The Chemistry of Life's Origins

NATO ASI Series

Advanced Science Institutes Series

A Series presenting the results of activities sponsored by the NATO Science Committee, which aims at the dissemination of advanced scientific and technological knowledge, with a view to strengthening links between scientific communities.

The Series is published by an international board of publishers in conjunction with the NATO Scientific Affairs Division

A Life Sciences **B Physics**	Plenum Publishing Corporation London and New York
C Mathematical and Physical Sciences **D Behavioural and Social Sciences** **E Applied Sciences**	Kluwer Academic Publishers Dordrecht, Boston and London
F Computer and Systems Sciences **G Ecological Sciences** **H Cell Biology** **I Global Environmental Change**	Springer-Verlag Berlin, Heidelberg, New York, London, Paris and Tokyo

NATO-PCO-DATA BASE

The electronic index to the NATO ASI Series provides full bibliographical references (with keywords and/or abstracts) to more than 30000 contributions from international scientists published in all sections of the NATO ASI Series.
Access to the NATO-PCO-DATA BASE is possible in two ways:

– via online FILE 128 (NATO-PCO-DATA BASE) hosted by ESRIN, Via Galileo Galilei, I-00044 Frascati, Italy.

– via CD-ROM "NATO-PCO-DATA BASE" with user-friendly retrieval software in English, French and German (© WTV GmbH and DATAWARE Technologies Inc. 1989).

The CD-ROM can be ordered through any member of the Board of Publishers or through NATO-PCO, Overijse, Belgium.

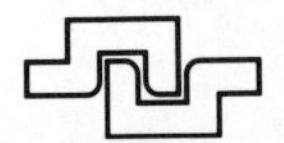

Series C: Mathematical and Physical Sciences - Vol. 416

The Chemistry of Life's Origins

edited by

J. M. Greenberg

Laboratory Astrophysics,
University of Leiden,
Leiden, The Netherlands

C. X. Mendoza-Gómez

Laboratory Astrophysics,
University of Leiden,
Leiden, The Netherlands

and

V. Pirronello

Istituto di Fisica,
Università di Catania,
Catania, Italy

Kluwer Academic Publishers

Dordrecht / Boston / London

Published in cooperation with NATO Scientific Affairs Division

Proceedings of the NATO Advanced Study Institute and
2nd International School of Space Chemistry
Erice, Sicily, Italy
20–30 October 1991

Library of Congress Cataloging-in-Publication Data

The Chemistry of life's origins / edited by J.M. Greenberg, and
C.X. Mendoza-Gómez, and V. Pirronello.
p. cm. -- (NATO ASI series. Series C, Mathematical and
physical sciences ; vol. 416)
Includes index.
ISBN 0-7923-2517-6 (alk. paper)
1. Cosmochemistry--Congresses. 2. Cosmology--Congresses.
3. Life--Origin--Congresses. I. Greenberg, J. Mayo (Jerome Mayo),
1922- . II. Mendoza-Gómez, C. X. III. Pirronello, V. (Valerio)
IV. Series: NATO ASI series. Series C, Mathematical and physical
sciences ; no. 416.
QB450.C546 1993
577--dc20 93-20895

ISBN 0-7923-2517-6

Published by Kluwer Academic Publishers,
P.O. Box 17, 3300 AA Dordrecht, The Netherlands.

Kluwer Academic Publishers incorporates the publishing programmes of
D. Reidel, Martinus Nijhoff, Dr W. Junk and MTP Press.

Sold and distributed in the U.S.A. and Canada
by Kluwer Academic Publishers,
101 Philip Drive, Norwell, MA 02061, U.S.A.

In all other countries, sold and distributed
by Kluwer Academic Publishers Group,
P.O. Box 322, 3300 AH Dordrecht, The Netherlands.

Printed on acid-free paper

Printed in the Netherlands

CONTENTS

PREFACE

This volume contains the lectures presented at the second course of the International School of Space Chemistry held in Erice (Sicily) from October 20 - 30 1991 at the "E. Majorana Centre for Scientific Culture". The course was attended by 58 participants from 13 countries.

The Chemistry of Life's Origins is well recognized as one of the most critical subjects of modern chemistry. Much progress has been made since the amazingly perceptive contributions by Oparin some 70 years ago when he first outlined a possible series of steps starting from simple molecules to basic building blocks and ultimate assembly into simple organisms capable of replicating, catalysis and evolution to higher organisms. The pioneering experiments of Stanley Miller demonstrated already forty years ago how easy it could have been to form the amino acids which are critical to living organisms.

However we have since learned and are still learning a great deal more about the primitive conditions on earth which has led us to a rethinking of where and how the condition for prebiotic chemical processes occurred. We have also learned a great deal more about the molecular basis for life. For instance, the existence of DNA was just discovered forty years ago.

It is becoming clearer and clearer from fossil evidence and from carbon isotopic distributions that life was extensively distributed as far back as 3.8 billion years ago. Just as in Darwinian theory, the ancestry of life as we know it today in all its complex and varied forms must have started with the simplest microscopic organisms. We still see all around us many simple bacteria and by observing how they function we try to work backwards to what their primitive ancestors did. It turns out that by coming to understand how all bacteria 'digest' carbon, the most basic element of life, we can detect the presence of living organisms on the earth simply by comparing (with carbonate deposits) how much of the light and heavy forms of carbon is found in the sediments containing the carbonaceous remains of these life forms. It has been a great surprise to find that life was already extensively present on the earth almost immediately after it was first possible to survive the last great bombardment 3.8 billion years ago. This is confirmed as well by the evidence for fossil microorganisms. The difficulty is that the impact record on the moon and other solar system bodies suggests that massive bombardment of the earth created an environment hostile to life for the first 700 million years of the earth's existence. With a total age of the earth of 4.5 billion years this severely limits the available time for life to have emerged. From molecules to living organisms must, in that case, have occurred not in hundreds of millions of years as earlier thought but perhaps only in thousands of years or even less. This raises some critical new questions about how life began. Was there enough time for building blocks to be formed on earth or were they imported from space? What *are* the building blocks? What are the minimum conditions or perhaps the ideal conditions for the emergence of life? Does the shortness of available time require a strongly localized favorable environment and set of prebiotic molecules rather than an extensively less favorable set of conditions? Is the emergence of life dependent on the preexisting chirality or handedness of the molecules which is a universal characteristic of living matter? Is there a minimum complexity threshold for the onset of autocatalytically closed sets of reactions to occur? Is racemization only inhibited with the advent of bounding semi-permeable membranes with chiral selectivity?

We have attempted to introduce the subject of life's origins by considering all possible sources of prebiotic molecules, from interstellar space to the protosolar nebula, to meteorites and, of course, from the earth itself. Many different kinds of questions must be addressed of how, when and where life began. This is primarily a matter of chemistry. But what kind of chemistry and *what environments*? Could life have started or existed on other planets of our solar system? And, if not, why not? Was there ever life on Mars? In what ways do other bodies of the solar system resemble or differ from the earth both past and present?

There is unquestionably a wide variety of earth and space environments in which prebiotic molecules *could* have been created as well as a wide variety of ways by which these molecules could have been considered as the initiators of life. The theory of the RNA world has received wide acceptance as providing the self-replicating requirement which is one of the keys to "living" beings. Is the term "world" an appropriate word or should we perhaps go back to the word "pond" as an initial state? Oligomerization certainly requires a high local concentration. The possibility that certain common clay minerals could have provided the catalytic reactions producing RNA oligomers is attractive because it is a highly localized process.

If we consider the earth as the source as well as the bearer of life we have to recognize that the oxidation state of the prebiotic atmosphere is now generally believed to have been a poor source of the molecules hydrogen cyanide, formaldehyde and ammonia which are key chemical intermediates to prebiotic chemical evolution and which are readily produced in an atmosphere which was originally presumed to be reduced. However, instead of the atmosphere, could the early ocean, which was in an obviously reduced state, have been the prime source of the endogenous prebiotic molecules? Interestingly, we now believe that the oceans were to a major extent brought in by comets so, perhaps one way or another, exogenous molecular sources were involved in the origins of life. Once the ocean existed, marine hydrothermal systems may have provided a source of prebiotic molecules. Other possible sources offered by the ocean could have been in the physical and chemical processing which can occur in its interface with the atmosphere where waves cause continuous mixing in sprays and bubbles. But the advantages offered by the ocean for prebiotic evolution are balanced by the disadvatages of dilution of the products. Does one, even with the ocean, require a specific local concentration or site of interaction for the prebiotic molecular pathway to succeed beyond the simplest form?

In recent years there have been many suggestions that the basic building blocks of life may have been created not on earth but rather in the vast regions of the space between the stars in our Milky Way and delivered to earth in sufficient quantities and in the required time frame to trigger rapid developments. The very small particles that make up the clouds of dust in space are very active chemical factories driven by the ultraviolet light of distant stars leading to complex organic molecules. This conjecture has been tested in the laboratory and has actually been confirmed by astronomical observation. It turns out that about 20% of all solid matter in our galaxy consists of largely prebiotic molecules and that comets are the repositor of such molecules in our solar system. Of course there are many, many unanswered questions about how these interstellar prebiotic molecules could have survived the impact of the comets and asteroids on earth. But at least as far as comets are concerned, the fact that they have been shown to be made up of very loosely held together small particles leads to the possibility that they would, upon impacting with the earth's atmosphere, break up into many fragments, many of which could then relatively gently floated to the earth. Of significance here is the requirement of a dense early atmosphere.The subject of life or of prebiotic type conditions in the solar system other than earth has also been discussed. Prebiotic synthesis in planetary environments has been reviewed with the introduction of many recent ideas which circumvent some of the problems created by the earth's atmosphere possibly not providing the simple moleculear ingredients thought to be needed for creating the amino acids. These take into account the evolution of the earth's atmosphere and ocean. The exogenous point of view of the origin of the prebiotic molecules is covered in part by a very detailed description of the analysis of the prebiotic products contained in meteorites. Whether the initial chemical composition of these meteorites was created in the protosolar nebula or in interstellar space can be examined by comparing the discussions in the chapter on protosolar chemistry and on the chemical evolution of interstellar dust with the analysis of the meteorites and the current knowledge of comet ingredients.

The scientists at the Ettore Majorana school for Space Chemistry on the Chemistry of Life's Origins addressed all these problems. Some considered the early atmospheres of the Earth and Mars. Others spoke of the present atmosphere of Titan as a useful global laboratory in which we can study some of the basic chemical processes in a primitive prebiotic atmosphere. Some discussed the amino acids found in meteorites.

It is important to realize that the methods used in the study of planetary atmospheres apply also to our own current problems of contamination of the atmosphere by carbon dioxide and its possible deleterious effects on the earth's climate. There are messages being deciphered that tell of the existence of living material in places as inhospitable as the Arctic and Antarctic as well as in geothermal vents. Were the prebiotic conditions for life to have evolved much more exotic than we had pictured? Planetary scientists, geo-chemists, astronomers and astrophysicists are searching for new answers.

Space probes such as the Halley, Giotto and Vega missions have provided us with proof of the abundance of organic molecules and interstellar dust in comets. Further space probes of the outer solar system, of Mars and of comets are aimed at understanding the origin both of life and of our solar system in general. Like all the other great problems that have fascinated and stimulated the minds of men through the ages, we believe the answers to the origin of life are there to be uncovered and it is a pleasure to acknowledge the opportunity provided by the support of NATO and of the "Ettore Majorana Centre for Scientific Culture" in Erice to bring together outstanding scientists from so many and varied disciplines to study the methodology and the hypotheses which are being applied so diligently to discover our origins.

A special debt of gratitude is owed to the staff of the Centre for their hospitality and constant support during the meeting.I would also like to express my sincere thanks to Dr. Celia Mendoza-Gómez and Professor Valerio Pirronello for their help in editing this volume.

J. Mayo Greenberg
University of Leiden
Laboratory Astrophysics
Niels Bohrweg 2
2300 RA Leiden
The Netherlands

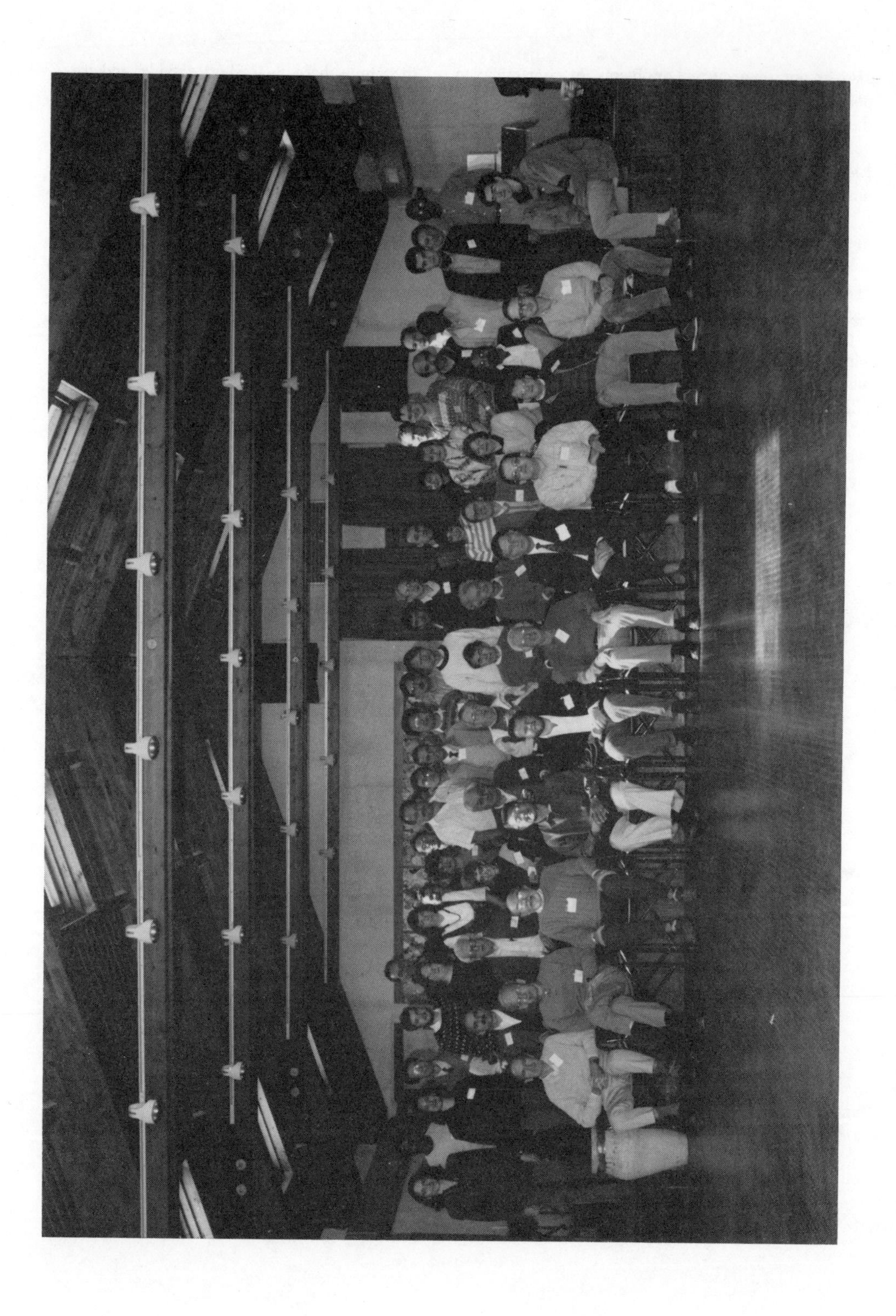

1. John Cronin
2. James P. Ferris
3. André Brack
4. Sherwood Chang
5. Valerio Pirronello
6. J. Mayo Greenberg
7. Vladimir Strelko
8. Wolfgang J. Duschl
9. Bruce Fegley
10. Jim Kasking
11. Peter Jenniskens
12. Mario di Martino
13. Charles Gehrke

27. Melinda Hutson
28. Carlos O. Della Védova
29. Julie Bartley
30. Stefan Pitsch
31. Sepideh Chakaveh
32. Konstantinos Fostiropoulos
33. Marvin Kleine
34. Zhang-fan Xing
35. Osama Shalabiea
36. Yong Bok Lee
37. Michael Wilson
38. Akira Kouchi
39. Eric Faure

14. Gözen Ertem
15. Yves Merle
16. Richard S. Young
17. Alan Schwartz
18. Jean Heidmann
19. Raik Mikelsaar
20. T. Kazaseva
21. Subhash Bhatia
22. Lina Tomasella
23. George Cooper
24. Paola Caselli
25. Jean Pierre J. Lafon
26. Alex Ruzicka

40. Stefan Rauch-Wojciechowski
41. Elizabeth Chacón B.
42. Jean-Baptiste Renard
43. Valérie Beaumont
44. Christine Lécluse
45. Hanno Sponholz
46. Michiel van Vliet
47. Ulrich Baumann
48. Tullio Aebischer
49. Joshua Obaleye
50. Andrea Petrocchi
51. Gennady Dolnikov
52. Vernon Morris

INTERSTELLAR DUST EVOLUTION: A RESERVOIR OF PREBIOTIC MOLECULES

J. MAYO GREENBERG AND
CELIA X. MENDOZA-GÓMEZ
Huygens Laboratory
Niels Bohrweg 2
P.O.Box 9504
2300 RA Leiden
The Netherlands

ABSTRACT. The physical and chemical evolution of interstellar dust is followed using a combination of observations, theory and laboratory analog studies. Laboratory analog experimental results on photochemical processes in interstellar dust mantles induced by ultraviolet radiation in space are analyzed. The organic molecular products consist of a wide variety of species, many of which are of obvious prebiotic significance.

1. Introduction

Interstellar dust is the ensemble of solid particles carried about by the gas which fills the space between the stars. The gas and dust clouds are the material out of which stars and solar systems are formed. The chemical composition of interstellar clouds is well represented in comets which are clearly the most primitive bodies of our solar system. If we are to understand comets it is first necessary to understand their primeval ingredients. The interstellar particles come in a variety of forms and sizes whose chemical identities and structures are provided by the combination of astronomical observations, laboratory studies and theoretical modelling. It is now recognized that a substantial fraction of the interstellar dust is in the form of complex organic molecules (Greenberg 1973, 1982a, Mendoza-Gómez 1992, Schutte 1988, Schutte and Greenberg 1986, Sandford et al 1991, Butchart et al 1986). The nature of these molecules depends to a large extent on how and where they are created. While it has often been hypothesized that a substantial amount of carbon bearing particles (or large molecules) found in the interstellar medium are formed in stellar atmospheres (see Tielens, 1991, and references therein) it appears likely that the most abundant source is the interstellar medium itself. A possible pathway to these molecules is via the energetic processing by ultraviolet photons and cosmic ray ions of simple molecules observed as mantles on the dust particles. The first laboratory experiments created to study this showed that so-called dirty ice (dominated by solid H_2O) dust mantles were the precursors of many organic molecules (Greenberg et al, 1972). Similar experiments by Khare and Sagan (1973) conducted at low temperatures but under somewhat less stringent conditions than those in interstellar space gave similar indications.

The aim of this chapter is to sketch the evolution of interstellar dust from simple to

J. M. Greenberg et al. (eds.), The Chemistry of Life's Origins, 1–32.

complex molecules by following some selected observational aspects of dust in a variety of environments, showing how its chemical composition may be studied using infrared spectroscopy and by relating these observations to laboratory analog studies which mimic the interstellar environments. The laboratory samples are also analyzed with gas chromatograph-mass spectroscopy and, for the larger molecules, this is supplemented with mass spectroscopy. One may, in this way, hope to arrive at a description of interstellar dust as it appears in the final stages of cloud contraction just preceding the formation of a protosolar system. In the next chapter these results will be applied to a prediction of the composition and morphology of comets and finally to how the complex organics created in interstellar space may be the precursors of life on the earth or in other solar systems.

2. Interstellar Dust Size and Growth by Accretion

The principal purpose of this section is to provide a basis for the evolutionary description of the particles. The wavelength dependence of the extinction and polarization leads to a definition of particle sizes. The cosmic abundance of the elements leads to a constraint on chemical components.

2.1. EXTINCTION AND POLARIZATION SIZE CONSTRAINTS

The shape of the continuum extinction by the dust grains is the factor determining their sizes, or more strictly, their size distributions. The light coming from stars is attenuated by the scattering and absorption by the intervening matter. The sum of the scattering and absorption is called extinction. In addition to this extinction there is frequently observed a certain degree of linear polarization. This polarization is caused by the fact that the two components of polarized light, into which the initial unpolarized star light may be decomposed, are attenuated differently by the particles. This can only occur if the particles are both nonspherical and aligned. We shall not give a detailed description of the how and why of the alignment mechanism but merely state here that it is almost always caused by magnetic effects such that the grains are spinning with their short axes (highest moment of inertia) directed along the magnetic field. The differential extinction by the particles whose long axes are along and perpendicular to the two polarized components results in a partial linear optical polarization perpendicular to the field direction. The maximum linear polarization varies somewhat in different regions but is characteristically about $\lambda^{-1} = 1.82\ \mu m^{-1}$. From the theory of the scattering of electromagnetic radiation by small particles (see e.g. van de Hulst 1957) it can be seen that this means that the particles responsible for the polarization have a mean size, a (semidiameter), such that $2\pi a/\lambda \approx 1$ for visible light ($\lambda \approx 500$nm). This tells us immediately that there must be a number of interstellar dust grains in the tenth micron size range. The full description of the sizes of the grains is actually deduced from the wavelength dependence of the extinction. In Figure 2.1 we show a *mean* extinction curve. We must note that variations in this extinction curve from star to star are observed and may be used to provide very important evidence for evaluating physical changes in the dust-gas cloud environments but these will not be taken up here. We wish here merely to demonstrate certain *characteristic* features and to show how these lead to a description of *characteristic* particle sizes.

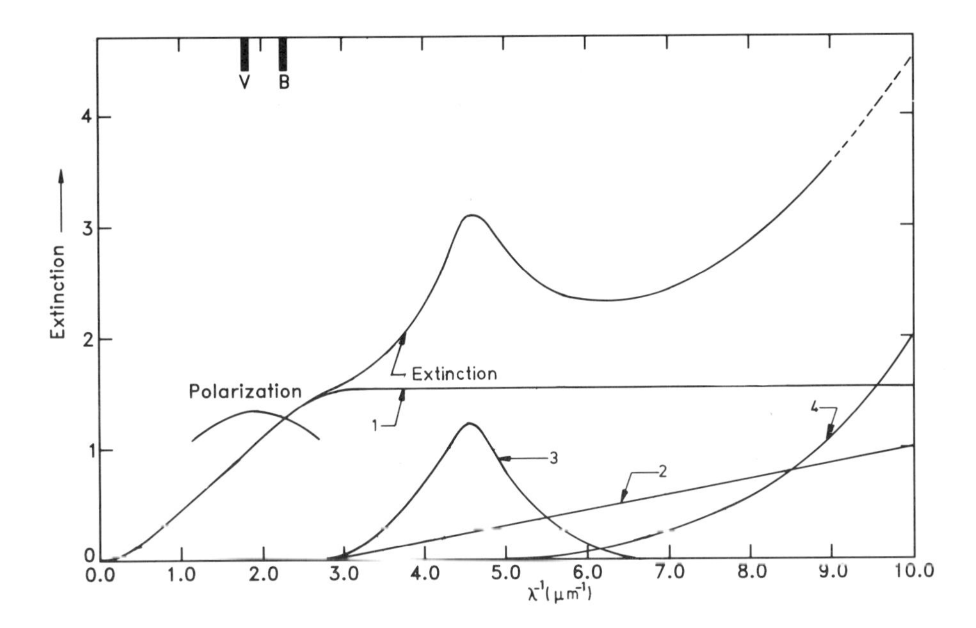

Figure 2.1. Schematic decomposition of the interstellar extinction curve. Shown also is the wavelength dependence of the polarization. The polarization is almost certainly produced by component (1) which accounts for the visual extinction. Component (2) is a linear term in the visual to ultraviolet extinction. Component (3) is the contribution by the "hump" particles. Component (4) is the far ultraviolet curvature.

The decomposition of the extinction curve shown in Fig. 2.1 is dictated by a combination of theoretical and observational (empirical) arguments.

Component (1) is produced by particles, whose mean size is $a \approx 0.1$ μm. It is a fundamental property of the extinction by solid particles that it starts at zero (for infinite wavelength), rises to a maximum at a certain value of the wavelength which depends on the size and optical properties of the particles and finally saturates; i.e., approaches asymptotically a constant value. The extinction efficiency, defined as the extinction divided by the geometrical cross section, approaches the value 2. In any case, beyond the first maximum the extinction can not rise again for any realistic grain material (see, for many examples, Greenberg, 1968). In fact the addition of absorbtion generally leads to a reduction in extinction in the ultraviolet for 0.1 μm particles. Thus the component (3) must be attributed to a different population of particles. Since the hump peak is at $\lambda \approx 0.2$ μm, this must be caused by particles whose mean size is $a \leq 0.03$ μm. A further constraint may be imposed by the fact that the hump particles are purely absorbing and this limits their size to $a \leq 0.01$ μm. The hump particles are almost certainly carbonaceous but their exact chemical composition is not yet established (Jenniskens, 1992, Greenberg et al, 1986, Onaka et al, 1992). The linear rise (component 2) starting at $\lambda^{-1} \approx 2.8 \mu m^{-1}$ is an *empirical* contribution for which we do not yet have a clear description in terms of particles

(Jenniskens, 1992). The rise after the dip is again due to very small particles whose size is < 0.01 μm and a possible candidate for this contribution may actually be thought of as large molecules generically described as polycyclic aromatic hydrocarbons. These particles, or large molecules, are being intensively and extensively studied (Allamandola et al, 1989, Leger et al, 1989) but are a secondary component in terms of the mass distribution in the interstellar dust population. Other supposed species for the small particles are quenched carbonaceous composite (Sakata et al, 1984). The major mass is in the 0.1 μm particles which we call the *large* particles. Furthermore, from the point of view of our interest in prebiotic molecules, they are the most likely to be the source of the relevant constituents.

2.2. COSMIC ABUNDANCE ELEMENTAL COMPOSITION CONSTRAINTS

The cosmic abundances of the major *condensable* elements in the interstellar medium are given in Table 2.1. Interstellar dust must therefore primarily be made up of these elements in various molecular combinations. One combination shows up immediately when the dust is observed in the infrared.

Table 2.1 Relative cosmic (solar system) abundances of the most common elements

	(1)	(2)	(3)
H	1	1	1
He	0.068	0.081	0.079
C	4.17(-4)	4.45(-4)	4.90(-4)
N	0.87(-4)	0.91(-4)	0.98(-4)
O	6.92(-4)	7.40(-4)	8.13(-4)
Mg	0.399(-4)	0.396(-4)	0.380(-4)
Si	0.376(-4)	0.368(-4)	0.355(-4)
S	0.188(-4)	0.189(-4)	0.162(-4)
Fe	0.338(-4)	0.331(-4)	0.467(-4)

(1) A.G.W. Cameron (1982) in C. Barnes et al (eds.), Elements and Nuclidic Abundances in the Solar System, Cambridge Univ.Press, p. 23
(2) E. Anders and Mitsuru Ebihara (1982), Geochim. Cosmochim. Acta 46, 2363-2380.
(3) N. Grevesse (1984), Phys. Scripta T8, 49-58.

This is the silicate component whose Si-0 stretch at about 9.7 μm is the dominant one in the infrared extinction towards the center of the Milky Way as observed by Willner and Pipher, 1982 (see Fig. 2.2). No other absorption feature is comparable in strength so that the first impression is that perhaps silicates consisting of Si, Mg, Fe (the *rockies*) are the main components of the dust, even though oxygen, carbon and nitrogen (the *organics*) are ten times as abundant. In fact, the abundances of the elements Si, Mg and Fe are indeed insufficient to fully explain the amount of extinction. Since we know the total extinction to the galactic center (total grain area) and we know the mean size of the dust grains we may calculate the total *volume* of dust per unit area along the line of sight.

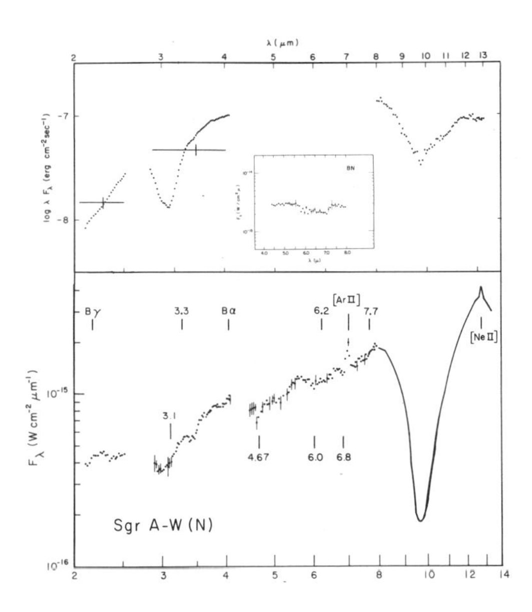

Figure 2.2. A comparison of the extinction to the B.N. object in Orion (the first observation of solid H_2O in molecular clouds) and to the galactic center (used as representative of diffuse (low density) regions). Note the absence of the H_2O ice band towards the galactic center but the presence of a relatively small absorption at about 3.4 μm.

It is generally observed that the dust and (gas) are well correlated so that, if we know the mean hydrogen density towards the galactic center, we can find the ratio of the volume of dust to the hydrogen number. If one then computes the number of *rocky* elements in the dust volume it turns out that this falls short of the required value by a factor of about 3 (Greenberg, 1991). This means that one requires the organics *along with* the *rockies*. A logical way to account for this is to note that the silicate particles which are injected into the interstellar medium by stars are condensation nuclei on which the abundant condensable atoms and molecules can form a mantle. Since the mean dust temperature in space is in the range T_d = 10 -20 K any atom or molecule which hits will stick and form a frost of organics. This will occur in the denser regions (molecular clouds) and we should then be able to observe such molecules as frozen H_2O. The growth rate of grains depends on the number density of the accreting atoms and molecules *and* on whether or not they stick. It is likely that the heavy atoms and molecules stick with a probability of unity but that hydrogen, while it may remain on the grain surface for some short time (d'Hendecourt, Allamandola, Greenberg, 1985), is finally returned to the gas. However, during the time it is travelling over the grain surface it has a probability of unity of encountering one of the heavy atoms, like oxygen, and hydrogenate it. The energy barrier for this reaction is zero and the energy-momentum balance is accommodated by the grain. This formation of H_2O is shown schematically in Figure 2.3. The rate of growth of an H_2O mantle is given by

$$d(H_2O)/dt = (1-\delta_O)\ [O/H]_{C.A.}\ n_H\ v_O\ \pi a_d^2$$

where δ_0 is the depletion of oxygen in the gas (because of the existence of the grains), v_O is the oxygen velocity, n_H = number density of hydrogen, a_d = dust radius, $[O/H]_{C.A}$= cosmic abundance of oxygen. The observed mean abundance of H_2O in grain mantles in molecular clouds has been derived by Whittet et al (1988) to be given, in terms of the H_2O absorption, by $\tau(H_2O)/A_v = 0.08$. Given an H_2O absorption per molecule as $\sigma_{H2O} = 5 \times 10^{-19} cm^2/mol$ (band strength 1.6×10^{-16} and half width 335 cm^{-1}), it may be shown that the time required to achieve this mantle is $t \approx$ 3-4 x 10^5 yr. Similar reactions will occur with carbon and nitrogen leading *initially* to a grain mantle consisting of H_2O, CH_4 and NH_3. It is, of course, well observed that CO in molecular clouds takes up about 20% of the available carbon so that it will also accrete along with the water, methane and ammonia making up a basic icy grain mantle as configured in Fig. 2.3. Surface reactions may also be considered to produce more complex molecules like formaldehyde and methanol (Tielens and Allamandola, 1987). It was by surface reactions of O, N and C with H that van de Hulst (1949) derived the original *dirty ice* model of interstellar dust.

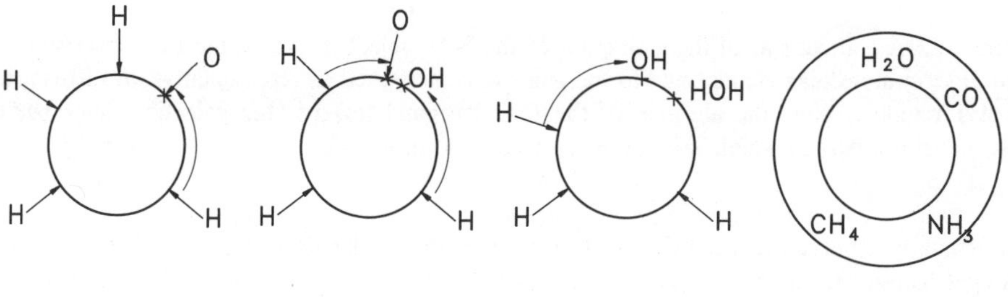

Figure 2.3. Schematic of the way H_2O mantles are formed by accretion and reactions of oxygen with hydrogen on grain surfaces. Note that a minimum grain size is required for mantle growth because for very small ($a \leq 0.01\mu m$) the reaction energy of O with H will cause the desorption of the molecules (Greenberg, 1968, 1980, Greenberg and Hong, 1976, Aannestad and Kenyon, 1979).

If the accretion process is allowed to continue unchecked, all the condensable species will be on the grain -and *out* of the gas- on a time scale of about

$$\tau_{ac} \approx 2 \times 10^9/n_0 \text{ yrs}$$

where n_0 is the number density of hydrogen in all forms ($n_0 = n_H + 2n_{H2}$). Since the molecular cloud lifetimes are of the order of several 10^7 years this would imply that in a moderately dense cloud with $n_0 = 10^4\ cm^{-3}$ there should be no observable molecules in the gas phase. This, of course, is not the case so that some desorption mechanism must exist which maintains some sort of equilibrium or, more precisely, steady state between the grain mantles and the gas. A number of such desorption mechanisms have been suggested but one which has been shown to be experimentally verified is an impulsive one triggering the chain reaction and heat generated by stored free radicals in grain mantles (Greenberg, 1973, 1979a, d'Hendecourt et al, 1982, Schutte and Greenberg, 1991). It depends on the fact of ultraviolet photoprocessing of the grain mantles even in dense clouds. Another form of energetic processing results from low energy cosmic rays and is described by Pirronello in this volume.

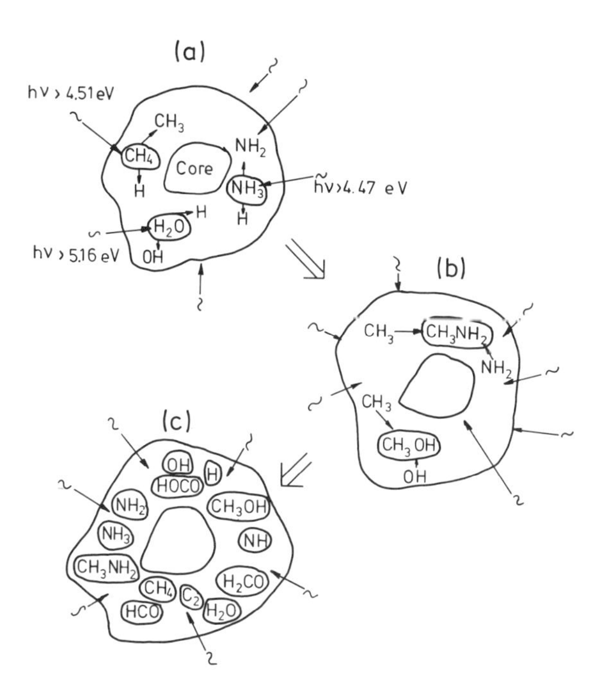

Figure 2.4. Various stages of photochemical grain evolution at low temperature. a) Initial simple accreted molecules and photolysis by ultraviolet photons to produce radicals: H_2O + hν → OH + H; b) Recombination of adjacent radicals producing new molecules; c) Resulting radicals and molecules at an intermediate stage of evolution.

3. Infrared Spectroscopy

The most direct way to observe the chemical composition of interstellar dust is by means of its infrared spectrum. Some of the principal transition frequencies of the characteristic molecular groups fall in a spectral region which is accessible to ground based telescopes. Although the atmosphere is opaque at many frequencies, it has enough windows that observations have been able to provide a substantial body of data of the major constituents of the dust. The sizes of the interstellar particles are small enough compared with the infrared absorption wave lengths that the particle spectra are not greatly distorted by the amount of particle scattering compared to the absorption; i.e., *particle extinction* closely follows the *material absorption.*

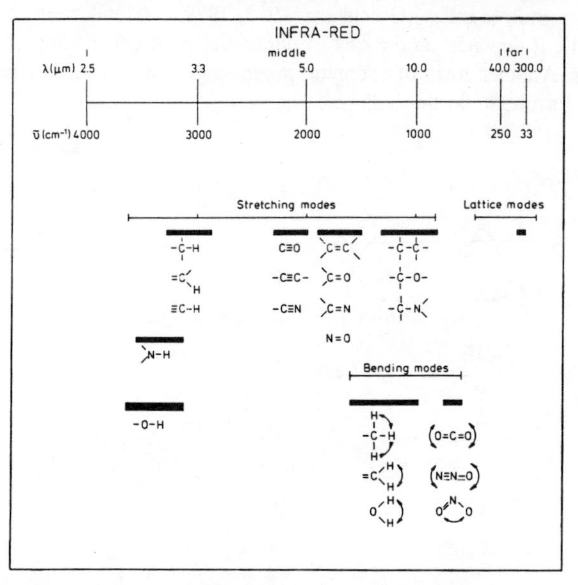

Figure 3.1. Characteristic vibration frequencies of various molecular groups.

The most abundant condensable (atomic) species are oxygen, carbon and nitrogen so that one can use Figure 3.1 to indicate where the vibration frequencies may be found for various combinations of these atoms with H or with each other. We are concerned here with spectral absorption or emission in the solid state. Stretching modes have the highest frequencies which decrease with the square root of the masses (μ) of the constituents and increase with the square roots of the bond strength (k) as predicted by classical oscillation theory $\nu = \sqrt{(k/\mu)}$. Bending modes and lattice modes are at the low frequency end.

Specific frequencies for different molecules are often found in the literature but in mixtures with other molecules these frequencies may be shifted and the absorption band strength as well as its shape may be modified. The determination of such effects as well as effects of temperature require laboratory studies (see Tielens et al, 1991, for a detailed discussion of CO absorption). A major feature of solid state spectra in comparison with spectra of gaseous molecules is the suppression of rotational structure. An example of this suppression is shown in Figure 3.2 where a solid state mixture spectrum is compared with the gas phase spectrum of exactly the same mixture taken at the same spectral resolution.

We see, for example, that the R and P branches of the gas phase rotational features of CO are absent in the solid and replaced by the pure vibrational feature (the Q branch) at 4.61μm which is totally absent in the gas phase. Similarly the very large H_2O solid ice band at 3.07 μm is absent in the gas phase and, of even greater importance for astronomy, has an integrated absorption strength and peak absorption strength far exceeding (by ≈100) that of the gas phase. This is a result of the strong dipole-dipole interactions of the water molecules with each other.This is a fortunate effect because without this amplification, knowledge of the chemistry of dust (at least the presence of H_2O grain mantles)would have been delayed by many years.

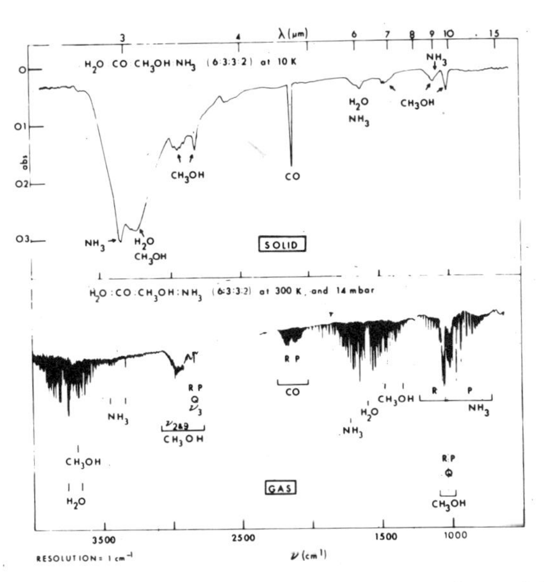

Figure 3.2. Comparison of infrared absorption spectrum of a solid and gaseous mixture $H_2O:CO:CH_3OH:NH_3$(6:3:3:2) taken at the same (1 cm^{-1}) resolution. The upper figure is for the amorphous solid at 10K; the lower figure is for the gas at room temperature

The assignment of various features seen in the infrared absorption spectra of interstellar dust to a range of molecules in grain mantles has advanced spectacularly in the last 3-4 years. Where before only the most abundant constituents like H_2O and CO were identified we now have a quite respectable list. One thing which must be pointed out is that, just as in the case of molecules in the gas, the relative abundances of solid state molecules depend on the local conditions as well as the past history of the dust grains. This caveat should be recalled not only in such obvious cases as the dependence of the relative abundance of H_2O and CO on temperature because of the large difference in vapor pressure but also in many more involved situations as occur (or have occurred) in regions of star formation. We summarize here now some examples of observed interstellar dust mantle molecules with particular relevance to the comet volatile composition because their relative abundances are so different from those derived from protosolar density (see Fegley, this volume).

Solid methanol was identified by and quantitatively described by Grim et al (1991) in a study of the C-H stretch at 3.53 μm in the wing of the ice band of W33A. Its abundance relative to H_2O was estimated at about 7%. This estimate was considerably less than the 55% or even as high as 80% derived by Tielens et al (1984) and by Tielens and Allamandola (1987). The situation appears to have been resolved by studies of the C-O stretching absorption near 9.8μm in W33A, NGC7538 IRS9, W3 IRS5 and AFGL2136 with the result that the CH_3OH abundance is probably in the 0.03-0.04 range with a strict upper limit between 0.06-0.18 (Schutte et al, 1991). While CO_2 (see section 5b) was expected to be comparable in abundance with CO, it was not detected until quite recently because its main feature lies at 4.27 μm in the 4-8 μm range which is inaccessible from the ground due to CO_2 in the atmosphere. The presence of CO_2 was derived using a difficult subtraction in the low resolution IRAS spectrum (d'Hendecourt and de Muizon, 1989) to arrive at a feature at 15.2 μm which could be compared with a laboratory spectrum. While NH_3 must certainly exist at some stage in interstellar dust its unambiguous identification has not been entirely successful. To round out this discussion I will summarize the situation on volatile molecules in grain mantles by noting that infrared features have been identified with OCS, OCN^-, H_2S, CH_4 (Lacy et al, 1991). Other molecules, like formaldehyde, which are probably present (Tielens and Allamandola, 1987), have earlier escaped positive identification for a variety of reasons. However a reexamination of the spectrum of the icy grains in the protostellar object GL 2136 reveals that about 3% H_2CO is in the mantles.

4. Ultraviolet Photoprocessing of Grain Mantles

Given the existence of icy grain mantles as pictured in Figure 2.3 and the presence of energetic ultraviolet photons we envisage initiating processes as illustrated in Fig. 2.4(a) An ultraviolet photon may dissociate H_2O into OH and H and, given enough energy, the H atom will escape from the grain leaving behind the frozen free radical OH. Should this radical be in physical contact with another radical it will *immediately* recombine to form a new molecule because radical-radical reactions occur with no energy barrier (zero activation energy). The process leads to new and more complex molecules and radicals (Fig. 2.4(c)). The storage time of radicals at the normal grain temperatures is long enough so that once having reached a maximum steady state they will survive for many millions of years at $T < 20K$ (Greenberg et al, 1992). We note that surface reactions may also lead to such radicals as HCO, OH etc. But within the grain as well as on the surface the radicals are produced by ultraviolet photons. The time scale for the photoprocessing depends on the ultraviolet flux.

The mean interstellar ultraviolet photon flux with energies ≥ 6 eV in the diffuse regions of space is about $\Phi_{hv} = 10^8$ cm^{-2} s^{-1}. In dust clouds this flux is attenuated so that deep in the interior it is probably reduced by a factor $\sim e^{-2Av}$ where A_v is the dust extinction in the visual (about 500 nm). A reduction of the UV flux by 10^{-4} occurs at 1 pc depth in a cloud whose density is of the order of $n_0 = 10^4$ cm^{-3}; i.e. a molecular cloud. At greater depths or at greater densities the *interstellar* UV flux penetration may become unimportant. But within such clouds there are two additional sources of ultraviolet: (1) ultraviolet generated by low energy cosmic rays ionizing hydrogen molecules (Prasad and Tarafdar, 1983), (2) shock generated ultraviolet by the birth of stars (Norman and Silk, 1980). It is likely that the lowest level of ultraviolet flux in the average molecular cloud is never less than $\Phi_{hv}^{M.C} \approx 10^4$ cm^{-2} s^{-1} (Greenberg, 1982b).

An estimate of the time scale for photoprocessing is derived from the time it takes for every molecular bond within the mantle to have absorbed a bond breaking photon. Assuming a grain radius of 0.1 μm and a molecular (bond) density of $n_{M.B.} = 2 \times 10^{22}$ cm^{-3} we get $\tau_{hv} = 5 \times 10^6$ yrs for $\Phi_{hv} = 10^4$ cm^{-3} and only 5×10^2 years in diffuse clouds. This implies that significant photoprocessing of grain mantles can take place during a molecular cloud lifetime of 5×10^7 yrs. It will turn out to be important to know the radical production time scale. It is obvious that nearest neighbor radicals are unstable so that the maximum concentration of radicals is ~10% corresponding to next nearest neighbors. In practice ~1% is about the *maximum* free radical concentration. The time to achieve this depends on the radical production efficiency which, measured in the laboratory, is about 0.5 (Schutte and Greenberg, 1991). The resulting time scale for achieving 1% radicals in the dust mantles in a molecular cloud is then only $\tau_{rad} \approx 10^5$ yrs. We note that this is about the accretion time scale and is also of the order of the time to grow a mean mantle thickness of H_2O as derived in section 2b.

As already alluded to, laboratory experiments must be used to study the photochemical evolution of dust in space. This will be discussed in the next section.

5. Laboratory Analogue Studies of Grain Evolution

Measuring the infrared spectra of low temperature ice mixtures as they may occur in interstellar space is one of the basic means by which one can study the evolution of interstellar dust. In the laboratory the dust core is simulated by a surface (the cold finger) cooled to 10 K in a vacuum chamber. Gas mixtures of simple molecules are deposited in a controlled way through a narrow inlet and condense on the cold finger. This icy mantle may then be irradiated by a vacuum ultraviolet hydrogen lamp and/or warmed up as occurs to the dust mantles in molecular clouds. A schematic of the laboratory is shown in Fig. 5.1.

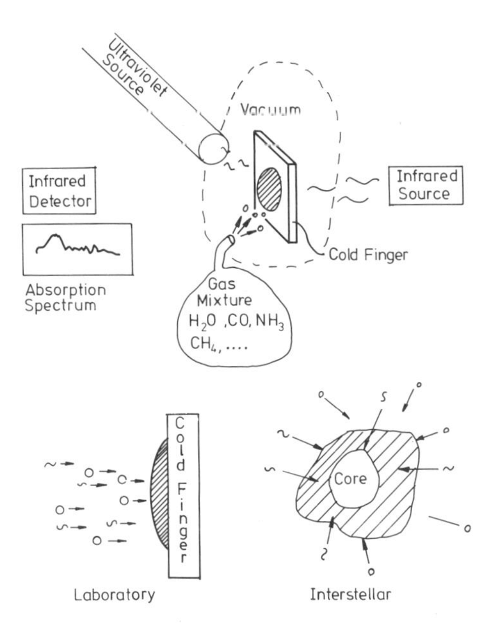

Figure 5.1. Schematic of photochemistry laboratory.
Note that one hour in the laboratory is equivalent to 10^3 - 10^7 yrs in space depending on whether the local environment corresponds to a low or high density cloud.

Details of the laboratory procedure may be found in Mendoza-Gómez (1992). The material on the cold finger is studied with infrared spectroscopy to follow changes in the molecular composition and to follow changes in the infrared band strengths and shapes as a function of the temperature and the matrix in which the particular molecule is imbedded. Further analytic techniques are those used to study the analog sample by mass spectrometry and gas chromatography-mass spectroscopy (GCMS) and high performance liquid chromatography (HPLC). When a sample is warmed up after irradiation some of it evaporates. In interstellar space, dust temperatures greater than 100 K are exceptional other than in regions of star formation and then, only near the stars. The molecules which may evaporate at $T > 100$ K are studied in the laboratory by being pumped off into a chamber whose contents are later examined with mass spectrometry. In the laboratory samples there always remains a residue at room temperature which is called organic refractory. Infrared spectra of such residues appear to be very similar to the spectra of interstellar dust observed in regions where the volatile mantles have been evaporated or eroded away. We will compare a few selected laboratory results with observation of interstellar dust in various types of astronomical environments. We will also present some results on the chemical and mass spectrometric analyses of various dust components which may be relevant to the origin of life. Figure 5.2 shows how theory, laboratory measurements and observations are coordinated.

Fig. 5.2 Schematic of how laboratory results and theoretical computations are combined with observations of infrared spectra to arrive at grain mantle models. (Courtesy of Edinburgh Observatory publications). The infrared satellite observation will provide data in the hitherto (almost) inaccessible wavelength range between 5 and 9 μm where most of the key signatures of complex organic molecules are found.

5.1. THERMAL PROCESSES

The strongest absorption feature in space (with the exception of the 9.7 μm Si-O stretch of silicates) is that of H_2O ice at ~ 3.1 μm. The ice bands at 6 μm and 13 μm are not as strong and are obscured by or mixed with other features. A series of spectra of an H_2O:CO ice mixture heated to various temperatures shows how the ice bands evolve - starting with amorphous ice which ultimately becomes crystalline. Simultaneously the CO band is seen to decrease, although the CO may remain trapped in the H_2O matrix well beyond the temperature at which it evaporates as a bulk solid (Schmitt et al, 1989, 1989a, 1989b). Comparison of the 3 μm ice band observed in two stellar objects shows these effects clearly. The star Elias 16 is situated beyond the Taurus cloud and the ice mantles are on cold dust grains along the full line of sight in the cloud. On the other hand HL Tau is known to have an accretion disk in which the local dust is heated. We see how the spectrum of each object indicates the fact of cool dust or of heated dust by the shape of the ice band (van de Bult et al, 1984). As expected, the CO band strength relative to the H_2O strength in HL Tau is extremely small as compared with that in Elias 16 (Geballe, 1986) and, in fact, it could be zero indicating that all or most of the CO has been evaporated. In molecular clouds the CO absorption is linearly proportional to the H_2O absorption (Whittet and Duley, 1992)

$$\tau_{4.67} = 0.69\ (\tau_{3.0} - 0.25)$$

the offset (0.25) probably being accounted for by the relative stability of the H_2O under either deposition or desorption conditions (Schutte and Greenberg, 1991). The ice band has been observed now in *many* objects and it is clearly the dominant species in the icy grain mantle (Tanaka et al, 1990, Whittet et al, 1988, Sato et al, 1990, Smith et al, 1988).

5.2. ULTRAVIOLET PHOTO-PROCESSING

In the laboratory at Leiden many spectra have been taken of irradiated and warmed up icy mixtures such as seen in Fig. 5.3 (d'Hendecourt, 1984).It is readily seen in Fig. 5.3 that upon irradiation CO_2 is abundantly produced but we here emphasize the short wavelength feature on the wing of the CO band. This feature, at 2170 cm^{-1}, persists up to very high temperatures (well above the 95 K shown here) while CO evaporates away.
In Fig. 5.4 we compare the spectrum of a protostellar object W33A with an irradiated warmed up icy mixture. Looking at the 4.61 μ region in W33A there is an *indication* of some remaining CO. But when compared with Fig. 5.3, it is immediately apparent that the 2170 cm^{-1} feature is, in fact, the dominant feature here. The feature has been identified as OCN^- (Grim and Greenberg, 1991). It is not always dominant as shown by comparison with another object NGC 7538 IRS9 in Fig. 5.5 (Tielens et al, 1991). The evidence for varying U.V. and thermal processing is becoming abundantly evident as astronomical spectra accumulate. The explosive ejection of molecules which balances the accretion in molecular clouds has been discussed most recently by Schutte and Greenberg (1991). This and the complex interaction between dust and gas are the subject of a continuing study using the laboratory ultraviolet processing results on various ices in combination with the atom, ion and molecular interactions in the gas phase (d'Hendecourt et al, 1985, Breukers, 1991).

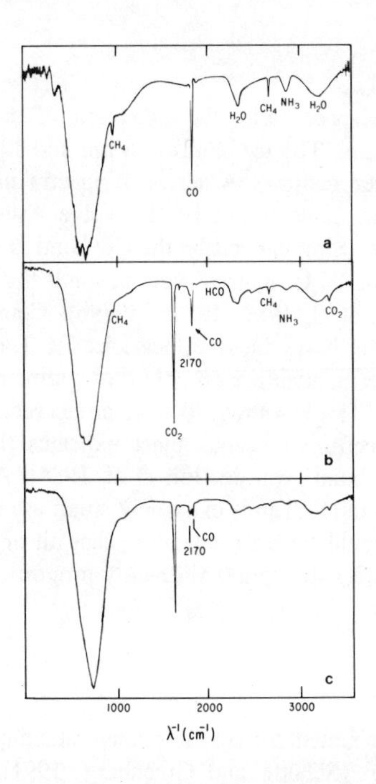

Figure 5.3 Infrared spectra of the mixture H_2O:CO:CH_4:NH_3: a) after deposition, no radiation, T=12K; b) after 300 h ultraviolet irradiation, T=12K; c) warmup to 95K.

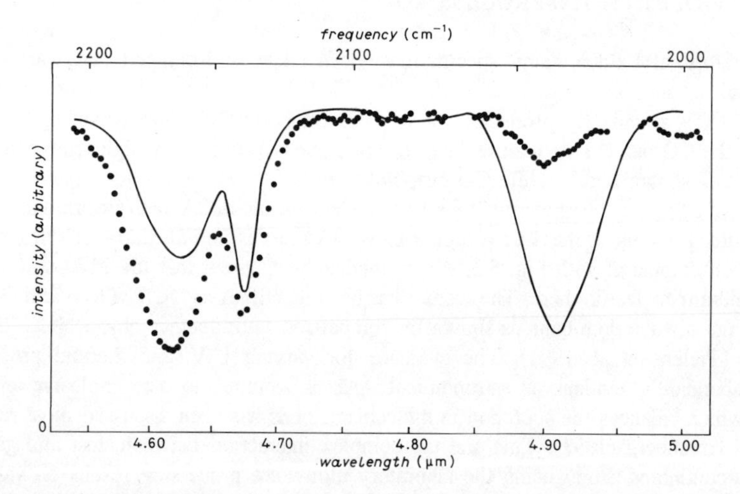

Figure 5.4. Comparison of the observed spectrum of the protostellar object W33A with a laboratory spectrum of an irradiated warmed up icy mixture containing H_2S along with H_2O, NH_3 and CO .

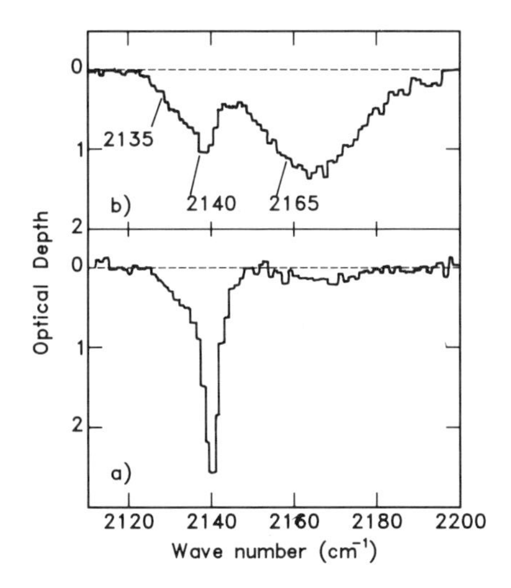

Figure 5.5. Comparison of NGC 7538 IRS9 spectra and W33A spectra.

5.3. ORGANIC RESIDUES - INFRARED

As already pointed out, there appears to be an insufficiency of rocky elements to produce the observed amount of interstellar extinction by silicates alone, so that mantles of the organics - OCN elements- must be involved. However, if one examines the spectrum toward the galactic center, the silicate feature far outweighs any evidence for mantles - certainly the ice band is absent. But the apparently insignificant (with respect to the "normally" strong ice bands) feature at 3.4 μm is *the* evidence for the organic mantle needed. Spectroscopically it compares well with that of the organic residue which results from long term photoprocessing of ices. It was predicted that such a mantle material should be observed wherever the ice band is not present (Greenberg 1973) and has now been observed not only towards the galactic centre but also towards the object for which it was originally predicted, namely VI Cygni #12 (Sandford et al, 1991). See Fig. 5.6 for a comparison of various 3.4 μm features in interstellar dust, and Fig. 5.7 for comparison with laboratory spectra. It was just difficult to observe earlier because the absorption strength per unit mass is much less than that of the H_2O and the detection techniques in the 70's were not yet good enough.

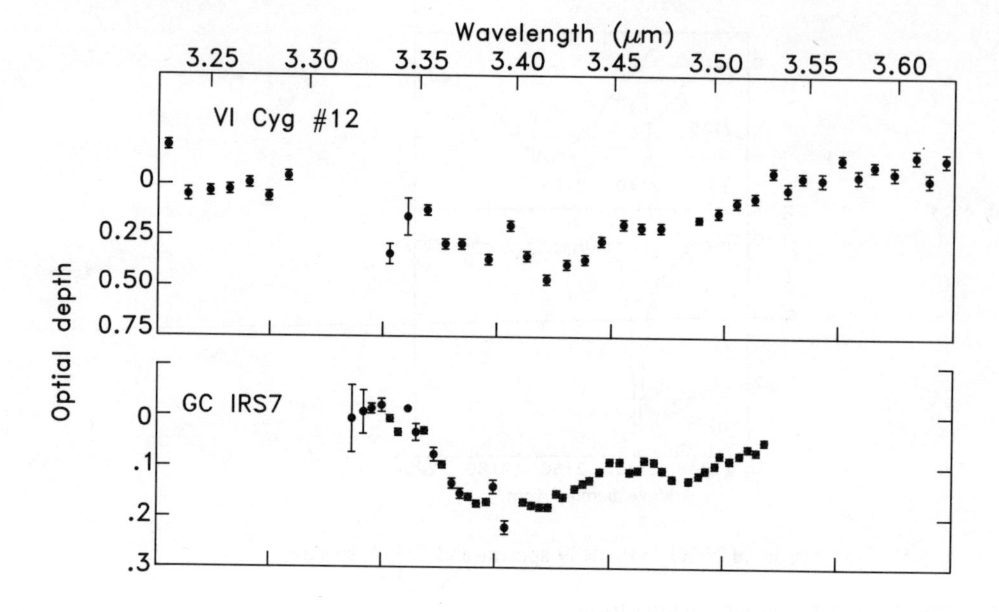

Figure 5.6. Comparison of various 3.4 μm features in interstellar dust.

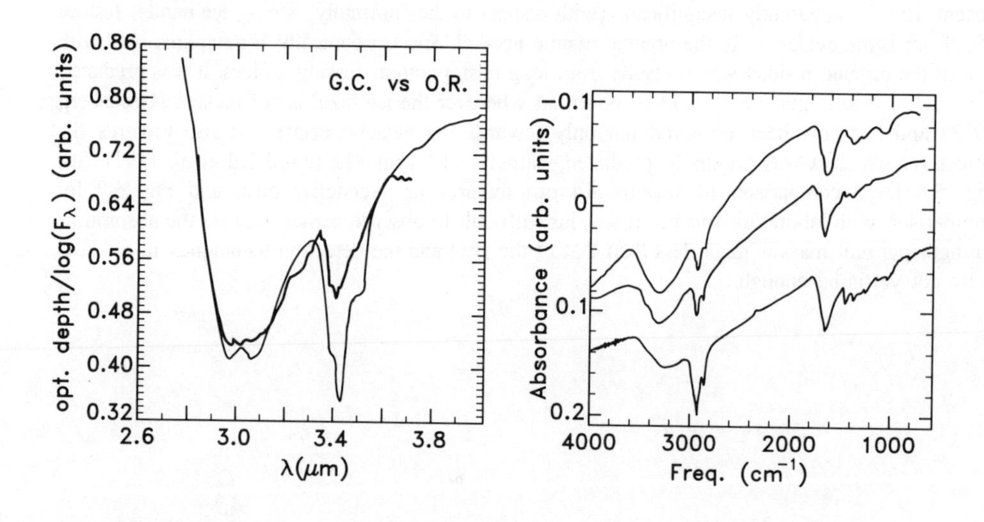

Figure 5.7.a) Comparison of the 3.4 μm feature in interstellar dust with laboratory spectra; b)Infrared spectra of three residues of photolyzed ices: Upper- H_2O:CO=0.71:0.29; Middle-H_2O:CO:NH_3=5:2:1; Lower-H_2O:CO:CH_4:NH_3=0.66:0.09:0.6:0.09.

In figure 5.7(b) we note that the spectrum of the organic refractories is rich between 5 and 8 μm (1800-1200 cm^{-1}) a region which is normally not accessible from the ground based observations. With the advent of the Infrared Satellite Observatory (ISO) it will be possible to study this spectral region and to relate its features to laboratory residues thus providing the possibility of a far more complete identification of the interstellar organics than can be obtained from the 3.4 μm feature alone. This latter gives only limited clues to the types of organics in space.

The presence of such organic refractory material as indicated by the 3.4 μm feature is also seen in meteorites (see Fig. 5.8). The possible connection between interstellar organics and those in meteorites is discussed in the chapter by Cronin and Chang. We shall further discuss it with reference to comets in the following chapter.

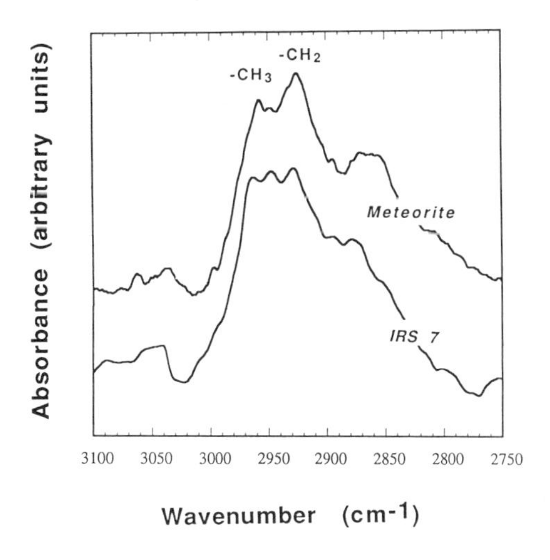

Figure 5.8. The 3.4 μm feature in a meteorite compared with the dust to the galactic center.

The major question to be answered is whether the limited 3.4 μm spectral region uniquely determines whether the organics in meteorites are (1) direct descendants of interstellar organics or (2) are modified forms of the interstellar organics or (3) have resulted from synthesis of simpler molecules within the parent asteroidal body. The latter has been suggested with, as a basis, the Strecker-cyanohydrin synthesis of amino-acids from aldehydes, HCN and NH_3 in water. A factor favoring this suggestion for the Murchison meteorite is the presence of hydrated silicates and carbonates in the Murchison meteorite, both following from the preservation of H_2O and the much more volatile CO_2 in the original aggregation process. However this preservation of volatiles implies the preservation of the interstellar organics in toto so that it would not be surprising if *all* the possible sources of organics in meteorites listed above should be represented. This makes meteoritic organics interesting but difficult to interpret uniquely.

As expected, therefore, the composition of the laboratory organic residues should provide a major source for comparison with meteorites of the carbonaceous chondritic classes CI1 and CM2.

5.4. ORGANICS - CHEMICAL AND MASS ANALYSES

Details of the following may be found in Mendoza-Gómez (1992). We summarize here some of the results of GCMS, HPLC and mass spectrometric analysis of the refractory (at room temperature) and semi-refractory (evaporates at T > 100 K) products of ultraviolet processed ices.

In Table 5.1 we list the products found for that part of the residues which are rinsed from the cold finger. The lactic acid is undoubtedly a contaminant because it does not have any variation with varying isotopically labelled molecules in the mixtures. All the others are well established (Briggs et al 1992). Amino acid analyses of one of our samples by Cronin (see chapter by Cronin and Chang, this volume) indicated evidence for many biological amino acids such as serine, glycine, alanine etc. with glycine being the most abundant. We note that photoprocessing of interstellar grain mantles leads to the presence of identifiable prebiotic mirror symmetry, molecules such as glyceric acid and glyceramide (Agarwal et al, 1985, Briggs et al, 1992). Note also that glycerol is made in good abundance and this may have substituted for ribose in a simpler nucleic acid polymer (Orgel, 1987) as pictured in Fig. 5.9.

Table 5.1 Products from the photolysis of $CO:H_2O:NH_2$ mixtures at 10 K[a] (Briggs et al, 1992).

	Compound	Formula	Amount[b](%)
1.	Ethylene glycol[c]	$HOCH_2CH_2OH$	4.6 ± 4.4
2.	Lactic acid[d]	$CH^3CH(OH)CO_2H$	21 ± 17
3.	Glycolic acid	$HOCH_2CO_2H$	24 ± 12
4.	3-Hydroxypropionic acid	$HOCH_2CH_2CO_2H$	0.76 ± 0.52
5.	Formamidine	$HCNH(NH_2)$	0.10 ± 0.18
6.	2-Hydroxyacetamide	$HOCH_2CONH_2$	17 ± 10
7.	Hexamethylene tetramine	$(CH_2)_6N_4$	1.2 ± 1.4
8.	Urea	NH_2CONH_2	2.2 ± 6.2
9.	Biuret	$NH_2CONHCONH_2$	0.50 ± 0.96
10.	Oxamic acid	NH_2COCO_2H	0.09 ± 0.20
11.	Ethanolamine	$HOCH_2CH_2NH_2$	0.38 ± 0.57
12.	Glycerol	$HOCH_2CH(OH)CH_2OH$	4.4 ± 2.5
13.	Glycine	$NH_2CH_2CO_2H$	0.27 ± 0.51
14.	Oxamide	$NH_2COCONH_2$	0.13 ± 8.6
15.	Glyceric acid	$HOCH_2CH(OH)CO_2H$	5.9 ± 5.6
16.	Glyceramide	$HOCH_2CH(OH)CONH_2$	15 ± 10

a The approximate ratio of $CO:H_2O:NH_3$ was 5:5:1 in the condensed phase.

b The average of 10 analyses. The yields given in percent based on the response of the mass spectrometer assuming the same sensitivity for each compound. The error is given at the 95% confidence level.

c It was not possible to determine if the ethylene glycol observed using ^{13}CO contains ^{13}C.

d Isotopic labelling studies have shown lactic acid to be a contaminant and not a photoproduct.

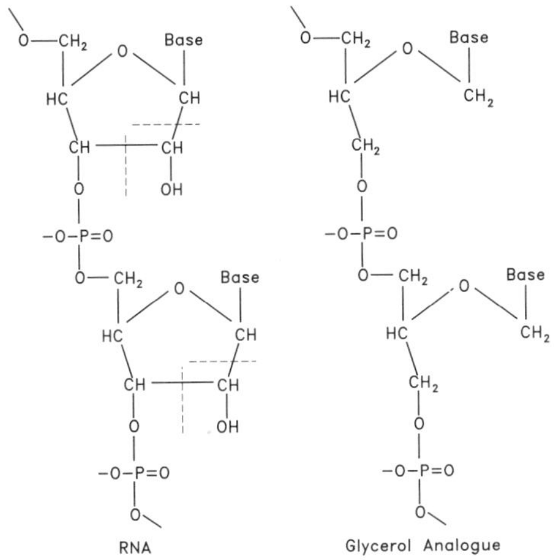

Figure 5.9. Nucleic acid polymers. Left, RNA contains ribose as a connecting unit. Right, glycerol may have been an ancestral analogue. From Orgel(1987).

In general the photoproduced molecules seem to be well suited as possible forerunners of a pre-RNA world (see Schwartz, this volume).

In Tables 5.2 and 5.3 are presented identified mass spectra products (Mendoza-Gómez, 1992). The residue mass spectra are quite complex (see Fig. 5.10) and only a few identifications are given here.

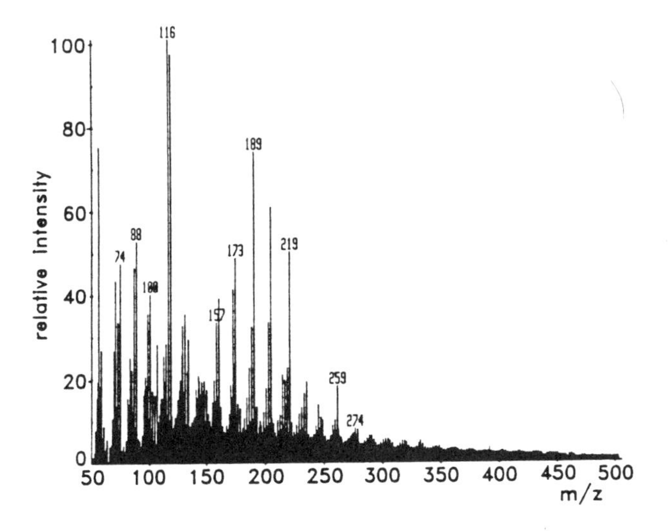

Figure 5.10. Example of the Mass Spectra obtained for an organic residue sample.

Analysis of the laboratory photoproduced organic refractories show that they can be separated into at least three groups: very complex polymers of at least 400 AMU, which have not been identified yet; a large hydrocarbon part, including minor oxygen and nitrogen containing material, in the region of 100 to 250 AMU, which are in the process of being fully characterized; and specific compounds in the region of less than 100 AMU, already confirmed to be real by isotopic labelling. These lower molecular weight compounds were studied in detail previously (see Briggs et al, 1992) by GC-MS and HLPC and were found to be C_2-C_3 hydroxy acids and hydroxy amides, glycerol, urea, hexamethylene tetramine, formamidine, ethanolamine, glycine and other amino acids. Many heterocyclic aromatic molecules seem also to be present and it has been suggested that the photoprocessing of grain mantles is required to be the source not only of the hump particles (Greenberg et al, 1986, Jenniskens et al, 1993), but the PAH's as well (Greenberg et al, 1992).

Three of the compounds identified in the organic refractory by GC-MS, glycine, glycolic acid and oxalic acid, have been detected in the Murchison meteorite (Cronin et al, 1988, Peltzer et al, 1984, Peltzer and Bada, 1978).

More recently, the middle molecular weight fraction (see Mendoza-Gómez et al, 1993 and Nibbering et al, 1993) of the laboratory photoproduced residues has been analyzed using several types of mass spectrometry. These new results indicate the presence of highly unsaturated, aromatic hydrocarbons (see table 5.2). This agrees qualitatively with what has been found for meteoritic material (Grady et al, 1983, Gilmour and Pillinger, 1985) and cometary material (Kissel and Krueger, 1987, Mukhin et al, 1990). The most refractory part of the organic residue material consists of relatively high molecular weight (≥ 400 AMU) polymers, which have not yet been completely characterized.

Table 5.2. High-resolution fast atom bombardment mass-spectral data of the organic refractory residues (U.V. irradiated H_2O:CO:NH_3=5:5:1)

-I-	-II-	-III-	-IV-
219	219.1394	219.1285050	$C_{14}H_{19}O_2$
203	203.1429	203.1435904	$C_{14}H_{19}O$
189	189.1285	189.1279403	$C_{13}H_{17}O$
173	173.1326	173.1330257	$C_{13}H_{17}$
159	159.1181	159.1173756	$C_{12}H_{15}$
157	157.1006	157.1017255	$C_{12}H_{13}$

I: FAB peak $[M+H]^+$
II: high-resolution experimentally obtained mass
III: theoretical mass of the proposed elemental composition
IV: proposed elemental composition

Finally, we also studied the volatile part by means of a gas collector (Mendoza-Gómez, 1992). Once the initial gas mixture (H_2O:CO:NH_3=5:5:1) was irradiated at 12 K, the sample was slowly warmed up, and the material coming off was trapped with a gas collector between given temperatures. The results found for material coming off from 100 K to room temperature are shown in Table 5.3. It is interesting to note how little nitrogen is contained in the intermediate masses shown here compared with the compounds discussed earlier (Briggs et al, 1992). In fact, nitrogen appears to be underabundant in comet volatiles (Wyckoff et al, 1991) and may possibly be embedded in the highly polymerized fraction which in our residue has not yet been analyzed. Indirectly these new results appear to confirm the comet origins as being interstellar.

It is possible that in some astrophysical situations, when the dust is heated in the presence of protostellar sources, these intermediate size molecules are evaporated and become mixed with the gas.

Table 5.3. High-resolution (H.R.) electron mass-spectral data of the gas material collected from 100 K to room temperature (same sample as in table 5.2)

-I-	-II-	-III-	-IV-
44	43.98993	43.9898293	CO_2
44	44.02936	44.0262148	C_2H_4O
70	70.04208	70.0418649	C_4H_6O
84	84.05819	84.0575150	C_5H_8O
98	98.07269	98.0731650	$C_6H_{10}O$
103	103.0431	103.0421992	C_7H_5N
115	115.1354	115.1360997	$C_7H_{17}N$
116	116.0615	116.0626003	C_9H_8

I: EI peak M^+
II: high-resolution experimentally obtained mass
III: theoretical mass of the proposed elemental composition
IV: proposed elemental composition

Some of the molecules identified using high resolution(elemental compositions), MS/MS mass spectral data and model compounds for direct comparison are shown below: Note how cyclic compounds seem to be well represented. This is encouraging because polycyclic aromatics appear to require a source in the interstellar medium (Greenberg et al 1993) rather than circumstellar as often suggested (Frenklach,1990) .

m/z 136 m/z 142

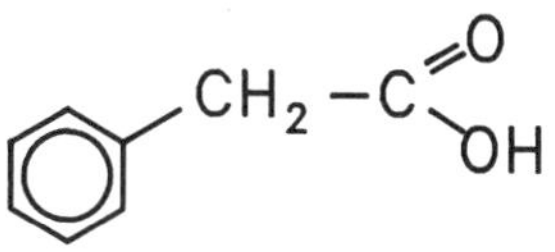

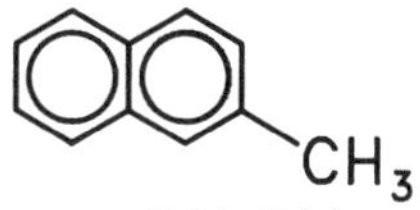

m/z 129  m/z 214

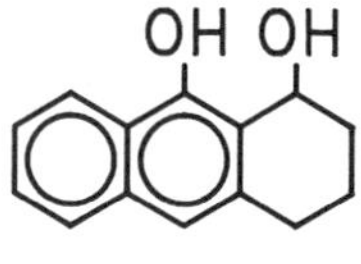

m/z 98 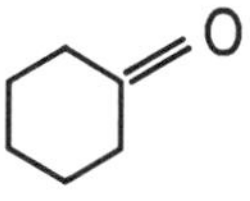 m/z 102

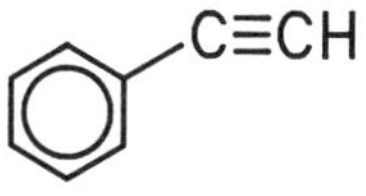

m/z 116 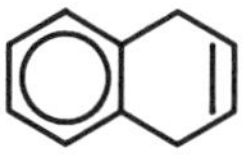 m/z 84

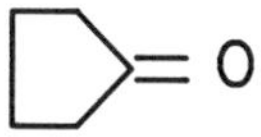

5.5. RELEVANCE TO THE ORIGIN OF LIFE

It has been proposed (Gilbert, 1986) that RNA was probably the first self-replicating macromolecule. The formose reaction might be the most plausible prebiological route to ribose synthesis. Ribose is, however, only a few percent at best of the total formose reaction product, and may not have been stable under primitive Earth conditions (Shapiro, 1988). Derivatives of glycerol and other similar polyalcohols have been proposed as substitutes for ribose in acyclic nucleotide analogs (Joyce et al, 1987, Joyce, 1989). These simple molecules may have combined with purine and pyrimidine bases and polymerized to form the first nucleic-acid like bio-macromolecules (see Fig. 5.9).

However, even though laboratory and observational results show that the organic matter in interstellar dust should contain many prebiotic molecules, a nagging question has remained as to how these molecules, which are formed racemically, were assembled into biological polymers with a single symmetry; i.e. selectively assembled with a handedness characteristic of all living organisms.

The assumption that "life can not and never could exist without molecular dissymmetry" (Terent`ev and Klabunovskii, 1959) is quite generally accepted. Recently Bonner (Bonner and Rubinstein, 1990, Bonner, 1991) has reviewed the status of this assumption. He has also reviewed all known mechanisms which could have produced the degree of chirality in prebiotic molecules sufficient to provide a basis for amplification on the earth and the subsequent formation of prebiotic polymers to initiate life. His conclusion is that no such mechanism exists on earth itself but that in interstellar space, the small (submicron) dust particles with mantles of organic molecules (Greenberg, 1984) are occasionally subjected to irradiation by circularly polarized ultraviolet light which can selectively destroy either right (D) or left (L) monomers in the mantle molecules. The source of such radiation is found along the magnetic axis of neutron stars which are the remnants of supernova explosions. Bonner's hypothesis is based primarily on the assumptions: (1) that prebiotic molecules with mirror symmetry exist on interstellar dust mantle molecules; (2) that clouds of such dust pass with sufficient frequency near neutron stars and (3) that circularly polarized light interacting with very cold systems (dust is generally at temperatures $T_d \approx 10$ K) has the same effects on mirror symmetry molecules as if they were at room temperature. Assumption (1) is justified by noting the molecules in Table 5.1. Assumptions (2) and (3) are being subjected by us to further scrutiny.

The analysis of our organic residues and the proposition that comets are primordial aggregates of interstellar dust suggest that the major components necessary for synthesis of nucleic-acid-like molecules (Loomis, 1988) could have been delivered by impacts of fluffy comets on the Earth (see Greenberg this volume). However, another possibility is direct delivery of the dust to the Earth as the solar system passes through the interstellar medium.

6. Interstellar dust on the Earth

As the Earth revolves around the sun, and the solar system as a whole revolves around the center of the Milky Way, we are continually passing through the gas and dust in space. The solar system at the current epoch is in the midst of the most tenuous kind of gas so that the interstellar dust can barely make its presence felt or observed and then only by measurements made outside of the Earth's orbital plane. Even so, a real detection of interstellar dust passing through the solar system at the distance of Jupiter has been made and suggests that the solar wind and magnetic field have less effect than had been expected (Grun et al,1993).

At quite a few times in the past, the solar system was immersed in dense molecular clouds from which it could have gathered large quantities of dust. Although including the gas phase molecules H_2CO and HCN can play a role in prebiotic chemistry, we also realize that the total fraction of available C, O, and N seen in these molecules is quite small compared to what we infer to be in the dust. Consequently, although we may use the gas phase molecules as an indicator, we will consider that the bulk of the organic molecules are bound in the dust.

We conservatively estimate that 0.1% of *all* the mass of our entire Milky Way is in the dust. For example, in a molecular cloud of 1 pc radius and a hydrogen density $n_H = 10^4$ cm^{-3} the complex organic molecules alone -of which about 1/4 are the very complex organic residues (yellow stuff)- account for a mass equal to that of about 100 times the mass of our sun.

However it is not enough to show that space is an enormous chemical factory producing more complex molecules than all possible planets. For example, if now or at any time since life began, a large quantity of this material were dumped on the planet earth while we passed through a dark cloud, it could hardly provide a competition for the already ongoing life process. If anything, it might be poisonous rather than life-giving. As an aside, it is interesting to speculate how the presence of a cloud around the solar system might appear. First of all, if we were in the center of a 1 pc radius, $n_H = 10^4$ cm^{-3} cloud, the Milky Way and all of the stars would disappear to the naked eye. In fact, all we would see would be the sun (insignificantly affected in brightness), the moon and the visible planets. On the other hand, the sky would light up very brightly -much brighter than the present zodiacal light or Milky Way- after sunset because of the scattered sunlight off the interstellar dust particles.

To return to the question of why we consider the prebiotic chemistry of space as a possible alternative for the origin of life to the prebiotic chemistry arising from atmospheric phenomena. Two reasons, as we see it, are that this chemistry could have been important earlier in the life of the Earth (see Kasting and Schidlowski chapters) and the state of chemical evolution of the dust molecules were sufficiently key life progenitors that upon deposition they were capable of further development at an already advanced level, including chirality as already mentioned.

6.1. ACCRETION OF INTERSTELLAR DUST BY THE PRIMITIVE EARTH.

There are many ways by which matter from space has been -and still is- deposited on the earth. For example, when the earth was first formed, there were enormous quantities of debris still remaining in the solar system which bombarded the earth's surface. Furthermore, since the solar system may be presumed to have formed and remained for several million years within a molecular cloud complex it would have continued to accrete large quantities of interstellar matter during this time. However, the current thinking about the state of the earth's surface at such an early epoch makes it highly unlikely that even abundant deposits of prebiotic material could have either survived or found a suitable environment for life to have evolved. On the other hand, since there now appears to be evidence for life having already been present on the earth some 3.8 billion years ago we must limit ourselves to the questions of when and how prebiotic matter could have been deposited on the earth's surface during this first 700 million years and actually perhaps in an even narrower time frame after the late bombardment era, between ~ 500 and 700 million years. We consider here only the accretion during passage of the solar system through dense clouds. Comets will be considered later (Greenberg this volume). Meteorites were also possible contributors to the earth's early organic environment (see chapter by Cronin and Chang).

It turns out that at the distance of the sun from the galactic center, the solar system is rotating about the galactic center at about twice the angular speed of the spiral pattern.

Therefore, since the pattern consists of two main spiral arms and since the galactic rotation period of the solar system is about 200 million years (Bok, 1981), the sun and earth pass through a relatively high concentration of dust and gas at the inner edges of spiral arms every 110 million years. This is shown schematically in Figure 6.1.

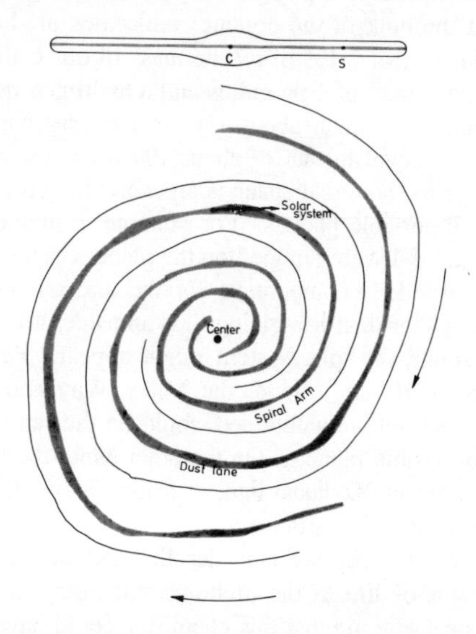

Fig. 6.1. Schematic diagram of a spiral galaxy seen edge-on and face-on showing the concentration of dust. The position and relative velocity of the solar system with respect to the spiral pattern is illustrated at a time when it was passing through a region where the dust clouds are concentrated.

We have thus passed through a spiral arm some 40 times since the formation of the solar system. It can be estimated (Talbot and Newman, 1977) that at the present epoch the collision probability of the earth with clouds of hydrogen density $n_H \sim 2 \times 10^3$ cm^{-3} is about 0.25 for each passage through a spiral arm. The number of clouds with density $n_H \sim 10^4$ cm^{-3} is somewhat less certain than that for the lower densities because of observational selection effects. However, a rough approximation to the fall-off in cloud number density is that it decreases inversely as matter density. We conclude then that during the 200 million year window the probability was of the order of unity that the earth passed through one very dense cloud ($n_H > 10^4$ cm^{-3}) and that the passage through clouds of density $n_H > 10^3$ cm^{-3} probably occurred as may as 3-4 times. We ignore the clouds of lower density because they may deposit material at too low a rate to be important. Perhaps this should be examined further because these clouds are far more abundant.

An estimate of the total mass of dust and molecules accreted during cloud passage is given essentially by a simple calculation based on sweeping up by the projected area of the earth. Effects of gas dynamics or of gravitational amplification or solar effects on this phenomenon appear to be relatively small. The rate at which the mass of *complex* molecules is accreted is:

$$dm_{org}/dt = \alpha \pi R_e^2 V \rho_{dm}$$

where α is a number of the order of unity depending on the orbits of the dust through the solar system near the earth, R_e = earth radius, V = cloud speed with respect to the solar system, ρ_{dm} = density of interstellar dust mantles. A mean relative azimuthal speed between the solar system and a cloud seems to be about 10 km s^{-1} (Spitzer, 1978). The density of the dust mantles has been shown to be about 0.5 percent of the hydrogen density. If we insert V = 10 km s^{-1}, $\rho_{dm} \approx$ o.005 $n_H m_H$ in the above equation we get $M \approx 5x10^8$ g yr^{-1} for $n_H \approx 2x10^3$ cm^{-3} and $M \approx 5x10^8$ g yr^{-1} for $n_H \approx 10^4$ cm^{-3}. Thus the rate of input of complex prebiotic type dust molecules during cloud passage is in the range of $5x10^8$ to $5x10^9$ grams per year.

Since the passage time through a typical cloud is 10^5 - 10^6 years, the total deposition is between 10^{14} g and 10^{16} g which is 10^{-14} to 10^{-12} of the earth's total mass. This is about 1% the current biomass of the earth and if partly chiral (Greenberg, this volume, Bonner 1991) could lead to chiral amplification and reproducible prebiotic polymers.

Since the surface to mass ratio of a dust grain is so high, modification of the chemical composition of the dust as it impinges on the earth through a dense atmosphere (see Kasting, this volume) would be small. Any existing atmosphere would act like a cushion slowing down these submicron particles without significantly heating them as occurs for much larger meteors and even the interplanetary dust (Brownlee, 1978). However, although the nonvolatile component would be relatively unaffected, the more volatile outer mantle constituents would be evaporated by heating from the sun and subsequent heating while penetrating the atmosphere.

7. Summary

It has been shown in this chapter that organic molecules of prebiotic significance are an abundant constituent of the interstellar dust. In Fig. 7.1 is shown a representative large grain as seen in low density clouds and as seen after all condensables have accreted as icy mantles and all small carbonaceous particles as inclusion in the mantles at the late stage of cloud contraction before the formation of the protosolar nebulae.

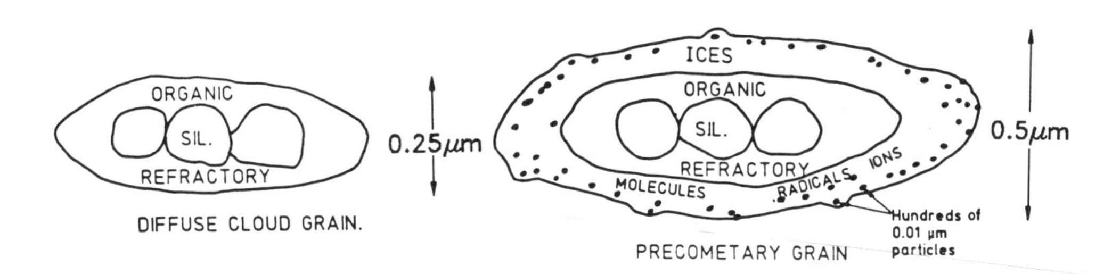

Figure 7.1. Dust model. A representative large grain as seen in low density clouds and as seen after all condensables have accreted as icy mantles and all small carbonaceous particles as inclusions in the mantles at the late stage of cloud contraction before the formation of the protosolar nebulae.

Table 7.1 gives a summary of the molecular species identified or inferred in each component of the average large presolar interstellar grain. Estimation of relative abundances are to be taken as indicative but not conclusive because they have to be based on an uncertain extrapolation of our knowledge of dust/gas chemistry in collapsing clouds.

Table 7.1. Molecules directly observed in interstellar grains and/or strongly inferred from laboratory spectra and theories of grain mantle evolution.

	Molecule	Comment*	
(1)[a]	H_2O	O	M2
(0.1)	CO	O	M2
(?)	H_2S	I	M2
(0.01)	NH_3	O	M2
(0.03)	H_2CO	O	M2
(?)	$(H_2CO)_n$	I	M2
(0.01)	OCN^-	O	M2
(0.01)	NH_4^+	O	M2
(0.05)	CH_3OH	O	M2
(0.01)	OCS	O	M2
(0.05)	CO_2	O	M2
(0.01)	CH_4	O	M2
(?)	S_2	I	M2
	complex organic	O	M1
	"Silicate"	O	C
	"Carbonaceous"	(O,I)	B
	PAH's	(O,I)	B

*: O= Observed; M1 = inner mantle; M2 = outer mantle, B = small, I = inferred; C = core
a: Estimation of relative abundances in interstellar dust *icy mantles.*

8. Acknowledgements

This work has largely been supported by NASA grant # NGR-33018148. Additional support has been provided by a grant from the Netherlands Space Research Organization (SRON). We thank Professor Nibbering and Dr. Fokkens (Inst. of M.S., Univ. of Amsterdam) for permission to quote their work liberally. Finally we thank Dr. Menno de Groot (Lab. Astrophysics, Univ. of Leiden) for his continued encouragement and his major contributions to making these experiments possible.

9. References

Aannestad, P.A. and Kenyon, S.J. (1979) 'Temperature fluctuations and the size distribution of interstellar grains', Astrophys. J. 230, 771-781.

Agarwal, V.K., Schutte, W., Greenberg, J.M., Ferris, J.P., Briggs, R. Connor, S., van de Bult, C.E.P.M. and Baas, F. (1985) 'Photochemical reactions in interstellar grains, photolysis of CO, NH_3 and H_2O', Origins of Life 16, 21-40.

Allamandola, L.J., Tielens, A.G.G.M. and Barker, J.R. (1989) 'Interstellar polycyclic aromatic hydrocarbons: the infrared emission bands, the excitation/emission mechanism, and the astrophysical implications', Astrophys. J. Suppl. Ser. 71, 733-775.

Bok, B.J. (1981) 'The Milky Way galaxy', Scientific American 244, March, 70-91.

Bonner, W.A. (1991), "The origin and amplification of biomolecular chirality" Origins of Life and Evolution of the Biosphere 11, 59.

Bonner, W.A., Rubinstein, F. (1990), in C. Ponnamperuma, F.R. Eirich (eds.), Prebiological Self Organization of Matter, p. 35.

Breukers, R.J.L.H. (1991) 'Thermal and chemical processes in the evolution of interstellar dust and gas', Thesis Leiden.

Briggs, R., Ertem, G., Ferris, J.P., Greenberg, J.M., McCain, P.J., Mendoza-Gómez, C.X. and Schutte, W. (1992) 'Comet Halley as an aggregate of interstellar dust and further evidence for the photochemical formation of organics in the interstellar medium', Origins of Life and Evolution of the Biosphere 22, 287-307.

Brownlee, D. (1978) 'Microparticle studies by sample techniques' in: 'Cosmic Dust', J.A.M. McDonnell (ed.), J. Wiley, N.Y., pp. 295-336.

Butchart, I., McFadzean, A.D., Whittet, D.C.B, Geballe, T.R. and Greenberg, J.M. (1986) 'Three mcron spectroscopy of the galactic centre source IRS 7', Astron. Astrophys. 154, L5-L7.

Chlewicki, G. (1985) 'Observational constraints on multimodal interstellar grain populations', Thesis Leiden.

Cronin, J.R., Pizzarello, S. and Cruikshank, D.F. (1988),"Organic matter in carbonaceous chondrites, planetary satellites, asteroids, and comets" in: J.F. Kerridge and M.S. Matthews (eds.), Meteorites and the Early Solar System, Univ. Arizona Press, p. 819.

Day, K.L. (1979) 'Mid-infrared optical properties of vapor-condensed magnetsium silicates', Astrophys. J. 234, 158-161.

Désert, F.X., Boulanger, F., Léger, A., Puget, J.L. and Sellgren, K. (1986) 'Nature of very small grains: PAH molecules or silicates?', Astron. Astrophys. 159, 328-330.

d'Hendecourt, L.B. (1984) 'Role of grains in interstellar molecule formation: integration of laboratory and theoretical studies', Thesis Leiden.

d'Hendecourt, L.B., Allamandola, L.J., Baas, F. and Greenberg, J.M. (1982) 'Interstellar grain explosions: molecule cycling between gas and dust', Astron. Astrophys. 109, L12-L14.

d'Hendecourt, L.B., Allamandola, L.J. and Greenberg, J.M. (1985) 'Time dependent chemistry in dense molecular clouds. I. Grain surface reactions, gas/grain interactions and infrared spectroscopy', Astron. Astrophys. 152, 130-150.

d'Hendecourt, L.B., and Jourdain de Muizon, M. (1989) 'The discovery of interstellar carbon dioxide', Astron. Astrophys. 223, L5-L8.

Frenklach, M. (1990), in: J.C. Tarter et al (eds), Carbon in the Galaxy: Studies from Earth and Space, NASA conf. publ. 3061, p. 259.

Geballe, T.R. (1986) 'Absorption by solid and gaseous CO towards obscured infrared objects', Astron. Astrophys. 162, 248-252.

Gilbert, W. (1986) 'The RNA world', Nature 319, 618.

Gilmour, I. and Pillinger, C.T. (1985), Org. Geochem. 8, 421.

Grady, M.M., Wright, I.P., Fallick, A.E. and Pillinger, C.T. (1983), in Proc. 8th symp. Antarctic Meteorites p. 289.

Greenberg, J.M. (1968) 'Interstellar grains' in B. Middlehurst and L.H. Aller (eds.), Stars and Stellar Systems, University of Chicago Press, pp. 221-364.

Greenberg, J.M. (1973) 'Chemical and physical properties of interstellar dust', in: M.A. Gordon and L.E. Snyder (eds), Molecules in the Galactic Environment, Wiley, p. 94-124.

Greenberg, J.M. (1978) 'Interstellar dust', in J.A.M. McDonnell (ed.), Cosmic Dust, Wiley, p. 187-294

Greenberg, J.M. (1979) 'Grain mantle photolysis: a connection between the grain size distribution function and the abundance of complex interstellar molecules', in B.E. Westerlund (ed.), Stars and Star Systems, Reidel, pp. 173-193.

Greenberg, J.M. (1980) 'Interstellar dust: a sink and source of molecules' in Les Spectres des Molecules Simples au Laboratoire et en Astrophysique, Universite de Liege, pp. 555-560.

Greenberg, J.M. (1982a) 'What are comets made of? - a model based on interstellar dust' in L.L. Wilkening (ed.), Comets, University of Arizona Press, pp. 131-163.

Greenberg, J.M. (1982b) 'Dust in dense clouds. One stage in a cycle' in J.E. Beckman and J.P. Phillips (eds.), Submillimetre Wave Astronomy, Cambridge University Press, pp. 261-306

Greenberg, J.M. (1984) 'The structure and evolution of interstellar grains', Scient. Amer. 250, 124-135.

Greenberg, J.M. (1991) 'Physical, chemical and optical interactions with interstellar dust' in J.M. Greenberg, V. Pirronello (eds.), Chemistry in Space, Kluwer, Dordrecht, pp. 227-261.

Greenberg, J.M., de Groot, M.S. and van der Zwet, G.P. (1986) 'Carbon components of interstellar dust', in The Comet Nucleus Sample Return Mission, Proc. workshop Canterbury, UK (15-17 July 1986), SP-249, pp. 143-151

Greenberg, J.M. and Hong, S.S. (1976) 'Infrared emission from grains with fluctuating temperatures' in M. Rowan-Robinson (ed.), Far Infrared Astronomy, Pergamon, p. 299.

Greenberg, J.M., Yencha, A.J., Corbett, J.W.and Frisch, H.L. (1972) 'Ultraviolet effects on the chemical composition and optical properties of interstellar grains', Memoires de la Societe Royale des Sciences de Liege, 6e serie, tome III, pp. 425-436.

Greenberg, J.M., Mendoza-Gómez, C.X., de Groot, M.S. and Breukers, R. (1992) 'Laboratory dust studies and gas-grain chemistry' in T.K. Millar and D.A.Williams (eds.), Dust and Chemistry in Astronomy, IOP Publ., pp. 265-288.

Grim, R.J.A., Baas, F., Geballe, T.R., Greenberg, J.M. and Schutte, W. (1991) 'Detection of solid methanol toward W 33A', Astron. Astrophys. 243, 473-477.

Grün, E., et al (1993) 'Discovery of Jovian dust streams and interstellar grains by the Ulysses spacecraft', Nature 362, 428-430.

Jenniskens, P. (1992) 'Organic matter in interstellar extinction', Thesis Leiden.

Jenniskens, P., Baratta, G.A., Kouchi, A., de Groot, M.S., Greenberg, J.M. and Strazzula, G. (1993) 'Carbon dust formation on interstellar grains', Astron.Astrophys. accepted for publication.

Joyce, G.F. (1989) 'RNA evolution and the origins of life', Nature 338, 217.

Joyce, G.F., Schwartz, A.W., Miller, S.L. and Orgel, L.E. (1987) 'The case for an ancestral genetic system involving simple analogs of the nucleotides', Proc. Nat. Acad. Sci. (USA) 84, 4398-4402.

Khare, B.N. and Sagan, C. (1973) 'Experimental interstellar organic chemistry: preliminary findings' in M.A. Gordon and L.E. Snyder (eds.), Molecules in the Galactic Environment. Wiley.

Kissel, J. and Krueger, F.R. (1987) 'The organic component in dust from comet Halley as measured by the PUMA mass spectrometer on board Vega 1', Nature 326, 775.

Lacy, J.H., Carr, J.S., Evans, N.J., Baas, F., Achtermann, J.M. and Arents, J.F. (1991) 'Discovery of interstellar methane: observations of gaseous and solid CH_4 absorption toward young stars in molecular clouds', Astrophys. J. 376, 556-560.

Léger, A., Verstraete, L., d'Hendecourt, L., Dutuit, O., Schmidt, W. and Lauer, J.C. (1989) 'The PAH hypothesis and the extinction curve' in L.J. Allamandola (ed.), Interstellar Dust, Kluwer, pp. 173-180.

Loomis, W.R. (1988) 'Four Billion Years', Sinauer Ass., Sunderland, MA.

Mendoza Gómez, C.X. (1992) 'Complex irradiation products in the interstellar medium', Thesis Leiden.

Mendoza-Gómez, C.X. and Greenberg, J.M. (1993) 'Laboratory simulation of organic grain mantles', Origins of Life and Evolution of the Biosphere 23, 23-28.

Mendoza-Gómez, C.X., Greenberg, J.M., Nibbering, N.M.M. and Fokkens R.H. (1993) *In preparation.*

Mitchell, G.F., Allen, M. and Maillard, J.-P. (1988) 'The ratio of solid to gas phase CO in the line of sight to W33A', Astrophys. J. 333, L55-L58.

Mukhin, L.M., Dikov, Y.P., Evlanov, E.N., Fomenkova, M.N., Nazarov, M.A., Priludsky, O.F., Sagdeev, R.Z. and Zubkov, B.U. (1989), in Lunar Planetary Science conf. XX, Houston, Texas, USA, p. 733.

Nibbering, N.M.M., Fokkens, R.H., Mendoza-Gómez, C.X. and Greenberg, J.M. (1993) *In preparation.*

Norman, C. and Silk, J. (1980) 'Clumpy molecular clouds: a dynamical model self-consistently regulated ty T Tauri star formation', Astrophys. J. 238, 158.

Onaka, T., Nakada, Y., Tanabe, T., Sakata, A. and Wada, S. (1986) 'A quenched carbonaceous composite (QCC) grain model for the interstellar 220 nm extinction hump', Astroph. Space Sc. 118, 411-413.

Orgel, L. (1987) 'Evolution of the genetic apparatus: a review' Cold Spring Harbor symp. Quant. Bio. 52, 9-16.

Peltzer, E.T., and Bada, J.L. (1978) 'α-Hydroxycarbolic acids in the Murchison meteorite', Nature 272, 443.

Peltzer, E.T., Bada, J.L., Schlesinger, G. and Miller, S.L. (1984), Adv. Space Res. 4, 69.

Prasad, S.S., and Tarafdar, S.P. (1980) 'UV radiation field inside dense clouds: its possible existence and chemical implications', Astrophys. J. 267, 603-609.

Sakata, A., Wada, S., Tanabe, T. and Onaka, T. (1984) 'Infrared spectrum of the laboratory-synthesized quenched carbonaceous composite (QCC): comparison with the infrared unidentified emission bands',. Astrophys. J. 287, L51-L54.

Sandford, S.A., Allamandola, L.J., Tielens, A.G.G.M., Sellgren, K., Tapia, M. and Pendleton, Y.

(1991) 'The interstellar C-H stretching band near 3.4 microns: constraints on the composition of organic material in the diffuse interstellar medium', Astrophys. J. 371, 607-620.

Sandford, S.A., Allamandola, L.J., Tielens, A.G.G.M., and Valero, G.J. (1988) 'Laboratory studies of the infrared spectral properties of CO in astrophysical ices', Astrophys. J. 329, 498-510.

Sato, S., Nagata, T., Tanaka, M. and Yamamoto, T. (1990) 'Three micron spectroscopy of low-mass pre-main-sequence stars', Astrophys. J. 359, 192-196.

Schmitt, B., Espinasse, S., Grim, R.J.A., Greenberg, J.M. and Klinger, J. (1989) 'Laboratory studies of cometary ice analogues' in Proc. Intern. workshop on Physics and Mechanics of Cometary Materials, ESA SP 302.

Schmitt, B., Greenberg, J.M. and Grim, R.J.A. (1989a) 'The temperature dependence of the CO infrared band strengths in CO:H_2O ices', Astrophys. J. 340. L33-L36.

Schmitt, B., Grim, R. and Greenberg, J.M. (1989b) 'Spectroscopy and physico-chemistry of CO:H_2O and CO_2:H_2O ices' in Proc. 22nd ESLAB symp. on infrared spectroscopy in astronomy, Salamanca, Spain, SP 290, pp. 213-219.

Schutte, W.A. (1988) 'The evolution of interstellar grain mantles', Thesis Leiden.

Schutte, W.A., Geballe, T.R., van Dishoeck, E.F. and Greenberg, J.M. (1993) 'Solid formaldehyde towards the protostar GL2136', in preparation.

Schutte, W.A. and Greenberg, J.M. (1986) 'Formation of organic molecules on interstellar dust particles', in F. Israel (ed.), Light on Dark Matter,Reidel, pp. 229-232.

Schutte, W.A. and Greenberg, J.M. (1991) 'Explosive desorption of icy grain mantles in dense clouds', Astron. Astrophys. 244, 190-204.

Schutte, W.A., Tielens, A.G.G.M. and Sandford, S.A. (1991) '10 micron spectra of protostars and the solid methanol abundance', Astrophys. J. 382-529.

Shapiro, R. (1988), Origins of Life 18, 71.

Smith, R.,G., Sellgren, K. and Tokunaga, A.T. (1988) 'A study of H_2O ice in the 3 micron spectrum of OH 213.8+4.2 (OH 0739-14)', Astrophys. J. 334, 209-219.

Spitzer, L. Jr. (1978) 'Physical Processes in the Interstellar Medium', J. Wiley & Sons, New York.

Talbot, R.J. Jr. and Newman, M.J. (1977) 'Encounters between stars and dense interstellar clouds', Astrophys. J. Supplement 34, 295.

Tanaka, M., Sato, S., Nagata, T. and Yamamoto, T. (1990) 'Three micron ice-band features in the ρ Ophiuchi sources', Astrophys. J. 352, 724-730.

Terent'ev, A.P., Klabunovskii, E.F. (1959), in F. Clark, L.M. Synge (eds), The Origin of Life on Earth, Pergamon, New York, p. 95.

Tielens, A.G.G.M. (1991) 'Characteristics of interstellar and circumstellar dust' in A.C. Levasseur-Regourd and H. Hasegawa (eds.), Origin and Evolution of Interplanetary Dust, Kluwer, pp. 405-412.

Tielens, A.G.G.M., and Allamandola, L.J. (1987) in G.E. Morfill and M. Scholer (eds.), Physical Processes in Dense Clouds, Reidel, p. 333.

Tielens, A.G.G.M., Allamandola, L.J., Bregman, J., Goebel, J., d'Hendecourt, L. and Witteborn, F.C. (1984) 'Absorption features in the 5-8 micron spectra of protostars, Astrophys.J. 287, 697-706.

Tielens, A.G.G.M., Tokunaga, A.T., Geballe, T.R. and Baas, F. (1991) 'Interstellar solid CO: polar and nonpolar interstellar ices', Astrophys. J. 381, 181-199.

Van de Bult, C.E.P.M., Greenberg, J.M. and Whittet, D.C.B. (1984) 'Ice in the Taurus molecular cloud: Modelling of the 3 micron profile', Monthly Not. RAS 214, 289-305

Van de Hulst, H.C. (1957) 'Light scattering by small particles', Wiley.

Van de Hulst, H.C. (1949) 'The solid particles in interstellar space', Rech. Astron. Obs. Utrecht 11, pt. 2.

Whittet, D.C.B., Bode, M.F., Longmore, A.J., Adamson, A.J., McFadzean, A.D., Aitken,D.K. and Roche, P.F. (1988) 'Infrared spectroscopy of dust in the Taurus dark clouds: ice and silicates', Mon. Not. R. astr. Soc. 233, 321-336.

Whittet, D.C.B. and Duley, W.W. (1992) 'Carbon monoxide frosts in the interstellar medium', Astron. Astrophys. Rev. 2, 167-189.

Willner, S.P., and Pipher, J.L. (1982), in G.R. Reigler and A.D. Blandford (eds.), The Galactic Center, Amer. Inst. Phys., p. 77.

Wyckoff, S., Tegler, S.C. and Engel, L. (1991) 'Ammonia abundances in four comets', Astrophys. J. 368, 279-286.

LABORATORY SIMULATIONS OF GRAIN ICY MANTLES PROCESSING BY COSMIC RAYS

Valerio Pirronello
Istituto di Fisica
Universita' di Catania
95125 Catania
Italy

ABSTRACT. Chemical alterations induced by cosmic rays in the icy mantles of grains inside dense interstellar clouds can produce prebiotic species that may be important to understand the seeds of the origin of life in the Universe. The widespread experimental effort to reproduce in the laboratory processes involving fast ions and ice mixtures of astrophysical interest in a sort of "simulation" of what may take place in space is reported.

1. Introduction

In recent years laboratory investigations of the principal effects of the interaction among energetic charged particles, frozen gases and mixtures of them have given clear evidence of the fact that a rich chemistry can take place, even at temperatures as low as 10 K, in regions few tens of Angstrom wide along the track of impinging ions.

On a complexity scale synthesized species range from the most simple, but important, molecule H_2 (Brown et al., 1982) to prebiotic species like formaldehyde (Pirronello et al., 1982; 1989) to polymeric materials (see e.g. Moore and Donn, 1982).

High confidence, even if till now not yet incontrovertible proof, exists that amino acids can be synthesized by the energy loss of fast projectiles in suitable ice mixtures containing hydrogen, carbon, nitrogen and oxygen.

The plan of this presentation is the following: first we will give a short description of the current knowledge about the energetic particle environment in which interactions relevant to our subject take place; we will then review the experimental results on the chemical modifications induced in ice mixtures by energetic impinging ions; some attention will be also paid to the mechanisms that are responsible for the formation of chemical species and special emphasis will be given to non thermal equilibrium processes of synthesis of molecules that usually go under the name of "Hot atom chemistry", a framework known to proceed under certain conditions in the gas phase that has been proposed by Roessler (1986) also for reactions induced by fast particles penetrating in solids.

J. M. Greenberg et al. (eds.), The Chemistry of Life's Origins, 33–53.

2. The Particle Environment

The particle environment, we will refer to mainly, is that of interstellar dense clouds, as described in this book by Mayo Greenberg. If no stellar sources are embedded the background energetic particles are substantially those belonging to the galactic component of cosmic rays (C.R.). Particles and ions that do not have enough energy to penetrate into the solids but can be important for gas phase reaction networks and surface chemistry on grains will not be considered here.

The existence of a continuous bombardment of our Earth by fast ions coming from external space was discovered by Victor Hess around 1912 during a balloon flight as is testified to by the rhyme that appears in one of the walls of the Majorana Center in Erice (Sicily)

Here in the Erice maze
cosmic rays are the craze
and this because a guy named Hess
ballooning up found more not less

Up to now investigations on cosmic rays are performed both from the ground, by balloon born missions and by space crafts, either in orbit around the earth or in interplanetary missions like Voyager and Ulysses. The composition of C.R. in the vicinity of the earth shows a relevant abundance of nucleons (about 98%) and a minority of electrons and positrons (about 2%). In Figure 1 (taken from Wefel, 1988) is shown a comparison of the relative abundance of elements with atomic number up to 30 for galactic cosmic rays (GCR) and solar energetic particles (SEP) measured at the Earth.

The composition and energy distribution of light cosmic rays measured near the earth (see Figure 2) is not always representative of the distribution of cosmic rays at large distances from the Sun and in particular is not representative of those cosmic rays that pervade our galaxy and that penetrate interstellar clouds inducing several interesting effects and in particular regulating the chemical reaction networks inside them through the ionization degree of chemical species. At energies higher than 10 GeV the C.R. spectrum is well fitted by a power law that goes as $E^{-2.7}$ while the energy spectrum below 1 GeV is almost completely unknown. This is because the solar magnetic field, carried by the solar wind regulates the electrodynamics inside a region, called the heliosphere, that extends till about one hundred A.U. Such an interplanetary magnetic field screens out the low energy light cosmic rays, and especially hydrogen, helium and electrons, from the inner solar system region. Theoretical effort to evaluate, from the measured spectrum of light C.R. the spectrum outside the heliosphere has been done for instance by Morfill, Voelk and Lee (1976). They, by means of the transport theory in an archimedean spirally shaped interplanetary magnetic field having some scattering irregularities, estimated a spectrum of cosmic rays outside the heliosphere that can be as high as thirty or forty times the measured one. Recently on the basis of an idea of Pirronello (1985), it has been proposed a method to obtain "observational" information by means of the detection in cometary comae of molecules synthesized by the cosmic ray irradiation of cometary nuclei when they still were in the Oort cloud (Pirronello and Lanzafame, 1989).

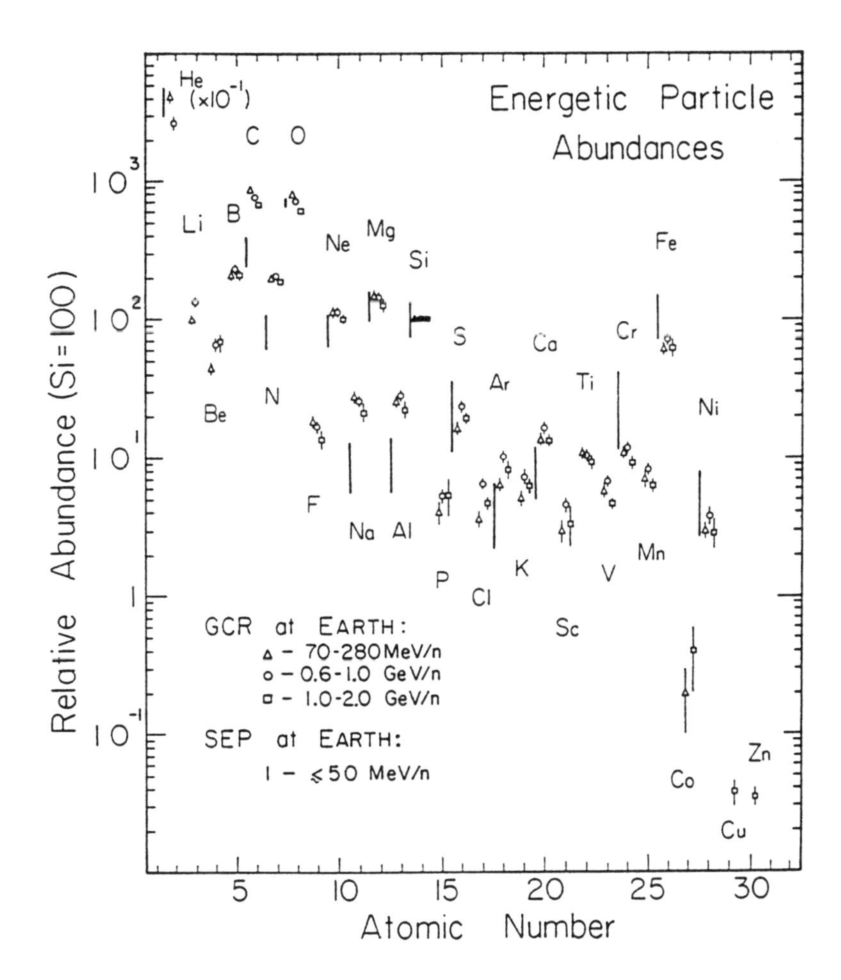

Figure 1. Relative abundances of galactic cosmic rays (GCR) and solar energetic particles (SEP) measured in the vicinity of the Earth in several energy intervals

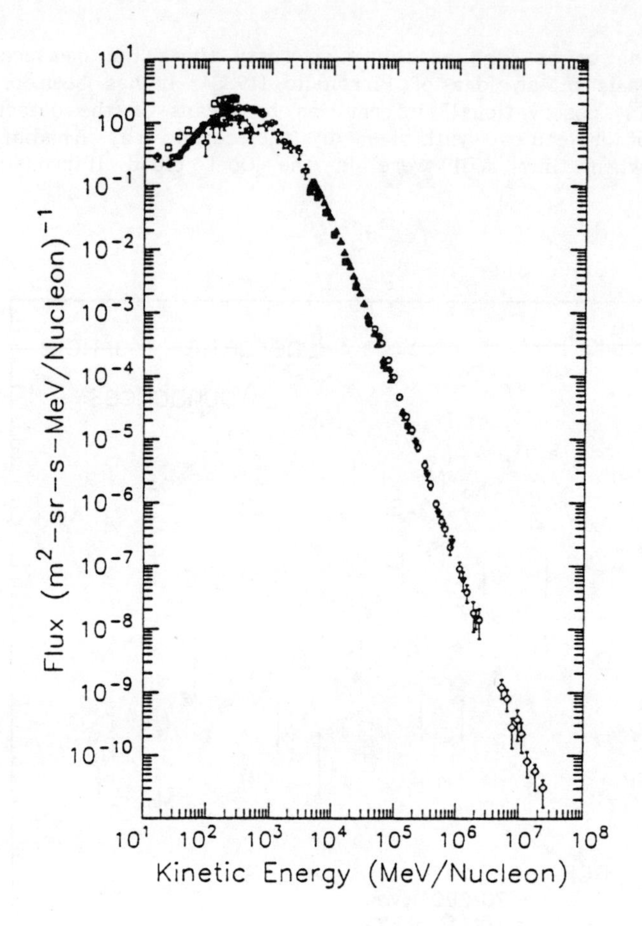

Figure 2. The differential energy spectrum for galactic cosmic rays

3. Molecule Formation

Fast particles penetrating a solid lose part of their energy by elastic and inelastic collisions with atoms and molecules of the target material. The energy lost per unit path length, the so called "stopping power" of the ion in the material **dE/dx**, is tabulated (Ziegler, 1980) and is due to the superposition of the "inelastic" $(dE/dx)_{in}$ and the "elastic" $(dE/dx)_{el}$ contributions, i.e.

$$(dE/dx)_{tot} = (dE/dx)_{in} + (dE/dx)_{el}$$

The first of the two contributions is due to elastic collisions between the impinging particle and the "core" (nucleus plus inner shell electrons) of the target atoms and is sometimes called "nuclear"; the other one is due to ionizations, excitations and charge exchange reactions induced by the impinging particle interacting with the electronic cloud of the target atoms and for this reason is often called "electronic". Important features of the elastic and inelastic contributions to the energy loss of particles in matter are:

a) both contributions (the nuclear one and the electronic one) first increase, go through a maximum and then decline with energy;
b) the nuclear stopping power is dominant at lower kinetic energies of the impinging ion, while the electronic one is dominant at higher energies;
c) approximately the electronic contribution becomes dominant over the nuclear contribution when the speed of the projectile is equal to or higher than the average orbital speed of the external electrons of atoms.

Several relevant effects are induced in solids by particle bombardment such as erosion (that consists in the ejection of atoms and molecules from the solid during irradiation), structural and optical properties changes. But the effect we want to pay more attention here is the **synthesis of new molecules** (some of which are particularly difficult to form in the gas phase in the extreme conditions encountered in the interstellar medium).

The possibility to perform laboratory simulations, reproducing conditions that are similar to those in which the interaction between particles and ices occur in space, gives to these studies a particular relevance.

3.1. LABORATORY SIMULATIONS

The type of equipment used to simulate in the laboratory the processes induced by the ion penetration in ices and also to measure quantitatively the effects of such interactions have been described in detail in Pirronello (1991) and will be only briefly recalled here. In general the experimental set up consists of an ultrahigh vacuum (UHV) cryo pumped chamber in which is situated a target that can be cooled to temperatures close to those occurring in dense interstellar clouds. According to the range of energy that it is intended to investigate, the generation of energetic ions occurs by means of ion guns, Van de Graaff accelerators, cyclotrons and so on.

Analyses of the effects and in particular of the chemical effects are essentially made either through mass spectrometry or by means of infrared spectroscopy. The first technique allows detection of species that are ejected during the bombardment and also species that remain trapped in the icy matrix when they are allowed to reach the gas phase by raising the temperature of the whole target. The second technique is able to detect only synthesized species still in the solid because those released during the irradiation never reach a high enough optical depth to be observed; the other drawbacks of this technique are that a column density of at least 10^{15} molecules synthesized per square centimeter is needed to overcome the threshold of detectability (while only few times 10^{12} molecules are enough for mass spectrometry) so that up to now it has been almost impossible to get quantitative results. This type of equipment allows "in situ" analyses (i.e. analyses that are made maintaining the irradiated sample still inside the scattering chamber where the interaction occurs); other "non in situ" analyses can be performed using either gas or liquid cromatography or X ray microscopes and so on. Naturally "in situ" techniques have to be considered the most important ones because they give the certainty that the effects observed are those really induced by the bombardment and not side ones introduced as an artifact through the analytical process, as can happen in the case of "non in situ" methods in which the irradiated material is manipulated, dispersed in gas or liquid, oxidized and so on.

A typical experiment of irradiation of molecular ices is carried out in the following way. After the scattering chamber has been evacuated and the required vacuum conditions have been reached, the target holder is cooled down to the temperature chosen and then the accretion of the icy film is allowed. Ice films, as thin as a couple of hundred monolayers, are deposited from the gas phase onto a substrate at about 10 K at a very low rate (to accrete an amorphous layer according to Venkatesh et al. (1974) and Rice (1975) prescriptions and according to Leger et al. (1979) and Hagen et al. (1981) with considerations on the nature of the icy mantle structure on interstellar grains).

Macroscopic non uniformities in the icy film can be avoided by providing the end of the steel pipe, used to deposit the gas mixture on the cold substrate, with a micron sized channel plate that will distribute the gas in a flat cylinder instead of in a conical shape; a movable "doser" placed at a short distance from the cold substrate also helps to obtain a well defined uniform ice layer.

Segregation (or layering) of species during a deposition of a gas mixture (for instance methane and water) can occur because different species have different sticking coefficients and hence different residence times on the walls of the pipe used to deposit the mixtures (Pirronello and Brown, unpublished). This produces a delay in the migration of some species and hence a layering of one component of the mixture with respect to the others. Such a problem can be avoided by maintaining the pipe itself at a temperature of about 400 K during deposition.

After the target has been prepared it can be irradiated with ions, using always a dose rate "δ" (number of impinging particles per second) in a range in which the yield of the effect monitored changes linearly with "δ", in order to measure single particle effects. This is a mandatory condition for applications to any astrophysical scenarios, where particle fluxes are extremely low when compared with those that can be attained in the laboratory.

When a Van de Graaff accelerator, or another source of ions of comparable energy, is available small, inverse biased, solid state detectors placed in the vacuum chamber close to the beam entrance allow measurement of the flux of particles of the beam scattered backwards from the target material to perform the so called Rutherford Backscattering analysis (RBS). This technique provides accurate measurements of the thickness of the bombarded layer, of its atomic composition, and of their changes with the dose of impinging particles. All these parameters are extremely important to maintain under controlled conditions during the whole experiment and to evaluate the erosion rate of the deposited icy film.

RBS analysis is performed by detecting light particles (usually helium ions and sometime protons) at MeV energies that are elastically scattered backwards from the nuclei of atoms belonging to the target. In the case of thin targets energy spectra of backscattered ions show peaks whose position is characteristic of each element and whose area is proportional to its abundance. Such a technique then allows one to obtain an accurate analysis of the elemental composition of the sample. Its accuracy increases for heavier elements and, naturally, the technique is not applicable in its usual way to detect the presence of elements as heavy as or lighter than the projectile.

RBS, to be sensitive also to the presence of hydrogen (an important component of almost all molecular ices of astrophysical interest), has then been performed in a modified way, as shown schematically in Figure 3.

On the left side of the upper part (a) of the figure is shown the silicon substrate, on which has been previously deposited a thin (250 A thick) gold layer (called the gold marker). When the substrate has been cooled to the required temperature, the ice layer to be irradiated is then accreted (with the above mentioned "caveat"). On the left side of the lower part (b) of the figure is shown a typical energy spectrum of backscattered ions. The sharp peak relative to ions that bounced off the gold atoms of the marker is seen as well as the broad shoulder relative to those backscattered by nuclei belonging to the "semi-infinite" silicon substrate, that appears at lower energies because silicon is lighter than gold. The important characteristic to be remarked is that the position of the gold peak in the spectra is shifted toward lower energies by an amount "ΔE_1" with respect to the position it would have had (see the arrow) if no ice film would have been accreted on the substrate. The origin of such a shift of the marker peak is due to the energy loss helium ions suffer in the ice layer both before and after bouncing against the Au nuclei. From a quantitative point of view the energy shift "ΔE" is related to the thickness "Δx" of the ice film through a function "$f(dE/dx)$" of the total stopping power (also due to the hydrogen component); a function that depends on the geometry involving the ion beam direction, the position of the target and that of the particle detector, i.e.

$$\Delta E = f(dE/dx)\ \Delta x$$

After the first thickness measurement, the sample is irradiated with the particles required (possibly the same helium beam, or protons or heavier ions and so on) until a certain fluence is reached. Another RBS spectrum is taken to measure again the thickness of the remaining layer

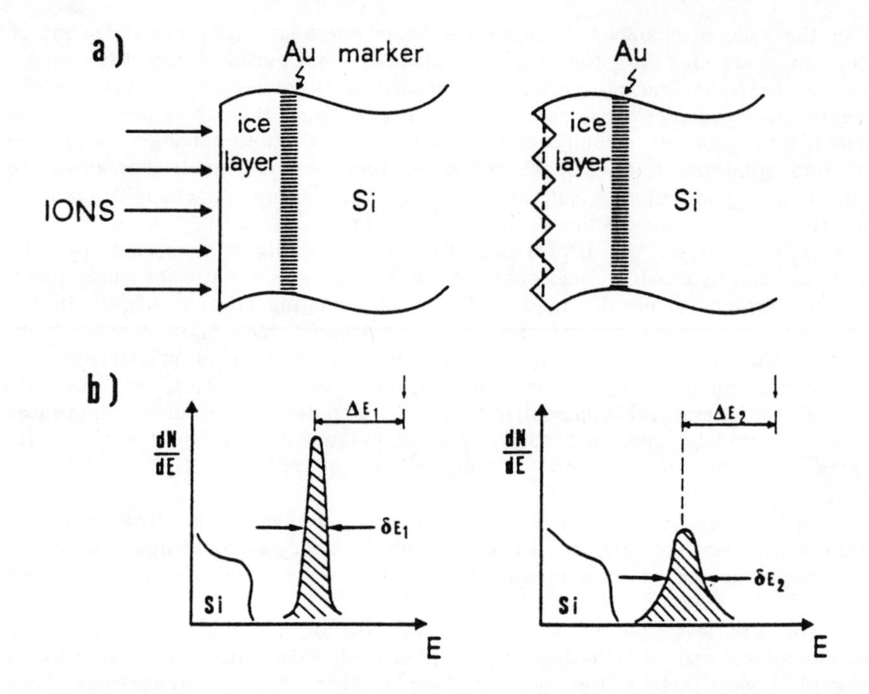

Figure 3 Schematic of RBS analysis of a molecular ice layer showing also structural changes of the film during irradiation

(Figure 3, lower part (b), on the right). As a rule, after the bombardment the shift of the Au marker peak "ΔE_2" (relative to the position it would have in the bare substrate, position indicated in the figure by the arrow) is smaller than the previous one "ΔE_1" because the thickness of the ice layer has diminished. From the difference in thickness, and knowing the dose of particles that bombarded the sample it is trivial to deduce the erosion yield characteristic of that particle at that energy at that temperature of the substrate.

As it can be seen from a comparison between the two spectra in the schematic, the Full Width at Half Maximum (FWHM) of the Au peak is different from one another: after a certain fluence of impinging projectiles it is wider than before, i.e.

$$\delta E_1 < \delta E_2$$

The explanation of the variation in the FWHM "δE" can be found in the upper part on the right of the same Figure 3. Such an increase is due to a change in the surface topography of the bombarded layer. Impinging ions that cross the ice layer where it has become thinner lose less energy and in the spectrum show up in the right wing of the peak, those crossing where the layer has remained thicker lose more energy and appear in the left wing of the peak broadening it. This effect, for films that are thin relative to the range of particles in the material, has been proved (Pirronello et al., 1981) to be proportional to the initial thickness of ice and so to involve all of it, changing its structure to a fluffier one.

3.2. RESULTS

Detection of molecules formed by the energy loss of impinging ions is now well established; several groups in several laboratories around the world have been able to synthesize chemical species that span from the simple molecular hydrogen to complex hydrocarbons. Only in very few cases, by the way, the production yield (the number of molecules synthesized by each impinging ion) has been measured quantitatively as a function of the many parameters it may depend on, such as: the energy of the impinging ion and its charge state, the temperature of the frozen layer, its degree of crystallinity and so on.

It was in 1982 that the group of Brown and co-workers at the Bell Laboratories had evidence that chemistry was going on along the track of fast ions in the irradiated solid (Brown et al., 1982). Bombarding heavy water ice layers with 1.5 MeV He^+ they detected and measured the release of D_2 and O_2. Their results are shown in Table 1 as a function of the temperature of D_2O ice.

TABLE 1. Yields of D_2 and O_2 ejected by D_2O ice upon bombardment with 1.5 MeV He^+ as a function of the substrate temperature

T(K)	D_2	O_2
60	0.2	0.6
80	0.5	0.8
100	1.0	1.2
120	2.9	2.1
140	7.5	4.0
160	30*	11*

*estimated

The process of production and release of these molecules, that involved the whole ice, occurred upon bombardment and (as is clear from Table 1) with a constant increase with temperature. Such a behavior is almost certainly due to the superposition of a temperature dependence in the formation of these new species and of a dependence of their mobility in the ice layer on temperature. This last effect, that might also be more relevant than the other one, is well known among solid state physicists and can be regarded as a typical example of a "radiation-enhanced" process.

Even more interesting chemical effects have been obtained by the irradiation of mixtures of species, prepared inside a steel bottle in the gas phase at room temperature and then deposited at the required slow rate onto the cold substrate to obtain an amorphous intimate ice mixture.

The first work on this subject was certainly made by Berger (1961) and Oro (1963), who irradiated layers of water, ammonia and methane and, using a *not in situ* analysis, observed the formation of several new molecules. After those efforts this type of investigation was abandoned and only in the early eighties Pirronello et al. (1982) restarted to irradiate frozen mixtures. They, unlike many others (see e.g. Moore et al., 1983), preferred to bombard only binary mixtures in the persuasion that it is better to obtain quantitative yields of the investigated processes rather then only qualitative information on the kind of molecules synthesized even if such choice would have meant to proceed much more slowly. With such a program in mind they first irradiated a frozen layer (1.8×10^{18} mol cm^{-2} thick) of $D_2{}^{16}O$ and $^{13}C^{18}O_2$ at 9 K with a total dose of the order of 10^{15} He^+ cm^{-2} at 1.5 MeV of and measured an average production yield of formaldehyde (either with one isotope of oxygen or with the other)

$$Y_{D_2CO} = 3.7$$

This total average yield is really the sum of the two average yields relative to $D_2C^{16}O$ and $D_2C^{18}O$. Both species were in fact detected, a fact that with the tracing of the origin of the oxygen atom can give important hints on the mechanism of formation of such species. In particular the abundance of $D_2C^{16}O$ is not easily explainable in the light of ordinary chemistry; in fact, looking at the structure of the molecule

```
D
      C  =  O
D
```

it is possible to see that in order to produce $D_2C^{16}O$ thermally a carbon atom, after loosing two ^{18}O atoms should recombine with two deuterium and one ^{16}O atom. As can be imagined, such a process is long and not very probable to occur. Perhaps a more efficient way to produce $D_2C^{16}O$ would be to **insert** a carbon in the water molecule.

The other important fact was that the synthesized formaldehyde molecules were released at the end of bombardment when the ice layer sublimed. In the light of the few experiments already described the chemical species, produced by the passage of an energetic particle are, once formed, either mobile enough to "percolate" through the ice layer, to reach the surface and to leave it, or not mobile enough to even start the first step of the process and hence to migrate outward.

Up-to-date a significant number of efforts on the formation of molecules under ion beam bombardment in frosts have been produced; among those worth mentioning either because of priority or because of the originality of their results are the following.

Moore et al. (1983) and Moore and Donn (1983) detected in the infrared the synthesis of carbon and nitrogen monoxide and dioxide in mixtures mainly composed of water, ammonia and methane and irradiated at the temperature of 20 K by 1 MeV H^+ (see for instance Figure 4, taken from their work, that shows the IR spectrum of the ice mixture before and after the proton bombardment). Moore (1984) furthermore observed the synthesis of SO_3 by irradiating frozen SO_2.

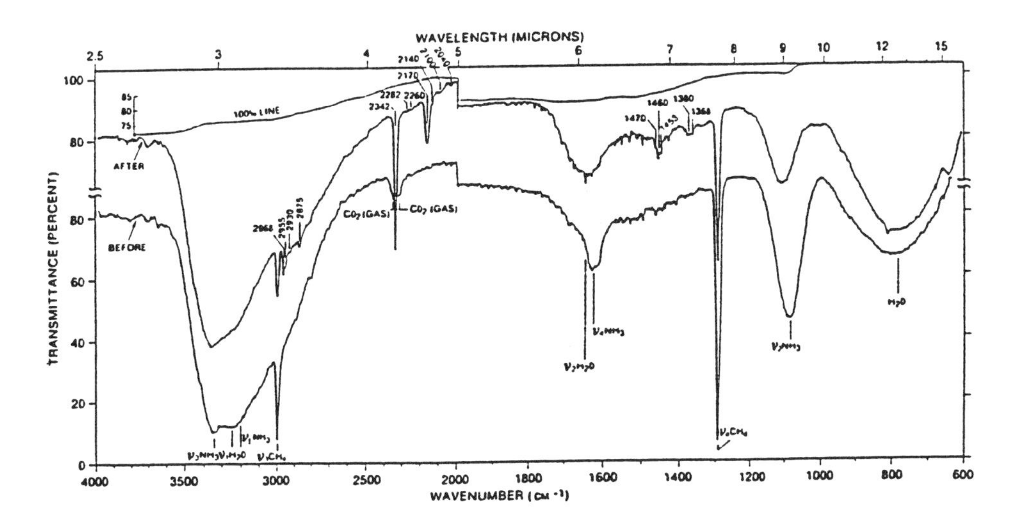

Figure 4. Infrared spectra of ice mixtures before and after bombardment with swift protons

Bar-Nun et al. (1985) irradiated water ice with H^+ and Ne^+ (using the range of energies in which the energy deposition in matter occurs predominantly through elastic processes) obtaining, as had Brown et al. (1982), molecular hydrogen and oxygen. Two major differences have to be mentioned: opposite to Brown et al. (1982) they found that the ejection of H_2 and O_2 during irradiation was more relevant than that of water molecules and furthermore they detected also an important atomic hydrogen contribution that could not be ascribed to the cracking fraction of water molecules or molecular hydrogen in the electron impact ionizer of the mass spectrometer.

Benit *et al* (1986, 1987, 1988) studied chemical processing in binary mixtures made of water and either carbon monoxide, carbon dioxide or ammonia by keV ions and more recently Benit and Brown (1990) investigated bilayers of isotopically labelled water films to study in great detail the formation of molecular oxygen.

A special quotation is deserved by the Julich group of Roessler

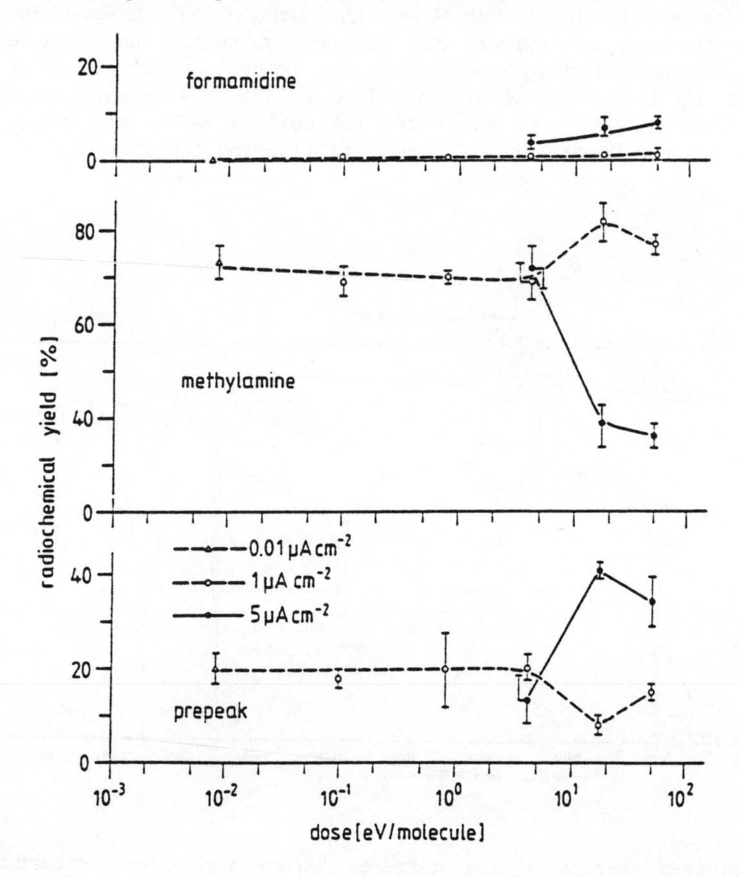

Figure 5. Radiochemical yields of species as a function of the dose

and collaborators because of their experimental work (see e.g. Nebeling et al., 1985; Roessler, 1986; Roessler, 1987; Armborst and Roessler, 1986; Roessler, 1988) but also, as we will see later, because of the framework they propose for the formation mechanisms of some molecules. In some important experiments they produced hot carbons through the nuclear reaction $N(p,\alpha)^{11}C$ and observed, for instance in NH_4Cl, together with some other simpler species, absorption bands that could be assigned to guanidine, formamidine and methylamine (see e.g. Figure 5, taken from Roessler (1986) where results are given in terms of the radiochemical yield, i.e. number of product species, either radicals or final products, per hundred eV).

Production of several radicals and molecules in pure CO, NH_3 and H_2O layers and also in H_2O/CO and NH_3/CO mixtures bombarded with keV ions at 20 K was obtained by Haring et al. (1983, 1984a,b,c). They also, by means of time of flight technique, measured the energy distribution of species ejected during irradiation.

Pirronello et al. (1988a,b) have been investigating chemistry in mixtures heavily isotopically labelled water and methane with 1.5 MeV He^+. They obtained and measured the formation of both mobile and non mobile species. An example of the results obtained is shown in Figure 6, for three

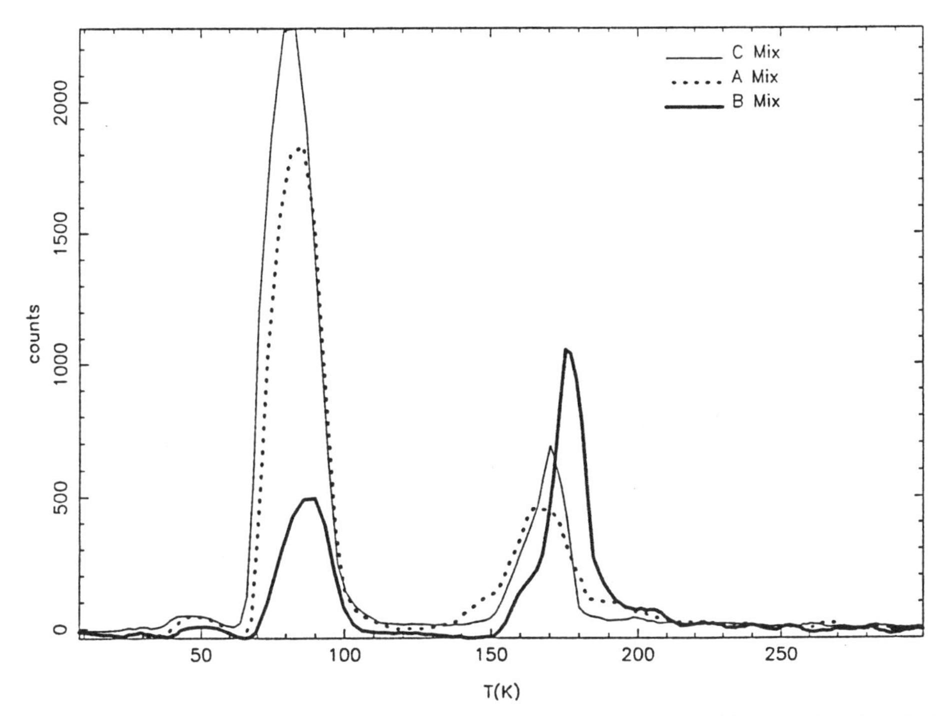

Figure 6. Mass 38 release with temperature in three different bombarded sample of $H_2O/^{13}CD_4$

different mixtures: "A" equal amounts of the two components, "B" richer in water, "C" richer in methane) of mass 38 that in the heavily isotopically labelled mixture could be methanol $^{13}CD_3OH$.

When an ice or an ice mixture formed by species that contain carbon atoms are irradiated by energetic particles till high doses, due to the formation and escape of volatile species like for instance H_2 and O_2, they undergo a deep and irreversible change in composition. The layer becomes richer and richer in carbon, C-H and C-O bonds are progressively destroyed (Tombrello, 1987) and are substituted by the stronger C-C bonds. As a consequence its boiling point increases till the point of becoming stable at room temperature as it is the yellow stuff of Mayo Greenberg (Greenberg, 1978). For very high doses of irradiation it can evolve into a layer of almost pure carbon with a degree of crystallinity that depends also on the substrate structure. These materials have received considerable attention (Cheng and Lanzerotti, 1978; Moore and Donn, 1982; Moore et al., 1983; Calcagno et al., 1983; Foti et al., 1984; Lanzerotti et al., 1985; Lanzerotti et al. 1987a) because of their importance for astrophysical applications. They have been analyzed both optically and by RBS. In pure frozen methane the rate at which the ratio H/C changes is quite relevant (with an effective initial cross section $\sigma \cong 5 \times 10^{-16}$ cm^2 that progressively diminishes with the decrease of the concentration of hydrogen.

3.3 MECHANISMS

In the energy range below 1 keV per amu, atoms and molecules of the target are energized via the direct transfer of momentum due to elastic collisions among the impinging ions and the nuclei plus core electrons of the target atoms in primary events (see e.g. Behrisch, 1981, 1983 and references therein); these atoms (called recoils), once displaced, collide with other atoms in neighboring molecules and produce a "cascade of collisions" that has been theoretically described in the linear regime by Sigmund (1969) and later by Falcone (1989) and in the non linear one by Vineyard (1976).

In the energy region in which it is dominant the energy loss via electronic processes spherical regions of energized material can be produced around primary events (that in this case are usually ionizations). If the energy released by the impinging ion remains confined for time intervals that are long enough it can be thermalized producing the so called thermal spike.

In both energy ranges atoms and radicals are set in motion and by interacting with nearby molecules can produce the formation of new species. Several of such chemical species formed under ion bombardment may be due just to the classical chemical reactions that occur in thermal equilibrium. For such reactions the kinetic energy of the partners is the one described by the Maxwell-Boltzmann distribution and the rate constants are in general those given by the Arrhenius law. However, together with the thermal classical reactions also "suprathermal" reactions, often referred as "Hot atom chemistry", may take place during irradiation of molecular solids. As mentioned before, the credit of proposing also for space chemistry occurring in the solid phase these types of reactions, that were already known to take place in the gas phase (see for instance the book edited by T. Matsuura in 1984), goes entirely to Kurt Roessler and his group (see e.g. Roessler, 1986; 1992). These reactions are characterized by the fact that at least one of the partners has a kinetic energy in the range between one and ten eV. Reactions of this type can occur more easily than thermal ones also when they have high activation energies and when they are endothermic. The peculiarity of these types of reactions when they go on in a solid consists in a much lower probability to have fragmentation of excited intermediate states with respect to the case in which they go on in the gas phase.

The most important "suprathermal" reactions are:

a) the abstraction of an atom from a molecule; it occurs when the bond energy of the new molecule to be formed is higher than that of the old one to be broken;
b) the insertion of an atom into bonds of the molecules belonging to the target; it occurs preferentially when polyvalent atoms are suprathermal and leads to intermediate products that may stabilize or fragment; while fragmentation is a probable fate for the intermediate products when this type of reaction occurs in the gas phase, stabilization is more probable in solids because many more ways of transferring energy exist in the solid phase than in the gas phase;
c) the collisional attack of a molecule from one hot atom that produces, as in the previous case, intermediate products that may decompose or stabilize;
d) the formation of a collision complex with two or more possible ligands.

A resume of these type of reactions is given in Table 2 taken from Roessler (1986).

The main reason why this kind of reaction should have great relevance (Roessler, 1992) is connected to the fact that rate constants for hot reactions, after a threshold, increase steeply to a plateau that can be many orders of magnitude higher than the rate constants relative to the thermal reactions of the same type. Such a high difference (many orders of magnitude) can be explained in terms of the little overlapping in energy between the Maxwell-Boltzmann distribution and the cross section curve and the better overlap with hot species. Molecules synthesized by ion bombardment in solids, if they are not mobile, remain trapped inside it; this is the case of species of this type formed in grain mantles by cosmic ray irradiation at least till when they are restored to the gas phase either by the chemical explosion of the mantle (Greenberg, 1973; 1978; d'Hendecourt et al., 1982) or through the interaction between icy mantles of grains and the heavy component of cosmic rays (Leger et al., 1985; Hedin et al., 1987; Johnson et al., 1991).

However species that are produced close to the surface of the solid or species that are mobile enough can be ejected in the gas phase during the irradiation itself. The yield of ejection (defined as number of ejected molecules per each impinging projectile) is a function of the

TABLE 2.
Mechanism of chemical reactions of hot atoms

1. ABSTRACTION

$X^{*} + HR \longrightarrow HH + R^{\cdot}$

$C^{*} + HR \longrightarrow CH + R^{\cdot} \rightarrow CH_2 \rightarrow CH_3 \rightarrow CH_4$

$N^{*} + HR \longrightarrow NH + R^{\cdot} \rightarrow NH_2 \rightarrow NH_3$

$O^{*} + HR \longrightarrow OH + R^{\cdot} \rightarrow OH_2$

$X^{*}(=\text{halogen}) + HR \longrightarrow XH + R^{\cdot}$

2. INSERTION OF POLYVALENT ATOMS

$C^{*} + H_2O \longrightarrow [HCOH] \rightarrow CO + H_2$; $[HCOH] \rightarrow H_2CO$

$C^{*} + NH_3 \longrightarrow [HCNH_2] \rightarrow H_2C=NH$

$C^{*} + CH_4 \longrightarrow [HCCH_3] \rightarrow HC\equiv HC + H_2$; $[HCCH_3] \rightarrow H_2C=CH_2$

$N^{*} + H_2O \longrightarrow [HNOH] \rightarrow NO + H_2$; $[HNOH] \rightarrow HNO + H$

$N^{*} + NH_3 \longrightarrow [HNNH_2] \xrightarrow{+H} H_2N-NH_2$

$N^{*} + CH_4 \longrightarrow [HNCH_3] \rightarrow CN + 2H_2$; $[HNCH_3] \xrightarrow{+H} H_2N-CH_3$

3. COLLISION

$C^{*} + OH_2 \longrightarrow C\cdot\cdot OH_2 \rightarrow CO + H_2$; $C\cdot\cdot OH_2 \rightarrow H_2CO$

$C^{*} + NH_3 \longrightarrow C...NH_3 \rightarrow HCN + 2H$; $C...NH_3 \rightarrow CNH + H_2$

IN LARGER MOLECULES:

SUBSTITUTION

$C^{*} + CH_3Br \longrightarrow CH_3Cl + Br$

4. COLLISON COMPLEX WITH TWO OR MORE POSSIBLE LIGANDS

$C^{*} + 2H_2O \rightarrow CO_2 + 4H$

$N^{*} + 2H_2O \rightarrow NO_2 + 4H$

$C^{*} + 2NH_2^{-} \rightarrow CN_2^{2-} + 4H$

stopping power dE/dx, either the nuclear one or the electronic one according to the energy range of the penetrating particle. In the case of cosmic rays irradiating grain mantles in dense interstellar clouds dominant effects are those produced by electronic processes. Three different regimes can be singled out according to the dependence of the yield on "dE/dx":

a) the linear one;
b) the quadratic one;
c) the cubic one.

The linear regime occurs when primary events (i.e. ionization processes), that give rise to the formation of the spherical regions in which molecules are energized, are well separated from one another. The probability that they intersect the surface allowing ejection of synthesized species then increases linearly with the stopping power (Brown et al., 1984) This is specifically the case of protons already at energies of few hundred keV or higher; for each primary event, in fact according to Fano (1946), the amount of energy released in the solid through ionizations, excitations and charge exchange reactions is roughly constant so they have to be well separated to give a low dE/dx.

The quadratic regime occurs (Johnson and Evatt, 1980) when the energized spherical regions overlap each other (this is the case of Helium ions at MeV energies); the cubic regime is obtained only in the case of heavy ions (see e.g. Salephour et al., 1986; Hedin et al., 1987; Johnson et al., 1989; Fenyo et al., 1990). In such a case, in fact, the above mentioned spherical regions overlap completely to give a deposition with a very high energy density. An explosive expansion of a whole chunk of material is the result of an enormous pressure build up that is able to release in the gas phase entire large molecules such as **leucine**.

4. STRUCTURAL AND OPTICAL PROPERTIES ALTERATIONS

Among the other relevant effects, that are of course related and occur simultaneously with erosion and change in the composition of the irradiated icy layer, there are also changes in its structure and in its optical properties.

That ion bombardment can cause both amorphization (disorder) in crystalline solids and vice versa recrystallization in amorphous materials is already well known. In the nuclear regime of the stopping power it is due to defect formation as a direct consequence of collision cascades. In the electronic regime it depends essentially on the rate at which the heat transferred to the target by the beam is removed. If such a rate is high enough amorphization occurs. If it is low, recrystallization takes place.

These structural changes can be studied using several techniques, for instance when MeV ion beams are available by means of channeling or RBS (it has already been mentioned that crystalline regrowth around the ion track can leave its signature in the increase of the FWHM, see e.g. Brown et al., 1978; Pirronello et al., 1981), but very impressively they can be investigated studying the changes of the optical properties of irradiated water ice.

Recently (O'Shaughnessy et al., 1988; Strazzulla et al., 1988; Sack et al., 1991) the reflectivity of water ice has been measured under ion bombardment (see Figure 15, taken from O'Shaughnessy et al., 1988). In particular Sack et al., (1991) have measured alterations in the UV-visible range in a wide interval of temperatures (between 20 K and 90 K) using light and heavy ions and for icy films that were accreted on the substrate with different methods and which therefore characterized by different structures.

In general irradiation of water films leads to to a decrease in reflectivity in the UV and in a slight brightening in the visible at very low irradiation dose. Even if they cannot exclude the presence of some absorption features (for instance embedded O_2 or OH) they think that the dominating factor affecting the optical properties of the layer is a change in scattering due to changes in the structure of the ice, especially in the concentration of internal scatterers and in the average grain size of the "polycrystalline amorphous" film they obtained at the rate of deposition they used. (Narten et al., 1976; Hagen et al., 1981).

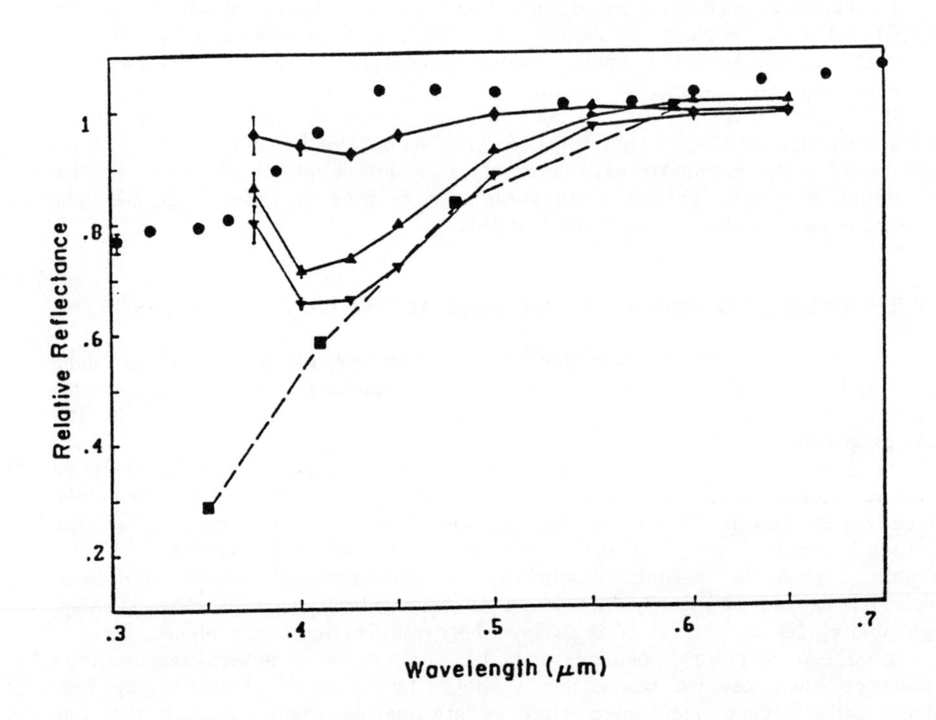

Figure 7. Reflectivity alteration of water ice films induced by ion irradiation (from O'Shaughnessy et al., 1988)

5. Conclusions

A brief description has been given of the results and the current ideas on the mechanisms responsible for the formation of simple and complex molecules in ices and ice mixtures, that are encountered in several important environments in space, irradiated with energetic ions. Before finishing it is worth mentioning that applications of the results (relative to the already discussed processes) to several astrophysical scenarios have been performed in the past ten years (see e.g. Pirronello, 1991). The interested reader will find an exhaustive review of the applications inside the solar system in Johnson (1990); among papers on applications of these effects in the interstellar medium and around stars one can mention Averna and Pirronello (1991), Di Martino et al. (1990), Johnson et al. (1991), Pirronello (1987), Pirronello and Averna (1988), Strazzulla (1985) and so on.

A final comment to conclude with is mandatory: up-to-date the importance of chemical processes induced by particle bombardment for the production of prebiotic or even biotic species has not yet been assessed and more experimental and theoretical efforts are needed in the future.

6. References

Armborst, E., Roessler, K.: 1986, Report-Jul 2089

Averna, D., Pirronello, V.: 1991, Astron. Astrophys. **245**, 239

Bar-Nun, A., Herman, A.G., Rappaport, M.L.: 1985, Surface Sci. **150**, 143

Behrisch, R. (ed): 1981, "Sputtering by Particle Bombardment I", Springer-Verlag, Berlin, Fed. Rep. of Germany

Behrisch, R. (ed): 1983, "Sputtering by Particle Bombardment II", Springer-Verlag, Berlin, Fed. Rep. of Germany

Benit, J., Bibring, J-P., Della Negra, S., Le Beyec, Y., Mendenhall, M., Rocard, F., Standing, K.: 1987, Nucl. Instr. Meth. **B19/20**, 838

Benit, J., Bibring, J-P., Della Negra, S., Le Beyec, Y., Rocard, F.: 1986, Radiation Effects **99**, 105

Benit, J., Bibring, J-P., Rocard, F.: 1988, Nucl. Instr. Meth. **B32**, 349

Benit, J., Brown, W.L.: 1990, Nucl. Instr. Meth. **B46**, 448

Berger, R.: 1961, Proc. Nat. Acad. Sci. USA **47**, 1434

Brown, W.L., Augustyniak, W.M., Marcantonio, K.J., Simmons, E., Boring, J.W., Johnson, R.E., Reimann, C.T.: 1984, Nucl. Instr. Methods **B1**, 307

Brown, W.L., Augustyniak, W.M., Simmons, E., Marcantonio, K.J., Lanzerotti, L.J., Johnson, R.E., Reimann, C.T., Boring, J.W., Foti, G., Pirronello, V.: 1982, Nucl. Instr. Methods **198**, 1

Brown, W.L., Lanzerotti, L.J., Poate, J.M., Augustyniak, W.M.: 1978, Phys. Rev. Lett. **40**, 1027

Calcagno, L., Foti, G., Strazzulla, G.: 1983, Lettere al Nuovo Cimento **37**, 303

Cheng, A.F., Lanzerotti, L.J.: 1978, J. Geophys. Res. **83**, 2597

d'Hendecourt, L.B., Allamandola, L.J., Baas, F., Greenberg, J.M.: 1982, Astron. Astrophys. **109**, L12

Di Martino, M., Mantegazza, L., Pirronello, V.: 1990, *On the Energetic Component of FU Ori Wind*, in Dusty Objects in the Universe, Bussoletti, E., Vittone, A., eds., p. 189, Reidel Publ. Co., Dordrecht, The Netherlands

Falcone, G.: 1990, La Rivista del Nuovo Cimento **13**, 1

Fano, U.: 1946, Phys. Rev. **70**, 44

Fenyo, D., Sundqvist, B.U.R., Karlson, A., Johnson, R.E.: 1990, Phys. Rev. **B42**, 1895

Foti, G., Calcagno, L., Sheng, K., Strazzulla, G.: 1984, Nature **310**, 126

Greenberg, J.M.: 1973, in "Molecules in the Galactic Environment", Gordon, M.A., Snyder, L.E., eds., p.93, John Wiley and sons

Greenberg, J.M.: 1978, *Interstellar Dust*, in Cosmic Dust, McDonnell, J.A.M. ed., p. 187, John Wiley and sons

Hagen, W., Tielens, A.G.G.M., Greenberg, J.M.: 1981, Chem. Phys. **56**, 367

Haring, R.A., Haring, A., Klein, F.S., Kummel, A.C., De Vries, A.E.: 1983, Nucl. Instr. Meth. **211**, 529

Haring, R.A., Kolfschoten, A.W., De Vries, A.E.: 1984a, Nucl. Instr. Meth. **B2**, 544

Haring, R.A., Pedrys, R., Oostra, D.J., Haring, A., De Vries, A.E.:, 1984b, Nucl. Instr. Meth. **B5**, 476

Haring, R.A., Pedrys, R., Oostra, D.J., Haring, A., De Vries, A.E.: 1984c, Nucl. Instr. Meth. **B5**, 483

Hedin, A., Hakansson, P., Salehpour, M., Sundqvist, B.U.R.: 1987, Phys. Rev. **B35**, 7377

Johnson, R.E.: 1990,

Johnson, R.E., Evatt, R.: 1980, Radiation Effects **52**, 187

Johnson, R.E., Pirronello, V., Sundqvist, B., Donn, B.: 1991, Astrophys. J. **379**, L75

Johnson, R.E., Sundqvist, B.U.R., Hedin, A., Feyno, D.: 1989, Phys. Rev. **B40**, 49

Lanzerotti, L.J., Brown, W.L., Johnson, R.E.: 1985, *Laboratory Studies of Ion Irradiation of Water, Sulfur Dioxide and Methane Ices*, in Ices in the Solar System, Klinger, J., Benest, D., Dollfus, A., Smoluchowski, R., eds., p. 317, Reidel Publ. Co. Dordrecht, The Netherlands

Lanzerotti, L.J., Brown, W.L., Marcantonio, K.J.: 1987, Astrophys. J. **313**, 910

Leger, A., Jura, M., Omont, A.: 1985, Astron. Astrophys. **144**, 147

Leger, A., Klein, J., de Cheveigne, S., Guinet, C., Defourneau, D., Belin, M.: 1979, Astron. Astrophys. **79**, 256

Moore, M.H.: 1984, Icarus **59**, 114

Moore, M.H., Donn, B.: 1982, Astrophys. J. **257**, L47

Moore, M.H., Donn, B., Khanna, R., A'Hearn, M.F.: 1983, Icarus **54**, 388

Morfill, G.E., Voelk, H.J., Lee, M.A.: 1976, J. Geophys. Res. **81**, 5841

Narten, A. H., Venkatesh, C. G., Rice, S. A.: 1976, J. Chem. Pys. **64**, 1106

Nebeling, B., Roessler, K., Stoecklin, G.: 1985, Radiochim. Acta **38**, 15

Oro, J.: 1963, Nature **197**, 971

O'Shaughnessy, D.J., Boring, J.W., Johnson, R.E.: 1988, Nature **333**, 240

Pirronello, V.: 1985, *Molecule Formation in Cometary Environment*, in Ices in the Solar System, Klinger, J., Benest, D., Dollfus, A., Smoluchowski,

R., eds., p. 261, Reidel Publ. Co., Dordrecht, The Netherlands

Pirronello, V.: 1987, Nucl. Instr. Meth. **B19/20**, 959

Pirronello, V.: 1991, *Physical and Chemical Effects Induced by Fast Ions in Ices of Astrophysical Interest*, in Chemistry in Space, Greenberg, J.M., Pirronello, V., eds., p. 263, Kluwer Acad. Publ., Dordrecht, The Netherlands

Pirronello, V., Averna, D.: 1988, Astron. Astrophys. **196**, 201

Pirronello, V., Brown, W.L., Lanzerotti, L.J., Lanzafame, G., Averna, D.: 1988a, *Formaldehyde Formation in $H_2O/^{13}CD_4$ Frozen Mixtures by Swift Ions*, in Dust in the Universe, Bailey, M.E., Williams, D.A., eds., p. 281, Cambridge Univ. Press, Cambridge, United Kingdom

Pirronello, V., Brown, W.L., Lanzerotti, L.J., Maclennan, C.G.: 1988b, *HD and CO Release during MeV Ion Irradiation of H_2O/CD_4 Frozen Mixtures at 9 K*, in Experiments on Cosmic Dust Analogues, Bussoletti, E. et al. eds., p. 287, Kluwer Acad. Publ., Dordrecht, The Netherlands

Pirronello, V., Brown, W.L., Lanzerotti, L.J., Marcantonio, K.J., Simmons, E.: 1982a, Astrophys. J. **262**, 636

Pirronello, V., Lanzafame, G.: 1989, Astrophys. J. **342**, 527

Pirronello, V., Strazzulla, G., Foti, G., Rimini, E.: 1981, Astron. Astrophys. **96**, 267

Rice, S.A.J.: 1975, Top. Curr. Chem. **60**, 109

Roessler, K.: 1986, Radiation Effects **99**, 21

Roessler, K.: 1987, Radiochim. Acta **42**, 123

Roessler, K.: 1988, Nucl. Instr. Meth. **B32**, 519

Roessler, K.: 1992, Nucl. Instr. Meth. (in press)

Sack, N.J., Boring, J.W., Johnson, R.E., Baragiola, R.A., Shi, M.: 1991, J. Geophys. Res. **96**, 17535

Salehpour, M., Hakannson, P., Sundqvist, B.U.R., Widdiyasekera, S.: 1986, Nucl. Instr. Meth. **B13**, 278

Sigmund, P.: 1969, Phys. Rev. **184**, 383

Strazzulla, G.: 1985, Icarus **61**,48

Strazzulla, G., Torrisi, L., Foti, G.: 1988, Europhys. Lett. **7**, 431

Tombrello, T.A.: 1987, Nucl. Instr. Meth. **B24/25**, 517

Venkatesh, C.G., Rice, S.A., Narten, A.H.: 1974, Science **186**, 927

Vineyard, G.H.: 1976, Radiation Effects **29**, 245

Wefel, J.P.: 1988, *An Overview of Cosmic Ray Research: Composition, Acceleration and Propagation*, in Genesis and Propagation of Cosmic Rays, Shapiro, M.M., Wefel, J.P., eds., p. 1, Reidel Publ. Co., Dordrecht, The Netherlands

Ziegler, J.F.: 1980, "Stopping Cross Sections for Energetic Ions in All Elements", Pergamon Press, New York

PHYSICS AND CHEMISTRY OF PROTOPLANETARY ACCRETION DISKS

WOLFGANG J. DUSCHL
Institut für Theoretische Astrphysik, Universität Heidelberg
Im Neuenheimer Feld 561, D-69120 Heidelberg, Germany

Interdisziplinäres Zentrum für Wissenschaftliches Rechnen, Universität Heidelberg
Im Neuenheimer Feld 368, W-69120 Heidelberg, Germany

and

INTERNET: wjd@platon.ita.uni-heidelberg.de

Abstract. First, we give a concise introduction to the formation, theory and stability of protoplanctary accretion disks In a second part, we present an overview of the physical processes that are currently thought to be of importance for the evolution of a protoplanetary disk. Finally, we discuss some new ideas of how chemical reactions might influence the global evolution of such disks.

There are many very good reviews available that summarize our present knowledge and ideas about protoplanetary disks. Consequently, I will not attempt to add just another review of the same kind to the literature. I rather prefer to discuss theoretical aspects that in my view bear great potential to foster the development of the topic in the next few years. In this respect, this written account deviates considerably from the oral version of the two 90-minute lectures. In the oral version, much emphasis was put on introducing the standard models for the protoplanetary disk. This encompasses only one chapter in the present version. The reasoning behind this is that most of the material that could be presented in section 4 is easily available in most libraries so that it might be more valuable to dwell on those things that are usually not discussed at great length (if at all) in this context and that bear great potential.

1. The formation of protostellar and protoplanetary disks

1.1. Star formation

The basic physical principles that lead to the formation of flat disk-like structures in the course of the collapse of a molecular cloud core with angular momentum are very simple to understand but immensely complicated to model. Even in the 1-dimensional case, i.e., without angular momentum, there are no satisfying models of the protostellar collapse available yet. The main reason for this is that severe numerical problems are still to overcome (Kürschner, 1992; Kürschner, 1993, in preparation).

Despite these fundamental problems, it is clear that the collapse of a rotating cloud will lead to the formation of a disk like structure simply due to the balance of angular momentum. This could be avoided only if there were a mechanism that could carry away angular momentum on a time scale that is considerably shorter than the ones of the collapse. Furthermore, there are many direct and indirect evidences for the formation of disks in the course of star formation, for instance

– the ecliptic plane in the solar system in which the orbits of all of the major

J. M. Greenberg et al. (eds.), The Chemistry of Life's Origins, 55–73.

planets (and even those of the majority of other objects in the solar system) is concentrated.

- direct observation of the dust and gas disk around β Pictoris (Smith and Terrile, 1984), a star that is just on its evolution towards the main sequence or just about to reach it (Paresce, 1991).
- the presence of disks in T Tauri- and FU Orionis-type variable young stars (Beckwith et al., 1990; Bouvier and Bertout, 1992; Calvet et al., 1991)

For a summary of the current status of the formation of viscous protostellar accretion disks, see Tscharnuter (1989).

1.2. What is a protoplanetary disk ?

In the course of star formation several phases are to be gone through. In the following we shall use the expression *protoplanetary (accretion) disk* only for those disks in which a major part of the total mass of the forming star and its planetary system is already accreted into the stellar core itself. This means that the disk is evolving in the potential well of the central protostellar core. In the earlier proto*stellar* phase, considerably more mass that has yet to be accreted onto the protostellar core, is still in the disk.

2. The basic physics of accretion disks

In our approach, we assume that the actual star formation process has already (almost) come to its end, and that consequently the mass of the remaining accretion disk is small compared to that of the newly formed star onto which accretion takes place. In the sense of our above given definition of a proto*planetary* disk (in contrast to a proto*stellar* disk), this assumption is fully justified. Furthermore, when developing the theoretical model, we neglect the influence of magnetic fields on the flow of matter. Magnetic fields, nonetheless, may play a rôle in such disks as small scale magnetic fields may be the cause of viscosity in accretion disks (see below). Under these assumptions, one can formulate[1]:

2.1. The standard model for accretion disks

The standard model for accretion disks, in its earliest form, dates back to Kant (1755). Interestingly enough, it was first formulated as an explanation for the solar system's origin.

This class of models reached a state of quantitative treatment when von Weizsäker (1943) and, especially, Lüst (1952) came back to ideas one and a half to two centuries old. von Weizsäcker's and Lüst's theoretical approach was re-invented (unfortunately without proper reference to the original papers) by Shakura and Sunyaev (1973) and Novikov and Thorne (1973). The basic features of these disks are:

[1] Despite the fact that protoplanetary systems and dwarf novae are very different systems in many respects, the underlying basic physics of the disk is remarkably similar in many respects. As the theory of accretion disks is perhaps best developed for the dwarf nova case, we will take examples from that case whenever it is justified.

- The disk's mass is small compared to the mass that creates the gravitational potential in which it evolves.
- The geometrical thickness (in the direction perpendicular to the disk plane) is small compared to the radius from the center of the disk, i.e., the disks are geometrically thin.
- Radial pressure and temperature gradients are negligible, i.e., if there are stationary solutions, these disks have to be centrifugally supported. Consequently, this means that the rotational velocity is (almost) Keplerian over which a very much smaller radial velocity component is superimposed.

In addition to these physical assumptions, there are usually some mathematical simplifications that are typical for the standard model:

- The disks are assumed to be rotationally symmetric, i.e., any structure in azimuthal direction is not taken into account. However, this assumption has been challenged recently in the context of angular momentum transport mechanisms. It was argued that spiral waves may develop that could act as a means of redistributing angular momentum and mass. While this idea, in principle, would give a very appealing way of getting rid of the viscosity problem, all the models that are discussed in literature do not really succeed in replacing local viscosity mechanisms yet (Matsuda et al., 1990).
- A mirror symmetry with respect to the disk plane is assumed. Recently it has been discussed that – at least for some kinds of objects – this might be an oversimplification. It seems as if warps might possibly occur in very hot disks that are capable of driving strong winds (Pringle, 1992; Horn and Meyer, 1993, in preparation), but this is not necessarily of importance in the case of protoplanetary disks.

2.2. Stationary accretion disks

2.2.1. Coordinate system

In principle, the geometry of the system has parts where a spherical coordinate system would be the best choice (namely for the star and its immediate vicinity), while in the disk itself a cylindrical coordinate system is much more suitable. When describing the disk, we shall always use a cylindrical coordinate system $\{s, z, \varphi\}$. The radial coordinate s is measured in the plane of the disk, the z-coordinate is taken perpendicularly to the s-direction. As mentioned above, because of symmetry assumptions, the azimuthal coordinate φ will not be important. Occasionally we will also use the spherical radial coordinate r $(= \sqrt{s^2 + z^2})$.

2.2.2. Gravitational force

As we assume the disk to be of negligible mass (i.e., massless), the gravitational potential in which it evolves is known and stationary. Thus, we do not have to solve Poisson's equation, but can write immediately for the gravitational acceleration $\vec{g}$:

$$\vec{g} = -\frac{GM}{r^2}\frac{\vec{r}}{r}. \tag{2.1}$$

G is the gravitational constant and M the mass of the central object that dominates the gravitational potential. In the case of protostellar disks, when there is still a considerable part of the system's total mass in the disk, the accretion process will change M considerably within a viscous time scale (see sect. 2.3 below). This has to be taken into account.

2.2.3. Balance of the radial forces

In the radial direction, the gravitational force is balanced by the centrifugal force, i.e., Kepler's 3rd law applies for the angular velocity ω:

$$\omega(s) = \sqrt{\frac{GM}{s^3}}. \tag{2.2}$$

Actually, instead of the (cylindrically) radial distance s, the (spherical) radial distance r would be required. Because of the fact that the disks are geometrically thin (i.e., $s \gg z$), we replace $r = \sqrt{s^2 + z^2} = s\sqrt{1 + \left(\frac{z}{s}\right)^2} \approx s$. Thus we neglect terms that are at least quadratic in z/s.

2.2.4. Balance of the vertical forces

In vertical direction, we have

$$\frac{d^2z}{dt^2} = g_z - \frac{1}{\rho}\frac{\partial P}{\partial z}; \tag{2.3}$$

g_z is the vertical component of the gravitational acceleration ($g_z = -GMz/s^3$), ρ the matter denisty, and P the pressure. As a material function, there has to be some prescription between P and the variables ρ and T (temperature). In the simplest case this could be the ideal gas law.

2.2.5. Surface density

For the convenience of discussing the radial structure of geometrically thin accretion disks, we introduce the surface density Σ:

$$\Sigma = \int_{-\infty}^{\infty} \rho dz. \tag{2.4}$$

2.2.6. Continuity equation

$$\dot{M} = -2\pi s \left(\Sigma v_s + \frac{s}{2}\dot{\Sigma}_{\text{in}}\right) \tag{2.5}$$

describes the connection between the mass flow rate $\dot{M}$, the surface density Σ, and the radial mass flow velocity v_s. Eq. 2.5 is the integrated version of the actual continuity equation

$$s\frac{\partial \Sigma}{\partial t} + \frac{\partial}{\partial s}(s\Sigma v_s) = s\dot{\Sigma}_{\text{in}}. \tag{2.6}$$

On the right hand side, $\dot{\Sigma}_{\text{in}}$ constitutes a source term. There can be two different kinds of source terms. One is a steady radial inflow into the disk at its outer edge. This is the standard case in close binaries with mass tranfer from the secondary into the accretion disk of the primary. Mathematically, in this case the mass flow rate $\dot{M}$ is given as a (variable or constant) integration constant while $\dot{\Sigma}_{\text{in}}$ is 0. The other kind of possible source term is material "raining" into the disk at all (or some) radii. In our case, this will be due to material from a surrounding cloud of matter. This source term definitely plays a important, perhaps even dominating rôle in the case of proto*stellar* disks, but is also of great interest in the present situation.

2.2.7. Balance of angular momentum

Similarly to the continuity equation, one can also formulate an equation of balance of angular momentum:

$$s\frac{\partial}{\partial t}\left(\Sigma s^2\omega\right) + \frac{\partial}{\partial s}\left[\left(s\Sigma v_s\right)\left(s^2\omega\right)\right] = \mathcal{G} + s\dot{I}_{\rm in}. \qquad 2.7$$

The quantity $\mathcal{G}$ on the right hand side describes the net effect of viscous torques between radially neighbouring matter (due to differential rotation in the disk [eq. 2.2] there is friction between radially neighbouring matter):

$$\mathcal{G} - -\frac{3}{2}\frac{\partial}{\partial s}\left(\nu\Sigma s^2\omega\right). \qquad 2.8$$

In eq. 2.7, the first term describes the local change of angular momentum surface density ($\Sigma s^2\omega$; $s^2\omega$ is the specific angular momentum) due to advection of matter with angular momentum (second term: $(\dot{M}/2\pi)\cdot s^2\omega$) and due to transport due to viscous torques (right hand side, see above). The last term on the right hand side again accounts for angular momentum that is added to the disk through the matter source term $\dot{\Sigma}_{\rm in}$. In eq. 2.8, ν is the coefficient of viscosity.

2.2.8. Frictional release of energy

The balance of energy under the present assumptions reads:

$$\frac{\partial T_{\rm c}}{\partial t} + \left\{v_s\frac{\partial T_{\rm c}}{\partial s}\right\} = \frac{2}{c_P\Sigma}\left(\frac{3}{4}\omega\int_{-\infty}^{\infty} t_{s\varphi}dz - \sigma T_{\rm eff}^4\right) + \left\{\nu_{\rm therm}\frac{1}{s}\frac{\partial}{\partial s}\left(s\frac{\partial T_{\rm c}}{\partial s}\right)\right\}. \qquad 2.9$$

Under almost all circumstances the radial advection term ($\{v_s\ldots\}$ on the left hand side) and the radial diffusion term ($\{\nu_{\rm therm}\ldots\}$ on the right hand side) are negligible within the framework of our approximations, and are indeed neglected in the standard model. Only in the case of unstable disks (see below) in the immediate vicinity of the region where the instability is occuring, these two terms become important.

In eq. 2.9, $T_{\rm c}$ denotes the disk's central (i.e., $z = 0$) temperature, $T_{\rm eff}$ its effective temperature (which is related with the emergent flux: $F = \sigma T_{\rm eff}^4$), σ the Stefan-Boltzmann constant, c_P the specific heat for constant pressure, $t_{s\varphi}$ the $s-\varphi$-component of the viscous stress tensor, and $\nu_{\rm therm}$ the thermal diffusivity.

Without the two terms in braces, it follows from eq. 2.9 that all the energy released through viscous friction is radiated locally.

In the stationary case, the equation reduces to

$$\frac{\partial F_z}{\partial z} = \frac{3}{2}t_{s\varphi}\omega; \qquad 2.10$$

F_z is the vertical energy flux at an altitude z.

2.2.9. Energy transport

Energy is transported either through radiation or convection:

$$\frac{\partial T}{\partial z} = - \begin{cases} \dfrac{3\kappa\rho}{4acT^3} F_z & \text{for radiative transport,} \\ g_z \rho \dfrac{T}{P} \nabla_{\text{conv}} & \text{for convective transport.} \end{cases} \qquad 2.11$$

κ is the opacity coefficient, and ∇_{conv} the quantity $d \log T / d \log P$ for convective stratification. For the radiative case, we made use of the diffusion approximation which is good only for optically thick cases. For the case of optically thin stratification, we refer the reader to Hubeny (1989).

2.2.10. Material functions

To close the set of equations, several material functions need to be known, namely the equation of state ($P(\rho, T)$), the viscosity ν (and – equivalently – the stress tensor element $t_{s\varphi}$), the opacity κ, the mean molecular weight μ, and the specific heat c_P. Finally, we have to make use of some model that describes the energy transport by convection.

The main problem in describing accretion disk arises from a fundamental lack of knowledge of how turbulence acts. This shortcoming plays a twofold rôle in the present questions. It makes description of both the convection and the viscosity uncertain. Only recently was it accomplished to at least find a consistent description for both phenomena in accretion disk. Usually, one makes the approach to describe convection by a mixing length model, and to parametrize viscosity. These so-called α-disk models use the following ansatz for a turbulent viscosity:

$$\nu = \frac{2}{3} \alpha h c_s; \qquad 2.12$$

c_s is the isothermal sound velocity, h the local vertical pressure scale height, and α an unknown parameter (that – unfortunately – contains practically all the physics; within the framework of our approximations, eq. 2.12 is equivalent to $t_{s\varphi} = \alpha P$). Surprisingly enough, it turns out that despite this very approximative approach disk models are much more successful than one would expect. The only two ways of determining the quantity α are either a consistent modelling of viscosity and convection, or a comparison of observed and predicted time scales.

2.3. Relevant time scales

Time scales enter the problem through eqs. 2.2, 2.3, 2.6, 2.7, and 2.9 (Kippenhahn and Thomas, 1982). Of those, the simplest one is the *Keplerian time scale* τ_{Kepler}:

$$\tau_{\text{Kepler}} = \frac{1}{\omega}. \qquad 2.13$$

The next one characterize the time it takes the vertical structure to reach hydrostatic equilibrium; from eq. 2.3 the *dynamical time scale* can be estimated:

$$\frac{h}{\tau_{\text{dyn}}^2} \approx \frac{P}{\rho h} \quad \Rightarrow \quad \tau_{\text{dyn}} = \frac{h}{c_s}, \qquad 2.14$$

i.e., it takes a sound travel time through the disk in vertical direction to establish equilibrium. By comparison with the other term of the equation, one finds

$$\tau_{\text{dyn}} = \tau_{\text{Kepler}}. \tag{2.15}$$

The *thermal adjustment time scale* may be estimated from eq. 2.9:

$$\tau_{\text{therm}} = \frac{c_P \Sigma T}{\omega t_{s\varphi} h} \approx \frac{1}{\alpha\omega}. \tag{2.16}$$

For α not too different from 1, the thermal, the dynamical, and the Keplerian time scales are of same order.

Finally, there is a time scale that describes how long it takes a mass element to be advected through the disk due to viscous torques. This *accretion* or *viscosity time scale* can be estimated:

$$\tau_{\text{visc}} = \frac{s^2}{\nu}. \tag{2.17}$$

In the framework of the thin disk assumptions, one can approximate this by

$$\tau_{\text{visc}} \approx \frac{1}{\alpha}\left(\frac{s}{h}\right)^2 \frac{1}{\omega}. \tag{2.18}$$

For the usual case of $\alpha \not\ll 1$, this leads to

$$\tau_{\text{Kepler}} \approx \tau_{\text{dyn}} \approx \tau_{\text{therm}} \ll \tau_{\text{visc}}. \tag{2.19}$$

2.4. The Evolution of Accretion Disks

The hierarchy of time scales as deduced in the previous section allows for one of two approximations in most cases, depending on the question one is interested to answer.

Either one follows only the very shortest time scales, i.e., the dynamical and – for α not too different from 1 – the thermal one. This is justified if one is interested in very short term changes in the disk. Then, one may assume that all processes involving longer time scales, especially the viscous evolution, take infinitely long to cause changes. In our case this would mean that any radial distribution of surface densities – be it in equilibrium or not – remains unaltered during the processes that are resolved in time, i.e., we approximate the longer time scales by infinity.

The opposite approach is justified when one is not so much interested in short term evolution but rather in the viscous one. Then one can approximate the situation by assuming that only the longest relevant time scale is resolved. Processes that involve shorter time scales are approximated by assuming that they run infinitely fast. In this context, the shorter time scales are approximated by 0.

In principle one could also isolate an intermediate time scale by approximating shorter one by 0 and longer ones by inifity. In our context we have only two time scales involved (again, if α is not too different from 1), so there is no intermediate time scale involved. If one analyzed the instability as discussed in section 3, one would find that there is actually a third, intermediate time scale, namely the one within which the information about an instability is communicated through the disk in radial direction. We will not follow this in more detail.

For all the time scale approximations just discussed it is crucial that there is a hierarchy of time scales. In the following, we will take into account only the viscous evolution. This means that we assume dynamical and thermal processes to happen instantaneously. So, we have to keep only time scales in the equations describing the balance of mass and angular momentum.

The time dependent continuity equation,

$$\frac{\partial \Sigma}{\partial t} + \frac{1}{s} \cdot \frac{\partial}{\partial s} \left(s \Sigma v_s \right) = \dot{\Sigma}_{\text{in}}, \tag{2.20}$$

together with the equation describing the balance of angular momentum,

$$\frac{\partial}{\partial t} \left(\Sigma s^2 \omega \right) + \frac{1}{s} \cdot \frac{\partial}{\partial s} \left(s \Sigma v_s s^2 \omega \right) = -\frac{3}{2} \cdot \frac{1}{s} \cdot \frac{\partial}{\partial s} \left(s^2 \nu \Sigma \omega \right) + \dot{I}_{\text{in}}, \tag{2.21}$$

may be combined to one equation of the type of the diffusion equation

$$\frac{\partial \Sigma}{\partial t} = \frac{3}{s} \cdot \frac{\partial}{\partial s} \left[\sqrt{s} \frac{\partial}{\partial s} \left(\nu \Sigma \sqrt{s} \right) \right] + f(\dot{\Sigma}_{\text{in}}, \dot{I}_{\text{in}}). \tag{2.22}$$

The second term on the right hand side allows for mass infall into the disk. The solution of this equation together with the stationary equations for energy release (eq. 2.9) and energy transport (eq. 2.11) allows to describe the viscous evolution of a geometrically thin Keplerian accretion disk. Here it is written for the case of a central point source that dominates the gravitational potential. It is straight forward to extend it to the case of an extended background potential (Duschl, 1988; von Linden, Duschl, and Biermann, 1992, in preparation). The situation changes considerably when selfgravity within the disk becomes important. While rough approximations are still possible (Duschl, 1988), the full problem still awaits a solution.

2.5. A critical review of the concept of Keplerian accretion disks

The standard model as introduced above assumes that the accretion disk is almost in a centrifugally supported equilibrium, i.e., matter moves with almost exactly the Keplerian circular velocity. On the other hand, at least in later stages of the star formation process, the protostellar core is rotating only slowly compared to Keplerian. As a consequence of this, there has to exist a transition zone between the domain where the gravitational forces are centrifugally balanced (i.e., the Keplerian disk) and that where pressure forces stabilize the structure (i.e., the protostellar core). In the standard models this region is assumed to be a boundary layer with infinitely small radial extent.

This is a mathematically as well as a physically very doubtful concept. Mathematically the assumption leads to a singularity, physically it leads to inconsistency. The inconsistency is introduced by the fact that in such a configuration, exactly half the energy available through accretion has to be radiated from an infinitely small volume. Both problems are usually "healed" by assuming that there is a finite but small radial extent of the boundary layer – small even compared to the geometrical dimensions of the star – and that the same amount of energy now comes from a small but finite volume. One has to "pay" for this by accepting the arbitrariness in the choice of the transition zone's radial width and by introducing another inconsistency. Indeed, if there is a finite radial extent of the transition region, the amount of energy radiated from this region has to be determined consistently which usually is not done.

As yet, there is no satisfying theoretical description of the transition region between star and disk available. From the observational side, spectra of T Tau stars strongly indicate that part of the inner disk is not a stationary Keplerian equilibrium disk (Bertout, Bouvier, Duschl, and Tscharnuter, 1993, in preparation). This is also manifested by the fact that the radial run of the effective temperature does not follow the simple $s^{-3/4}$-law as is required by the standard models (cf. observations by Beckwith et al. 1990). Deviations from this law due to the inconsistent boundary layer are confined to very small radial regions. The reason for this deviation might be either a non-stationary disk, or an extended boundary layer or both.

3. Stability of Keplerian accretion disks

In the first part of this section, we deduce the condition for stability of an accretion disk against thermal-viscous instability. This is by no means the only possible instability but it is the one that plays the most important rôle in the context that is discussed here. In the second section, we use a one-zone model for the vertical structure of the disk and deduce for this approximation which physical processes determine whether a disk is stable or not.

3.1. A CONDITION FOR THERMAL-VISCOUS INSTABILITY

In the regime of time scales which is of interest in the context that is discussed here, the evolution of the disk is described by eq. 2.22. That is mathematically equivalent to a diffusion equation with a variable diffusion coefficient that itself is a function of the solution of the diffusion equation. This equation was derived in the previous section. From its linearized form ($\delta\Sigma$'s are small deviations from the stationary solution Σ_0),

$$\frac{\partial \delta\Sigma}{\partial t} = \frac{3}{s} \cdot \frac{\partial}{\partial s}\left[\sqrt{s}\frac{\partial}{\partial s}\left\{\sqrt{s}\left.\frac{\partial \nu\Sigma}{\partial \Sigma}\right|_0 \delta\Sigma\right\}\right] \quad \Rightarrow \quad \frac{\partial b}{\partial t} = \frac{3}{4s}\frac{\partial \nu\Sigma}{\partial \Sigma}\frac{\partial^2 b}{\partial \sqrt{s}^2}, \qquad 3.1$$

(b is the quantity in braces on right hand side of the first equation) (Meyer and Meyer-Hofmeister, 1981) one can deduce the stability condition

$$\frac{\partial(\nu\Sigma)}{\partial\Sigma} > 0. \qquad 3.2$$

This is the stability condition for geometrically thin Keplerian accretion disks against a thermal-viscous instability.

This instability is well known to take place in accretion disks. It is known to be of special importance in accretion disks around white dwarfs in dwarf nova systems. There, this instability is the cause for the frequent outbursts of the systems. As some of the dwarf novae are also eclipsing binaries, it is not only possible to observe the spatially integrated emission of energy but also it is possible to gain spatially resolved information about the evolution of the instability itself. The observed features are in agreement with the theoretical picture of the evolution of outbursts. In section 5, we will discuss in how far the same kind of instability – altough for quite a different combination of physical parameters – may be an important phase in the evolution of protoplanetary disks.

3.2. The physical background of the instability

It is very instructive to isolate the processes that determine the stability of accretion disks. This is done in a one-zone approximation in vertical direction, i.e., we do not calculate the vertical structure of the disk but replace the respective quantities by appropriate averages, and replace derivatives in vertical direction ($\partial/\partial z$) by division through the disk characteristic vertical extent (h). Additionally, we assume that changes of the viscosity parameter α and the molecular weight μ do not play an important rôle. A full treatment, including $\dot{\Sigma}_{\rm in}$ and variable α and μ will be given elsewhere (Duschl and Tscharnuter, 1993, in preparation).

In the following, we analyze the disk structure at some fixed radius s from the central object. First, we assume that the equation of state is modelled to a sufficient accuracy by the laws of an ideal gas:

$$P = \frac{\rho \Re T}{\mu}, \tag{3.3}$$

(μ: mean molecular weight; $\Re$: universal gas constant) and introduce a surface density Σ

$$\Sigma = 2\rho h. \tag{3.4}$$

Hydrostatic equilibrium in vertical direction,

$$\frac{\partial P}{\partial z} = -g_z \rho, \tag{3.5}$$

with the gravitational acceleration in vertical direction, g_z,

$$g_z = -\frac{GM}{s^2} \cdot \frac{z}{s}, \tag{3.6}$$

leads to

$$P \sim \Sigma h. \tag{3.7}$$

Here, we made use of one of the approximations of the one-zone description in vertical direction:

$$\left| \frac{\partial P}{\partial z} \right| \approx \frac{P}{h}. \tag{3.8}$$

Together with the equation of state, this leads to

$$T \sim h^2. \tag{3.9}$$

This equation gives also the reason why and under which circumstances accretion disks are geometrically thin ($h \ll s$): A disk is thin if it is sufficiently cool.

Viscous heating causes a vertical gradient of the energy flux F:

$$\frac{\partial F}{\partial z} = -t_{s\varphi} s \frac{d\omega}{ds}. \tag{3.10}$$

In the thin disk approximations (and this holds true in general, not only for the one-zone models), the s-φ component of the viscous stress tensor, $t_{s\varphi}$ is reduced to

$$t_{s\varphi} = \alpha P. \tag{3.11}$$

Replacing again the derivative in vertical direction by a division by h and introducing a quantity W that describes how much of the energy is transported by radiation:

$$W = \frac{F_{\text{rad}}}{F}, \tag{3.12}$$

one gets

$$F_{\text{rad}} \sim WhP. \tag{3.13}$$

F_{rad} itself is approximated by the diffusion approximation, i.e., the stratification is assumed to be optically thick (∇_{rad}: radiative gradient ∇; κ: Rosseland mean opacity; a: radiation constant; c: velocity of light):

$$F_{\text{rad}} = \frac{2acT^4}{3\Sigma\kappa}\nabla_{\text{rad}}. \tag{3.14}$$

This leads to

$$\Sigma^2\kappa W \sim h^6\nabla_{\text{rad}}. \tag{3.15}$$

Integrating the stationary equation of transport of angular momentum (eqs. 2.7 and 2.8), setting the integration constant to 0, and replacing integrals in vertical direction by multiplications by h, one gets

$$\dot{M} \sim \Sigma h^2 \tag{3.16}$$

and

$$W\Sigma^5\kappa \sim \dot{M}^3\nabla_{\text{rad}}. \tag{3.17}$$

On the other hand, for geometrically thin disks, we have

$$\nu \sim T, \tag{3.18}$$

and thus

$$\dot{M} \sim \nu\Sigma. \tag{3.19}$$

Putting everything together, we get for the stability condition

$$\frac{\partial \log \dot{M}}{\partial \log \Sigma} > 0. \tag{3.20}$$

On the other side, eq. 3.17 gives a relation between $\dot{M}$ and Σ that is only depending on material functions. We regard the density ρ and the temperature T (or its appropriately averaged values in vertical direction) as the independent variables and model the dependence by a power law ansatz

$$\frac{W\kappa}{\nabla_{\rm rad}} \sim \rho^A T^B. \qquad 3.21$$

This leads to

$$\frac{W\kappa}{\nabla_{\rm rad}} \sim \rho^A T^B \sim \left(\frac{\Sigma}{h}\right)^A T^B \sim \left(\frac{\Sigma}{\left(\dot{M}/\Sigma\right)^{1/2}}\right)^A \left(\frac{\dot{M}}{\Sigma}\right)^B \sim \frac{\Sigma^{\frac{3A}{2}-B}}{\dot{M}^{\frac{A}{2}-B}} \qquad 3.22$$

and to

$$\frac{\partial \log \dot{M}}{\partial \log \Sigma} = \frac{5+\frac{3A}{2}-B}{3+\frac{A}{2}-B} = \frac{10+3A-2B}{6+A-2B}. \qquad 3.23$$

Condition 3.20 then defines two domains of instability:

$$2B-6 < A < \frac{2}{3}B - \frac{10}{3}, \qquad 3.24$$

and

$$\frac{2}{3}B - \frac{10}{3} < A < 2B-6. \qquad 3.25$$

It is important to note that the instability depends only on the exponents but not on the absolute values of the independent quantities. Usually, i.e., in the context of close binaries, one finds that especially strong gradients in the opacities κ can cause instabilities.

We emphasize this as in most of the standard models that are discussed in the next section, the question whether the disk is stable or not is simply neglected. This is not necessarily in contradiction with the results presented in these papers. While for showing the crucial quantities that determine stability or instability of the disks, the vertical one-zone approximation is a good and convenient tool, for actually detecting the instability in the models itself, a proper determination of the vertical structure is indispensable. This is very well known from accretion disk models in dwarf novae, i.e., close binaries systems with mass transfer. It was for quite some time known that an instability like the one discussed here may exist in the accretion disks in these systems. But only when one abandoned the approximations of the one-zone model and when then for the first time complete models of the vertical structure were calculated, was the instability found (Meyer and Meyer-Hofmeister, 1981).

In section 4, we will briefly compile an overwiew over presently discussed models for the protoplanetary disk and its evolution. In section 5, we will present some preliminary thoughts about a modification of the existing models. This modification relies on the instability of the disk and on the chemical reactions in the disk that only the higher temperatures due to the instability causes.

4. Standard models for protoplanetary disks and observational constraints

As was already stated at the beginning of our paper, it is not our aim to give here a complete review of all the models for protoplanetary disks that are currently under consideration, we rather prefer to give a guide to easily accessible literature. We will comment on some models from which progress is expected in the near future.

While the first ideas about accretion disks already date back already a quarter of a millenium ago, the first quantitative models were developed in the late 40s and early 50s of this century. The era of "solar nebula model" started with Cameron's (1962) paper. In that paper and a subsequent one along the same line of thought (Cameron and Pine, 1973) it was assumed that the collapse of interstellar material to form the protoplanetary disk was almost an instantaneous process.

At the end of the 60s, following Larson's (1969) landmark paper on how star formation proceeds in a highly non-homologous way, it became clear that the evolution of the protosolar nebula has to be described in the framework of a viscous accretion disk that may be "fed" by continuously infalling material. The latter is required as the time scales over which the disk is capable of rearranging itself by transporting matter and angular momentum ($\tau_{\mathrm{visc}} \approx 10^3 yrs$) is considerably shorter than the time scale over which the disk itself grows through matter falling from the star forming cloud ($\approx 10^5 yrs$). In view of the uncertainties in our knowledge of how the protostellar collapse really proceeds (Tscharnuter, 1989), this has to be regarded as a preliminary requirement.

In the standard picture of the evolution of a viscous protoplanetary accretion disks one expects basically three phases (Morfill, 1985; Wood and Morfill, 1988):

- Phase 1 is the phase during which the disk is being formed. In this phase, the infall into the disk ($\dot{\Sigma}_{\mathrm{in}}$ in eqs. 2.6, 2.20, and 2.22) is more efficient than the transport through the disk to the central body. The net effect of this phase is that the dimensions of the disk are growing while the transport of angular momentum and mass becomes more efficient. In terms of time scale arguments, the first phase is the one during which the viscous time scale is considerably longer than the time scale over which a mass comparable to the disk mass is falling into the disk.
- In a subsequent phase, sort of a "quasi-stationary" disk evolution could be reached. This is the case when the disk is capable of reacting to variations in the mass infall on time scales that are not very much longer than the variation time scales. During this phase, the mass of the disk is approximately constant.
- Phase 3 is characterized by a cease in the infall of matter in the disk. When mass no longer is replenished from outside, the disk enters a new phase of evolution as now the dominant process is no longer a quasi-stationary processing of material through the disk but rather a clearing of the disk. How this proceeds is highly dependent on the viscosity prescription (discussed below). In the standard models one assumes that during this phase, accretion becomes gradually less efficient as the disks cool and become optically thin and radiatively stratified. If turbulence ceases, this phase could be identified with the one during which dust grains begin settleing into the disks plane and as a consequence of this, planetesimal accretion could occur.

In much more detail, the currently discussed solar nebula models are reviewed by Wood and Morfill (1988). In the years since then especially the papers by Mizuno, Markiewicz, and Völk (1988), Ruden and Pollack (1991), Völk and Morfill (1991) and Morfill, Spruit and Levy (1992) deserve attention.

Two topics that are of very high importance for the evolution of the protoplanetary nebula but still have to be adressed in more extent are:

– What is the cause for viscosity?
– Are the model solutions stable at all?

In the following, we will make comments on both questions, and will deal with the second topic in even more extent in the next section.

4.1. Viscosity in protoplanetary nebulae

In principle, there are three possible mechanisms one can think of when looking for a driving agent for gas and dust disks:

– global mechanisms,
– ◇ gravitational torques
– ◇ magnetic fields
– local mechanisms
– ◇ turbulent viscosity

The most natural and best understood form of viscosity, namely molecular viscosity, is excluded beyond any doubts as the evolution time scales resulting from a pure molecular viscosity would be several orders of magnitude longer than the Hubble time. In other words: accretion disks where only molecular viscosity is at work do not evolve over the Hubble time. The fact that we are here and do not live right in the central plane of a protoplanetary disks immediately excludes molecular viscosity.

The only kind of accretion disks that are observable directly to quite some extent and for which detailed models exist are those in dwarf nova systems. There, in principle also all three mechanisms could play an important rôle. It has turned out that one gets the best theoretical representation of what is observed by assuming that a turbulent viscosity is at work. This is also the assumed in most of the presently discussed protoplanetary models (with the exception of Hayashi, 1981, who argued for magnetic forces as the driving agent). Despite this, the importance of magnetic fields in all kinds of accretion disks may be underestimated (Tout and Pringle, 1992).

It is known that in both kinds of accretion disks (protoplanetary ones and dwarf nova disks) convective energy transport is of importance at least during part of the evolution. Only recently has it been shown for dwarf nova disks that one can model convection and turbulent viscosity in those regions where convection is the dominant means of energy transport in a self-consistent way (Duschl, 1989). In these calculations, turbulence was described following a mixing length picture. It turned out that then the resulting viscosity parameter α is in the range of $10^{-1\ldots-3}$. This is approximately the parameter range that is also in discussion for protoplanetary disks.

In protoplanetary disk models it is usually assumed that turbulence dies out when convection dies out. This seems to be a reasonable approach when argued that there is only one turbulence underlying both convection and viscosity. Nonetheless, in the case of dwarf novae, this is in clear contradiction with what is observed. Simply from time scale arguments, one can deduce that in the phases when energy is not transported through convection, the viscosity is even more efficient. So, there must be some – yet unknown – mechanism present that not only keeps viscosity from dying out together with convection but rather makes it even more efficient.

In this respect, most of the presently discussed protoplanetary nebula models take a very different approach by assuming that without convection the redistribution of angular and thus accretion of mass, i.e., the depletion of mass, also becomes very inefficient. The fact that this contradicts what we know from dwarf novae disks is not necessarily a problem as we are talking about quite a different parameter range. But nonetheless it should be clarified how protoplanetary disks evolve with considerable viscosity even in non-convective regions (although one does not really want the disk to evolve faster in later stages as this would shorten the time that is available for the formation of the planets).

4.2. Stability of protoplanetary nebulae

Under certain circumstances, viscous accretion disks may turn unstable (see section 3). As was shown in Section 3.2., the opacity κ is a crucial agent that can decide about stability or instability. The question whether an instability may occur or not does not depend so much on the absolute values of κ but rather on the derivatives $\partial \log \kappa / \partial \log \rho|_T$ ($\hat{=}$ A in eq. 3.22) and $\partial \log \kappa / \partial \log T|_\rho$ ($\hat{=}$ B in eq. 3.22). In the temperature range that is of relevance for protoplanetary disks, there is quite a number of features in the opacities each of which might give rise to an instability of the disk (see, e.g., Pollack et al., 1986). Whether these features really lead to instabilities has to be clarified by consistent model calculations of the disks' vertical structure. As is known from accretion disks in close binaries, a proper 1+1-dimensional model is the minimum prerequisit for deciding whether the disks are stable or not. First attempts to analyze the behaviour of unstable disks were undertaken by Lin and Papaloizou (1985) and Clarke et al. (1990). In the following section we will introduce some new ideas that might give an even more complicated and involved picture of the evolution of protoplanetary disks. Detailled model calculations are under way and will be published later (Duschl, Gail, Tscharnuter, 1993, in preparation).

5. The rôle of chemical reactions in the evolution of protoplanetary disks

T Tauri and FU Orionis stars are young stellar systems with circumstellar material. In these systems, most of the mass is already concentrated in the central stellar core. It is suspected that both types of stars are actually only the manifestations of the same underlying evolutionary status. FU Orionis-stars may be identified with the active phase of what is otherwise recognized as a T Tauri star.

There are strong indications that accretion disks play an important rôle (Bertout, 1989; Beckwith et al., 1990; Bouvier and Bertout, 1992). Recently (Marsh and Mahoney, 1992) there were even claims that there could be planets already present in the disk. However, as uncertain as the latter statement may be, many hints point in the direction that the T Tauri and FU Orionis phenomenon is closely related to many of the above discussed features of protoplanetary disks.

The activity of FU Orionis variables consititutes itself in a rise of luminosity by a factor of up to 10,000 within a few months, and a subsequent decline that lasts at least for many decades. Actually none of the FU Orionis stars that have been observed since the prototype FU Orionis itself was detected about half a century ago has reached its pre-outburst luminosity since, but also none has ceased declining in luminosity yet. The idea that the outbursts of FU Orionis-stars are due to a disk instability of the above discussed type is well established. Nonetheless, even the best models due not yet succeed to reproduce the time scales correctly (Clarke et al., 1990).

In the following, we summarize a new idea about chemical reactions that may play a rôle in protoplanetary accretion disks and that may be very closely related to FU Orionis outbursts (Duschl, Gail, and Tscharnuter, 1992).

As was shown above the onset and evolution of disk instabilities is mainly governed by the density and temperature dependence of the opacity of the disk's material. At low temperatures, the opacity is mainly due to dust and molecules, in particular CO and H_2O. During an outburst, temperature rises above the vaporization temperature of dust and the dissociation temperature of most of the molecules. In the later stage of the outburst evolution, when temperature falls, molecules will form again first, and later dust will condense out of the gas phase. It is by no means clear that this leads to a chemical composition that is identical to the one at the onset of the outburst. If the chemical composition is changed, one has to expect that the evolution is not governed by a limit cycle as one is used to in *ordinary* accretion disk outburst (like in dwarf novae) but by a limit cycle that is altered through the evolution of the outburst.

Before a (first) outburst, most of the carbon atoms that are present in the disk are bound in soot particles which vaporize in the course of the outburst. When the disk cools, carbon is not likely to form soot again, but rather to remain in the form of CO, in the gas. This means an irreversible change of the chemical composition of the gas and dust phase between before and after the outburst. This, in turn, alters the opacities considerably, thus changing the limit cycle that governs the outburst.

The evolution of such chemically modified accretion disks will be modelled in a combination of a standard 1+1-dimensional time-dependent accretion disk code that additionally (and in contrast to standard codes) incorporates two sets of transport reaction equations (Duschl, Gail, and Tscharnuter, 1993, in preparation).

Molecules: The time scales of molecule dissociation and formation is governed on the one hand by characteristic reaction time scales and on the other hand by the time scales within which macroscopic parameters (e.g., pressure and temperature) change. In the disks' outer zones, the densities are so low that the reaction time scales and the hydrodynamic ones will be of the same order. This requires a kinetic treatment of the chemical processes.

For this, the rate equations

$$\frac{\partial n_i}{\partial t} + \nabla (n_i \mathbf{v}) = R_i \qquad 5.1$$

have to be solved for all species i that contribute significantly to the gas opacity and that participate in their dissociation or formation processes.

The main contributers to the opacity are H_2, H_2O, and TiO, for higher temperature also SiO and CO are of importance. This means that the reaction network mainly has to include the chemistry of the elements H, C, O, Si, and Ti. As in the disk C is less abundant than O, for the relevant temperatures and pressures all C is in CO as long as there are molecules present at all. Under these circumstances, C-H compounds will not form.

Dust formation and vaporization: The time scale for vaporization is short compared to the hydrodynamic one. This allows to treat the problem of the vaporization of dust during an increase of temperature in the approximation of *instantaneous* dust vaporization. The vaporizing material has to be taken into account as an additional source term R in the molecules' rate equations.

The problem of dust condensation during cooling of the gas can be separated into the question of the formation of condensation seeds and the one of the growth of macroscopic dust particles. It can be shown that the domain of cluster sizes between the critical cluster for seed formation and macroscopic cluster is crossed so quickly that this intermediate phase of particle growth does not play a crucial rôle. This allows to treat all particles that are larger than the critical one as macroscopic.

For computing the dust contribution to the total opacity only the integral quantity

$$\kappa_\nu(t) \propto \int \kappa_\nu(N) \cdot f(N,t) \qquad 5.2$$

is required. $f(N,t)$ is the distribution function of cluster sizes. In general, the dust absorption coefficient can be represented by the moments of the distribution function:

$$K_i(t) = \int f(N,t) N^{\frac{i}{3}} dN. \qquad 5.3$$

For these, a closed system of transport equations can be derived:

$$\frac{\partial K_0}{\partial t} + \nabla \mathbf{v} K_0 = J_*$$

$$\frac{\partial K_i}{\partial t} + \nabla \mathbf{v} K_i = \frac{i}{3} \frac{1}{\tau_w} K_{i-1} \quad (1 \leq i \leq 3) \qquad 5.4$$

($\mathbf{v}$: velocity). Here, J_* is the rate of seed formation and

$$\frac{1}{\tau_w} = \sum_i \alpha_i n_i v_i \sigma \qquad 5.5$$

denotes the growth time scale (σ: surface per atom in the dust particle; v average velocity of gas particles; n_i particle density of the molecules that are responsible for the growth; α_i: sticking coefficient).

The consumption of molecules due to dust growth is taken into account through the rate

$$R = -\frac{2}{3} \alpha n v \sigma K_2 \qquad 5.6$$

in the respective rate equations.

Under the conditions present in the disk (slow cooling), the seed formation is quasi-stationary. The nucleation rate is given by

$$J_* = \frac{1}{\tau_w} n_{cr} \qquad 5.7$$

(n_{cr} equilibrium particle density of the least abundant cluster on the path from the molecule to dust; τ_w growth time scale from one cluster size to the next larger one on the reaction path). The equation for J_* together with one for n_{cr} closes the coupled system for molecule and dust formation and destruction.

As a consequence of this, we expect the disk instability to be altered through its own action, and consequently the time scales to be modified. It remains to be seen whether this can account for the observations that the ratio between the ris and the decline time scale in FU Orionis-stars is considerably smaller than the standard disk outburst models (i.e., without chemical reactions) predict.

6. Summary

It seems as if the processes that lead to the formation of protostellar and protoplanetary disks were well identified. During and after the formation, a description as a viscous gas and dust accretion disks is the most reasonable approach. During the last two decades, several models have been developed that describe the evolution of such disks. Within the last decade, protoplanetary disk became also accessible to observations (T Tauri-stars, FU Orionis-outbursts, β Pictoris, ...). It is of great help that in other fields of astrophysics, disks also play a rôle and are accessible to direct observations (especially in dwarf nova systems, of which some are even eclipsing binaries that thus allow for a spatial resolution of the disk structure). From the disk models that have been demonstrated in such systems one may suspect that not all processes that can play a rôle in protoplanetary accretion disks are already taken into account in the necessary way (viscosity in domains where energy is mainly transported by radiation; accretion disk in-/stabilities). Finally one can identify processes that can influence the evolution of protoplanetary disks considerably that are not yet taken into account but that might have the potential of being intimately related to directly observable features in young stellar objects (chemical reactions and its relation to FU Orionis-outbursts).

7. References

Beckwith, S.V.W, Sargent, A.I., Chini, R.S, and Güsten, R., 1990: *Astronomical J.* **99,** 924

Bertout, C., 1989: *Ann. Rev. Astron. Astrophys.* **27,** 351

Bouvier, J. and Bertout, C., 1992: *Astron.&Astrophys.* in press

Calvet, N., Hartmann, L., and Kenyon, S.J., 1991: *Astrophys. J.* **383,** 752

Cameron, A.G.W, 1962: *Icarus* **1,** 13

Cameron, A.G.W. and Pine, M.R., 1973: *Icarus* **18,** 377

Clarke, C.J., Lin, D.N.C., and Pringle, J.E., 1990: *Monthly Notices Roy. Astron. Soc.* **242,** 439

Duschl, W.J., 1988: *Astron.&Astrophys.* **194,** 43

Duschl, W.J., 1989: *Astron.&Astrophys.* **225,** 105

Duschl, W.J., Gail, H.-P., and Tscharnuter, W.M., 1992: *Bull. American Astron. Soc.* **24,** 798

Hayashi, C., 1981: *Prog. Theoret. Phys. Suppl.* **70,** 35

Hubeny, I., 1989, in: F. Meyer, W.J. Duschl, F. Frank, and E. Meyer-Hofmeister (eds.) *"Theory of accretion disks"*, Kluwer Academic Publishers, Dordrecht, The Netherlands, p.445

Kant, I., 1755: *"Allgemeine Naturgeschichte und Theorie des Himmels"*, bey Johann Friederich Petersen, Königsberg und Leipzig

Kippenhahn, R. and Thomas, H.-C., 1982: *Astron.&Astrophys.* **114,** 77

Kürschner, R., 1992: *"Die numerische Totalenergieerhaltung bei der Simulation des protostellaren Kollaps"*, Diploma thesis, University of Heidelberg, Germany.

Larson, R.B., 1969: *Monthly Notices Roy. Astron. Soc.* **145,** 271

Lin, D.N.C. and Papaloizou, J., 1985, in: D.C. Black and M.S. Matthews (eds.) *"Protostars and Planets II"*, The University of Arizona Press, Tucson, AZ, U.S.A., p.981

Lüst, R., 1952: *Zeitschr. Naturforschung* **7a,** 87

Marsh, K.A. and Mahoney, M.J., 1992: *Astrophys. J. Letters,* submitted

Matsuda, T., Sekino, N., Shima, E., Sawada, K., and Spruit, H, 1990: *Astron.&Astrophys.* **235,** 211

Meyer, F. and Meyer-Hofmeister, E., 1981: *Astron&Astrophys.* **104,** L10

Mizuno, H., Markiewicz, W.J., and Völk, H.J., 1988: *Astron.&Astrophys.* **195,** 183

Morfill, G.E., 1985, in: R.A. Lucas, A. Omont, and R. Stora (eds.) *"Birth and Infancy of Stars"*, North Holland, Amsterdam, The Netherlands, p.693

Morfill, G.E., Spruit, H., Levy, E., 1992, in: J. Lunine and E. Levy (eds.) *"Protostars and planets III"*, The University of Arizone Press, Tucson, AZ, U.S.A., in press

Novikov, I.D. and Thorne, K.S., 1973, in: C. DeWitt and B.S. DeWitt (eds.) *"Black Holes"*, Gordon and Breach Science Publishers, New York, NY, U.S.A., p.343

Paresce, F., 1991: *Astron&Astrophys.* **247,** L25

Pollack, J.B., McKay, C.P. and Christofferson, B.M., 1986: *Icarus* **64,** 471

Pringle, J.E., 1992: *Monthly Notices Roy. Astron. Soc* in press

Ruden, S.P. and Pollack, J.B., 1991: *Astrophys. J.* **375,** 740

Shakura, N.I. and Sunyaev, R.A., 1973: *Astron.&Astrophys.* **24,** 337

Smith, B.A. and Terrile, R.J., 1984: *Science* **226,** 1421

Tout, C.A. and Pringle, J.E., 1992: *Monthly Notices Roy. Astron. Soc* in press

Tscharnuter, W.M., 1989, in: F. Meyer, W.J. Duschl, F. Frank, and E. Meyer-Hofmeister (eds.) *"Theory of accretion disks"*, Kluwer Academic Publishers, Dordrecht, The Netherlands, p.113

Völk, H.J. and Morfill, G.E., 1991: *Space Sci. Rev.* **56,** 65

Weizsäcker, C.F. von: 1943, *Zeitschr. Astrophys.* **22,** 319

Wood, J.A. and Morfill, G.E., 1988, in: J.F. Kerridge and M.S. Matthews (eds.) *"Meteorites and the early solar system", The University of Arizona Press, Tucson, AZ, U.S.A., p.329*

CHEMISTRY OF THE SOLAR NEBULA

BRUCE FEGLEY, JR.
Department of Earth & Planetary Sciences
and McDonnell Center for the Space Sciences
Washington University
St. Louis, MO 63130-4899 USA

ABSTRACT. We review theoretical models of thermochemical processes in the solar nebula which consider the effects of nebular dynamics on the chemistry of the abundant, chemically reactive volatile elements H, O, C, N, and S. Specific, testable predictions of these models are described. We also use the theoretical models to interpret the latest available data on the abundances and molecular speciation of volatiles in comet P/Halley.

1. Introduction

The thermochemical interactions between gases and grains in the solar nebula played a central role in establishing the observed volatile element inventories of the planets, their satellites, and the small bodies in the solar system (e.g., the asteroids, comets, and meteorites). For example, the oxidation and sulfurization of Fe metal grains, the synthesis of organic compounds by grain catalyzed chemistry between $CO + H_2$, the extent of evaporation and thermal reprocessing of presolar grains are all different types of thermochemical interactions which ultimately influenced the volatile element content of the solid planet-forming materials in the solar nebula.

In this paper we review thermochemical processes in the solar nebula with an emphasis on the chemistry of the abundant, chemically reactive volatiles H, O, C, N, and S. We begin by discussing the elemental abundances in solar composition material and then move on to consider the types of interstellar materials accreted by the solar nebula. Next we examine the isotopic evidence for inefficient thermal processing in the nebula. After reviewing the results of thermochemical equilibrium models, which are the foundation for any discussion of nebular chemistry, we describe the basis of thermochemical kinetic models of nebular chemistry which attempt to consider the effects of nebular mixing and dynamics on the chemical processes taking place. We begin reviewing the results of these models with detailed descriptions of the gas phase and grain catalyzed chemistry of carbon and nitrogen, then move on to the sulfurization and oxidation of Fe metal, and the mechanisms for water retention in solid grains. The effects of this high temperature chemistry on the composition of low temperature condensates are then reviewed. At this point we present current observational data on the abundances of volatiles in comet Halley and interpret these data in terms of the nebular chemistry models discussed earlier.

2. Elemental Abundances in the Solar Nebula

A knowledge of the elemental abundances in the solar nebula (i.e., in solar composition material) is essential for any discussion of nebular chemistry. An understanding of how the solar abundances of the chemical elements have been determined is also useful for anyone interested in cosmochemical modelling of the solar nebula because the chondritic meteorites, which provide much of the abundance data, originally formed in the solar nebula and their elemental compositions have been little altered since that time.

J. M. Greenberg et al. (eds.), The Chemistry of Life's Origins, 75–147.

The first attempts to determine the abundances of the elements probably date back to Clarke (1889) who attempted to find periodicities in the relative abundances of the elements in the Earth's crust. However, as we now know, the relative elemental abundances in the terrestrial crust have been modified by planetary differentiation and weathering processes and generally are not representative of the abundances of the elements in solar composition material. Thus, Clarke's attempts were doomed to failure.

Instead, it is necessary to analyze samples of material which formed in the solar nebula and have retained their elemental composition without alteration since that time. Obviously, spectroscopic determination of the elemental abundances in the Sun can provide the necessary information with the exception of deuterium and a few light elements such as Li, B, and Be which have been partially or totally consumed by thermonuclear fusion reactions. In fact this approach has been taken and is very useful provided that the relevant inputs for the data analysis are well known. For example, it is necessary to know the oscillator strengths for the spectral lines of interest and it is also necessary to realistically model physical conditions in the solar photosphere where the lines are formed. Any possible elemental fractionation processes which affect the elemental abundances in the solar photosphere must also be understood. The topic of solar photospheric abundances is reviewed by Grevesse (1984).

Another highly successful approach has been to determine elemental abundances in the chondritic meteorites (so named because they contain small, rounded glass beads known as chondrules), which were also formed in the solar nebula. In the modern era, chemical analyses of the elemental abundances in chondrites was first done by Goldschmidt, the Noddacks, and their colleagues in the 1920s and 1930s. This work was critically assessed by Goldschmidt (1937, 1954) who compiled the first table of elemental abundances in meteorites. This tabulation showed that to first approximation, the abundances of the non-volatile elements in meteorites and in the Sun were similar. It also served as a stimulus for the seminal work of Suess (1947a,b) who postulated that the abundances of the nuclides, and especially the odd mass number nuclides, are a smooth function of mass number.

Suess (1947a,b) used this postulate to adjust the elemental abundances to produce a smooth variation of abundance with mass number. In some cases, such as Re, adjustments of up to a factor of 100 were made and shown to be correct by subsequent chemical analyses of meteorites. Many of the details of this curve are reviewed by Woolum (1988).

A later paper by Suess and Urey (1956) carried this approach even further and produced an influential table of solar elemental abundances. One outcome of their table was the pioneering studies of stellar nucleosynthesis mechanisms by Burbidge et al (1957). Another outcome was an increasing number of high quality analytical studies of elemental abundances in chondritic meteorites. Much of this work is summarized in the compilation edited by Mason (1971). Much of the later work since that compilation was assembled is reviewed by Mason (1979) and in the papers by Anders and Ebihara (1982) and Anders and Grevesse (1989). It is safe to say that in the intervening 35 years since the publication of Suess and Urey (1956), the improvements in chemical analyses of meteorites, in the understanding of stellar nucleosynthesis mechanisms, and in astronomical observations of elemental abundances in the Sun and other stars have led to vast improvements in our knowledge of the solar abundances of the elements.

The result of all these efforts is displayed in Table 1, based on Anders and Grevesse (1989), which summarizes present knowledge of the solar abundances of the elements. It shows the atomic abundances of the elements in CI chondrites (normalized to 10^6 atoms of Si), in the solar photosphere (normalized to 10^{12} atoms of H), and elemental abundances by mass in the Orgueil CI chondrite. The CI chondrites are chosen as an abundance standard because their elemental composition matches that of the Sun more closely than that of any other meteorite class. The Orgueil meteorite is the most widely distributed and frequently analyzed CI chondrite.

Perhaps the single most important point illustrated by the data in Table 1 is that the chemistry of solar composition material and of the solar nebula is overwhelmingly dominated by hydrogen. The third most abundant element overall and the second most abundant chemically reactive element, oxygen, has only about 0.1% of the H atomic abundance in solar matter. Excluding He, the atomic abundances of all other elements combined sum up to about 70% of the oxygen elemental abundance. This situation is in stark contrast to the elemental abundances at the surface of the Earth where O is the most abundant element in the crust and H is a trace element contained in a thin oceanic veneer. The contrast is even more severe on Venus where H is depleted by about a factor of 10^5 relative to its observed surficial abundance on the Earth. Clearly one of the most important problems facing cosmochemists is how to explain the evolution of the volatile inventories of Venus, Earth, and Mars from the compositionally very different solar nebula. This problem is still unresolved.

A second important point demonstrated by the solar elemental abundances is that the very similar abundances of carbon and oxygen results in an intimate coupling of their chemistry. In the Anders and Grevesse (1989) tabulation of solar elemental abundances the atomic C/O ratio is ~0.42, while in the earlier tabulations of Cameron (1973, 1982) the C/O ratio was 0.57 and 0.60, respectively. The more recent data of Grevesse et al (1991) on the photospheric carbon abundance correspond to a C/O ratio of 0.47. One of the most stable molecules observed in nature is CO. The similarity of the C and O abundances (C/O ~0.4-0.6) dictates that over a wide range of P,T conditions in the solar nebula CO is the most abundant carbon gas and is also either the first or second most abundant oxygen gas, depending on the degree of dissociation of water vapor. Changes in the gas phase abundance of CO and H_2O, for example by freezing out the water vapor (Stevenson and Lunine 1988), can then alter the total C/O ratio in the gas phase. In turn, changing this ratio alters the oxygen fugacity of the solar nebula and the major element mineralogy of grains formed from the nebular gas (e.g., Larimer 1975; Larimer and Bartholomay 1979). Alteration of this mineralogy has first order consequences such as the formation of carbide, nitride, and sulfide minerals that are stable at high temperature, thus leading to efficient retention of these important volatile elements in rocky material. There are also important second order consequences such as the condensation of cohenite Fe_3C instead of Fe metal, which may influence the nature of grain catalyzed chemistry which can proceed.

The similarity of the C/O ratio is also important for determining the amount of water ice in volatile-rich bodies that formed in the outer solar nebula. If all carbon remained as CO at the low temperatures in the outer nebula, then the water ice abundance was decreased below the amount which could condense if CO had already been converted to CH_4. On the other hand, if CO were efficiently converted to CH_4 and/or other hydrocarbons, then a sizeable fraction of the total O was released from CO and was available for formation of water ice. As we shall see later, the ice/rock mass ratios in "icy" bodies formed in the solar nebula, where CO was the dominant carbon gas are predicted to be lower than the ice/rock ratios in "icy" bodies formed in giant protoplanetary subnebulae, where CH_4 was the dominant carbon gas.

Another significant point shown by the elemental abundance data is that carbon chemistry was an important facet of nebular chemistry. This stems from the fact that C is the third most abundant chemically reactive element. In fact, several questions such as whether or not CO can be converted to CH_4 in any appreciable quantities with decreasing temperature and the extent to which it undergoes grain catalyzed reactions with H_2 to form organic compounds are fundamental in determining the share of volatiles inherited by the Earth and the form in which these volatiles were provided.

TABLE 1. Abundances of the Elements in CI Chondrites and in the Solar Photosphere.

Atomic Number	Element Name & Chemical Symbol	Abundance in CI Chondrites (Si = 10^6 atoms)	Abundance (by mass) in Orgueil CI chondrite	Abundance in Solar Photosphere (H = 10^{12} atoms)
1.	Hydrogen (H)	2.79×10^{10}	2.02%	1.00×10^{12}
2.	Helium (He)	2.72×10^{9}	56 nL/g	9.77×10^{10}
3.	Lithium (Li)	57.1	1.49 μg/g	14.45
4.	Beryllium (Be)	0.73	24.9 ng/g	14.13
5.	Boron (B)	21.2	870 ng/g	398
6.	Carbon (C)[a]	1.01×10^{7}	3.45%	3.98×10^{8}
7.	Nitrogen (N)[b]	3.13×10^{6}	3180 μg/g	1.00×10^{8}
8.	Oxygen (O)	2.38×10^{7}	46.4%	8.51×10^{8}
9.	Fluorine (F)	843	58.2 μg/g	3.63×10^{4}
10.	Neon (Ne)	3.44×10^{6}	203 pL/g	1.23×10^{8}
11.	Sodium (Na)	5.74×10^{4}	4900 μg/g	2.14×10^{6}
12.	Magnesium (Mg)	1.074×10^{6}	9.53%	3.80×10^{7}
13.	Aluminum (Al)	8.49×10^{4}	8690 μg/g	2.95×10^{6}
14.	Silicon (Si)	1.00×10^{6}	10.67%	3.55×10^{7}
15.	Phosphorus (P)	1.04×10^{4}	1180 μg/g	2.82×10^{5}
16.	Sulfur (S)	5.15×10^{5}	5.25%	1.62×10^{7}
17.	Chlorine (Cl)	5,240	698 μg/g	3.16×10^{5}
18.	Argon (Ar)	1.01×10^{5}	751 pL/g	3.63×10^{6}
19.	Potassium (K)	3,770	566 μg/g	1.32×10^{5}
20.	Calcium (Ca)	6.11×10^{4}	9020 μg/g	2.29×10^{6}
21.	Scandium (Sc)	34.2	5.83 μg/g	1,259
22.	Titanium (Ti)	2,400	436 μg/g	9.77×10^{4}
23.	Vanadium (V)	293	56.2 μg/g	1.0×10^{4}
24.	Chromium (Cr)	1.35×10^{4}	2660 μg/g	4.68×10^{5}
25.	Manganese (Mn)	9,550	1980 μg/g	2.45×10^{5}
26.	Iron (Fe)[c]	9.00×10^{5}	18.51%	3.24×10^{7}
27.	Cobalt (Co)	2,250	507 μg/g	8.32×10^{4}
28.	Nickel (Ni)	4.93×10^{4}	1.10%	1.78×10^{6}
29.	Copper (Cu)	522	119 μg/g	1.62×10^{4}
30.	Zinc (Zn)	1,260	311 μg/g	3.98×10^{4}
31.	Gallium (Ga)	37.8	10.1 μg/g	759

Atomic Number	Element Name & Chemical Symbol	Abundance in CI Chondrites (Si = 10^6 atoms)	Abundance (by mass) in Orgueil CI chondrite	Abundance in Solar Photosphere (H = 10^{12} atoms)
32.	Germanium (Ge)	119	32.6 µg/g	2,570
33.	Arsenic (As)	6.56	1.85 µg/g	------
34.	Selenium (Se)	62.1	18.2 µg/g	------
35.	Bromine (Br)	11.8	3.56 µg/g	------
36.	Krypton (Kr)	45	8.7 pL/g	------
37.	Rubidium (Rb)	7.09	2.30 µg/g	398
38.	Strontium (Sr)	23.5	7.80 µg/g	794
39.	Yttrium (Y)	4.64	1.53 µg/g	174
40.	Zirconium (Zr)	11.4	3.95 µg/g	398
41.	Niobium (Nb)	0.698	246 ng/g	26.3
42.	Molybdenum (Mo)	2.55	928 ng/g	83.2
44.	Ruthenium (Ru)	1.86	714 ng/g	69.2
45.	Rhodium (Rh)	0.344	134 ng/g	13.2
46.	Palladium (Pd)	1.39	556 ng/g	49
47.	Silver (Ag)	0.486	197 ng/g	8.7
48.	Cadmium (Cd)	1.61	680 ng/g	72.4
49.	Indium (In)	0.184	77.8 ng/g	45.7
50.	Tin (Sn)	3.82	1680 ng/g	100
51.	Antimony (Sb)	0.309	133 ng/g	10
52.	Tellurium (Te)	4.81	2270 ng/g	------
53.	Iodine (I)	0.9	433 ng/g	------
54.	Xenon (Xe)	4.7	8.6 pL/g	------
55.	Cesium (Cs)	0.372	186 ng/g	------
56.	Barium (Ba)	4.49	2340 ng/g	135
57.	Lanthanum (La)	0.446	236 ng/g	16.6
58.	Cerium (Ce)	1.136	619 ng/g	35.5
59.	Praseodymium (Pr)	0.1669	90 ng/g	5.1
60.	Neodymium (Nd)	0.8279	463 ng/g	31.6
62.	Samarium (Sm)	0.2582	144 ng/g	10
63.	Europium (Eu)	0.0973	54.7 ng/g	3.2
64.	Gadolinium (Gd)	0.33	199 ng/g	13.2
65.	Terbium (Tb)	0.0603	35.3 ng/g	0.8

Atomic Number	Element Name & Chemical Symbol	Abundance in CI Chondrites (Si = 10^6 atoms)	Abundance (by mass) in Orgueil CI chondrite	Abundance in Solar Photosphere (H = 10^{12} atoms)
66.	Dysprosium (Dy)	0.3942	246 ng/g	12.6
67.	Holmium (Ho)	0.0889	55.2 ng/g	1.8
68.	Erbium (Er)	0.2508	162 ng/g	8.5
69.	Thulium (Tm)	0.0378	22 ng/g	1
70.	Ytterbium (Yb)	0.2479	166 ng/g	12
71.	Lutetium (Lu)	0.0367	24.5 ng/g	5.8
72.	Hafnium (Hf)	0.154	108 ng/g	7.6
73.	Tantalum (Ta)	0.0207	14.0 ng/g	------
74.	Tungsten (W)	0.133	92.3 ng/g	12.9
75.	Rhenium (Re)	0.0517	37.1 ng/g	------
76.	Osmium (Os)	0.675	483 ng/g	28.2
77.	Iridium (Ir)	0.661	474 ng/g	22.4
78.	Platinum (Pt)	1.34	973 ng/g	63.1
79.	Gold (Au)	0.187	145 ng/g	10.2
80.	Mercury (Hg)	0.34	258 ng/g	------
81.	Thallium (Tl)	0.184	143 ng/g	7.9
82.	Lead (Pb)	3.15	2430 ng/g	70.8
83.	Bismuth (Bi)	0.144	111 ng/g	------
90.	Thorium (Th)	0.0335	28.6 ng/g	1.3
92.	Uranium (U)	0.009	8.1 ng/g	<0.34

Elemental abundance compilations do not list the following radioactive elements which have no stable isotopes and are not found in meteorites: Technetium (43), Promethium (61), Polonium (84), Astatine (85), Radon (86), Francium (87), Radium (88), Actinium (89).

The abbrevations used for abundances in the Orgueil CI chondrite have the following meanings: % = mass %, nL/g = 10^{-9} liters/gram, µg/g = 10^{-6} grams/gram, pL/g = 10^{-12} liters/gram, ng/g = 10^{-9} grams/gram

The abundances in Table 1 are generally from Anders and Grevesse (1989) with the following values also being included:

[a]The photospheric carbon abundance is from Grevesse et al (1991).

[b]The photospheric nitrogen abundance is from Grevesse et al (1990).

[c]The photospheric iron abundance is from Biémont et al (1991) and Holweger et al (1990).

Finally, a fourth point illustrated by the abundances in Table 1 is that the chemistry of the solar nebula is essentially the chemistry of 8 elements (H, O, C, N, Mg, Si, Fe, S). To a lesser extent Al, Ca, Na, Ni, Cr, P, and Mn are also important by virtue of their abundances. However, several of the elements in this second list (Ni, Cr, Mn) simply alloy with Fe without engaging in any other chemistry. Thus, to a good first approximation, the elemental abundances dictate that the chemistry of solar material is the chemistry of only a handful of the naturally occurring elements. Therefore the rest of this review will emphasize the chemistry of these 12 elements while discussing chemistry in the solar nebula.

3. Composition and Preservation of Presolar Gases and Grains

The abundances in the preceding section tell us how much of each element was present in the solar nebula but do not tell us what form(s) each element was in when it was accreted by the solar nebula accretion disk. As discussed below, this information is potentially important because thermal processing in the solar nebula was not 100% efficient. Thus a knowledge of the initial gas and grain composition is useful for modelling the nebular reactions affecting the accreted interstellar gas and dust.

Qualitatively we expect that the interstellar gas and grains were thermally and chemically reprocessed to varying degrees depending on several factors such as the distance of the accreted material from the proto-Sun, the type of grains (e.g., rocky, organic, icy, etc.), whether the interstellar material was accreted at an early or late stage of nebular evolution, the rate of radial transport in the solar nebula relative to the rate of equilibrating reactions in the gas parcel, and to the rate of accretion of small grains into larger clumps. Obviously some grains in the solar nebula experienced complex histories such as cycles of evaporation, condensation, re-evaporation, and re-condensation. Other grains were presumably totally destroyed while some others were incorporated into meteorite parent bodies essentially unaffected by nebular processes. The work by Cameron and Fegley (1982) on the position of grain evaporation fronts in the solar nebula is a quantitative attempt to model the survivability of different types of rocky and metallic grains as a function of their radial and vertical position in the solar nebula. Related work has also been done by Morfill and colleagues (e.g., Morfill and Volk 1984).

A variety of sources provide information on the probable composition of the gases and grains accreted by the solar nebula. In recent years radio astronomy has detected a diverse suite of molecules in interstellar space and in the circumstellar shells of carbon stars (e.g., Irvine and Knacke 1989; Ormont 1991). At present over 70 molecules ranging in complexity from OH to $HC_{10}CN$ have been detected. Tables 2 and 3 schematically summarize a small subset of these data on the composition of interstellar and circumstellar gas and grains, indicate the possible effects of nebular chemistry on the accreted gas and grains, and also list possible preservation sites for the presolar species.

Several important points are indicated by these two tables. Perhaps the most significant is that the identification of presolar material in primitive objects (e.g., comets and some types of asteroids and meteorites) may be ambiguous because several important interstellar molecules are also predicted to form in the solar nebula. This problem may be most severe for CO and N_2. If complete chemical equilibrium were attained, all interstellar carbon species would be converted to CO in the high temperature regions of the solar nebula and all interstellar nitrogen species would be converted to N_2. In the absence of a potentially diagnostic isotopic ratio, it is then impossible to distinguish interstellar CO and N_2 from nebular CO and N_2. Likewise, interstellar H_2S, unless it is isotopically distinctive in some way, is impossible to distinguish from nebular H_2S.

TABLE 2. Representative Interstellar Molecules and Their Possible Fates in the Solar Nebula[a]

Molecule(s)	Comments on Nebular Chemistry	Possible Preservation Site(s)
H_2	Major gas, o-p ratio reset to high T value	Does not condense in nebula
CO	Also stable high T form of C in nebula, use C & O isotopes to distinguish nebular from interstellar CO?	Trapped in clathrate hydrate or condensed as CO ice in comets or other icy bodies in outermost nebula
N_2	Also stable high T form of N in nebula, use N isotopes to distinguish nebular from interstellar N_2?	Trapped in clathrate hydrate or condensed as N_2 ice in comets or other icy bodies in outermost nebula
HDO	Isotopic exchange will equilibrate D/H with nebular H_2 at sufficiently high temperatures in inner nebula	HDO ice in comets & icy satellites, HDO in hydrated silicates on asteroids?
HCN	Also produced by shock chemistry in solar nebula & Jovian protoplanetary subnebulae, use isotopes to distinguish source?	HCN ice in comets and outer solar nebula bodies?, conversion to organic matter on meteorite parent bodies
H_2CO	Thermal decomposition and/or photolysis to H_2 + CO, polymerization to POM?	H_2CO & POM observed in comets
CH_3C_2H	Oxidation to CO at high temperatures	Cometary ices?
CH_3CN	Thermal decomposition	Cometary ices?
SO_2	Reduction to H_2S + HS at high temperatures	Cometary ices? (UV spectra give an upper limit of <0.1% of solar S abundance in several comets)[b]
COS	Thermal decomposition and/or photolysis to CO + S, subsequent conversion of S to H_2S	Cometary ices? (radio data give an upper limit of <10% of solar S abundance in Comet Levy)[c]
H_2S	Also stable high T form of S in nebula	Probably difficult to preserve & distinguish from nebular H_2S
H_2CS	Thermal decomposition to CO + H_2S at fairly low temperatures in nebula	Cometary ices? (radio data give an upper limit of about 50% of solar S abundance in Comet Levy)[c]

(a) The most abundant gases listed by Irvine and Knacke (1989), excluding radicals and ions. The latter species were probably chemically reprocessed on a rapid timescale.

(b) Kim and A'Hearn 1991

(c) Crovisier et al 1991

TABLE 3. Major Components of Interstellar Dust and Their Fate(s) in the Solar Nebula[a]

Component	Comments on Nebular Chemistry	Possible Preservation Site(s)
Silicates	Amorphous material will be annealed and equilibrated, crystalline material will attempt to equilibrate at sufficiently high temperatures	Low T meteorite matrices, but difficult to distinguish from more abundant nebular constituents unless isotopically anomalous
Graphite	Oxidized at high T to form CO, hydrogenated at low T to form CH_4	Circumstellar graphite grains observed in primitive meteorites
Polycyclic Aromatic Hydrocarbons (PAHs)	Oxidized at high T for form CO + H_2, isotopic equilibration with nebular vapor, partial pyrolysis to more C-rich matter	D-rich organics observed in low T meteorite matrices & interplanetary dust particles, not conclusively identified as PAHs
Amorphous Carbon	Oxidized at high T to form CO, hydrogenated at low T to form CH_4	Low T meteorite matrices
Icy Grain Mantles	Evaporation of ices, chemical & isotopic equilibration with nebular gas, photolysis in outer nebula	Comets & other icy bodies in outer solar nebula?
Organic Refractory Grain Mantles	More resistant to processes which destroy organics	D-rich organics observed in low T meteorite matrices & interplanetary dust particles
SiC	Oxidation (e.g., by O, OH, H_2O) to silica & eventually silicates	Circumstellar SiC grains observed in primitive meteorites
MgS	Oxidation (e.g., by O, OH, H_2O) to Mg oxides & silicates	Low T meteorite matrices?, sensitive to water & O_2

(a) Major components as reviewed by Tielens and Allamandola (1987).

A similar problem exists for distinguishing presolar silicates from those produced by thermal processing (e.g., condensation, evaporation) in the solar nebula. Again, isotopic differences are one way to distinguish exotic material. If this is not possible, for example by virtue of small sample size which prohibits isotopic measurements, then other approaches may prove useful. Recently Bernatowicz et al (1991) discovered small (7-21 nm) crystals of TiC inside isotopically anomalous graphite grains which are clearly of presolar origin. Thus, the included TiC must also be presolar. Analogously, finding a silicate inside another grain, which by virtue of its mineralogy and/or isotopic composition is clearly presolar, may also be used as proof of a presolar origin. However, because the average composition of the gas and grains accreted by the solar nebula has solar chemical and isotopic composition, by definition, then it is clear that not all presolar material has to be isotopically distinctive from nebular materials.

On the other hand, some interstellar molecules are either less likely to be produced by nebular chemistry or less likely to be produced in amounts as large as those observed in the interstellar medium. It is generally believed that deuterium-rich species fall into this category. Table 4 lists the D/H ratios in molecules in interstellar clouds and in different solar system objects. There is a wide range of D/H ratios from the protosolar value of about 3×10^{-5} to values of a few percent in molecules in interstellar clouds.. As discussed later, it is difficult to elevate the protosolar D/H ratio to such high values solely by thermochemical isotopic exchange reactions in the solar nebula because these reactions are kinetically inhibited at the low temperatures where the exchange has to occur (Grinspoon and Lewis 1987). However, ion-molecule reactions are facile at such low temperatures and provide a means for enhancing the D/H ratios in various hydrides.

Likewise, some presolar grains have distinctive compositions which are not representative of thermal processing in the solar nebula. The SiC, TiC, diamond, and graphite grains found in primitive meteorites (Amari et al 1990; Bernatowicz et al 1987, 1991; Lewis et al 1987; Tang and Anders 1988; Virag et al 1992; Zinner et al 1987) are probably the best examples of this situation. As discussed by Fegley (1988), SiC, TiC, and diamond are thermodynamically unstable in the solar nebula and would be oxidized to other compounds at the high temperatures in the inner solar nebula. As discussed later, graphite is stable at low pressures and low temperatures in a solar composition gas, but is thermodynamically unstable under all other conditions. But even in these relatively clear cut cases there is still room for some ambiguity. Silicon carbide and graphite are also predicted to be thermodynamically stable under the highly reducing conditions necessary for formation of the enstatite chondrites (Larimer 1975; Larimer and Bartholomay 1979). Diamond is not thermodynamically stable under these conditions but diamonds formed by impacts are also present in the ureilites, carbon-rich achondrites. Even in these cases, the identification of the SiC, graphite, and diamond grains as presolar rests upon their exotic isotopic compositions.

4. Isotopic Evidence for Inefficient Thermal Processing in the Solar Nebula

The discussion in the preceding section explicitly assumed that thermal processing in the solar nebula was not 100% efficient. What is the evidence that this is the case? Suess (1965) felt compelled to state that "Among the very few assumptions which, in the opinion of the writer, can be considered well justified and firmly established, is the notion that the planetary objects, i.e. planets, their satellites, asteroids, meteorites, and other objects of the solar system, were formed from a well-mixed primordial nebula of chemically and isotopically uniform composition. At some time between the time of formation of the elements and the beginning of condensation of the less volatile material, this nebula must have been in the state of a

TABLE 4. D/H Ratios in Interstellar Space and the Solar System[a]

Material	D/H Ratio	Reference(s)
C_3HD/C_3H_2	0.03-0.15	Gerin et al 1987; Bell et al 1988
NH_2D/NH_3	0.003 - 0.14	Olberg et al 1985; Walmsley et al 1987
C_2D/C_2H	0.01-0.05	Herbst et al 1987
DNC/HNC	0.01 - 0.04	Brown and Rice 1981
DCN/HCN	0.002 - 0.02	Wootten 1987
DC_5N/HC_5N	0.02	Schloerb et al 1981; MacLeod et al 1981
DC_3N/HC_3N	~0.02	Langer et al 1980; Wootten 1987
$HDCO/H_2CO$	~0.01	Guelin et al 1982
HDO/H_2O on Venus	0.016-0.019	Donahue et al 1982; DeBergh et al 1991
IDPs	$(1.2\text{-}16) \times 10^{-4}$	Zinner 1988
Ordinary chondrites	$(0.8\text{-}10.5) \times 10^{-4}$	Zinner 1988
HDO/H_2O on Mars	$(7.8\pm0.3) \times 10^{-4}$	Bjoraker et al 1989; Owen et al 1988
Carbonaceous chondrites	$(0.8\text{-}6) \times 10^{-4}$	Zinner 1988
Comet Halley HDO/H_2O	$(0.6\text{-}4.8) \times 10^{-4}$	Eberhardt et al 1987b
Terrestrial D/H (SMOW)	1.557×10^{-4}	Hagemann, Nief and Roth 1970
D/H in Titan's atm.	$\sim15 \times 10^{-5}$	Coustenis et al 1989; Owen et al 1986
Meteoritic Phyllosilicates	$\sim14 \times 10^{-5}$	Yang and Epstein 1983
Neptunian atm. D/H	$\sim12 \times 10^{-5}$	DeBergh et al 1990
Uranian atm. D/H	$\sim9 \times 10^{-5}$	DeBergh et al 1986
Nebular HD/H_2	$\sim3 \times 10^{-5}$	Anders and Grevesse 1989
Jovian atm. D/H	$(2.6\pm1.0) \times 10^{-5}$	Fegley and Prinn 1988
Saturnian atm. D/H	$(1.7\pm1.0) \times 10^{-5}$	Fegley and Prinn 1988
Interstellar HD/H_2	$(0.8\text{-}2) \times 10^{-5}$	Boesgaard and Steigman 1985

(a) Modified from Irvine & Knacke (1989) and Zinner (1988)

homogeneous gas mass of a temperature so high that no solids were present. Otherwise, variations in the isotopic composition of many elements would have to be anticipated." Thus, the formation of the solar nebula was believed to have erased the previously existing record of interstellar chemistry, and the chemical fractionations in meteorites and planetary material were subsequently produced largely by chemical equilibrium processes in the monotonically cooling solar nebula (e.g., see Anders 1968; Barshay and Lewis 1976; Grossman 1972; Grossman and Larimer 1974; Larimer 1967; Lewis 1972a; Lewis 1974).

However, this picture changed dramatically with the discovery by Clayton, Grossman and Mayeda (1973) of oxygen isotopic anomalies in meteorites. This was significant for several reasons: (1) the anomalies in the $^{16}O/^{17}O$ and $^{16}O/^{18}O$ ratios are not due to mass fractionation processes, (2) the anomalies are present in the most abundant element in rocky material, and (3) the anomalies are present in the third most abundant element overall and the second most abundant chemically reactive element in solar composition material.

It is true that prior to this work, there were reports of isotopic anomalies in some noble gases in meteorites, notably the presence of ^{129}Xe and of Ne-E, essentially pure ^{22}Ne (Reynolds 1960; Jeffery and Reynolds 1961; Black and Pepin 1969), and of large D/H ratios in some carbonaceous chondrites (Boato 1954; Briggs 1963). However, the significance of the large D/H ratios was not generally appreciated and because the known isotopic anomalies were otherwise restricted to the chemically unreactive noble gases, their existence did not have an important influence on models of nebular chemistry. However, once it became apparent that non-mass fractionated oxygen anomalies existed in meteoritic material, and were substantial effects (at the several % level), then the model of a solar nebula in which all accreted material was totally vaporized had to be modified because this scenario could not explain the preservation of ubiquitous anomalies in the most abundant element in stony meteorites.

Subsequently, the discovery (or rediscovery) of isotopic anomalies in many other elements reinforced the point that physical and chemical processes in the solar nebula did not completely erase the prior record of interstellar chemistry. This evidence essentially falls into four categories: (1) Large deviations of the D/H, $^{12}C/^{13}C$, $^{14}N/^{15}N$ ratios from their terrestrial values in some chondrites (e.g., Pillinger 1984; Zinner 1988). As mentioned earlier, these are difficult to explain solely by mass fractionation processes because chemical reactions between neutral species are probably kinetically inhibited at the low temperatures that are required to explain the observed effects by thermochemical isotopic exchange reactions. Instead either inheritance from interstellar chemistry involving ion-molecule reactions proceeding at low temperatures or from circumstellar chemistry in the atmospheres of certain types of stars is apparently required. (2) Non-mass fractionated isotopic anomalies in all elements analyzed (e.g., O, Mg, Si, Cr, Sr, Ba, Nd, Sm) in a small suite of Ca, Al-rich inclusions (CAIs) in carbonaceous chondrites (e.g., Lee 1988). These CAIs are known as FUN CAIs because of the Fractionated and Unknown Nuclear effects observed in them. (3) Non-mass fractionated isotopic anomalies in Ti in CAIs and meteorites (Lee 1988; Niederer, Papanastassiou, and Wasserburg 1981; Niemeyer and Lugmair 1984). These anomalies are attributed to the incomplete mixing and homogenization of Ti isotopes from at least four different stellar sources. (4) Anomalies in ^{26}Mg, ^{53}Cr, ^{60}Ni, ^{109}Ag, ^{129}Xe, ^{142}Nd, and ^{244}Pu-fission Xe which indicate the presence of the now extinct radionuclides ^{26}Al, ^{53}Mn, ^{60}Ni, ^{109}Pd, ^{129}I, ^{146}Sm, and ^{244}Pu in the early solar system (Wasserburg 1985; Podosek and Swindle 1988; Shukolyukov and Lugmair 1992). (5) The ubiquitous non-mass fractionated oxygen isotopic anomalies (Thiemens 1988).

Details of the different isotopic anomalies are reviewed in the articles cited above. However, the important point here is that thermal processing was not 100% efficient at homogenizing the interstellar gas and grains accreted by the solar nebula. In some cases, such as the light elements, large deviations of the isotopic ratios are observed in bulk meteorites or in phases separated from them. In other cases such as the more refractory elements, smaller effects corresponding to deviations of one part in 10^4 to one part in 10^3 from the terrestrial ratios are observed. The anomalies in oxygen can be at the percent level but may represent chemical processes which operated at some point during nebular history and may not necessarily require inefficient thermal processing of presolar material. However, inefficient homogenization after the production of the anomalies is still required for their preservation.

5. Thermochemical Equilibrium Models of Nebular Chemistry

The observational evidence reviewed above indicates that it is desirable to incorporate the influence of both the prevailing P,T conditions and of dynamics on the chemical reactions between gases and grains in the solar nebula. However, most models of nebular chemistry are based solely on thermochemical equilibrium and assume that only the P,T conditions control the nature and abundance of the gases and grains present in a system of solar composition. The inputs to thermochemical equilibrium models are the total pressure, the temperature, and the overall elemental composition. The outputs of the thermochemical equilibrium models are the temperature- and pressure dependent molecular speciation of all elements in the code, the condensation temperatures of condensed phases (pure solids & liquids, and solid & liquid solutions), and the abundances of the condensed phases as a function of temperature and pressure. Depending on the complexity of the particular code used, not all of these outputs may be calculated. The thermodynamic data used in the codes come from standard compilations such as the JANAF (Joint Army-Navy-Air Force) Tables, which contains data for about 1200 species. The P,T conditions assumed in the calculations come from physical models of the solar nebula (e.g., Boss, Morfill and Tscharnuter 1989; Cameron 1978, 1985; Lin and Papaloizou 1985; Ruden and Lin 1986; Wood and Morfill 1988). In general, current nebular models predict pressures and temperatures of about 10^{-2} to 10^{-4} bars and 1000-2000 K in the inner solar nebula dropping to 10^{-7} to 10^{-9} bars and 20-100 K in the outer solar nebula. The exact pressure and temperature vary as a function of radial distance in the solar nebula. Although nebular accretion disk models do not specify adiabatic structure, many of the P,T profiles do resemble adiabats, perhaps with a discontinuity (or several discontinuities) due to opacity changes induced by grain condensation. A representative P,T profile, initially based on the work of Lewis (1974) is shown in Fig. 1. The same profile is used throughout the rest of this paper.

Thermochemical equilibrium models have enjoyed great popularity because they give a first order description of the chemical and mineralogical composition of the chondritic meteorites. However, as we will discuss in more detail later the great appeal of thermochemical equilibrium models is also their Achilles heal because no consideration is given to either the rate at which chemical equilibrium might be approached or to the pathway by which this occurs. This problem is most severe for chemical reactions predicted to occur at low temperatures in the solar nebula and for chemical reactions involving carbon and nitrogen.

For the present, we will keep these limitations in mind while we review the major results of thermochemical models of nebular chemistry. These results are important for several reasons: (1) they provide the foundation upon which the more detailed and sophisticated thermochemical kinetic models of nebular chemistry are built, (2) as stated above they provide a first order description of the chemical and mineralogical composition of the chondritic meteorites, and (3) they provide important information about the types of rocky and metallic grains that are potential catalysts for heterogeneous chemistry involving C and N compounds.

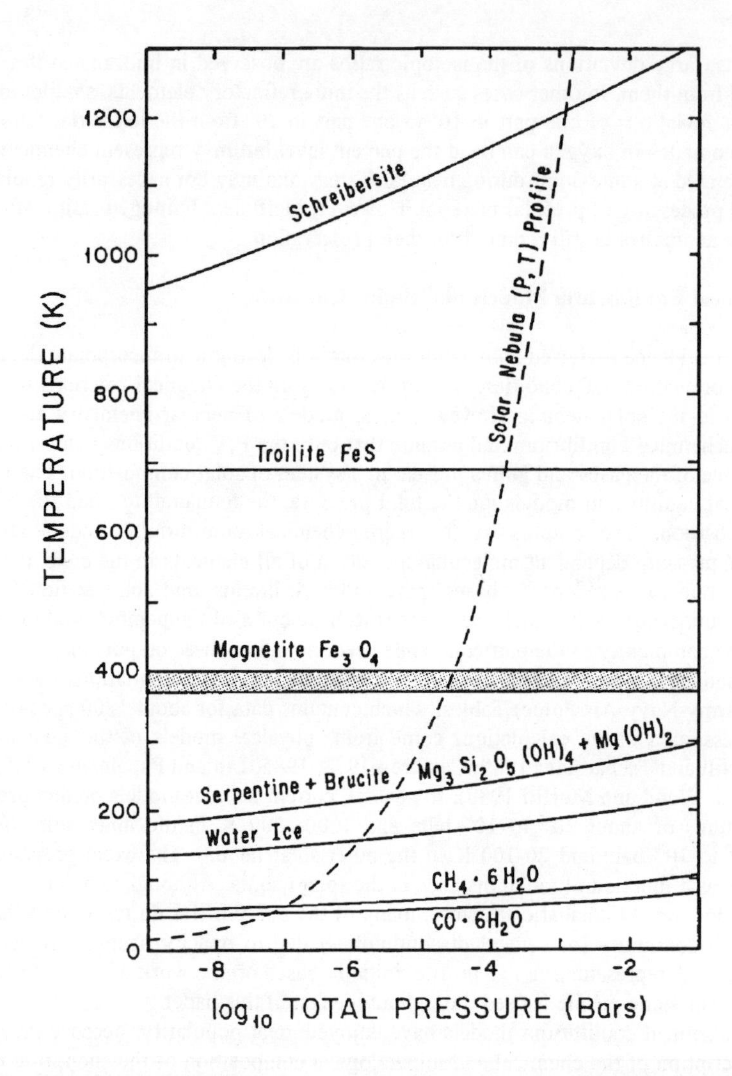

Figure 1. The thermodynamic stability fields for several exemplary gas-grain chemical reactions in the solar nebula. The P,T profile shown is originally based on the work of Lewis (1974), and has also been adopted in several subsequent publications (e.g., Barshay 1981; Fegley 1988; Fegley and Prinn 1989; Prinn and Fegley 1989). The shaded region for magnetite illustrates the range of formation temperatures for the case of all carbon remaining as CO (lower line) and for all carbon being found as CH_4 (upper line). CO clathrate will form if CO remains the major carbon gas in the nebula while CH_4 clathrate will form if CH_4 is present, as in the giant protoplanetary subnebulae. Modified from Fegley (1988).

Table 5 and Figs. 1-3 provide a convenient summary of the major results of thermochemical equilibrium models of nebular chemistry. Table 5 lists the major gases, initial condensates, and condensation temperatures for each of the naturally occurring elements. The data in Table 5 are for an assumed total nebular pressure of 10^{-4} bars. This is because many of the results of condensation calculations in meteoritics and cosmochemistry are done at a constant total pressure, which is generally taken as either 10^{-4} or 10^{-3} bars. Figs. 2-3 illustrate the thermodynamic stability fields and abundances for condensates of the major elements from 150-1800 K in the inner regions of the solar nebula. These results, which are from Barshay (1981), give a detailed picture of the predicted condensate composition as a function of temperature throughout the region where the terrestrial planets and chondrites formed. Although not as comprehensive, a complementary illustration of condensate stability fields is provided by Fig. 1 which shows how the condensation temperatures of selected phases vary as a function of the assumed total pressure in the solar nebula. The important points demonstrated by the results of thermochemical equilibrium models are as follows.

The phases stable at the highest temperatures (either because they are the first condensates or the last phases to evaporate) are Ca-, Al-, Ti-bearing minerals such as hibonite ($CaAl_{12}O_{19}$), corundum (Al_2O_3), perovskite ($CaTiO_3$), spinel ($MgAl_2O_4$), gehlenite ($Ca_2Al_2SiO_7$), and åkermanite $Ca_2MgSi_2O_7$, The refractory minerals are predicted to contain the less abundant refractory lithophiles (elements which geochemically prefer to be in a silicate phase) in solid solution. The refractory lithophiles include the REE, Ti, Zr, Hf, V, Nb, Ta, Sc, Y, Sr, Ba, Th, U, and Pu. Also predicted to be stable at these high temperatures are metallic alloys composed of the Pt-group elements (Ru, Os, Rh, Ir, Pt), Mo, W, and Re. Details of the major element condensation calculations, and descriptions of the comparisons of the predicted and observed mineral assemblages are given by Grossman and Larimer (1974), Kornacki and Fegley (1984), Larimer (1988), and Palme and Fegley (1990). The condensation calculations for the refractory trace lithophiles and for the refractory metals (also known as the refractory siderophiles) are discussed by Fegley and Kornacki (1984), Fegley and Palme (1985), Palme and Wlotzka (1976), and Kornacki and Fegley (1986). Descriptions of the mineralogy of CAIs, of the refractory lithophile abundance patterns and of the refractory metal nuggets are given in several references (Bischoff and Palme 1987; Blum et al 1988; El Goresy et al 1978; Fegley and Ireland 1991; Fuchs and Blander 1980; MacPherson et al 1988).

The refractory minerals predicted by the condensation calculations are observed in the CAIs in carbonaceous chondrites and the observed mineral assemblages are more or less similar to the predicted mineral assemblages in the condensation calculations. Likewise, the refractory lithophiles are observed in solid solution in several of these minerals, with abundance patterns which are plausibly explained by the condensation calculations. The predicted refractory metal nuggets are also found in the CAIs and have compositions consistent with those predicted by the chemical equilibrium models. Although many aspects of the formation of CAIs are still controversial, the overall similarity of the predictions and observed phase assemblages and trace element abundance patterns suggests that vapor-solid fractionation processes played an important role in the formation of the CAIs, and by implication were important in at least some regions of the solar nebula at some time (or times).

Another important point for the present discussion is that none of the high temperature phases are considered to be good catalysts for carbon and nitrogen chemistry in the nebular environment. Industrial experience shows that the oxide and silicate minerals found in CAIs are generally not good catalysts for reduction of CO and N_2 or for organic compound formation. The Pt metal nuggets are of course potentially quite good catalysts, but their availability is limited to high temperature regions where they remain exposed to the nebular gas. However, the

TABLE 5. Equilibrium Condensation Chemistry of the Elements in the Solar Nebula

Atomic Number & Chemical Symbol	Condensation T (K) (P = 10^{-4} bars)	Initial Condensate in Solar Nebula	Major Gases in Solar Nebula	Notes & Sources
1. H	180	$H_2O(s)$	H_2	A, 1.
2. He[a]	<5	He(s)	He	A, 1.
3. Li	1225	Li_2SiO_3 in $MgSiO_3$	LiCl, LiF	MV, 2.
4. Be[b]	1400 (10^{-3} bars)	$BeAl_2O_4$ in $MgAl_2O_4$	$Be(OH)_2$, BeOH	RL, 3.
5. B[b]	745-759 (10^{-3} bars)	$NaBO_2(s)$	$NaBO_2$, KBO_2, HBO_2, H_3BO_3	MV, 3.
6. C[c]	78	$CH_4 \cdot 6H_2O(s)$	CO, CH_4	A, 1.
7. N[d]	120	$NH_3 \cdot H_2O(s)$	N_2, NH_3	A, 1.
8. O[e]	---	---	CO, H_2O	A
9. F	736	$Ca_5(PO_4)_3F$	HF	MV, 4.
10. Ne[a]	~5	Ne(s)	Ne	A, 1.
11. Na	970 (50%)	$NaAlSi_3O_8$ in feldspar	Na, NaCl	MV, 4.
12. Mg	1340 (50%)	$Mg_2SiO_4(s)$	Mg	ME, 5, 13.
13. Al	1670	$Al_2O_3(s)$	Al, AlOH, Al_2O, AlS, AlH, AlO, AlF	RL, 6.
14. Si[f]	1529	$Ca_2Al_2SiO_7(s)$	SiO, SiS	ME, 6.
15. P	1151 (50%)	$Fe_3P(s)$	PO, P, PN, PS	MV, 4, 7.
16. S	684	FeS(s)	H_2S, HS	MV, 4, 12
17. Cl	863 (50%)	$Na_4[AlSiO_4]_3Cl(s)$	HCl, NaCl, KCl	MV, 4.
18. Ar	50	$Ar \cdot 6H_2O(s)$	Ar	A, 14.
19. K	1000 (50%)	$KAlSi_3O_8(s)$ in feldspar	K, KCl, KOH	MV, 4.
20. Ca	1634	$CaAl_{12}O_{19}(s)$	Ca	RL, 6.
21. Sc	1652 (50%)	$Sc_2O_3(s)$	ScO	RL, 8.
22. Ti	1600	$CaTiO_3(s)$	TiO, TiO_2	RL, 6.
23. V	1455 (50%)	diss. in $CaTiO_3(s)$	VO_2, VO	RL, 8.
24. Cr	1301 (50%)	diss. in Fe alloy	Cr	MV, 9.
25. Mn	1190 (50%)	Mn_2SiO_4 in olivine	Mn	MV, 2.
26. Fe	1337 (50%)	Fe alloy	Fe	ME, 7, 9.
27. Co	1356 (50%)	diss. in Fe alloy	Co	RS, 9.
28. Ni	1354 (50%)	diss. in Fe alloy	Ni	RS, 9.
29. Cu	1170 (50%)	diss. in Fe alloy	Cu	MV, 2.

Atomic Number & Chemical Symbol	Condensation T (K) ($P = 10^{-4}$ bars)	Initial Condensate in Solar Nebula	Major Gases in Solar Nebula	Notes & Sources
30. Zn	684 (50%)	ZnS diss. in FeS	Zn	MV, 2.
31. Ga	918 (50%)	diss. in Fe alloy	GaOH, GaCl, GaBr	MV, 10.
32. Ge	825 (50%)	diss. in Fe alloy	GeS, GeSe	MV, 10.
33. As	1012 (50%)	diss. in Fe alloy	As	MV, 10.
34. Se	684 (50%)	FeSe diss. in FeS	H_2Se, GeSe	MV, 2.
35. Br[b]	~350	$Ca_5(PO_4)_3Br(s)$	HBr, NaBr	HV, 4.
36. Kr	54	$Kr \cdot 6H_2O(s)$	Kr	A, 14.
37. Rb[b]	~1080	diss. in feldspar	Rb, RbCl	MV, 5, 13.
38. Sr	1217 (50%)	diss. in $CaTiO_3(s)$	Sr, $SrCl_2$, $Sr(OH)_2$, SrOH	RL, 8.
39. Y	1622 (50%)	$Y_2O_3(s)$	YO	RL, 8.
40. Zr	1717 (50%)	$ZrO_2(s)$	ZrO_2, ZrO	RL, 8.
41. Nb	1517 (50%)	diss. in $CaTiO_3(s)$	NbO_2, NbO	RL, 8.
42. Mo	1595 (50%)	refractory metal alloy	MoO, Mo, MoO_2	RS, 9.
44. Ru	1565 (50%)	refractory metal alloy	Ru	RS, 9.
45. Rh	1392 (50%)	refractory metal alloy	Rh	RS, 9.
46. Pd	1320 (50%)	diss. in Fe alloy	Pd	MV, 9.
47. Ag	993 (50%)	diss. in Fe alloy	Ag	MV, 2.
48. Cd[b]	430 (10^{-5} bars)	CdS in FeS	Cd	HV, 11.
49. In[b]	470 (50%)	InS in FeS	In, InCl, InOH	HV, 11.
50. Sn	720 (50%)	diss. in Fe alloy	SnS, SnSe	MV, 2.
51. Sb	912 (50%)	diss. in Fe alloy	SbS, Sb	MV, 10.
52. Te	680 (50%)	FeTe diss. in FeS	Te, H_2Te	MV, 2.
53. I	?	?	I, HI	MV/HV?
54. Xe	74	$Xe \cdot 6H_2O(s)$	Xe	A, 14.
55. Cs	?	?	CsCl, Cs, CsOH	MV/HV?
56. Ba	1162 (50%)	diss. in $CaTiO_3(s)$	$Ba(OH)_2$, BaOH, BaS, BaO	RL, 8.
57. La	1544 (50%)	diss. in $CaTiO_3(s)$	LaO	RL, 8.
58. Ce	1440 (50%)	diss. in $CaTiO_3(s)$	CeO_2, CeO	RL, 8.
59. Pr	1557 (50%)	diss. in $CaTiO_3(s)$	PrO	RL, 8. .
60. Nd	1563 (50%)	diss. in $CaTiO_3(s)$	NdO	RL, 8.
62. Sm	1560 (50%)	diss. in $CaTiO_3(s)$	SmO, Sm	RL, 8.
63. Eu	1338 (50%)	diss. in $CaTiO_3(s)$	Eu	RL, 8.

Atomic Number & Chemical Symbol	Condensation T (K) ($P = 10^{-4}$ bars)	Initial Condensate in Solar Nebula	Major Gases in Solar Nebula	Notes & Sources
64. Gd	1597 (50%)	diss. in $CaTiO_3$(s)	GdO	RL, 8.
65. Tb	1598 (50%)	diss. in $CaTiO_3$(s)	TbO	RL, 8.
66. Dy	1598 (50%)	diss. in $CaTiO_3$(s)	DyO, Dy	RL, 8.
67. Ho	1598 (50%)	diss. in $CaTiO_3$(s)	HoO, Ho	RL, 8.
68. Er	1598 (50%)	diss. in $CaTiO_3$(s)	ErO, Er	RL, 8.
69. Tm	1598 (50%)	diss. in $CaTiO_3$(s)	Tm, TmO	RL, 8.
70. Yb	1493 (50%)	diss. in $CaTiO_3$(s)	Yb	RL, 8.
71. Lu	1598 (50%)	diss. in $CaTiO_3$(s)	LuO	RL, 8.
72. Hf	1690 (50%)	HfO_2(s)	HfO	RL, 8.
73. Ta	1543 (50%)	diss. in $CaTiO_3$(s)	TaO_2, TaO	RL, 8.
74. W	1794 (50%)	refractory metal alloy	WO, WO_2, WO_3	RS, 9.
75. Re	1818 (50%)	refractory metal alloy	Re	RS, 9.
76. Os	1812 (50%)	refractory metal alloy	Os	RS, 9.
77. Ir	1603 (50%)	refractory metal alloy	Ir	RS, 9.
78. Pt	1411 (50%)	refractory metal alloy	Pt	RS, 9.
79. Au	1284 (50%)	Fe alloy	Au	MV, 2.
80. Hg	?	?	Hg	MV/HV?
81. Tl[b]	448 (50%)	diss. in Fe alloy	Tl	HV,11.
82. Pb[b]	520 (50%)	diss. in Fe alloy	Pb, PbS	HV, 11.
83. Bi	472 (50%)	diss. in Fe alloy	Bi	HV, 11.
90. Th	1598 (50%)	diss. in $CaTiO_3$(s)	ThO_2	RL, 8.
92. U	1580 (50%)	diss. in $CaTiO_3$(s)	UO_2	RL, 8.

Notes to Table 5

The condensation temperatures either indicate where the condensate first becomes stable or where 50% of the element is condensed and 50% is in the gas. The 50% condensation temperature is generally used when solid solutions are formed.

The major gases vary as a function of temperature and total pressure. The gas chemistry in Table 5 is generally valid at the condensation temperature of each element, and was either taken from the original references below or calculated as part of this work.

Sources cited in Table 5: (1) Lewis 1972; (2) Wai & Wasson 1977; (3) Cameron et al 1973; (4) Fegley & Lewis 1980; (5) Grossman & Larimer 1974; (6) Kornacki & Fegley 1984; (7) Sears1978; (8) Kornacki & Fegley 1986; (9) Fegley & Palme 1985; (10) Wai & Wasson 1979; (11) Larimer 1973; (12) Larimer 1967, (13) Wasson 1985, (14) Sill & Wilkening 1978.

(a) This temperature is below cosmic background and condensation will not occur.

(b) The condensation temperature and initial condensate are uncertain and need to be re-evaluated.

(c) As discussed in the text, kinetic inhibition of the CO to CH_4 conversion yields either $CO \cdot 6H_2O(s)$ or CO(s) as the initial condensate.

(d) As discussed in the text, kinetic inhibition of the N_2 to NH_3 conversion yields either $N_2 \cdot 6H_2O(s)$ or $N_2(s)$ as the initial condensate.

(e) Oxygen is the most abundant element in rocky material and a separate condensation temperature is meaningless. The bulk of oxygen condenses as water ice; the remainder is present as CO or in rocky material.

(f) Most Si condenses when the silicates $MgSiO_3$ and Mg_2SiO_4 form (e.g., 1340 K at 10^{-4} bars). See the $MgSiO_3$ condensation curve in Figure 2.

Key to abbreviations used for cosmochemical classification of the elements: A = atmophile, HV = highly volatile, ME = major element, MV = moderately volatile, RL = refractory lithophile, RS = refractory siderophile

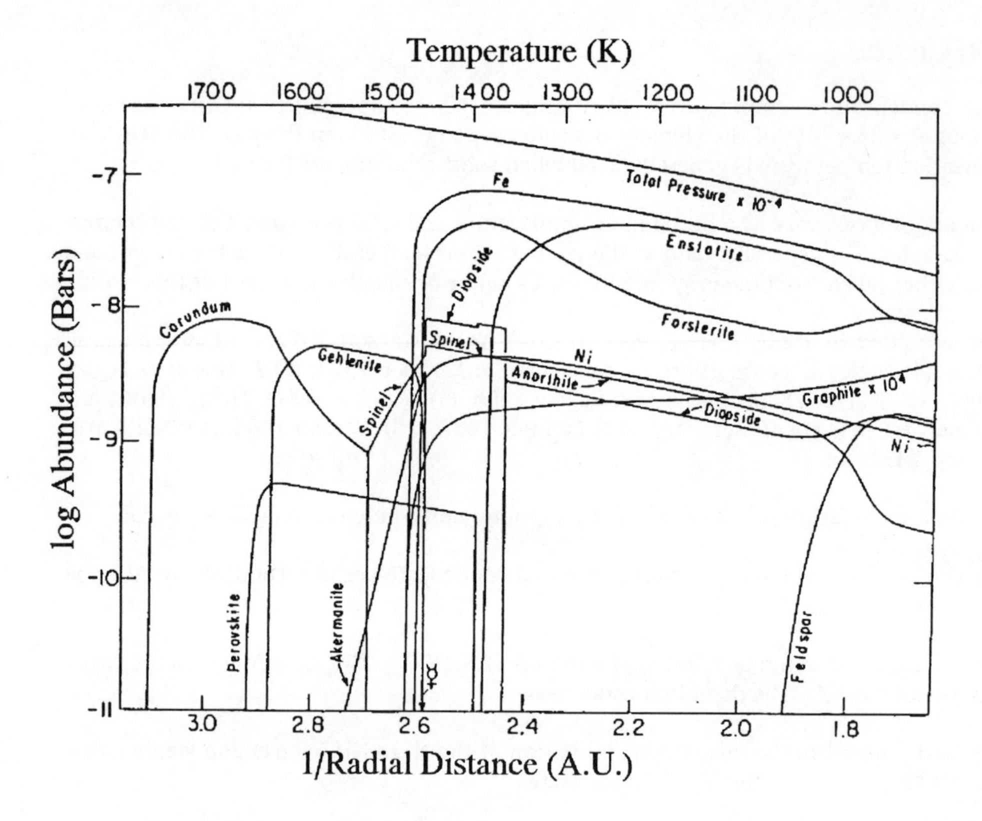

Figure 2. The chemical equilibrium condensation sequence from 900 to 1800 K in the inner regions of the solar nebula. The abundances of the different condensates stable at complete chemical equilibrium are displayed along the P,T profile illustrated in Fig. 1. The condensation of Fe metal at about 1450 K marks the appearance of a potentially important catalyst for reactions involving carbon and nitrogen compounds The line labelled graphite shows the thermodynamic activity of carbon dissolved in Fe metal under equilibrium conditions. The astrological symbols for Mercury, Venus, Earth, and Mars are shown at the appropriate places on the distance scale which is in inverse astronomical units. Fig. 3 shows the condensate stability fields and abundances at lower temperatures. Modified from Barshay (1981).

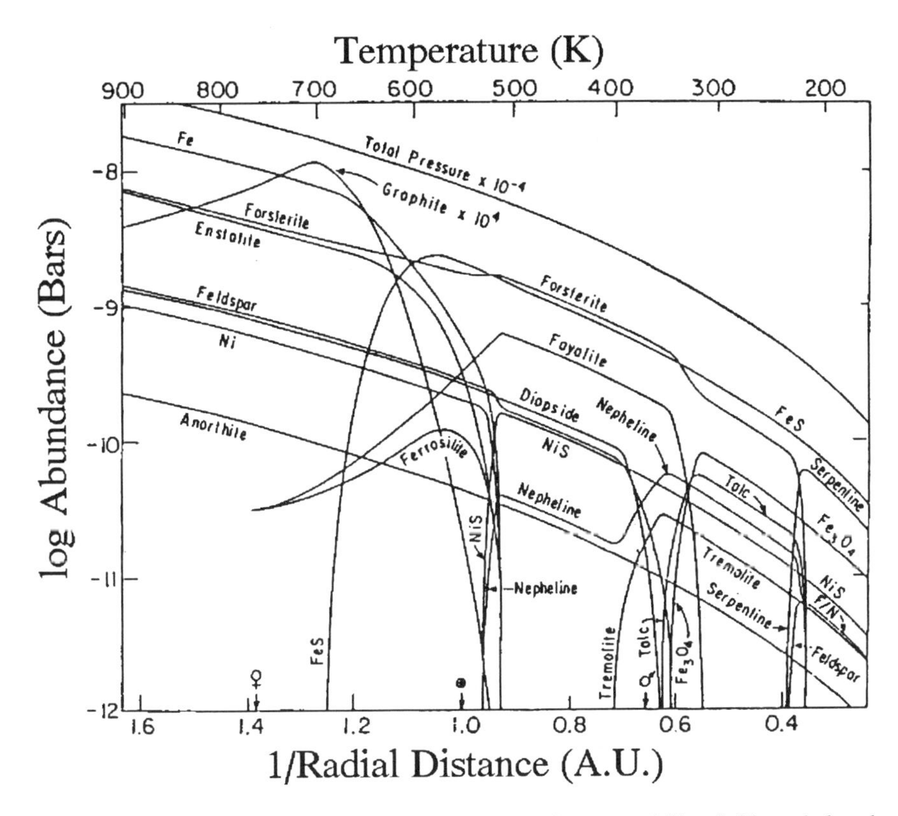

Figure 3. As in Fig. 2 from about 200 to 900 K. The condensate stability fields and abundances are indicated along the nebular P,T profile illustrated in Fig. 1. The astrological symbols for Mercury, Venus, Earth, and Mars are shown at the appropriate places on the distance scale. Note the formation of FeS at 690 K which could potentially deactivate Fe grains as catalysts by coating them with a sulfide layer.The Fe-bearing silicates fayalite Fe_2SiO_4 and ferrosilite $FeSiO_3$, which form in solid solution with their magnesian counterparts, first have appreciable abundances at about 800 K. Formation of these two minerals may be kinetically inhibited because of slow solid state diffusion between Fe metal grains and silicates. All Fe metal which has not already been converted to troilite is removed by magnetite formation at about 400 K. Modified from Barshay (1981).

observational evidence shows that the Pt metal nuggets are physically sequestered inside CAIs, and thus are removed from exposure to the nebular gas at fairly high temperatures by condensation of the refractory oxide and silicate minerals. Any remaining Pt metal nuggets which escape this fate will of course be dissolved in the much more abundant Fe metal alloy, once it condenses. Thus, by the time that the nebular temperature has dropped to the point where organic compounds can be formed by processes such as Fischer-Tropsch-type (FTT) reactions, the Pt metal nuggets are no longer available as catalysts.

At lower temperatures, Fe alloy containing other siderophile (metal-loving) elements in solid solution. , forsterite (Mg_2SiO_4), and enstatite ($MgSiO_3$) are predicted to form. Together, these phases make up the bulk of the condensible rocky material in solar composition gas. The condensation (or evaporation) curve for Fe-Ni alloy is an important phase boundary for carbon and nitrogen chemistry because the metal alloy is potentially the most abundant and most effective catalyst for reactions involving these two elements. At high temperatures some of the Fe alloy can react with P gases to form schreibersite Fe_3P. However, because of the mass balance constraints imposed by solar elemental abundances (see Table 1), this reaction only consumes a minor amount of the total available alloy. Much more of the Fe metal is removed by formation of troilite (FeS), which occurs at the pressure independent temperature of 690 K. This involves the reaction of $H_2S(g)$ with the metal grains to form more Ni-rich metal grains coated by FeS. The reaction is pressure independent because the overall stoichiometry involves consumption of one mole of $H_2S(g)$ and production of one mole of $H_2(g)$. At still lower temperatures of 370-400 K, the remaining metal alloy can be attacked by nebular water vapor to form magnetite (Fe_3O_4). Again, this reaction is pressure independent because as the same number of moles of $H_2O(g)$ and $H_2(g)$ are consumed and produced, respectively.

If the Fe-Ni remains well mixed with the nebular gas throughout its stability field, it is a potential catalyst throughout this entire temperature range. However, if the metal grains do not remain well mixed with the nebular gas (e.g., by settling to form much larger metal chunks in the nebular midplane on a time scale that is rapid with respect to nebular cooling) then their catalytic efficiency will be correspondingly diminished. Likewise, if troilite formation at 690 K results in FeS coatings on all metal grains, then no metallic surface will be left exposed to catalyze reactions at lower temperatures. In later sections when we discuss Fe metal grain catalyzed reactions of CO and N_2 we assume that the metal grains in fact remain well mixed with the nebular gas at the solar Fe/H_2 abundance ratio, and that the exposed surfaces are not coated by FeS and remain catalytically active down to 370-400 K where magnetite formation occurs.

Two other major events are predicted to occur below 690 K where troilite forms. One is the formation of hydrated silicates by the reaction of nebular water vapor with anhydrous silicates. However, as we discuss later, hydration of anhydrous silicates by nebular water vapor probably did not actually occur. Qualitatively, the process is analogous to hydrating rock in a near vacuum. Quantitatively, as shown later, the time scale for hydrated silicate formation in the solar nebula is orders of magnitude longer than the lifetime of the nebula itself. The other event is the condensation of water ice. Fig. 1 schematically illustrates the pressure-dependent temperatures at which both processes are predicted to occur. Hydrated silicate formation is exemplified by the formation of serpentine [$Mg_3Si_2O_5(OH)_4$] + brucite [$Mg(OH)_2$] by the reaction of forsterite and water vapor. The water ice condensation line shown in Fig. 1 simply represents the line along which the partial pressure of water vapor in the solar nebula becomes equal to the water vapor pressure over H_2O(ice). If nebular pressures ever became high enough, the water ice condensation curve would cross the freezing point (273 K), and liquid water would form. This is unlikely.

TABLE 6. Summary of Hydrated Silicate Condensation Calculations[a]

Condensation Temp. (K)[b]	Hydrated Phase(s) and Chemical Formula(s)	Water Content (w/o)	Sources
~500	Tremolite [$Ca_2Mg_5Si_8O_{22}(OH)_2$]	2.2	1
<470[c]	Na phlogopite [$NaMg_3AlSi_3O_{10}(OH)_2$]	4.5	2
~460[d]	Hydroxyapatite [$Ca_5(PO_4)_3OH$]	1.8	3
~400[e]	Serpentine [$Mg_3Si_2O_5(OH)_4$]	13	4
~230[e,f]	Serpentine [$Mg_3Si_2O_5(OH)_4$]	13	5
<274[g]	Serpentine [$Mg_3Si_2O_5(OH)_4$] + Brucite [$Mg(OH)_2$]	16.1	2
~225	Serpentine [$Mg_3Si_2O_5(OH)_4$] + Brucite [$Mg(OH)_2$]	16.1	6
~400[e]	Talc [$Mg_3Si_4O_{10}(OH)_2$]	4.8	1
~340[e,f]	Talc [$Mg_3Si_4O_{10}(OH)_2$]	4.8	5
~160-280[h]	Talc [$Mg_3Si_4O_{10}(OH)_2$]	4.8	7
~250	Talc [$Mg_3Si_4O_{10}(OH)_2$] + Brucite [$Mg(OH)_2$]	8.3	6
~160[i]	Water Ice [H_2O]	100	8

(a) Modified from Prinn & Fegley 1989.

(b) The condensation temperature is the highest temperature at which the hydrated phase is stable along the solar nebula P,T profile shown in the figures.

(c) Na phlogopite forms at 470 K at 10^{-3} bars and will form at lower temperatures along the solar nebula P,T profile used here.

(d) The water content is calculated on the basis of 1 H_2O molecule per 2 hydroxyapatites.

(e) This result is predicated on solid-solid chemical equilibrium which is unlikely at these low temperatures (see text).

(f) Barshay's calculations used a feldspar-nepheline buffer for silica at T<600 K.

(g) This assemblage forms at 274 K at 10^{-3} bars and will form at lower temperatures along the solar nebula P,T profile used here.

(h) Larimer & Anders 1967 list 2 possible talc formation reactions: $3MgSiO_3(s) + SiO_2(s) + H_2O(g) \rightarrow Mg_3Si_4O_{10}(OH)_2(s)$ and $3Mg_2SiO_4(s) + 5SiO_2(s) + 2H_2O(g) \rightarrow 2Mg_3Si_4O_{10}(OH)_2(s)$ with condensation temperatures of 160-280 K at total pressures of ~ 10^{-6}-10^{-3} bars.

(i) The shaded region for water ice condensation in Fig. 10 illustrates the range of condensation temperatures assuming that all carbon is either CH_4 (higher T) or CO (lower T).

Sources: (1) Lewis 1972a; (2) Hashimoto & Grossman 1987; (3) Fegley & Lewis 1980; (4) Lewis 1974; (5) Barshay 1981; (6) Prinn & Fegley 1989; (7) Larimer & Anders 1967; (8) Lewis 1972b

At even lower temperatures below the water ice condensation curve, the highly volatile gases such as CO, N_2, CH_4, NH_3, etc. are predicted to condense. This may occur by formation of a distinct hydrate with water ice, as in the case of $NH_3 \cdot H_2O(s)$ or by trapping the gas in the water ice crystalline lattice as in the case of clathrate hydrate formation. The clathrate hydrates which could have formed involve CO, N_2, and CH_4. Their ideal chemical formulas are $CO \cdot 6H_2O(s)$, $N_2 \cdot 6H_2O(s)$, or $CH_4 \cdot 6H_2O(s)$. Finally, at even lower temperatures around 20 K, the pure ices of these gases may have condensed. However, it is very unlikely that nebular temperatures were ever low enough (< 20 K) to condense Ne(s) or $H_2(s)$.

6. Thermochemical Kinetic Models of Nebular Chemistry

The central problem with chemical equilibrium models of nebular chemistry was probably recognized first by Urey, who pioneered the applications of physical chemistry in general and chemical thermodynamics in particular to cosmochemistry. While discussing the applications of chemical thermodynamic methods and data to nebular modelling Urey (1953) noted that "Our data in this field give much information relative to possible reactions, and at higher temperatures they certainly give us practically assured knowledge of the chemical situations due to the high velocities of reactions, at least in homogeneous systems, providing the data are adequate, which is unfortunately not always the case. At lower temperatures, thermodynamic equilibrium may not be reached even in periods of time that are long compared to the age of the universe, and at these temperatures the kinetics of thermal reactions or of photochemical reactions become important."

It is somewhat ironic that Urey's early recognition of the potential problems with purely chemical equilibrium models of nebular chemistry was not acted on for nearly 30 years (e.g., Lewis and Prinn 1980; Prinn and Fegley 1981). However, one decade later, it is generally acknowledged that thermochemical reactions in the solar nebula were coupled with and influenced by nebular dynamics, although the exact extent and nature of this coupling and influence are currently a matter of great debate. At one extreme there is the view that nebular mixing was very inefficient at transporting thermally reprocessed material outward from the hotter nebular regions inward of a few A.U. to the colder nebular regions at distances of a few to a few tens of A.U. (Stevenson 1990). At the other extreme there is the opposite view that nebular mass transport outward was extremely efficient and that the high temperature speciation formed in the hotter inner regions dominated the chemical composition of the entire solar nebula (Prinn 1990). At present, the efficiency of nebular mass transport for polluting the outer nebula with thermochemical products formed in the inner nebula is still controversial. In the following discussion we adopt the view that nebular thermochemistry played the dominant role in controlling the molecular speciation throughout the entire solar nebula due either to mass transport from the inner nebula, or due to processing in the hot, dense giant protoplanetary subnebulae postulated around the gas giant planets during their formation.

7. Carbon and Nitrogen Chemistry in the Solar Nebula

Carbon and nitrogen chemistry in the solar nebula are complex topics which are still incompletely understood. We begin by discussing the gas phase equilibrium chemistry and then consider the effects of kinetic inhibition of the important gas phase thermochemical reactions responsible for interconverting CO & CH_4 and N_2 & NH_3 in the solar nebula. We then review the role of grain catalyzed reactions, such as Fischer-Tropsch-type (FTT) reactions which may have been important for the production of organic matter found on primitive bodies such as asteroids, comets, interplanetary dust particles, and meteorites.

7.1 Gas Phase Chemistry

To a good first approximation, the gas phase chemistry of carbon in a H_2-rich gas with solar elemental ratios is dominated by CO, and CH_4, and to an even better approximation the gas phase chemistry of nitrogen is dominated by N_2 and NH_3. Neglecting for a moment several complicating factors CO is the dominant carbon gas at high temperatures and low pressures, while CH_4 is the dominant carbon gas at low temperatures and high pressures in solar composition material (Urey 1953; Lewis, Barshay, and Noyes 1979). Likewise N_2 is the dominant nitrogen gas at high temperatures and low pressures, while NH_3 is the dominant nitrogen gas at low temperatures and high pressures (e.g., Fegley 1983). The oxidized (i.e., CO and N_2) gases are converted into the reduced (i.e., CH_4 and NH_3) gases by the net thermochemical reactions:

$$CO(g) + 3H_2(g) = CH_4(g) + H_2O(g)$$

$$N_2(g) + 3H_2(g) = 2NH_3(g).$$

CO is also converted into CO_2 by the net thermochemical reaction

$$CO(g) + H_2O(g) = CO_2(g) + H_2(g).$$

The thermochemical conversions of the oxidized C and N gases to their reduced counterparts proceed to the right with decreasing temperature at constant pressure. The conversion of CO into CO_2 also proceeds to the right with decreasing temperature. Thus, if chemical equilibrium were maintained as the gas cooled, the CO/CH_4, CO/CO_2, and N_2/NH_3 ratios would all decrease as the temperature decreased. This behavior is illustrated in Fig. 4 where these ratios are plotted as function of 1/T along the solar nebula P,T profile shown earlier.

Clearly, the extent to which the homogeneous gas-phase conversions of the oxidized to reduced C and N gases can proceed with decreasing temperature is dependent on the rates of the elementary reactions that make up the reaction pathway relative to the rates of nebular mixing and overall cooling. If the chemical conversion is much faster than the rate at which the gas parcel is mixed outward to cooler nebular regions (or than the rate at which overall cooling of the nebula proceeds) then chemical equilibrium will be closely approached as cooling proceeds. However if the chemical conversion is much slower than the rate at which mixing outward to cooler regions (or overall cooling) occurs, then chemical equilibrium will be frozen in, or quenched at some point, and the gas phase composition in cooler regions will be the same as that established at the quench point. If we denote the chemical lifetime for conversion of the oxidized to the reduced gases by t_{chem} and the nebular mixing (or overall cooling) time by t_{mix}, the former case is represented by the inequality $t_{chem} < t_{mix}$ and the latter case is represented by the inequality $t_{chem} > t_{mix}$. The intermediate case which occurs at the quench temperature (T_Q) is represented by the equality $t_{chem} = t_{mix}$. Once expressions for the t_{chem} and t_{mix} are available, it is then possible to calculate the position of the quench temperature T_Q in a solar nebula model.

Data from the chemical kinetic and chemical engineering literature can be used to calculate the t_{chem} values for the $CO \rightarrow CH_4$, $CO \rightarrow CO_2$, and $N_2 \rightarrow NH_3$ conversions (e.g., Lewis and Prinn 1980; Prinn and Fegley 1981, 1989). The t_{mix} values can be bounded by considering the

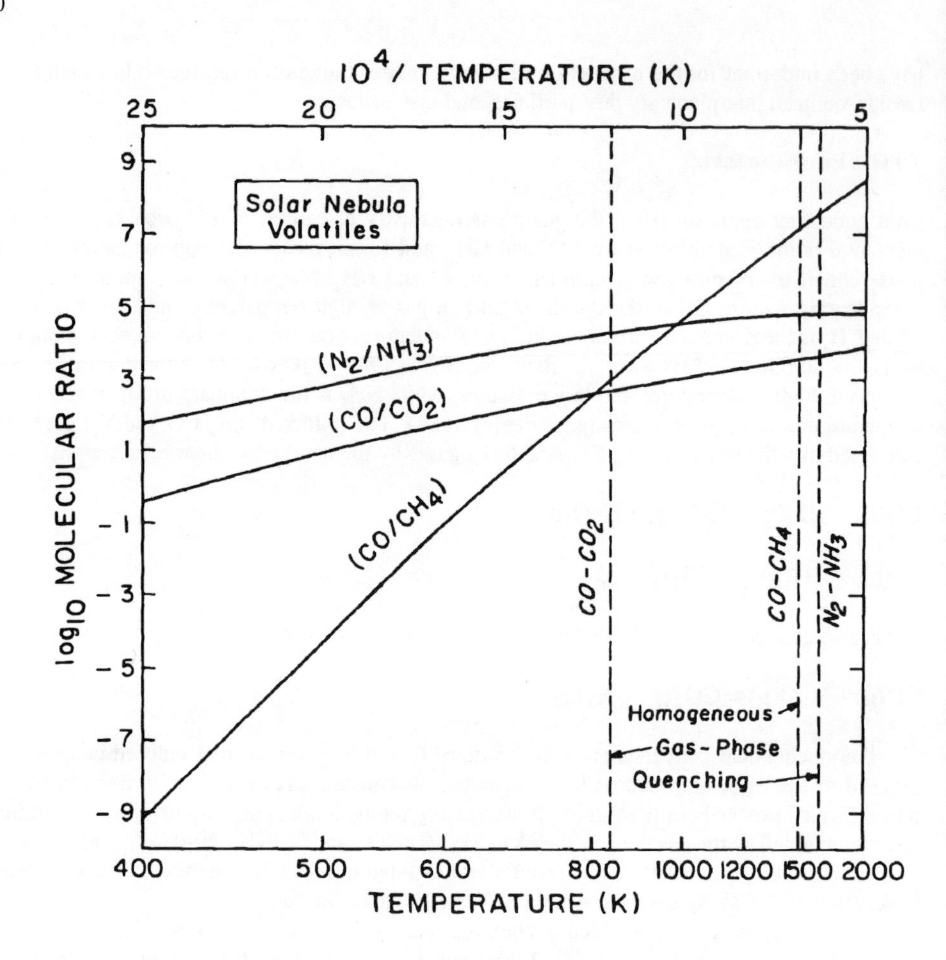

Figure 4. Changes in the equilibrium CO/CH_4, CO/CO_2, and N_2/NH_3 ratios as a function of temperature along the solar nebula P,T profile illustrated in Fig. 1. The vertical dashed lines indicate the temperatures at which the homogeneous gas phase conversions of CO to CH_4, CO to CO_2, and N_2 to NH_3 are quenched in the solar nebula if the conversions can proceed over the entire nebular lifetime of 10^{13} seconds. If mixing of hot gas to cooler regions occurs more rapidly, then the quench temperatures will be higher. Conversely, if the reactions could proceed for longer times, the quench temperatures will be lower. However, the chemical conversions clearly cannot proceed for longer than the nebular lifetime. The effects of Fe metal grains on the rates of these conversions are discussed in the text. Modified from Fegley and Prinn (1989).

fastest and slowest possible nebular mixing times. Following Cameron (1978), the fastest possible mixing time has generally been estimated as $t_{mix} \sim 3H/V_S \sim 10^8$ sec, where H is the radial density scale length and V_s is the sound speed in the solar nebula. The slowest possible mixing time has generally been equated to the lifetime of the solar nebula. On the basis of astronomical observations of disks around young stellar objects (e.g., Strom, Edwards, and Strom 1989) this lifetime is $\sim 10^{13}$ sec. The arguments below are unaltered if longer lifetimes of about $10^{14.5}$ sec (10 million years) consistent with ^{129}Xe measurements of the time interval between the formation of chondrules and meteorite matrix are used.

The results of comparing the t_{chem} and t_{mix} values are as follows. Homogeneous gas phase reactions with $t_{chem} > 10^8$ sec may be quenched in a turbulent, rapidly mixed region of the solar nebula, but may well proceed down to fairly low temperatures in more sluggishly mixed nebular regions. However, without making any assumptions about nebular mixing rates, it is clear that reactions with $t_{chem} \geq 10^{13\text{-}14.5}$ sec will certainly be quenched because this time corresponds to the longest possible mixing time, the lifetime of the nebula itself. Longer times are simply not possible (unless either the astrophysical or cosmochemical evidence has greatly underestimated the lifetime of the solar nebula).

In fact, when the t_{chem} values are calculated using kinetic data from the literature, we find that the quench temperature for the $CO \rightarrow CH_4$ conversion is about 1470 K. As illustrated in Fig. 5, this temperature is well inside the CO stability field at a point where the CO/CH_4 ratio is $\sim 10^7$. Likewise the calculated quench temperature for the $N_2 \rightarrow NH_3$ conversion is about 1600 K. As shown in Fig. 6, this temperature is also well inside the N_2 stability field and corresponds to a N_2/NH_3 ratio of $\sim 10^5$. The quench temperature for the conversion of CO to CO_2 is shown in Fig. 7. It is also inside the region where CO is dominant at a temperature of about 830 K and a CO_2/CO ratio of $10^{-2.7}$. The $CO \rightarrow CO_2$ conversion is thus more facile than the conversions of the oxidized to reduced species. This is a consequence of the relatively rapid interconversions between oxidized carbon species such as CO, CO_2, H_2CO, etc.

These quench temperatures are calculated for the representative nebular P,T profile shown earlier. The variation of the quench temperatures T_Q with the total assumed nebular pressure are also plotted in Figs. 5-7 where it is seen that T_Q drops with increasing pressure. Lower quench temperatures bring the point where equilibrium is frozen in closer to the boundaries in P,T space where the CO/CH_4 and N_2/NH_3 ratios are unity. However, total pressures of about 0.3 bars (at 1000-1100 K) are required for T_Q to intersect the CO/CH_4 phase boundary. Somewhat larger pressures of about 80 bars are required to intersect the N_2/NH_3 boundary. These pressures are several orders of magnitude higher than those calculated in many currently accepted nebular models. Such high pressures are probably inconsistent with the mineralogy of chondrites because complex multiphase liquids would condense instead of solid minerals under these high pressures (e.g., see Grossman and Larimer 1974), and the trace element abundance patterns in the CAIs in carbonaceous chondrites would also be much different than what is actually seen. Thus, it is very safe to state that the thermochemical kinetic models of homogeneous gas phase chemistry show that CO and N_2 were the dominant C and N gases throughout the solar nebula.

However, the situation is predicted to have been dramatically different in the higher density environments, known as the giant protoplanetary subnebulae, which existed around the gas giant planets during their formation. A representative P,T profile taken from the work of Prinn and Fegley (1981) on the Jovian subnebula is illustrated in Figs. 5-7. At a given temperature, the expected pressure in the Jovian (or Saturnian, etc.) subnebula is several orders of magnitude higher than that in the surrounding solar nebula. This difference has two important consequences. One is that the Jovian subnebula P,T profile lies within the thermochemical equilibrium stability fields of CH_4 and NH_3. The other is that the kinetics of the $CO \rightarrow CH_4$,

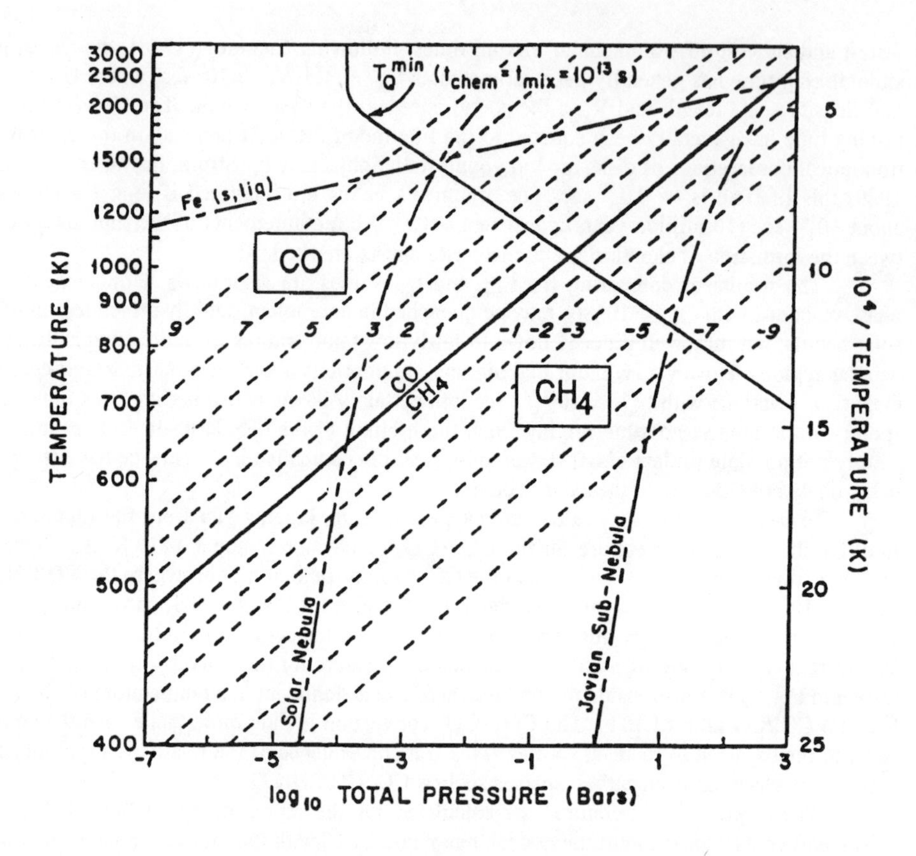

Figure 5. Gas phase equilibrium chemistry of carbon in a solar composition gas. Carbon monoxide is the dominant carbon gas at high temperatures and low pressures and methane is the dominant carbon gas at low temperatures and high pressures. The heavy solid diagonal line is the phase boundary between CO and CH_4. The two gases have equal abundances along this line. The dashed diagonal lines indicate different CO/CH_4 ratios. For example, the dashed line labelled 1 indicates $CO/CH_4 = 10$, and the dashed line labelled -2 indicates $CO/CH_4 = 0.01$. Also shown are the condensation curve for Fe(s, liq), the adopted P,T profiles for the solar nebula and the Jovian protoplanetary subnebula, and the line showing the quench temperature T_Q for the homogeneous gas phase conversion of CO to CH_4. This quench temperature is calculated assuming that the time available for the conversion (t_{chem}) is the same as the nebular lifetime (t_{mix}) of 10^{13} seconds. As mentioned previously, longer conversion times are not physically possible and shorter conversion times, corresponding to more rapid mixing, yield higher quench temperatures. The intersection of the T_Q line with the nebular and subnebular P,T profiles shows the CO/CH_4 ratio at the quench point. As discussed in the text, quenching yields a CO-rich solar nebula and a CH_4-rich subnebula. Finally, as first shown by Urey (1953) and later confirmed by Lewis, Barshay and Noyes (1979) there is a graphite stability field off to the lower left of this figure. The maximum temperature for graphite stability is about 470 K and the maximum pressure is about $10^{-7.6}$ bars. Modified from Fegley and Prinn (1989).

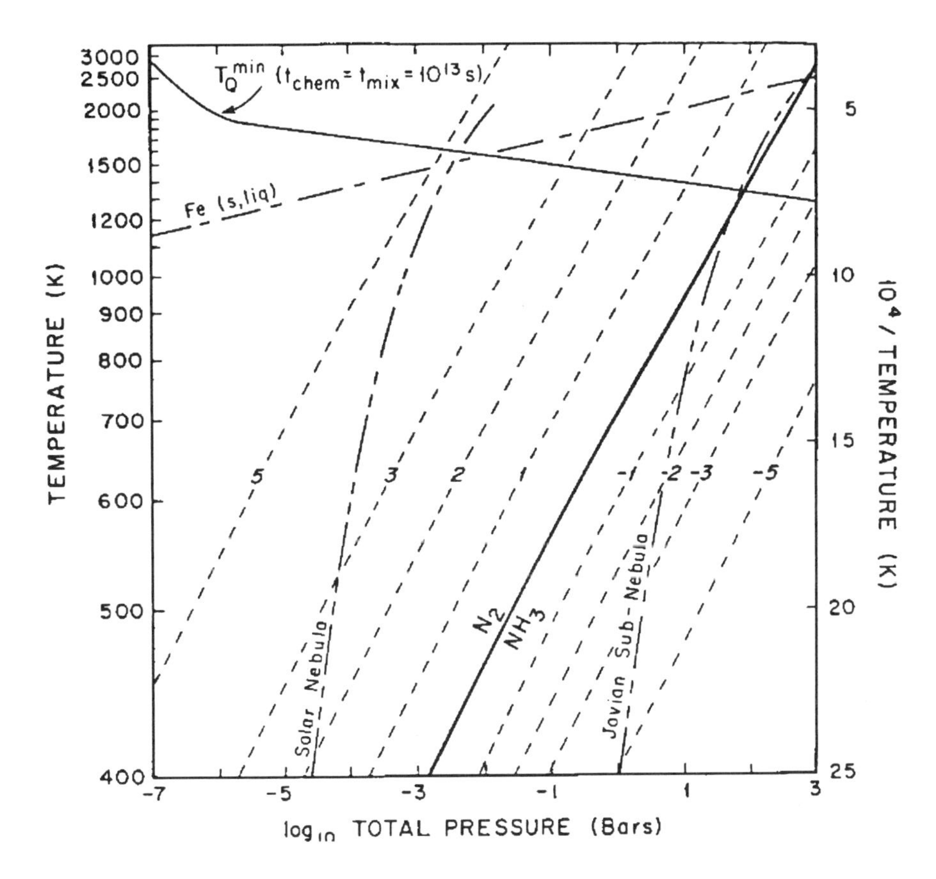

Figure 6. Gas phase equilibrium chemistry of nitrogen in a solar composition gas. Molecular nitrogen is the dominant nitrogen gas at high temperatures and low pressures and ammonia is the dominant nitrogen gas at low temperatures and high pressures The heavy solid diagonal line is the phase boundary between N_2 and NH_3. The two gases have equal abundances along this line. The dashed diagonal lines indicate constant ratios of N_2/NH_3 with a line labelled 2 indicating a ratio of 100 and a line labelled -3 indicating a ratio of 0.001. As in Fig. 5 the adopted P,T profiles for the solar nebula and the Jovian protoplanetary subnebula are also shown along with the quench temperature T_Q line for t_{chem} = t_{mix} = 10^{13} seconds. Likewise, the intersection of the T_Q line with the P,T profiles shows the N_2/NH_3 ratio at the quench points. Quenching yields a N_2-rich solar nebula and a NH_3-rich Jovian protoplanetary subnebula. Modified from Prinn and Fegley (1989).

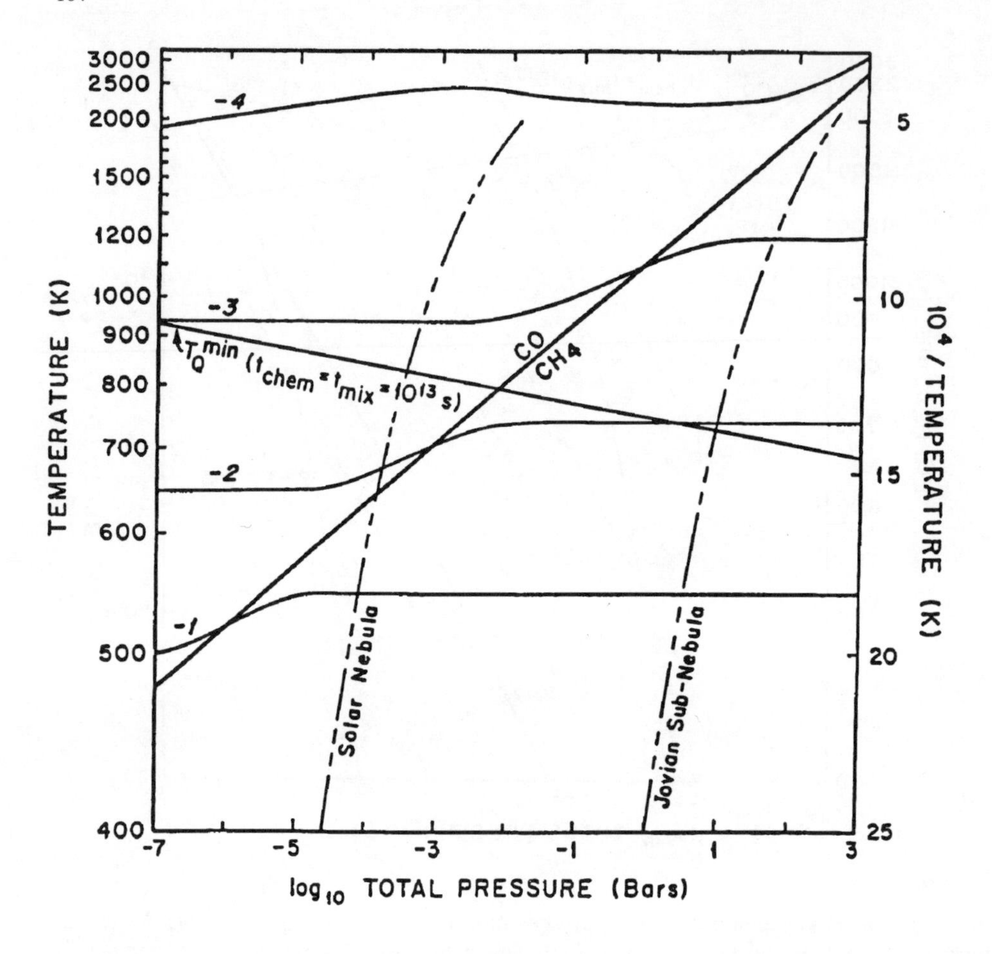

Figure 7. Chemical equilibrium ratios of CO_2/CO in a solar composition gas are plotted as a function of temperature and pressure along with the adopted P,T profiles for the solar nebula and the Jovian protoplanetary subnebula, and the T_Q line for the CO to CO_2 homogeneous gas phase conversion. Note that the CO to CO_2 conversion is relatively facile, leading to lower quench temperatures than for the reduction of CO and N_2. The phase boundary between CO and CH_4 is also shown for reference. The inflections in the contours of CO_2/CO are due to the change in the water abundance at this phase boundary. As shown by Lewis, Barshay, and Noyes (1979) there is a field off to the bottom left of this figure where CO_2 is the dominant carbon gas. This occurs inside the graphite stability field at temperatures below about 375 K and pressures below about 10^{-11} bars. Modified from Fegley and Prinn (1989).

$CO \rightarrow CO_2$, and $N_2 \rightarrow NH_3$ conversions are much more favorable in these environments because of the higher densities at a given temperature. As a result the conversions proceed down to lower temperatures where the reduced gases (or CO_2) are relatively much more abundant. In particular, for the Jovian subnebula P,T profile illustrated in Figs. 5-7, quenching of the $CO \rightarrow CH_4$ conversion is predicted at 840 K and $CH_4/CO \sim 10^6$, quenching of the $N_2 \rightarrow NH_3$ conversion is predicted at 1370 K and $NH_3/N_2 \sim 1$, and quenching of the $CO \rightarrow CO_2$ conversion is predicted at 740 K and $CO_2/CO \sim 0.01$ (Prinn and Fegley 1989). The dramatic differences between the predicted chemistry in the Jovian protoplanetary subnebula and the solar nebula are vividly demonstrated by the fact that the subnebular CH_4/CO ratio is $\sim 10^{13}$ times larger than the solar nebula ratio and that the subnebular NH_3/N_2 ratio is $\sim 10^5$ times larger than the solar nebula ratio. The giant protoplanetary subnebulae are thus predicted to be very efficient thermochemical processing plants for the production of reduced carbon and nitrogen gases. As we shall see later, this dramatic difference between C and N chemistry in the solar nebula and protoplanetary subnebulae has important implications for the chemistry of volatiles in the coma of comet P/Halley.

7.2 Grain Catalyzed Chemistry

The results above are specifically for homogeneous gas phase chemistry. However, as mentioned earlier, Fe metal, which is the most abundant transition metal in solar composition material, is predicted to be stable over a wide temperature range in the solar nebula. Industrial experience shows that Fe alloys are catalysts for the production of ammonia from the elements via the Haber process and for the production of synthetic fuels from $CO + H_2$ (Bond 1962; Dry 1981; Biloen and Sachtler 1981). Thus, the presence of Fe metal grains in the solar nebula may modify the results described above by accelerating the rates of the $CO \rightarrow CH_4$, $CO \rightarrow CO_2$, and $N_2 \rightarrow NH_3$ conversions and by converting a substantial portion of nebular CO to organic compounds via Fischer-Tropsch-type (FTT) reactions.

As we can see from Fig. 1, the lowest temperature at which Fe grains are expected to be present in the solar nebula is in the range of 370-400 K where magnetite is predicted to form. The Fe grains may be removed from the gas phase at higher temperatures if they agglomerate together and settle to the nebular midplane at a rate more rapid than that at which the nebula cools. Alternatively the Fe metal grains may remain dispersed, but catalytically inactive in the nebular gas at temperatures below 690 K if they are coated with FeS. In fact, petrographic studies of the unequilibrated ordinary chondrites, which are generally believed to preserve a record of nebular processes, show FeS-rimmed metal grains which may be nebular condensates (Rambaldi and Wasson 1981, 1984). Because neither situation can be predicted with certainty, we adopt the most conservative view possible, namely that Fe metal grains remain dispersed in the nebular gas with an abundance equal to the Fe/H_2 atomic ratio, at temperatures down to 370-400 K where magnetite is predicted to form. However, below this temperature, it is likely that the Fe metal grains will be rendered catalytically inactive by the formation of magnetite coatings, which chemically isolate them from the nebular gas.

By adopting this conservative viewpoint we can set a firm upper limit to the NH_3 abundance and a firm lower limit to the N_2/NH_3 ratio in the solar nebula because the magnetite formation temperature is taken as the quench temperature for the $N_2 \rightarrow NH_3$ conversion. From Fig. 6 we then see that taking $T_Q \sim$ 370-400 K yields $N_2/NH_3 \sim 100$ in the solar nebula. Of course, this result is applicable to the P,T profile assumed for our modelling. However, many, if not all of the currently accepted nebular models have P,T profiles which lie to the left of the one adopted here and will thus also yield $N_2/NH_3 \gg 1$. It is important to note that this treatment of the Fe grain catalyzed $N_2 \rightarrow NH_3$ conversion is independent of the actual reaction rate and of the assumed Fe grain size. A model dependent estimate of the solar nebula N_2/NH_3 ratio has also been made by (1) making the reasonable assumption that the typical Fe metal grain size (100 μm radius) in primitive chondritic meteorites is relevant to the solar nebula and (2) taking kinetic data from the chemical engineering literature for the rate of the Fe catalyzed conversion. This approach yields a slightly higher quench temperature of 530 K and a N_2/NH_3 ratio of about 170 if the $N_2 \rightarrow NH_3$ conversion is allowed to proceed for the entire nebular lifetime of 10^{13} seconds (Lewis and Prinn 1980). Therefore, the possibility of grain catalyzed NH_3 formation in the solar nebula does not alter the conclusion reached earlier from consideration of the homogeneous gas phase reaction kinetics, namely that N_2 was the dominant nitrogen gas throughout the entire solar nebula.

Of course Fe grain catalyzed NH_3 formation is also possible in the giant protoplanetary subnebulae around Jupiter and the other gas giant planets during their formation. As previously mentioned, the P,T profiles for these higher density environments are expected to lie within the NH_3 stability field. Thus, if catalysis is indeed effective, it will simply lead to the production of even more NH_3 and thus strengthen the conclusions reached earlier. In fact, taking the pressure-independent magnetite formation temperature of 370-400 K as the absolute minimum quench temperature yields $NH_3/N_2 \sim 10^5$. The model dependent estimate using the same kinetic data from the literature and the maximum conversion time of 10^{13} sec results in $T_Q \sim 495$ K and $NH_3/N_2 \sim 2000$ (Prinn and Fegley 1989). This ratio is applicable to the Jovian subnebula P,T profile shown in the figures. We would expect that the P,T profiles for the Saturnian subnebula and for any subnebulae of the other gas giant planets would lie at slightly lower pressures at a given temperature and thus have slightly lower NH_3/N_2 ratios. However, these ratios are still expected to be $\gg 1$.

Now we can consider how the presence of Fe metal grains affects the kinetics of the $CO \rightarrow CH_4$ conversion. Again, an absolute lower limit can be set by assuming that the Fe metal grains remain catalytically active and well mixed with the nebular gas down to the magnetite formation temperature. In this case, we obtain $CO/CH_4 \sim 10^{-9}$, in other words all carbon in the solar nebula is in the form of methane. The lower density P,T profiles found in other nebular models lead to larger CO/CH_4 ratios, but they are all still much less than unity. However, it is important to ask whether or not we actually expect Fe metal grains to catalyze the $CO \rightarrow CH_4$ conversion down to such low temperatures.

This point has been examined by Prinn and Fegley (1989) who used literature data on the rate of the heterogeneously catalyzed $CO \rightarrow CH_4$ conversion on ultra-clean, high purity, specially prepared Fe surfaces (Vannice 1975, 1982). The literature rate equation for the $CO \rightarrow CH_4$ conversion on metallic Fe particles is

$$\frac{d}{dt}[CH_4] = -\frac{d}{dt}[CO] = [sites]k_{site}P_{H_2}$$

where [i] is the gaseous molecular number density per cm^3, [sites] is the number density of all catalytically active sites on the surfaces of all Fe grains in each cm^3 of the solar nebula, P_{H2} is the H_2 partial pressure in bars, and k_{site} is the experimental rate constant expressed as the number of CH_4 molecules produced per active site per second. The value for k_{site} was taken as

$$k_{site} \sim 2.2 \times 10^7 \exp(-21,300/RT)$$

for the $CO \rightarrow CH_4$ conversion on clean Fe surfaces. The chemical time constant t_{chem} for the Fe grain catalyzed $CO \rightarrow CH_4$ conversion is then given by

$$t_{chem} = -[CO]/\frac{d}{dt}[CO].$$

Taking the shortest feasible mixing times of about 10^8 sec based on Cameron's (1978) estimate of transport at 1/3 sound speed, Prinn and Fegley (1989) calculated $T_Q \sim 900$ K and $CO/CH_4 \sim 10^{3.8}$. On the other extreme, taking the longest possible mixing time equal to the nebular lifetime of 10^{13}sec gave $T_Q \sim 520$ K and $CO/CH_4 \sim 10^{-3.5}$. The results of their calculations are displayed in Fig. 8. Mendybayev et al (1986) made an analogous model of Fe grain catalyzed CO reduction in the solar nebula, but used a slightly different rate constant in their calculations. They computed a quench temperature of 750 K and $CO/CH_4 \sim 10$ assuming that $t_{chem} = t_{mix} \sim 10^{9.5}$ seconds.

At this point it is important to emphasize that both the results of Prinn and Fegley (1989) and of Mendybayev et al (1986) are based on the rate constants measured for the $CO \rightarrow CH_4$ conversion on ultra-clean, high purity, specially prepared Fe surfaces, which do not exist in the solar nebula. This is important for two reasons. First, once Fe metal condenses in the nebula it begins acting as a solute, or host phase, for other elements that dissolve in it. Although the other transition metals which can dissolve in the Fe will probably not drastically change its catalytic activity, other predicted solutes such as P, S, C, N, and O (Kozasa and Hasegawa 1988; Fegley and Lewis 1980) will probably have more deleterious effects. Indeed, sulfur is a well known catalyst poison. Second, the laboratory studies of Fe catalyzed CO reduction show that the metal surface is rapidly inactivated by the forming of carbonaceous coatings as the reaction proceeds (e.g., Krebs, Bonzel and Gafner 1979; Vannice 1982). This effect is potentially relevant to the solar nebula because carbonaceous coatings and "tar balls" are commonly observed in interplanetary dust particles or IDPs (e.g., Bradley, Brownlee, and Fraundorf 1984; Bradley and Brownlee 1986). The model results are therefore probably best viewed as upper limits to the efficiency of Fe catalyzed CO reduction and as lower limits to the solar nebula CO/CH_4 ratio.

However, another, more likely course of events, is that Fe grains catalyzed organic compound formation from nebular $CO + H_2$ via FTT reactions. As noted previously (Fegley 1988, 1990; Fegley and Prinn 1989; Prinn and Fegley 1989), this pathway is more likely for the following reasons: (1) "tar balls", which are associations of carbonaceous material and Fe-bearing grains such as carbide, metal, or oxide, are commonly observed in chondritic IDPs (Bradley, Brownlee, and Fraundorf 1984; Bradley and Brownlee 1986; Bradley, Sandford, and Walker 1988; Christoffersen and Buseck 1983), (2) the "tar balls" are qualitatively similar to the carbonaceous deposits found on Fe-based FTT catalysts in the laboratory (Vannice 1982; Krebs, Bonzel, and Gafner 1979), (3) graphite is predicted to be stable at low T and P in a solar gas (Lewis, Barshay, and Noyes 1979), but its precipitation may be kinetically inhibited and hydrocarbons may form instead, (4) the low albedos and reflection spectra of many outer solar system objects implies the presence of dark carbonaceous material on the surfaces of these bodies (e.g., Himalia, Elara, Parisphae, Carme, Sinope, Lysithea, Ananke, Leda, Phoebe, the Uranian rings, part of the surface of Iapetus, the nucleus of comet P/Halley, etc.), (5) theoretical estimates of the carbon budget of the outer solar system indicate that about 10% of nebular CO was converted into organic matter (Simonelli et al 1989), and (6) carbonaceous material analogous to FTT reaction products is found in primitive chondritic meteorites (e.g., Hayatsu and Anders 1981; Studier, Hayatsu, and Anders 1968).

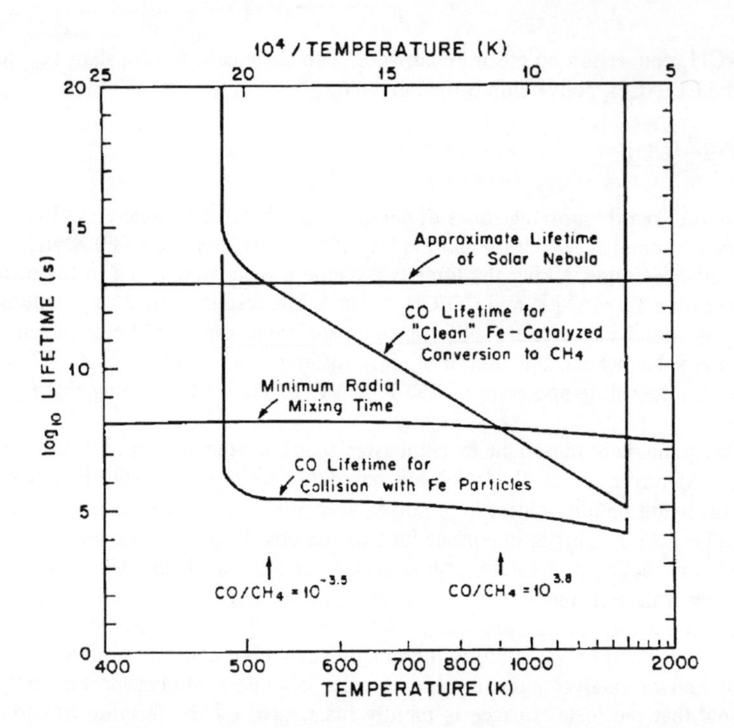

Figure 8. An illustration of the different chemical and physical timescales associated with the Fe grain catalyzed CO to CH_4 conversion in the solar nebula. The two horizontal lines show the estimated lifetime of the solar nebula (which constrains the longest possible time to do chemistry) and the estimated minimum radial mixing time (which fixes the shortest possible time to do chemistry). These two timescales are compared to the collision lifetime for CO to collide with 100 μm radius Fe grains and the chemical lifetime for the Fe grains to catalyze conversion of CO to CH_4. These times are calculated as described in the text. The results of the calculations show that if the Fe grains in the nebula are as efficient as high purity, specially prepared Fe catalysts in the laboratory, then a substantial fraction of CO may be converted to CH_4. However, as argued in the text, the more likely outcome of Fe grain catalyzed carbon chemistry is probably the formation of organic compounds via Fischer-Tropsch-type (FTT) reactions. Modified from Prinn and Fegley (1989).

Although some of the evidence presented by Anders and colleagues in favor of a FTT origin for organic compounds in meteorites is currently disputed (e.g., see Cronin's paper in this volume, and Shock and Schulte 1990a,b), these points taken together argue strongly that organic compounds were reasonably abundant in the solar nebula. The most plausible source of this organic matter is Fe grain catalyzed formation from CO, the most abundant carbon feedstock in the solar nebula.

The Fischer-Tropsch syntheses of alkanes, alkenes, and alcohols from nebular CO + H_2 are exemplified by reactions such as

$$(2n+1)H_2 + nCO = C_nH_{2n+2} + nH_2O$$

$$(n+1)H_2 + 2nCO = C_nH_{2n+2} + nCO_2$$

$$2nH_2 + nCO = C_nH_{2n} + nH_2O$$

$$nH_2 + 2nCO = C_nH_{2n} + nCO_2$$

$$2nH_2 + nCO = C_nH_{2n+1}OH + (n-1)H_2O$$

$$(n+1)H_2 + (2n-1)CO = C_nH_{2n+1}OH + (n-1)CO_2.$$

Fegley (1988) estimated the quench temperatures and time constants for FTT reactions in the solar nebula by using the kinetic theory of gases to calculate the rate of the grain catalyzed reactions. This approach is based on the assumption that the initial grain catalyzed reaction rate will depend upon the collision rate of CO (the less abundant reactant gas) with the grain surfaces (Fe metal grains). The grains are further assumed to be covered with sorbed H_2, which is reasonable since it is by far the most abundant nebular gas. Then, the collision rate (σ_i) of CO with the grains is given by

$$\sigma_i = 10^{25.4}[P_i/(M_iT)^{\frac{1}{2}}]$$

where σ_i has units of molecules cm^{-2} sec^{-1}, P_i is the CO partial pressure in bars, M_i is the CO molecular weight in g mole^{-1}, and T is the temperature in Kelvins. The total number of collisions of the CO molecules with all Fe grains in each cm^3 of the solar nebula is expressed as

$$\nu_i = \sigma_i A$$

where A is the total surface area of all Fe grains per each cm^3 of the nebula. For the purposes of the calculations, the Fe grains are assumed to be monodisperse, spherical grains that are fully dense and uniformly distributed at solar abundance in the nebular gas. Grain radii from 0.1 μm to 100 μm were used in the model calculations.

A collision time constant t_{coll} which is the time required for all CO molecules to collide with all Fe grains in each cm^3 of the solar nebula, can be calculated from the equation

$$t_{coll} = [CO]/\nu_i.$$

Now, if every collision of a CO molecule with an Fe grain led to organic compound synthesis, then t_{coll} would also be the chemical time constant t_{chem} for CO destruction. However, only a small fraction of collisions that have the necessary activation energy E_a lead to a chemical reaction. This fraction of reactive collisions is given by

$$f_i = \nu_i \exp(-E_a/RT)$$

where R is the ideal gas constant. Then the chemical time constant for CO destruction is

$$t_{chem} = [CO]/f_i = t_{coll}/\exp(-E_a/RT).$$

On the basis of E_a values from the meteoritic and chemical engineering literature (Dry 1981; Hayatsu and Anders 1981), Fegley (1988) took E_a = 90 kJ mole^{-1}. Over the nebular lifetime of 10^{13} sec, iron grains comparable in size to those in chondritic meteorites (100 μm radius) could convert ~ 10% of all CO into organic compounds before quenching occurred at about 510 K. On the other hand, much smaller Fe grains comparable in size to those in IDPs (0.1 μm radius) could convert this much CO into organics down to 440 K. However, these theoretical estimates are probably upper limits because H_2O sorption onto the Fe grain surfaces, thus reducing the amount of reactive sorbed H_2, was neglected. In any case, the estimated quench temperatures of 440-510 K are consistent with the laboratory results of Anders and colleagues (e.g., Anders, Hayatsu, and Studier 1973) showing that catalysis of FTT reactions was effective down to about 375 K. The estimated conversion of about 10% of CO is also consistent with the independent theoretical estimates of Simonelli et al (1989) that about 10% of the nebular CO was indeed converted into organic matter.

However, many important questions about the importance of FTT reactions in the solar nebula still remain to be answered. Prinn and Fegley (1989) emphasized that no systematic studies were available on the effects of process parameters such as the total pressure, temperature, time, CO/H_2 ratios, chemical and physical properties of the Fe metal catalyst, the presence of catalyst poisons such as H_2S, etc. In the absence of such data obtained under conditions relevant to the solar nebula, they used existing studies of FTT reactions under a range of conditions to provide some guidelines about the probable course of events in the solar nebula. Figure 9, taken from their paper, shows the effects of variable CO/H_2 ratios on the composition of hydrocarbons formed at a low total conversion of CO to organics. Not surprisingly, as the CO/H_2 ratio decreases (i.e., a more H_2 rich gas), the products in the single pass flow system become more enriched in CH_4 and C_2 hydrocarbons. Conversely, the product spectrum from an industrial system used to produce synthetic fuel is C_1 (2%), C_2 (2%), C_3 (5%), C_4 (5%), C_5-C_{11} (18%), C_{12}-C_{18} (14%), C_{19}-C_{23} (7%), C_{24}-C_{35} (21%), and >C_{35} (26%) (Dry 1981). This composition is rich in large molecules because the reactor is run to high total conversion of CO to organics and the outlet gas is recycled through the reactor to promote synthesis of larger molecules (Dry 1981). The FTT synthesis experiments conducted by Studier et al (1968) in static systems also gave larger molecules (as expressed by the maximum number of C atoms per molecule) than experiments run in flow systems under similar conditions. These results apparently indicate that FTT reactions on meteorite parent bodies may have been more important than similar reactions in the nebular environment, but no firm conclusions can be drawn at present.

8. Water Retention by Solid Grains

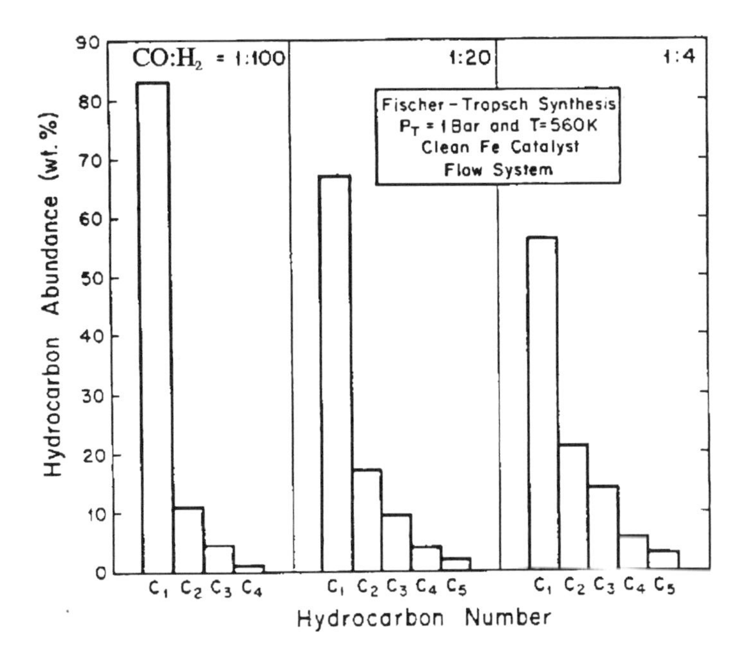

Figure 9. A bar graph showing the hydrocarbons produced by the Fischer-Tropsch synthesis on a Fe catalyst over a range of CO/H_2 ratios from 1:100 to 1:4. The results are for the initial stages of hydrocarbon synthesis (i.e., low total conversion of CO to products) in a flow system (Krebs et al 1979). Note that the lowest CO/H_2 ratio of 1:100 is still about 14 times larger than the solar CO/H_2 ratio from Table 1. Modified from Prinn and Fegley (1989) where the ratios are incorrectly shown as H_2/CO ratios.

The presence of oceans on the Earth but not on Venus illustrates one of the classic problems of planetary science, namely the mechanism for water retention by solid grains during planetary formation. Latimer (1950) and Urey (1952, 1953) suggested that hydrated silicates were responsible for water retention by the terrestrial planets. Despite the lack of accurate thermodynamic data which prevented them from calculating hydrated silicate stability fields in a solar gas, Urey (1953) was able to calculate the conditions under which water would be retained as brucite [$Mg(OH)_2$], water ice, and ammonia monohydrate $NH_3 \bullet H_2O(s)$. Urey (1953) concluded "that water and ammonia would have condensed in the solar nebula and the existence of the carbonaceous chondrites containing up to 10% of water confirms this conclusion..." Of course, the kinetic inhibition of the $N_2 \rightarrow NH_3$ conversion modifies Urey's original conclusion regarding ammonia hydrate stability.

During the intervening 40 years there have been several studies of hydrated silicate formation in the solar nebula. The results of these are summarized in Table 6, which lists condensation temperatures along the adopted nebular P,T profile, and in Fig. 10 which displays the pressure dependence of the hydrated silicate and water ice condensation curves. Fig. 10 also shows the calculated thermochemical isotopic fractionation for D/H between nebular H_2 and H_2O as a function of temperature. Several specific predictions result from these calculations. As the temperature decreases the first hydrous minerals to form are relatively water-poor phases such as hydroxyapatite, tremolite, and Na phlogopite. These minerals become stable in the 460-500 K range. However, these phases are seldom, if ever, seen in chondrites, and thus their existence in the solar nebula is problematic. Further decreases in temperature lead to the formation of more water-rich minerals such as brucite, talc, and serpentine, at temperatures of 400 K and below. These minerals are the major hydrous phases which are observed in water-bearing (e.g., CI, CM2, unequilibrated ordinary) chondrites. Decreasing temperature also leads to systematic variation of the D/H ratio in nebular water vapor. The net thermochemical equilibrium

$$HD(g) + H_2O(g) = H_2(g) + HDO(g)$$

proceeds to the right with decreasing temperature and thus nebular water vapor becomes preferentially enriched in D as the temperature drops. This enrichment is expressed both as a fractionation factor (simply the ratio of D/H in HDO to that in HD) and also as a δD value relative to Standard Mean Ocean Water (SMOW) in Fig. 10. Assuming that minimal isotopic fractionation occurs between nebular water vapor and the condensed phases, these results can then be used as a guide to the predicted D/H ratio in the hydrated silicates and in water ice. Thus, the hydrous phases formed at higher temperatures would be relatively D-poor and have a D/H ratio about 1-2 times the protosolar value of 3×10^{-5}, while the hydrous phases formed at lower temperatures would be relatively D-rich. In this regard it is interesting to note that the terrestrial D/H ratio of 1.6×10^{-4} (i.e., the D/H of SMOW), is not attained until $T \sim 200$ K.

The predictions of the thermochemical equilibrium calculations look relatively straightforward, but they are based on the attainment of solid-solid, gas-solid, and gas phase chemical equilibrium at low temperatures in the low density nebular environment. However, there are several good reasons to doubt that complete chemical and isotopic equilibrium are in fact attained under these conditions. We can illustrate some of the potential problems involved by considering serpentine formation via the vapor phase hydration of forsterite and enstatite

$$Mg_2SiO_4(s) + MgSiO_3(s) + 2H_2O(g) \rightarrow Mg_3Si_2O_5(OH)_4(s).$$

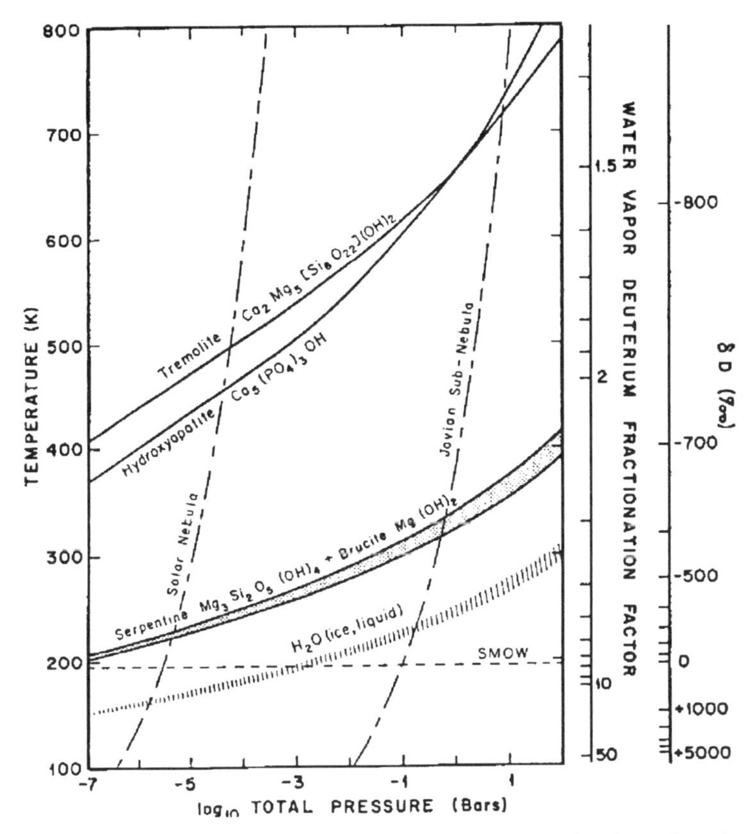

Figure 10. The thermodynamic stability fields of important water-bearing minerals and water ice over a range of temperatures and pressures applicable to the solar nebula and giant protoplanetary subnebulae. From top to bottom the hydrated silicate stability fields are from Lewis (1974) for tremolite, Fegley and Lewis (1980) for hydroxyapatite, and Fegley (1988) for serpentine + brucite and water ice. The shaded regions for serpentine + brucite and water ice illustrate the effects of having either CO (bottom curves) or CH_4 (top curves) as the dominant carbon gases. If CO remains the dominant carbon gas, then the amount of oxygen available for incorporation into water vapor is decreased. The amount of oxygen incorporated into rocky material, modelled as MgO + SiO_2, was also taken into account in these calculations. The thermochemical equilibrium fractionation factors (Richet, Bottinga, and Javoy 1977) for temperature dependent partitioning of D between HD(g) and H_2O(g) are also shown. These fractionation factors are independent of the absolute D/H ratio in either phase. However, the δD values , defined as $\delta D = 10^3[(D/H)_{water} - (D/H)_{SMOW}]$, do depend on the absolute D/H value chosen for H_2 in the solar nebula (taken as 2×10^{-5}). In this expression SMOW = Standard Mean Ocean Water with a D/H ratio of 1.557×10^{-4} (Pillinger 1984). In the text it is argued that both the gas-solid equilibration of water vapor with anhydrous silicates to form hydrous phases and the isotopic equilibration of HD and H_2O are kinetically inhibited at low temperatures in the solar nebula, but not in the giant protoplanetary subnebulae. Modified from Prinn and Fegley (1989).

This reaction requires the diffusion of Mg and Si between forsterite and enstatite at temperatures of 200 to 400 K. The literature data on cationic and oxygen diffusion in silicates can be combined with a model of hydrated silicate formation in the solar nebula to estimate the kinetics of this process. This has been done in several papers (Fegley 1988; Fegley and Prinn 1989; Prinn and Fegley 1989). The two silicate reactants, forsterite and enstatite are assumed to be in intimate contact with each other (i.e. in the same grain) for the entire lifetime of the solar nebula. The composite forsterite + enstatite grains are also assumed to be monodisperse spheres with radii of 0.1 μm. This assumed grain size is comparable to that observed in fine-grained matrix in chondritic meteorites and silicate grains in IDPs, but is smaller than that observed for most silicate grains in chondrites.

The rate determining step for hydrated silicate formation is then assumed to be solid-state diffusion in the composite grain. Either cationic (Mg-Si) or oxygen diffusion may be rate determining. Literature data show that anionic diffusion is in fact much slower. The characteristic diffusion time t_d can be estimated from the scaling relationship $t_d \sim r^2/D$ where r is the grain radius and D is the diffusion coefficient ($cm^2 sec^{-1}$). Fegley (1988) and Prinn and Fegley (1989) assumed that Mg-Si diffusion was rate determining and in the absence of diffusion constant data for this cationic pair assumed that it could be modelled using the available data on Fe-Mg diffusion in olivine (e.g., from Misener 1974). In this case t_d is $\sim 10^{23}$ sec at 400 K and $\sim 10^{55}$ sec at 200 K. On the other hand, Fegley and Prinn (1989) assumed that the much slower oxygen diffusion in olivine was rate limiting and estimated $t_d \sim 10^{41}$ sec at 400 K and $\sim 10^{89}$ sec at 200 K. These times are clearly much longer than the age of the solar system itself 10^{17} sec.

As noted by Fegley (1988), the actual situation in the solar nebula was probably not as favorable as that assumed in the models. Forsterite and enstatite were probably not in intimate contact throughout the entire lifetime of the solar nebula, accretion and coagulation produced grains much larger than 0.1 μm over the nebular lifetime, and solid-state diffusion (bulk and grain boundary) may have been quenched at these low temperatures. Simply put, it appears very probable that solid state chemistry was quenched at the low temperatures where it is needed to produce hydrated silicates.

As a consequence we are left with the vapor phase hydration of monomineralic silicate grains as the only available pathway for making hydrated silicates in the solar nebula. This pathway is exemplified by the two reactions

$$2Mg_2SiO_4(s) + 3H_2O(g) \rightarrow Mg_3Si_2O_5(OH)_4(s) + Mg(OH)_2(s)$$

$$4MgSiO_3(s) + 2H_2O(g) \rightarrow Mg_3Si_4O_{10}(OH)_2(s) + Mg(OH)_2(s).$$

However, as first noted by Fegley (1988) this pathway is also too slow to have occurred in the solar nebula. This is easily demonstrated by using the gas-grain kinetic model described earlier in connection with Fe grain catalyzed CO reduction. In this case the reactant gas is water vapor instead of CO, the grains are silicates instead of Fe metal, and the activation energy E_a is taken as ~ 70 kJ $mole^{-1}$. This is the activation energy measured by Layden and Brindley (1963) and Bratton and Brindley (1965) for the vapor phase hydration of MgO(s) to $Mg(OH)_2$(s). It is used because no activation energy data are available for the vapor phase hydration of silicates under conditions even close to those expected in the solar nebula. Again taking the most favorable assumptions of monodisperse, fully dense, spherical grains with radii of 0.1 μm we find that the chemical time constant t_{chem} for forsterite hydration in the solar nebula is much longer than the nebular lifetime. In fact $t_{chem} \sim 10^{18}$ sec is calculated. This calculation is illustrated in Fig. 11 where t_{chem} is plotted as a function of temperature along with the collision lifetime t_{coll} for colli-

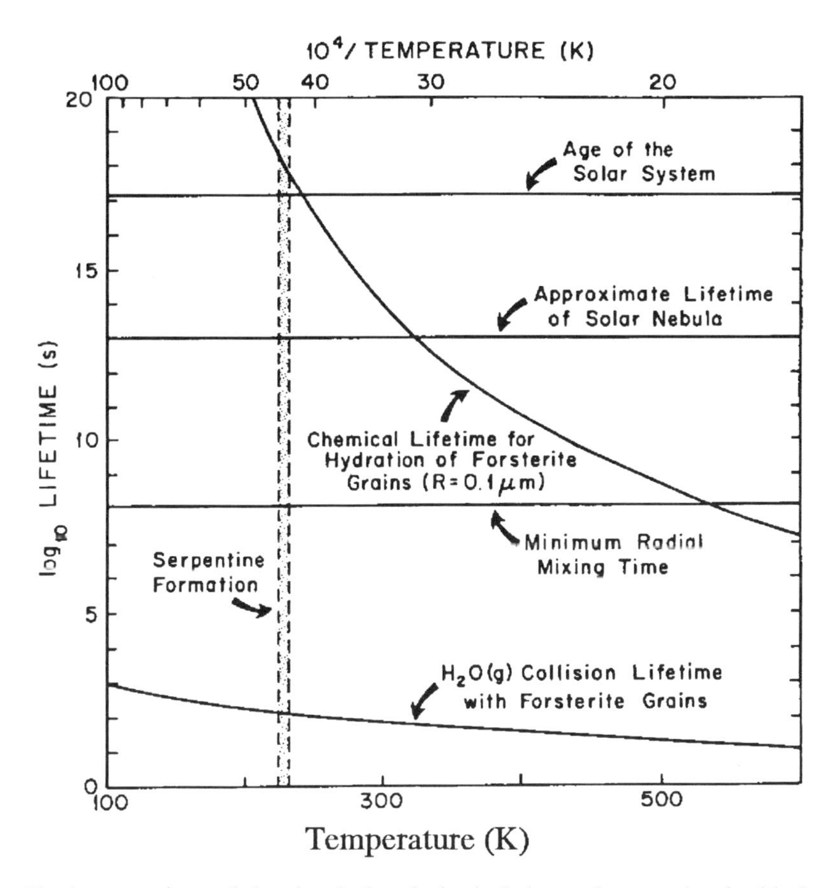

Figure 11. A comparison of the chemical and physical timescales associated with the vapor phase hydration of forsterite to different timescales in the solar nebula. The three horizontal lines represent the age of the solar system, the estimated lifetime of the solar nebula, and the estimated minimum radial mixing time (Cameron 1978) in the solar nebula. The vertical shaded region shows the maximum temperatures at which serpentine is thermodynamically stable in the solar nebula. The upper end of this range corresponds to all carbon being present as CH_4 and the lower end of this range corresponds to all carbon being present as CO. In these calculations, which are described in the text, forsterite is taken as a proxy for the more diverse suite of Mg silicates expected to be present at low temperatures in the solar nebula (e.g., see Fig. 3). Although the collision lifetime for H_2O molecules to collide with small forsterite grains (assumed to be 0.1 μm radius spheres) is short, only a small fraction of the collisions possess the necessary activation energy to hydrate the grains. Thus, the chemical lifetime for hydration of forsterite is much longer than the nebular lifetime at the low temperatures where serpentine first becomes thermodynamically stable. Modified from Prinn and Fegley (1989).

sions of H_2O molecules with forsterite grains. At the temperatures where serpentine finally becomes thermodynamically stable, t_{chem} is 10^5 times longer than the nebular lifetime and is about 10 times longer than the age of the solar system.

Once again however, these large values are probably only a lower limit to the actual times required for vapor phase hydration of silicates in the nebula. The activation energy for MgO hydration to $Mg(OH)_2$ is almost certainly lower than that required for silicate hydration because more bond breaking and solid-state diffusion is involved in the silicates than in MgO. Also, it is again unlikely that the 0.1 μm grains assumed in the calculation were actually present throughout the entire nebular lifetime. Finally, once the anhydrous silicates became covered by a thin layer of hydrated material, the reaction rate probably decreased due to solid-state diffusion and was not as rapid as that estimated from the water vapor collision frequency with forsterite grains. Thus, the conclusion that silicate hydration was kinetically inhibited in the solar nebula is robust.

On the other hand, similar models indicate that hydrated silicate formation was probably kinetically favorable in the higher density giant protoplanetary subnebulae around the gas giant planets during their formative stages. There are two complementary reasons for this state of affairs. One, which is illustrated by Fig. 10, is that the positive dT/dP for condensation curves leads to hydrated silicate formation at higher temperatures at the higher pressures expected in the giant protoplanetary subnebulae. Thus, for example, serpentine formation becomes thermodynamically favorable at about 325 K in the Jovian subnebula, versus about 225 K in the solar nebula. The higher total pressure in the subnebulae also leads to higher collision frequencies with the silicate grains. The combination of both factors decreases t_{chem} for forsterite hydration to 10^9 sec, or about 0.01% of the nebular lifetime for 0.1 μm grains. Much larger grains with radii of 1000 μm could also be hydrated over the nebular lifetime of 10^{13} sec if the rate remains controlled by the gas-grain collision frequency instead of solid-state diffusion. These calculations predict that hydrated silicate formation was favorable in the subnebulae around the gas giant planets.

There are several major implications of these calculations. One is that the hydrated silicates found in carbonaceous and unequilibrated ordinary chondrites are parent body, rather than nebular products. This theoretical conclusion is supported by extensive petrographic evidence showing that hydrated silicates in meteorites were produced by aqueous alteration on the meteorite parent bodies (e.g., Barber 1985; Boström and Fredriksson 1966; Bunch and Chang 1980; Tomeoka and Buseck 1985). Aqueous activity on the CI and CM parent bodies is also required to explain the formation of sulfate- and carbonate-bearing veins (e.g., DuFresne and Anders 1962; Fredriksson and Kerridge 1988; Richardson 1978). In addition, unidirectional oxidation of troilite in an aqueous environment provides a plausible explanation for the observed sulfur isotopic composition of FeS, elemental S, and the sulfate veins (Pillinger 1984). Finally, Clayton and Mayeda (1984) have shown that the oxygen isotopic compositions of CM2 chondrites require $T < 293$ K and liquid water volume fractions >44% to produce the hydrated silicates in them.. The oxygen isotopic data for CI chondrites require alteration in an even wetter and warmer environment to produce the hydrated silicates in these meteorites.

The second implication of the calculations is that the water currently found in hydrated silicates in CI and CM2 chondrites was originally retained on their parent bodies as water ice. The water ice condensation curve in Fig. 10 shows that condensation will take place at about 160 K ($R \sim 3.4$ A.U.) for this nebular P,T profile which assumes $T \propto R^{-1}$ and takes $T \sim 550$ K at 1 A.U. (Lewis 1974). This distance is about at the outer edge of the main asteroid belt. Because of the presence of water ice on the Galilean satellites Europa, Ganymede, and Callisto (Pilcher, Ridgway, and McCord 1972), the water ice front is generally placed at 5 A.U. in nebular models. However, the position of this front obviously varies with distance during the thermal evolution of the solar nebula (e.g., Ruden and Lin 1986) and first moves outward as the nebula warms up and later moves inward as the nebula cools. Also, the water ice front, like other condensation/evaporation fronts (Cameron and Fegley 1982) is closer to the proto-Sun at higher and lower regions off the midplane of the nebula. The inner boundary will be set by dispersal of the gas phase or by exhaustion of the available water vapor. The vertical structure and the time-dependent behavior of the water ice front are clearly important topics meriting further study, but the results presented here are consistent with water ice retention by the carbonaceous chondrite parent bodies.

Finally, a third implication of the calculations is that the hydrogen isotopic composition of the hydrated silicates in CI and CM2 chondrites should reflect that of the parent water ice, perhaps with some small modification due to the thermochemical isotopic fractionation that may take place during the aqueous alteration process. As noted in Table 4, the inferred D/H ratio in meteoritic hydrated silicates is about 14×10^{-5} ($\delta D \sim -100$ ‰) (Yang and Epstein 1983). Assuming no modification of the D/H ratio due to aqueous alteration, this corresponds to the D/H ratio of water ice at about 200 K (see Fig. 10). This would be consistent with water ice condensation at a pressure of about 10^{-3} bars. The D/H ratio of the water ice itself may either have been established by thermochemical isotopic exchange or by photochemical enrichment (Yung, et al 1988). However the former process is generally believed to be kinetically inhibited at such low temperatures (but see Lecluse and Robert (1992) for new experimental results that apparently indicate otherwise), while the latter process requires a dust-free nebula to allow penetration of the UV light driving the enrichment process.

9. Troilite and Magnetite Formation

Two other important gas-grain reactions are the formation of troilite and magnetite. Both proceed by the reaction of Fe metal grains with nebular vapor:

$$\mathrm{Fe(metal)} + \mathrm{H_2S(g)} \rightarrow \mathrm{FeS(s)} + \mathrm{H_2(g)}$$

$$\mathrm{3Fe(metal)} + \mathrm{4H_2O(g)} \rightarrow \mathrm{Fe_3O_4(s)} + \mathrm{4H_2(g)}.$$

As stated earlier, both reactions are pressure independent because the total pressure factors out of the equilibrium constant expressions, which depend only on the temperature and on the H_2S/H_2 and H_2O/H_2 mixing ratios. Fegley (1988) applied the same gas-grain kinetic model with the same basic assumptions to these systems as well. In the case of FeS an activation energy of about 105 kJ mole^{-1} from the work of Worrell and Turkdogan (1968) on FeS formation from Fe was used. In the case of Fe_3O_4, no activation energy data were available and $E_a \sim 80$ kJ mole^{-1} from the work of Turkdogan, McKewan, and Zwell (1965) on wüstite formation from water vapor and Fe was used instead.

The results of the kinetic calculations are displayed in Fig. 12. Troilite formation is predicted to be a rapid process with a chemical time constant of $\sim 10^{10}$ sec (about 0.1% of the nebular lifetime). If the rate of troilite formation stays controlled by gas-grain collisions, it can proceed on a time scale less than the nebular lifetime down to 525 K. On the other hand, solid-state diffusion constraints may take over at some point, but this does not change the basic result that FeS formation at 690 K is rapid. The results of these kinetic calculations are thus in accord with the intuitive expectation that "tarnishing" of Fe grains is a rapid process. The prediction that FeS formation was kinetically favorable in the solar nebula is also in accord with the petro-

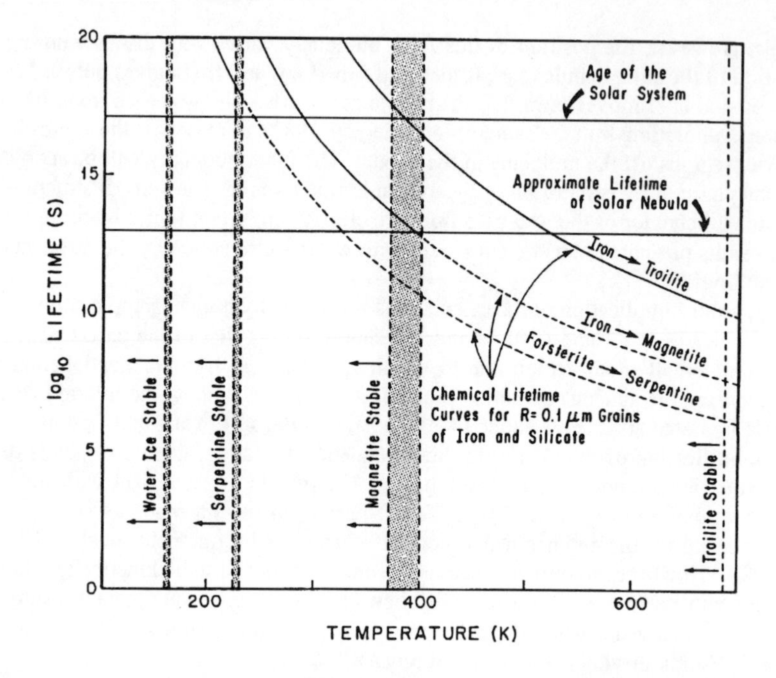

Figure 12. As in Fig. 11 but for troilite, magnetite, and serpentine formation in the solar nebula. The calculations, which are described in the text, show that troilite formation is kinetically favorable in the solar nebula because it proceeds in a short time relative to the nebular lifetime (about 0.1% of the estimated nebular lifetime). Magnetite formation is apparently slower, and may proceed on a timescale comparable to the nebular lifetime. However, intuition suggests that "rusting" of Fe grains by water vapor will be a rapid process and petrographic studies of magnetites in carbonaceous chondrites (Lohn and El Goresy 1992) support a nebular origin. The water ice condensation curve is also shown for comparison. From Fegley (1988).

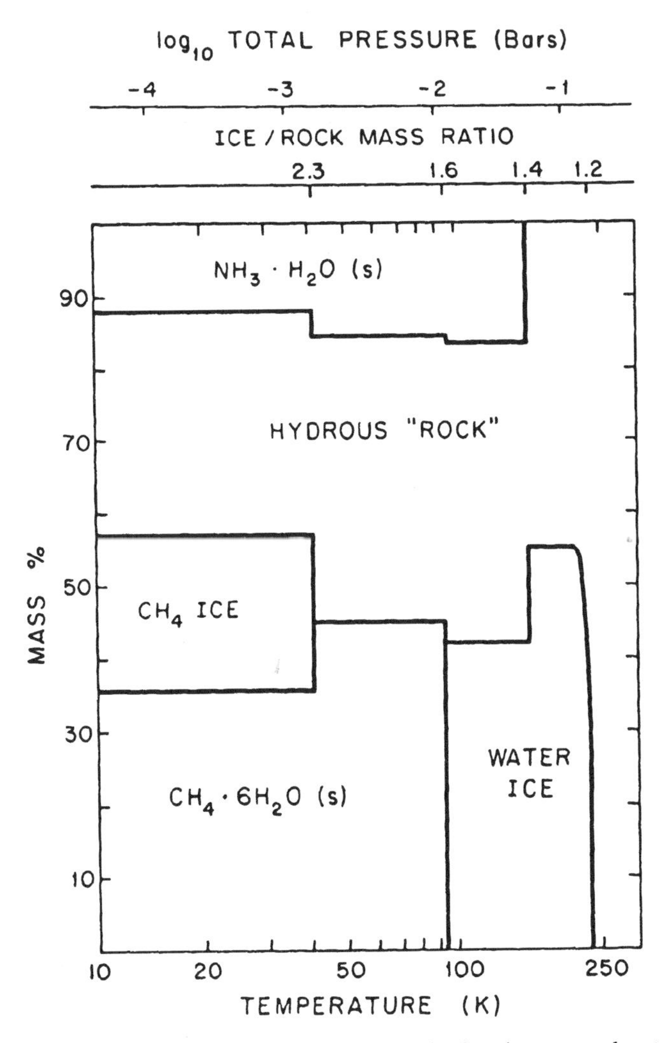

Figure 13. The low temperature condensation sequence in the giant protoplanetary subnebulae. As discussed in the text, the formation of CH_4, NH_3, hydrated rock, and clathrate hydrates is predicted to be kinetically and thermodynamically favorable in the giant protoplanetary subnebulae. The pressure scale along the top corresponds to the Jovian protoplanetary subnebula P,T profile adopted throughout this paper. Note that the ice/rock mass ratio is significantly higher than that predicted for the solar nebula (Fig. 14). Minor condensates such as HCN are not shown in this figure, but are nevertheless potentially important for organic compound synthesis in the atmosphere of Titan. Modified from Fegley and Prinn (1989).

graphic evidence of FeS-rimmed metal grains in the unequilibrated ordinary chondrites (Rambaldi and Wasson 1981, 1984). The calculated t_{chem} of about 10^{13} sec at 400 K for magnetite formation indicates that bulk magnetite formation may experience some kinetic inhibition, especially if CO remains the dominant carbon gas and magnetite is formed at 370 K. On the other hand, intuition suggests that the "rusting" of Fe metal grains by nebular water vapor is probably a rapid process. Furthermore, recent petrographic work by Lohn and El Goresy (1992) on magnetites in CI and CM2 chondrites suggests that the magnetites are nebular products. However, prior petrographic studies by Kerridge, MacKay, and Boynton (1979) concluded that the magnetites in CI chondrites were parent body products. Because of these petrographic disagreements and because the estimated chemical time constant for magnetite formation is based on the activation energy for wüstite formation, it is premature to draw firm conclusions regarding the kinetic favorability of the $Fe \rightarrow Fe_3O_4$ conversion in the solar nebula. This must await further experimental measurements and combined petrographic and isotopic studies.

10. Low Temperature Chemistry

Low temperature chemistry in these two environments is qualitatively different from the high temperature chemistry taking place above the water ice condensation curve. Above this curve, which is illustrated in Fig. 1, predominantly "rocky" and metallic grains are condensing from and interacting with the nebular gas, while below this curve predominantly "icy" grains are condensing from and interacting with the gas.

The nature of the "icy" grains which are formed both in the solar nebula and in the giant protoplanetary subnebulae depends to a large extent on the gas phase and gas-grain chemistry of carbon and nitrogen at higher temperatures. The two extremes are clearly the case of complete chemical equilibrium and the case of complete kinetic inhibition. Each will be examined in turn.

If complete chemical equilibrium is reached, then all carbon is present as CH_4 and all nitrogen is present as NH_3 at low temperatures. This case, or more precisely a close approach to it, is relevant to the giant protoplanetary subnebulae. The low temperature condensation sequence (neglecting the noble gases) is H_2O(ice) at 235 K, $NH_3 \bullet H_2O$(s) at 160 K, $CH_4 \bullet 6H_2O$(s) at 94 K, and CH_4(s) at 40 K (Prinn and Fegley 1989). The condensation temperatures given are specific to the Jovian protoplanetary subnebula P,T profile adopted in earlier figures, but the condensation sequence itself is relevant over a wide pressure range. Fig. 13 illustrates the resulting composition of the low temperature condensate as a function of temperature along the Jovian protoplanetary subnebula P,T profile. Initially, the condensate is totally composed of "rocky" material, including hydrous silicates. However, first the condensation of water ice and subsequently the condensation of ammonia hydrate and the other "icy" materials dramatically alter the condensate composition so that the "icy" materials are more abundant than the rock. Obviously, not all of these low temperature condensates will form in all environments because the temperature profile of the protoplanetary subnebula will blend into that of the surrounding solar nebula. In the Jovian subnebula ammonia hydrate is probably the most volatile condensate which will form while in the Saturnian subnebula methane clathrate hydrate $CH_4 \bullet 6H_2O$(s) is probably the most volatile condensate formed. Further out, in the Uranian and Neptunian subnebulae (if these existed) it is possible that solid methane ice also condensed.

On the other hand, if CO and N_2 reduction were both completely inhibited, then all carbon is present as CO and CO_2 and all nitrogen is present as N_2. The low temperature condensation sequence in this case, which may be a good analog to the solar nebula, is H_2O(ice) at 150 K, either NH_4HCO_3(s) at 150 K or NH_4COONH_2(s) at 130 K, CO_2(ice) at 70 K, CO clathrate hydrate, ideally $CO\cdot6H_2O(s)$ at 60 K, N_2 clathrate hydrate, ideally $N_2\cdot6H_2O(s)$ at 55 K, and residual CO and N_2 as the pure ices at 20 K (Prinn and Fegley 1989). Once again the noble gases have been neglected. The condensation temperatures are specifically for the adopted solar nebula P,T profile but the overall sequence remains the same over a wide pressure range. Because only a small fraction of total carbon is expected to be converted to CO_2 , and because only a small fraction of total nitrogen is expected to be converted to NH_3, the masses of condensates containing these species constitute $< 1\%$ of total C and N. It is also important to realize that in a CO-rich nebula, the amount of water ice which is available is insufficient to completely enclathrate all CO and N_2. At present the thermodynamic data for these two clathrates is also somewhat uncertain. In the case of CO clathrate hydrate, the only available experimental data are the dissociation pressure at the clathrate-liquid water-gas triple point at 273 K (Davidson et al 1987). In the case of N_2 clathrate hydrate, Miller (1961, 1969) has reported two different vapor pressure equations. Therefore it is difficult to specify exactly how much of each clathrate will form because the volatilities of each are poorly constrained. Fig. 14 displays the composition of low temperature condensate in the solar nebula as a function of temperature. In this instance, the rock is expected to be anhydrous because of the kinetic inhibition of hydrated silicate formation in the solar nebula. Anhydrous rock remains the dominant constituent of the low temperature condensate down to low temperatures where CO and N_2 ices condense. This is due to the diminished abundance of water ice in the CO-rich solar nebula. If temperatures never were lower than 20 K in the solar nebula, then anhydrous rock remained the major constituent of the low temperature condensate.

A third, intermediate case is somewhat more complex and less well constrained. In this scenario, which is relevant to the solar nebula, some fraction of CO has been converted into hydrocarbons by grain catalyzed FTT reactions. As a result the low temperature condensation sequence in order of decreasing temperature is as follows: anhydrous rock, "refractory" organic compounds analogous to the "tar balls" in IDPs, H_2O(ice), either $NH_4HCO_3(s)$ or $NH_4COONH_2(s)$, light hydrocarbons and their clathrate hydrates, CO_2(ice), $CO\cdot6H_2O(s)$, $N_2\cdot6H_2O(s)$, and residual CO and N_2. The condensation temperatures cannot be specified in this case without first knowing the amount of CO converted into hydrocarbons.

Additional complications are introduced by consideration of the kinetics of clathrate hydrate formation in the two different nebular environments. Briefly, clathrates are cage compounds in which the guest molecule is trapped inside a cage formed by the crystalline lattice of the host species. Clathrate hydrates are simply clathrates in which water ice is the host. Their formation has long been recognized as a potentially important mechanism for retention of carbon and nitrogen in ice-rich objects in the outer solar nebula (e.g., Delsemme and Swings 1952; Miller 1961; Delsemme and Miller 1970; Delsemme and Wenger 1970; Lewis 1972b; Delsemme 1976; Sill and Wilkening 1978; Lewis and Prinn 1980; Lunine 1989; Engel et al 1990). However, until very recently very little attention was paid to the kinetic feasibility of clathrate formation in the low temperature, low pressure environment of the outer solar nebula (e.g., Lewis and Prinn 1980; Lunine and Stevenson 1985; Fegley 1988).

As illustrated in Fig. 1 clathrates only become thermodynamically stable at very low temperatures in the solar nebula. For example, $CO\cdot6H_2O(s)$ becomes stable at about 60 K along the solar nebula P,T profile shown in the figure. Now in order for this clathrate to form, CO(g) in the solar nebula must collide with and then diffuse into water ice grains in the nebula. Fegley (1988) modeled this process using the gas-grain kinetic model described earlier. This is equivalent to assuming that the rate at which CO molecules collide with water ice grains is the rate determining step for clathrate formation. If fresh water ice is continually exposed to the nebular gas, for example by low velocity collisions as advocated by Lunine and Stevenson (1985), this

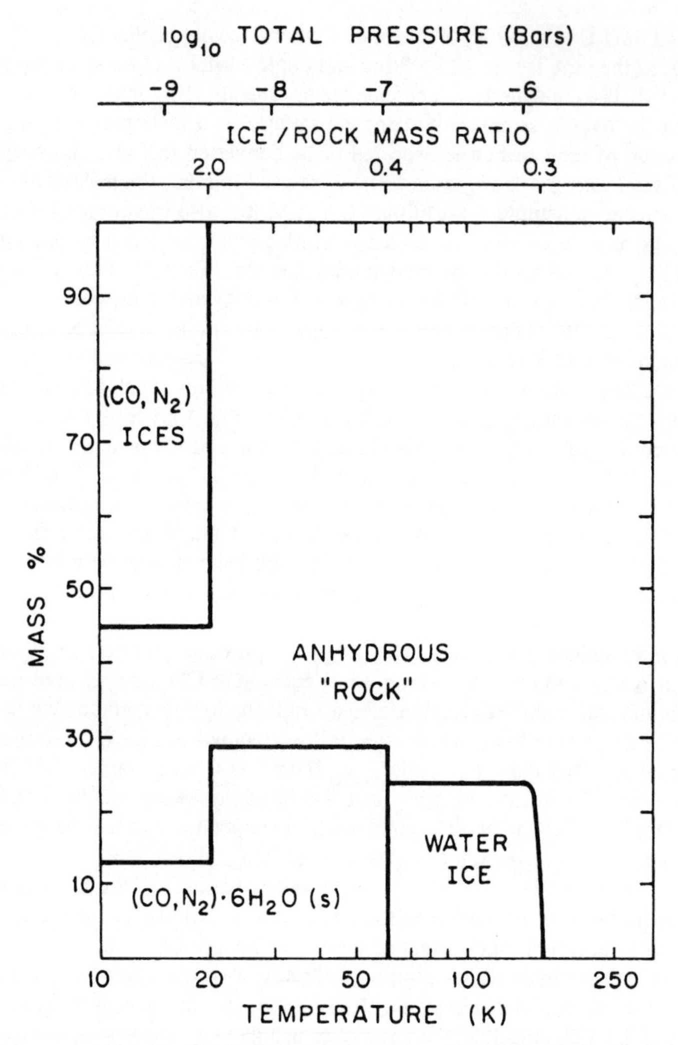

Figure 14. The low temperature condensation sequence in the solar nebula. As discussed in the text, the formation of CH_4, NH_3, and hydrated "rock" are predicted to be kinetically inhibited in the solar nebula. The formation of CO and N_2 clathrate hydrates is also probably kinetically inhibited. The pressure scale along the top corresponds to the nebular P,T profile adopted throughout this paper. Note that the ice/rock mass ratio in the CO-rich solar nebula is significantly lower than that in the CH_4-rich giant protoplanetary subnebulae. This difference is potentially diagnostic of the formation conditions of icy bodies in the outer solar system. Modified from Fegley and Prinn (1989).

may be the case. However, it is clear that solid-state diffusion may also become rate limiting, which will probably decrease the clathrate formation rate. For now, if we assume that the gas-grain collisions are rate limiting, we can estimate an upper limit to the rate and a lower limit to the time for clathrate formation in the solar nebula and in the giant protoplanetary subnebulae.

In the solar nebula, the time required for 6% of all CO, which is the maximum amount that can be enclathrated before running out of water ice, to collide with 1 μm radius, spherical, monodisperse water grains is about 40,000 seconds for the P,T profile shown in Fig. 1. The fraction of these collisions that have the necessary activation energy will be chemically reactive and lead to $CO \cdot 6H_2O(s)$ formation. Now in order for CO clathrate formation to take place within the nebular lifetime of 10^{13} sec, this activation energy must be ≤ 8 kJ $mole^{-1}$ (Fegley 1988). Higher activation energies will lead to longer clathrate formation times that exceed the lifetime of the solar nebula. However, 8 kJ is a relatively low activation energy by comparison with the 19 kJ $mole^{-1}$ required for HF diffusion through ice (Haltenorth and Klinger 1969) or with the 24.7 kJ $mole^{-1}$ required for CO_2 clathrate formation (Miller and Smythe 1970). If the activation energy for CO clathrate formation were 19 kJ $mole^{-1}$ the time required would be about 10^{21} sec, while if about 25 kJ $mole^{-1}$ is taken by analogy with CO_2 clathrate, the time required rises to even larger values. Nitrogen clathrate hydrate becomes stable at a similar temperature, and the same arguments apply in this case as well. Thus, it is apparent that clathrate formation will be kinetically inhibited in the low temperature, low pressure environment where it is thermodynamically feasible (but see Lunine et al 1991 for a dissenting viewpoint).

However, as pointed out by Fegley and Prinn (1989), clathrate formation is predicted to be kinetically favorable in the higher pressure environments of giant planet subnebulae. In this case, CH_4 is the dominant carbon gas and CH_4 clathrate formation becomes thermodynamically feasible a higher temperatures. As noted above, $CH_4 \cdot 6H_2O(s)$ becomes stable at about 95 K and 0.01 bars for the Jovian protoplanetary subnebula P,T profile. This pressure is approximately 5 orders of magnitude higher than the corresponding solar nebula pressure at the $CO \cdot 6H_2O(s)$ formation temperature of 60 K and thus leads to higher CH_4 gas collision rates with water ice grains. For example, the time for 22% of all CH_4 (which is the maximum amount that can be enclathrated before using up all the available water ice) to collide with $r = 1$ μm spherical, monodisperse ice grains is only 0.1 seconds. In this case, the activation energy for formation of $CH_4 \cdot 6H_2O(s)$ can be as large as 25 kJ $mole^{-1}$ to have the process occur during the lifetime of the solar nebula. Therefore, the results of these first order calculations predict (in accord with intuition) that CO and N_2 clathrate formation will be kinetically inhibited in the solar nebula but that CH_4 clathrate formation will be kinetically feasible in giant planet subnebulae. However, as previously stressed by Fegley (1988, 1990) and Fegley and Prinn (1989), experimental studies of the kinetics of clathrate formation are required for a comprehensive understanding of the kinetic constraints on clathrate formation in the solar nebula and in giant planet subnebulae. These experiments should focus on CH_4, CO, and N_2 clathrates and should be suitably designed so that the dependence of the rate on gas partial pressures, ice particle sizes, and temperature can be quantitatively measured.

Thus, two qualitatively different low temperature condensation sequences are expected in the solar nebula and in the giant protoplanetary subnebulae as a result of the differences in C and N chemistry at higher temperatures. In the solar nebula condensates of oxidized carbon and nitrogen compounds are important, formation of clathrate hydrates is kinetically inhibited, anhydrous rock is more abundant than water ice, and anhydrous rock remains the dominant constituent of the low temperature condensate down to the point where CO and N_2 ices condense. In the giant protoplanetary subnebulae condensates of reduced carbon and nitrogen compounds are important, clathrate hydrate formation is thermodynamically and kinetically favorable, hydrous rock is less abundant than water ice, and ices are the dominant constituent of the low temperature condensate. Finally, another important difference, first pointed out by Prinn and Fegley (1981), is the condensation of small, but nevertheless significant amounts of HCN in the giant protoplanetary subnebulae. Hydrogen cyanide is an important precursor for the abiotic synthesis of complex organic molecules (Oró and Kimball 1961; Abelson 1966). It is produced by the thermochemical conversion of some of the CH_4 and NH_3 into HCN,

$$CH_4 + NH_3 \rightarrow HCN + 3H_2$$

with the exact amount of conversion being kinetically controlled by the chemical reaction and radial mixing rates. For example the HCN abundances predicted by Prinn and Fegley (1981) range from about 1 to 100 μg/g depending on the assumed strength of radial mixing. As noted by these authors, "HCN even at these low predicted abundances could be extremely important as a starting material for the production of more complex organic compounds in the atmosphere of an icy satellite such as Titan."

11. Abundances Of C-H-O-N-S Compounds In Comet P/Halley and Other Comets

The preceding discussion has concentrated on theoretical predictions arising from current models of nebular chemistry. At this point it is therefore interesting to compare these predictions with observations in order to determine how useful the models are for understanding the volatile element inventories of solar system bodies. Detailed comparisons of the volatile inventories of the terrestrial planets, the chondritic meteorites and asteroids, the gas giant planets, icy satellites of the gas giant planets, and of Pluto with predictions of nebular chemistry models have already been presented by Prinn and Fegley (1989) and Fegley and Prinn (1989). Instead of repeating these discussions here, we will concentrate instead on a comparison of the model results with the volatile element abundances and molecular speciation in comet P/Halley and other recent comets. This is instructive for several reasons: (1) Earth-based observations and spacecraft flybys have provided a large amount of data on the chemical compositions of Halley and other recent comets, (2) comets are arguably the best preserved samples of nebular material, having presumably undergone less thermal processing subsequent to their formation than any type of chondrite, (3) comets are generally believed to have formed in the outer regions of the solar nebula and thus their compositions should shed light on the relative contributions of interstellar material and of thermally processed nebular material to chemistry in the outer solar nebula, and (4) significant new data have been published since the last such comparison was done by Fegley and Prinn (1989).

11.1 Water Vapor

As Weaver (1989) has noted, prior to the return of Halley observers had constructed a strong circumstantial case for the dominance of water vapor in the volatiles emitted by comets (e.g., see Delsemme 1982). However, water vapor was first observed directly in comet P/Halley using the Fourier Transform Infrared Spectrometer (FTIR) on the Kuiper Airborne Observatory (Mumma et al 1986) and was later observed in comet Wilson (1986l) with the same apparatus (Larson et al 1989). Subsequent measurements by the Neutral Mass Spectrometer (NMS) on Giotto showed that H_2O comprised $\geq 80\%$ of the volatiles emitted by Halley (Krankowsky et al 1986) and that the water vapor has a D/H ratio in the range of $0.6 - 4.8 \times 10^{-4}$ (Eberhardt et al 1987b). Finally, more recent measurements by Mumma, Weaver, and Larson (1987) and Mumma et al (1988) have provided data on the ortho-to-para ratio of water vapor emitted by Halley and by Wilson(1986l). For Halley, the derived ortho/para ratio is 2.3±0.2 while for Wilson(1986l) the derived ortho/para ratio is 3.2±0.2. If these data are taken at face value, the Halley ortho/para ratio implies a nuclear spin temperature of 25 K while the Wilson(1986l) ortho/para ratio implies statistical equilibrium ($T \geq 50$ K). However, Bockelée-Morvan and Crovisier (1990) retrieved different ortho-para ratios from the same data and have recently questioned the analysis of Mumma and colleagues. At present this discrepancy between the two groups is unresolved.

11.2 Carbon Compounds

The second most abundant group of volatiles (after water vapor) emitted from Halley are the oxidized carbon gases CO, CO_2, H_2CO, and CH_3OH. Five different measurements provide information on the abundance and distribution of CO emitted from Halley. The Giotto NMS data for mass 28, which is probably dominated by CO (see the discussion for N_2 below), has been interpreted as indicating a comet nucleus source for CO having $CO/H_2O \leq 0.07$ and an extended source in the inner coma for CO having $CO/H_2O \leq 0.15$ (Eberhardt et al 1987a). Infrared measurements from the IKS experiment on the Vega space probes yield $CO/H_2O \sim 0.05$ for a comet nucleus source (Combes et al 1988). Pioneer Venus Orbiter Ultraviolet Spectrometer (PVOUS) measurements of resonance UV emission from atomic hydrogen, oxygen, and carbon in the coma of Halley yield nominal H:O:C atomic ratios of 1:0.7:0.07, which are consistent with $CO/H_2O \sim 0.25$ or with $CO_2/H_2O \sim 0.14$ (Stewart 1987). Rocket-borne ultraviolet spectrometer measurements of resonance UV emission from atomic oxygen and carbon in the coma of Halley also yield similar CO/H_2O ratios of 0.20±0.05 for the one flight and 0.17±0.04 for a second flight (Woods et al 1986). Finally, measurements from the International Ultraviolet Explorer (IUE) satellite also yield a rough estimate for the CO/H_2O ratio of 0.1-0.2 for Halley (Festou et al 1986). This set of observations is generally interpreted as indicating a nucleus source for CO having $CO/H_2O \sim 0.02 - 0.07$ (by number) and a dispersed source which accounts for the balance of the observed CO (e.g., see Eberhardt et al 1987a; Weaver 1989).

Carbon monoxide has also been detected in comet Austin (1989c1), comet Levy (1990c), comet West (1976 VI) with $CO/H_2O \sim 0.3$ and in comet Bradfield (1979 X) with $CO/H_2O \sim 0.02$ (Weaver 1989; Mumma et al 1992, and references therein). The CO/H_2O ratios in Austin and Levy are in the range of 1-7% (Mumma et al 1992). As both Mumma et al (1992) and Weaver (1989) note, the "high" CO abundance in Halley and in comet West (1976 VI), and the "low" CO abundance in comet Bradfield (1979 X), comet Austin (1989c1), and comet Levy (1990c) may be explained as follows. The observations made with large fields of view of comets with large dust production rates (e.g., the rocket UV observations of Halley and West) yield larger apparent CO abundances because CO production from the nucleus and also from evaporating organic grains is being observed, while the observations made with smaller fields of view (e.g., IUE observations of Austin, Halley, and Levy) or observations made of comets with low dust production rates yield smaller apparent CO abundances because only CO production from the nucleus is being observed.

Carbon dioxide was observed with both the NMS experiment on Giotto and the IKS experiment on Vega. The NMS data yield $CO_2/H_2O \sim 0.04$ (Krankowsky et al 1986) while the IKS data yield a ratio of about 0.03 (Combes et al 1988). Both ratios are appropriate for CO_2 emitted from the nucleus and together with the adopted values of $\sim 0.02 - 0.07$ for CO/H_2O from the nucleus yield $CO/CO_2 \sim 0.50 - 2.3$. In other words, roughly equal amounts of CO and

CO_2 are being emitted from the nucleus of Halley. As discussed below, this rough equality has important implications for the origin of carbon-bearing gases and grains in Halley.

Formaldehyde has also been detected in Halley. The H_2CO/H_2O ratio obtained from the IKS measurements is ~ 0.04 (Combes et al 1988; Mumma and Reuter 1989) while a slightly lower value of ~ 0.02 was derived from radio wavelength observations by Snyder et al (1989). The analysis of the Giotto NMS data by Meier et al (1991) gave an upper limit of a few percent for H_2CO/H_2O. Both the radio wavelength observations and the Giotto NMS data apparently indicate a distributed source for at least some of the H_2CO (Krankowsky 1991; Snyder et al 1989). However, the IKS measurements supposedly refer to a nucleus source of H_2CO and not to formaldehyde released from the decomposition of POM in dust grains (Combes et al 1988). Formaldehyde has also been observed at radio wavelengths in comet Machholz (1988j) with a production rate an order of magnitude larger than that in Halley (dePater et al 1991) and in comet Austin (1989c1) with a H_2CO/H_2O ratio of 0.006 (Bockelée-Morvan et al 1991).

The detection of polyoxymethylene or POM $(H_2CO)_n$ in comet P/Halley was reported by Huebner (1987), Huebner, Boice and Sharp (1987), and Mitchell et al (1987). These authors interpreted regularly spaced peaks in ion mass spectra from the PICCA instrument (Korth et al 1986) as the cracking pattern of POM released from the dust grains. The POM was suggested to make up 2% (by mass) of the dust emitted from Halley (Mitchell et al 1987). However, Mitchell et al (1989) later proposed that a mixture of complex hydrocarbons containing nitrogen could also explain the observed mass spectra.

Infra-red spectroscopic observations of comets Austin (1989c1), Levy (1990c), Halley, and Wilson (1987 VII) showed the presence of CH_3OH and gave ratios of ~ 0.01-0.05 relative to water (Hoban et al 1991). Bockelée-Morvan et al (1991) also reported the radio wavelength detection of CH_3OH in comet Austin (1989c1) at a level corresponding to CH_3OH/H_2O ~ 0.01. Geiss et al (1991) interpreted Giotto IMS data as showing a CH_3OH/H_2O ratio of 0.003-0.015, and Eberhardt et al (1991) obtained a similar value of ~ 0.01 for the CH_3OH/H_2O ratio from the Giotto NMS data.

Carbon suboxide C_3O_2 has also been claimed in comet Halley at a level of 0.03-0.04 relative to water (Huntress, Allen, and Delitsky 1991). Photolysis of this species yields CO and either atomic C or C_2O, depending on the wavelength of the UV photons. Thus, it is an alternative to explain the distributed source of CO observed in Halley.

The CH_4/H_2O ratio in the volatiles emitted by Halley is currently controversial. Modeling of the Ion Mass Spectrometer data of Balsiger et al (1986) solely in terms of gas phase chemistry by Allen et al (1987) yields a CH_4/H_2O ratio of ~ 0.02. However, more recent modeling of the same data by Boice et al (1990) in terms of the decomposition of POM grains gives a lower CH_4/H_2O ratio of $\leq$ 0.005. Infrared observations from the Kuiper Airborne Observatory by Drapatz, Larson, and Davis (1987) gave an upper limit for $CH_4/H_2O \leq 0.04$ while IR observations at Cerro Tololo by Kawara et al (1988) gave CH_4/H_2O ~ 0.002 - 0.01 for assumed rotational temperatures of 50 to 200 K. The value adopted here for the CH_4/H_2O ratio is ~ 0.01 - 0.05 from the review by Weaver (1989). This value is adopted despite the more recent work of Boice et al (1990) for 2 reasons: (1) their analysis of the IMS data is based upon estimated rate coefficients for the POM chemistry and the sensitivity of their results to variations in the estimated rate data is not specified, (2) their analysis does not give a good fit to the CH peak at 13 amu without invoking an unidentified grain source for this peak. Together with the adopted CO/H_2O ratio of ~ 0.02 - 0.07 and the adopted CO_2/H_2O ratio of ~ 0.03 - 0.04, the adopted CH_4/H_2O ratio leads to CO/CH4 ~ 0.4 - 7.0 and CO_2/CH_4 ~ 0.6 - 4.0.

11.3 Nitrogen Compounds

In contrast to the carbon compounds discussed above, neither N_2 nor NH_3 has been observed in Halley. In both cases the inferred abundances of the parent molecules are deduced from observations of daughter molecules presumably produced by photolysis of the parents.

Allen et al (1987) originally derived a NH_3/H_2O ratio of ~ 0.01 - 0.02 from their analysis of the Giotto IMS data. However, a subsequent reanalysis of the same data by Marconi and Mendis (1988), who unlike Allen et al (1987) assumed a highly elevated solar UV flux at the time of the Halley spacecraft encounters, led to the conclusion that $NH_3/H_2O < 0.01$ and indeed may even be zero. However, the total absence of NH_3 in Halley is extremely unlikely given the Earth-based observations of NH_2 (Tegler and Wyckoff 1989; Wyckoff et al 1988; Wyckoff, Tegler and Engel 1989, 1991a) which is most plausibly produced by the photodissociation of NH_3. Tegler and Wyckoff (1989) derived $NH_3/H_2O = 0.005\pm0.002$ in comet P/Halley, which was later lowered to $NH_3/H_2O = 0.002\pm0.0014$ by Wyckoff, Tegler, and Engel (1991a). Ip et al (1990) derived NH_3/H_2O ~0.005 from an analysis of Giotto IMS data. The value adopted here for the NH_3/H_2O ratio is ~ 0.005 - 0.02. Wyckoff , Tegler, and Engel (1989, 1991a) have also observed NH_2 emission from comet P/Borrelly, comet Hartley-Good, and comet Thiele. The derived NH_3/H_2O ratios reported in their 1991 paper are 0.09±0.06% for Borrelly, 0.08±0.06% for Hartley-Good, and 0.16±0.11% for Thiele. However, as Weaver et al (1991) note, the only direct observation of NH_3 in a comet is a marginal detection of a radio line in comet IRAS-Araki-Alcock (1983 VII) by Altenhoff et al (1983).

Until recently only upper limits were available for the N_2/H_2O ratio in Halley. Wyckoff and Theobald (1989) observed N^+_2 in Halley and calculated a N_2/CO ratio ~ 0.002. Using the adopted value of ~ 0.02 - 0.07 for the CO/H_2O ratio leads to $N_2/H_2O \sim 4 \times 10^{-5}$ to 1×10^{-4}. A higher N_2/H_2O ratio of ~ 2×10^{-4} was derived by Wyckoff, Tegler, and Engel (1991b), but the difference is not significant for our arguments. In any case, the low N_2/CO ratio derived by Wyckoff and Theobald (1989) indicates that most of the mass 28 peak observed in the Giotto NMS is due to CO rather than to N_2. The adopted values for the N_2/H_2O and NH_3/H_2O ratios correspond to N_2/NH_3 ~ 0.002 - 0.025 while the value for N_2/NH_3 calculated by Wyckoff and colleagues on the basis of their own observational data is ~ 0.1.

Finally, HCN has also been detected at radio wavelengths in comet Kohoutek (Huebner et al 1974), in Halley at HCN/H_2O ~ 0.001 (Schloerb et al 1987; Despois et al. 1986), in comet Austin (1989c1) at HCN/H_2O ~ 0.0004, and in comet Levy (1990c) (Bockelée-Morvan et al 1991). Ip et al (1991) interpreted the Giotto IMS data as indicating $HCN/H_2O \leq 0.0002$. The reasons for the disagreement with the radio wavelength observations are unclear. Wyckoff et al (1989) used high resolution spectra of CN emitted from Halley to derive a $^{12}C/^{13}C$ ratio of 65±9, which is significantly lower than the terrestrial value of 89. However, subsequent work by Wyckoff and colleagues (Kleine et al 1991) showed that their initial carbon isotope determination was incorrect and that the $^{12}C/^{13}C$ ratio in Halley is in fact the same as the terrestrial value within observational error. Prior $^{12}C/^{13}C$ determinations for comets are 70±15 for Ikeya 1963 I (Stawikowski and Greenstein 1964), 100±20 for Tago-Sato-Kosaka 1969 IX by Owen (1973), 115(+30, -20) and 135(+65, -45) for Kohoutek 1973 XII by Danks et al (1974), and 100(+20, -30) for Kobayashi-Berger-Milon 1975 IX from Vanýsek (1977). All of the other cometary carbon isotopic ratios are derived from observations of C_2.

11.4 Sulfur Compounds

The sulfur species CS, S, S_2, and H_2S have been spectroscopically observed in comets; however, the only one of these four which is a "parent" molecule is H_2S. The solar elemental S/O abundance ratio of 0.021 (Anders and Grevesse 1989) suggests that sulfur species should be fairly abundant in comets, if a sizeable fraction of the total sulfur is in volatile form and not sequestered as FeS or otherwise locked up in the "rocky" component. The CS radical has been observed in many comets with abundances relative to water of about 10^{-3} (Feldman 1991). Although it has not been observed, CS_2 is generally believed to be the parent molecule for CS. Ultraviolet observations of atomic S indicate its presence in several comets, although its abundance relative to water is uncertain because of uncertainties in the atomic S line intensities and oscillator strengths (Crovisier et al 1991). Diatomic sulfur S_2 was observed in comet IRAS-Araki-Alcock (1983 d) by A'Hearn et al (1983). According to the recent analysis of Kim et al (1990) the S_2/H_2O abundance ratio is 0.0002 in this comet. Several groups have very recently reported the presence of H_2S in comets. Crovisier et al (1991) detected H_2S in comet Austin (1989c1) and comet Levy (1990c) at H_2S/H_2O ratios of 0.0027 and 0.0020, respectively. Marconi et al (1990) reported H_3S^+ in the Giotto Positive Ion Cluster Composition Analyzer (PICCA) data and estimated $H_2S/H_2O \sim 0.005$-0.022. Using newer data for H_2S photodissociation, Crovisier reanalyzed these data and found $H_2S/H_2O \sim 0.013$. Meier et al (1991) also reported evidence for H_2S from the Giotto mass spectrometer data and derived $H_2S/H_2O \sim 0.003$. The revised photodissociation lifetimes for H_2S and the radial ion profile observed in the NMS data are consistent with its production from the nucleus of comet P/Halley, and a dust particle source may not need to be invoked.

12. Interpretation of the Observed Molecular Abundances in Comet P/Halley

12.1 WATER

A fundamental question is whether or not the water in comet P/Halley and other comets is from the solar nebula or from the interstellar medium. If comet P/Halley is an assemblage of interstellar material, as proposed by some, then water, the most abundant volatile, should also have an interstellar origin. On the other hand, if comet P/Halley represents an assemblage of nebular and/or nebular plus interstellar material, then the water could be from the solar nebula. The source of the water has important implications for the source of the other, less abundant volatiles, especially for those which may be trapped in water ice as clathrate hydrates. This point can be illustrated by considering the following scenarios.

In the first scenario, we imagine that the water in comet P/Halley is relict interstellar ice. In this case, the water in Halley was never exposed to nebular temperatures high enough for it to evaporate or to isotopically exchange D and oxygen with nebular H_2 and O-bearing gases. The nuclear spins of the protons may also be representative of the interstellar medium, although this need not necessarily be the case, because spin exchange can occur at fairly low temperatures if the necessary catalysts are present. An important implication of this scenario is that clathrate hydrates would be absent from Halley because it is difficult to imagine how these compounds could have formed in the extremely cold, tenuous interstellar medium. We would then expect that the other volatiles in Halley are all present as pure ices. We would also expect that the chemical and isotopic composition of the volatiles in Halley could have little or no relation to that of average solar system material. In this case, the composition of Halley would be important for understanding the composition of the interstellar medium at the time that the solar system formed, but would provide little or no information about nebular processes.

In the second scenario, we imagine that the water in comet P/Halley is nebular in origin. This water could either be interstellar water that had evaporated and recondensed in the solar nebula, or it could be water that was formed in the solar nebula or in the giant protoplanetary subnebulae from chemical reactions such as:

$$CO + 3H_2 = CH_4 + H_2O.$$

In this scenario the water in comet P/Halley bears the signature of nebular processes and potentially provides information on these processes. Furthermore, in this case, it is possible that clathrate hydrates would have been formed if the water were chemically produced by CO hydrogenation in the giant protoplanetary subnebulae, where clathrate hydrate formation is kinetically favorable. Other important implications of a nebular origin for the water in Halley are that the isotopic composition could have equilibrated with the D/H ratio of nebular H_2 and O-bearing gases, and that the nuclear spins of the protons were equilibrated at the high temperature value. In this case the molecular composition of Halley provides very little information about interstellar speciation but provides a wealth of information about nebular speciation and nebular chemical processes. Another implication of this scenario is that other species, more volatile than water ice, are unlikely to have an interstellar origin.

The crystalline state (i.e., amorphous vs. crystalline), isotopic composition (D/H, oxygen isotopes), and nuclear spin state (ortho/para ratio) can potentially all shed light on this question. However, no data are available on the crystalline state of water ice in comet P/Halley. Furthermore, the interpretation of the oxygen isotopic data, which show a terrestrial $^{16}O/^{18}O$ ratio within observational errors, is nonunique given the independent evidence for large oxygen isotopic variations in the chondritic meteorites. Thus, only the D/H ratio and the ortho/para ratio may be potentially diagnostic of the origin of the water in comet P/Halley. We consider the D/H ratio first.

The observed D/H ratio of water in comet P/Halley is $0.6 - 4.8 \times 10^{-4}$ (Eberhardt et al 1987b). As noted earlier, this is the same as the terrestrial value of about 1.6×10^{-4} within the observational uncertainties. The equilibration temperature corresponding to a given D/H ratio in water can be calculated using the thermochemical isotopic fractionation factors tabulated by Richet, Bottinga, and Javoy (1977). The lower end of the range of D/H values derived for Halley (D/H $\sim 0.6 \times 10^{-4}$) corresponds to a temperature of ~ 465 K, while the upper end of the range of values ((D/H $\sim 4.8 \times 10^{-4}$) corresponds to a temperature of ~ 148 K. The classical picture of isotopic exchange envisions increasing deuterium enrichment in hydrides such as H_2O, CH_4, etc. with decreasing temperature (Geiss and Reeves 1981). In this picture, the D/H ratio represents the lowest temperature at which nebular water vapor and hydrogen exchanged deuterium. However, there are several reasons for believing that D/H exchange could not occur at low temperatures in the solar nebula. The first reason is that kinetic models of the D/H exchange rate between hydrogen and hydrides such as H_2O, CH_4, etc. in the atmospheres of the Jovian planets show that these reactions take longer than the age of the solar system at low temperatures (e.g., Fegley and Prinn 1988). It is therefore unlikely that these reactions will proceed any more rapidly in the much lower density nebular environment. Second, calculations by Grinspoon and Lewis (1987) show that the grain catalyzed D/H exchange rate is also probably too slow for isotopic equilibration to occur at low temperatures. But, as mentioned earlier, the recent experimental work of Lecluse and Robert (1992) apparently shows that D/H exchange between HD and H_2O will be rapid enough to produce Earth-like D/H ratios in nebular water.

Assume for the moment that the isotopic exchange temperatures are the maximum temperatures at which the water in Halley last exchanged deuterium with nebular H_2. If this view is taken, then the exchange process is viewed as a back-reaction in which D-rich water is losing deuterium to the surrounding nebular H_2. This process may occur as a consequence of reactions driven by thermochemistry (e.g., in the subnebulae surrounding the giant planets) or as a consequence of reactions driven by the interstellar radiation field impinging on the outer layers of the primitive solar nebula. The latter possibility is essentially the reverse of the scheme proposed by Yung et al (1988). Whether or not a purely thermochemical or a photochemical mechanism is envisioned, the temperature range of 148-465 K corresponding to the observed D/H ratio indicates significant amounts of thermal processing within the solar nebula for the water in comet P/Halley. Thus, the observed D/H ratio is compatible with a nebular origin for the water in Halley, or at least with nebular thermal processing at temperatures in the 148-465 K range.

On the other hand, if the ortho/para ratio of 2.3±0.2 deduced by Mumma and colleagues is taken at face value, the situation appears slightly more complicated. This low ratio implies a nuclear spin equilibration temperature of about 25 K, which is below the minimum temperatures predicted in most solar nebula models. An interstellar origin for the water in Halley is clearly implied from this result, unless a way can be found to retain the low temperature ortho/para ratio while isotopically exchanging D up to higher temperatures. As noted previously, Bockelée-Morvan and Crovisier (1990) retrieved different ortho-para ratios from the same data and questioned the analysis of Mumma and colleagues. In the absence of other information we do not use either ortho/para ratio as a constraint. Without compelling evidence in favor of an interstellar origin we regard it as highly probable that the water in comet P/Halley either originated (or was last thermally processed) in the solar nebula.

12.2 Carbon Compounds

The observational data reviewed earlier give the following ratios for the major carbon gases observed in comet Halley: CO/CO_2 ~ 0.5-2.3, CO/H_2CO ~ 0.5-1.8, CO/CH_3OH ~ 2.0-7.0, CO/C_3O_2 ~ 0.5-2.3, CO/CH_4 ~ 0.4-7.0, or in other words roughly equal abundances of all carbon gases within the uncertainties of the data. The elemental mass balance calculations of Delsemme (1988) further indicate that Halley contains the solar complement of carbon with about 25% of the total being found in the gaseous compounds and the remaining 75% being found in the CHON grains. Fegley (1990) pointed out that this carbon mass balance for Halley could be interpreted in two ways. The first approach, which we adopt here, is that the carbon mass balance indicates that both the volatile carbon gases and the more refractory carbon compounds in the CHON grains originated from the same reservoir which was fractionated into two groups of carbon compounds by some suite of chemical processes. The second approach, which was taken by Lunine (1989), is that the carbon mass balance is merely a coincidence and the volatile carbon gases and the refractory carbon in CHON grains have separate and decoupled origins.

Proceeding with the first approach we note that the 3:1 ratio between refractory and volatile carbon in Halley is qualitatively similar to the ratio of $P_{CO} \sim P_{CO2} \sim P_{CH4} \sim (1/3)A_{gr}$(the graphite abundance) predicted for the abundance of graphite and carbon gases at low temperatures and pressures ($T<470$ K, $P< 10^{-7.6}$ bars) inside the graphite stability field (Lewis, Barshay, and Noyes 1979). As noted earlier, grain catalyzed chemistry may proceed down to ~370-400 K where magnetite formation will deactivate the Fe grain catalyst. Furthermore, because of the difficulties in precipitating graphite from a solar gas, it is probably best to regard it as a proxy for organic matter. Formaldehyde, methanol, and carbon suboxide were not included in the work of Lewis, Barshay, and Noyes (1979) but the relatively rapid interconversions between oxidized carbon compounds such as CO-CO_2-H_2CO-CH_3OH (e.g., Warnatz 1984) suggest that all of these species can be formed at fairly low temperatures. We note that Fe catalyzed FTT reactions also produce alcohols, aldehydes, and CO_2 under laboratory conditions (Bond 1962; Dictor and Bell 1986;Dry et al 1972; Renshaw, Roscoe and Walker 1970). Thus, all of the volatile carbon species (except C_3O_2) can potentially be produced by grain catalyzed reactions. On the basis of calculations by Fegley (1988), Lunine (1989) and Engel et al (1989) also proposed that the abundances of the volatile gases CO, CO_2, and CH_4 in Halley are the result of grain catalyzed chemistry, but they regarded the involatile CHON grains as having a separate and decoupled origin from the carbon gases and did not discuss its origin in the light of the work by Lewis, Barshay, and Noyes (1979). However, as originally proposed by Fegley (1990) grain catalyzed chemistry may be responsible for both the volatile carbon gases and the refractory organic matter.

Many of the points mentioned earlier in connection with our discussion of grain catalyzed carbon and nitrogen chemistry argue for this model for the origin of carbon compounds in Halley. Briefly, these include: (1) the high abundance of Fe, the third most abundant cation (after Mg and Si) in rocky material, (2) the known catalytic activity of Fe in industrial processes, (3) the observed association of organic material with Fe phases (Fe alloy, carbides, and oxides) in interplanetary dust particles, and (4) the suspected association of some IDPs and comets. On the other hand, the problems mentioned earlier, such as the possible catalytic deactivation of Fe grains by troilite formation at 690 K, and the lack of laboratory studies under conditions relevant to the solar nebula, also apply here. The promising agreement between the Halley data and the laboratory results on FTT reactions point out the need for further experimental work in this important area.

Alternatively, some researchers have proposed an interstellar origin for the volatile carbon species in Halley because the observed CO/CH_4 ratio of $\sim$ 0.4-7.0 in Halley is similar to the CO/CH_4 ratios in solid grains in different molecular clouds: $\sim$ 0.4 in W33A, $\sim$ 0.3 in 7538 IRS1, and ~7.5 in 7538 IRS9 (Lacy et al 1991). The production of CH_4 and other hydrides by laboratory UV irradiation of simulated interstellar ices (e.g., Greenberg 1992) has also been used to argue for an interstellar origin of carbon gases in Halley. However, an interstellar origin for the CO and CH_4 in Halley requires preservation of the pristine interstellar grains. This is apparently contradicted by the arguments above showing that the water in Halley was processed to temperatures in the range of 148-465 K in the solar nebula. The more volatile CO and CH_4 ices would totally evaporate at these temperatures. An important implication of the interstellar model is that high D/H ratios are expected in CH_4, NH_3, and other hydrides formed on grains in molecular clouds (Tielens 1983). In contrast, the D/H ratios expected in CH_4 formed by grain catalyzed chemistry should reflect those of the starting HD, presumably the protosolar value of $\sim 3 \times 10^{-5}$. Thus, isotopic analyses of the volatiles emitted by comets will provide an important test of interstellar versus nebular origins.

It is also important to note, as emphasized by Prinn and Fegley (1989) and Fegley and Prinn (1989) that the CO/CH_4 ratio in Halley is compatible with nebular chemistry. Their model, which is schematically illustrated in Fig. 15, is a two component mixing model in which CO-rich nebular material is mixed with a smaller amount of CH_4-rich giant protoplanetary subnebular material. The mixing process may occur via collisions of icy bodies formed in the two environments after the nebular gas has dissipated, or chemical exchange between the two environments. Prinn and Fegley (1989) and Fegley and Prinn (1989) favored a collisional mixing process but did not model the dynamics of this mechanism. However, photogeologists have interpreted some of the cratering features observed in Voyager images of the icy satellites of Jupiter, Saturn, and Uranus in terms of bombardment by planetesimals during the early evolution of the solar system (e.g., Plescia 1987). Potentially testable implications of the two component mixing model include: (1) low D/H ratios in CH_4 because it originates from CO hydrogenation in the giant protoplanetary subnebulae, (2) a heterogeneous structure for the comet nucleus because it is a physical mixture of materials from two different nebular environments, (3) the

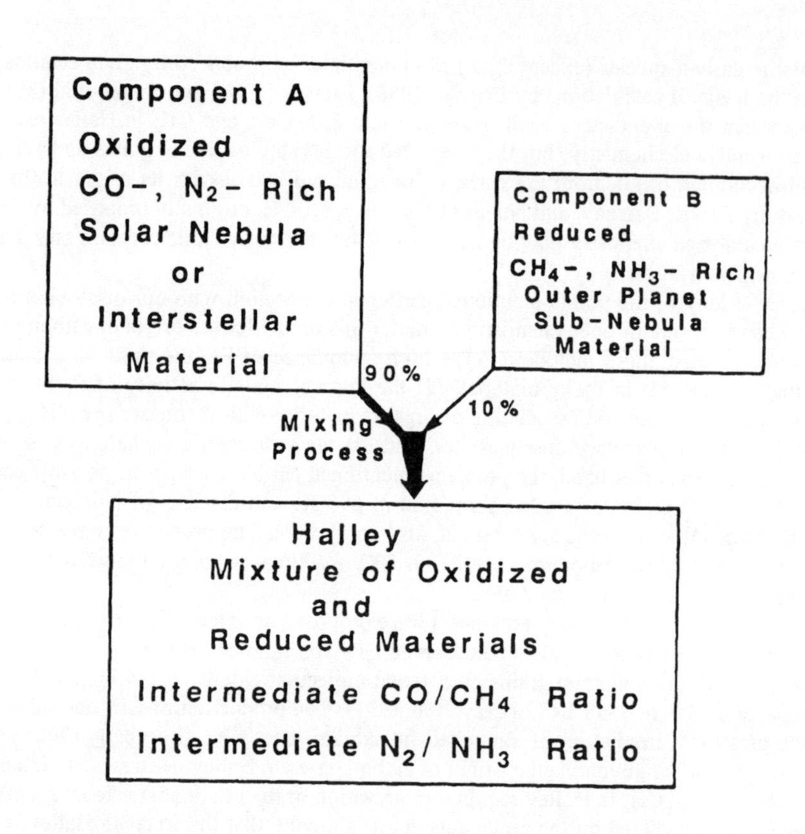

Figure 15. A schematic diagram illustrating the two component mixing model for the origin of volatiles in comet P/Halley (Prinn and Fegley 1989; Fegley and Prinn 1989). The inferred CH_4/CO and NH_3/N_2 ratios in the coma of Halley are intermediate between those expected in the solar nebula and in the giant protoplanetary subnebulae. This implies that the volatiles emitted from Halley are a mixture of material from the solar nebula (and/or interstellar medium) and the giant protoplanetary subnebulae. The additional data on abundances of H_2CO, CH_3OH, H_2S, and POM, which have become available since this model was originally proposed, also can be interpreted in terms of grain catalyzed chemistry in the solar nebula (Fegley 1990, this paper). As discussed in the text, there is no firm evidence showing that comet P/Halley represents pristine interstellar material. Modified from Fegley and Prinn (1989).

presence of both clathrate hydrates formed in the giant protoplanetary subnebula and of pure ices formed in the solar nebula, and (4) the presence of both anhydrous rock from the solar nebula and hydrated silicates from the giant protoplanetary subnebulae.

12.3 Nitrogen Compounds

Now we consider the interpretation of the observed abundances of nitrogen compounds in comet Halley. Prinn and Fegley (1989) and Fegley and Prinn (1989) pointed out that the NH_3/H_2O ratio of ~ 0.005-0.02 and the N_2/NH_3 ratio of ~ 0.002-0.025 in Halley are incompatible with either a solely solar nebula (N_2/NH_3 ~ 170) or a solely subnebula source (N_2/NH_3 ~ 0.0005). However, like the CO/CH_4 ratio, the N_2/NH_3 ratio can be explained by the mixing of material from these two different environments. On the other hand, the observed NH_3 abundances in the hot core of the OMC-1 molecular cloud (Blake et al 1987) have led Lunine (1989) and Engel et al (1989) to suggest an interstellar source for the NH_3 in Halley. Wyckoff, Tegler, and Engel (1991a) have also proposed an interstellar source on the basis of the very similar ammonia abundances they derived for four comets. They argued that a two component mixing process would be unlikely to give similar abundances in four comets with different dynamical characteristics. Again, the two alternatives can probably be tested by measuring the D/H isotopic ratio in the NH_3 because the interstellar source should give a substantially higher D/H ratio than the giant protoplanetary subnebula source.

The origin of HCN in Halley is also interesting. Fegley (1990) discussed the possibility that the HCN in Halley originated from lightning induced shock chemistry in nebular environments. This is an intriguing mechanism because nebular lightning has been suggested as a mechanism for chondrule formation (e.g., Cameron 1966; Whipple 1966). However, Prinn and Fegley (1989) noted that nebular lightning may also have distinctive chemical consequences. The high temperatures (several times 10^3 K) reached in lightning discharges lead to increasing degrees of molecular dissociation, atomization, and ionization with increasing temperatures. The recombination of these simple fragments during the rapid cooling of the shocked gas leads initially to the production of more complex fragments, then to thermally stable molecules such as HCN. Sufficiently rapid cooling quenches these stable molecules at their high temperature abundances, which are generally enhanced over their equilibrium abundances at much lower temperatures. Lightning is a potentially significant source of disequilibrium products, especially in the outer nebula beyond the water ice condensation front where comets are generally believed to have originated.

Prinn and Fegley (1989) modelled lightning induced shock chemistry in the solar nebula and in the giant protoplanetary subnebulae as adiabatic shock heating and predicted that the maximum HCN concentrations in these two environments occurred near temperatures of 3000-4000 K where ~ (0.2-6) × 10^{18} HCN molecules are formed per mole of shocked gas. This corresponds to maximum conversions of ~ 0.3% (in the solar nebula) and ~ 6.5% (in the Jovian protoplanetary subnebula) of total nitrogen into HCN. This compares favorably with Halley where the NH_3/H_2O ratio of ~ 0.005-0.02 and the HCN/H_2O ratio of ~ 0.001 correspond to a HCN/NH_3 ratio of ~ 0.05-0.2. Thus, as proposed by Fegley (1990) the HCN in Halley can be explained on the basis of lightning induced shock chemistry in either nebular environment, but only small dilution factors (<20 times in the solar nebula and <70 times in the Jovian subnebula) of the shocked gas by the unshocked gas are implied. Fegley (1990) noted that the low $^{12}C/^{13}C$ ratio of 65±9 derived for CN emitted by Halley (Wyckoff et al 1989) posed a problem for this model, but with the upward revision of the carbon isotopic ratio to an Earth-like value (Kleine et al 1991) this problem disappears.

12.4 Sulfur Compounds

Here we focus on H_2S and S_2. The H_2S/H_2O ratio in Halley is in the range of 0.003-0.022, which is similar to the NH_3/H_2O ratio of ~ 0.005-0.02. This suggests that the two gases are produced in equimolar amounts by the decomposition of NH_4SH. Ammonium hydrosulfide is a fairly volatile compound which can be formed in the solar nebula if not all sulfur reacts with Fe metal grains to form troilite. This could occur if accretion were sufficiently rapid to remove the metal from contact with the gas or if large chunks of metal reacted inefficiently with the nebula gas Fegley and Lewis (1980) calculated that NH_4SH would form at 134 K if a pure heterogeneous accretion model was adopted. Grains of NH_4SH emitted from the nucleus of comet Halley will decompose according to the reaction

$$NH_4SH = NH_3(g) + H_2S(g)$$

which has the equilibrium constant (Kelley 1937)

$$\log K = \log(P_{NH_3} \bullet P_{H_2S}) = 15.20 - 4823/T$$

Thus, the NH_4SH source explains the approximately equimolar amounts of NH_3 and H_2S observed in Halley. It also alleviates the need to postulate an extremely low temperature to preserve H_2S ice (e.g., Crovisier et al 1991). Ammonium hydrosulfide also provides a nebular source for H_2S at an abundance level lower than the solar S/H_2O ratio, because most of the sulfur has already reacted with Fe at higher temperatures to form troilite. The amounts of sulfur left to form NH_4SH will clearly depend on the efficiency with which troilite formation occurs, but this is not easily quantified. Another attractive feature of NH_4SH is that it is a potential source of the S_2 observed in comet IRAS-Araki-Alcock (1983d) by A'Hearn et al (1983). Grim and Greenberg (1986) have shown that S_2 can be produced by the UV irradiation of sulfur-bearing simulated interstellar ices. However, the work of Lebofsky and Fegley (1976) on the UV irradiation of NH_4SH led to the production of more complex sulfur compounds, such as polysulfides. They did not specifically look for S_2, but it is conceivable that it is also produced by UV irradiation of NH_4SH at sufficiently low temperatures.

Acknowledgments

This work was supported by grants from the NASA Origins of Solar Systems Program and the Planetary Atmospheres Program to Washington University (B. Fegley, P.I.).

References

Abelson, P.H. 1966. Chemical events on the primitive Earth. *Proc. Natl. Acad. Sci. (USA)* **55**, 1365-1372.

A'Hearn, M.F., P.D. Feldman, and D.G. Schleicher 1983. The discovery of S_2 in comet IRAS-Araki-Alcock 1983d. *Astrophys. J. Lett.* **274**, L99-L103.

Allen, M., M. Delitsky, W. Huntress, Y. Yung, W.H. Ip, R. Schwenn, H. Rosenbauer, E. Shelley, H. Balsiger, and J. Geiss 1987. Evidence for methane and ammonia in the coma of comet Halley. *Astron. Astrophys.* **187**, 502-512.

Altenhoff, W.J., W. Bartla, W.K. Huchtmeier, J. Schmidt, P. Stumpff, and M. Walmsley 1983. Radio observations of comet 1983d. *Astron. Astrophys.* **187**, 502-512.

Amari, S., E. Anders, A. Virag, and E. Zinner 1990. Interstellar graphite in meteorites. *Nature* **345**, 238-240.

Anders, E. 1968. Chemical processes in the early solar system, as inferred from meteorites. *Acc. Chem. Res.* **1**, 289-298.

Anders, E., and M. Ebihara 1982. Solar system abundances of the elements. *Geochim. Cosmochim. Acta* **46**, 2363-2380.

Anders, E., and N. Grevesse 1989. Abundances of the elements: Meteoritic and solar. *Geochim. Cosmochim. Acta* **53**, 197-214.

Anders, E., R. Hayatsu, and M.H. Studier 1973. Organic compounds in meteorites. *Science* **182**, 781-790.

Balsiger, H., K. Altwegg, F. Bühler, J. Geiss, A.G. Ghielmetti, B.E. Goldstein, R. Goldstein, P. Hemmerich, G. Kulzer, A.J. Lazarus, A. Meier, M. Neugebauer, U. Rettenmund, H. Rosenbauer, R. Schwen, E.G. Shelley, E. Ungstrup, and D.T. Young 1986. Ion composition and dynamics at comet Halley. *Nature* **321**, 330-334.

Barber, D.J. 1985. Phyllosilicates and other layer-structured minerals in stony meteorites. *Clay Minerals* **20**, 415-454.

Barshay, S.S. 1981. Combined condensation-accretion models of the terrestrial planets. Ph.D. thesis, Massachusetts Institute of Technology. 67pp.

Barshay, S.S. and J.S. Lewis 1976. Chemistry of primitive solar material. *Ann. Rev. Astron. Astrophys.* **14**, 81-94.

Bell, M.B., L.A. Avery, H.E. Matthews, P.A. Feldman, J.K.G. Watson, S.C. Madden, and W.M. Irvine 1988. A study of C_3HD in cold interstellar clouds. *Astrophys. J.* **326**, 924-930.

Bernatowicz, T., G. Fraundorf, T. Ming, E. Anders, B. Wopenka, E. Zinner, and P. Fraundorf 1987. Evidence for interstellar SiC in the Murray carbonaceous chondrite. *Nature* **330**, 728-730.

Bernatowicz, T., S. Amari, E.K. Zinner, and R.S. Lewis 1991. Interstellar grains within interstellar grains. *Astrophys. J.* **373**, L73-L76.

Biémont, E., M. Baudoux, R.L. Kurucz, W. Ansbacher, and E.H. Pinnington 1991. The solar abundance of iron: a "final" word! *Astron. Astrophys.* **249**, 539-544.

Biloen, P. and W.M.H. Sachtler 1981. Mechanism of hydrocarbon synthesis over Fischer-Tropsch catalysts. in *Advances in Catalysis,* ed. D.D. Eley, H. Pines, and P.B. Weisz, Academic Press, NY, pp. 165-216.

Bischoff, A. and H. Palme 1987. Composition and mineralogy of refractory-metal-rich assemblages from a Ca,Al-rich inclusion in the Allende meteorite. *Geochim. Cosmochim. Acta* **51**, 2733-2748.

Bjoraker, G.L., M.J. Mumma, and H.P. Larson 1989. The value of D/H in the Martian atmosphere: Measurements of HDO and H_2O using the Kuiper Airborne Observatory. in *Abstracts of the Fourth International Conference on Mars*, pp. 69-70.

Black, D.C. and R.O. Pepin 1969. Trapped neon in meteorites. II. *Earth Planet. Sci. Lett.* **6**, 395-405.

Blake, G.A., E.C. Sutton, C.R. Masson, and T.G. Phillips 1987. Molecular abundances in OMC-1: The chemical composition of interstellar molecular clouds and the influence of massive star formation. *Astrophys. J.* **315**, 621-645.

Blum, J.D., G.J. Wasserburg, I.D. Hutcheon, J.R. Beckett, and E.M. Stolper 1988. Origin of opaque assemblages in C3V meteorites: Implications for nebular and planetary processes. *Geochim. Cosmochim. Acta* **53**, 543-556.

Boato, G. 1954. The isotopic composition of hydrogen and carbon in the carbonaceous chondrites. *Geochim. Cosmochim. Acta* **6**, 209-220.

Bockelée-Morvan, D. and J. Crovisier 1990. in *Asteroids, Comets, and Meteors III*, ed. C.I. Lagerkvist et al, Uppsala University Press, Uppsala, pp. 263-265.

Bockelée-Morvan, D., P. Colom, J. Crovisier, D. Despois, and G. Paubert 1991. Microwave detection of hydrogen sulphide and methanol in comet Austin (1989cl). *Nature* **350**, 318-320.

Boesgaard, A.M. and G. Steigman 1985. Big bang nucleosynthesis: Theories and observations. *Ann. Rev. Astron. Astrophys.* **23**, 319-379.

Boice, D.C., W.F. Huebner, M.J. Sablik, and I. Konno 1990. Distributed coma sources and the CH_4/CO ratio in comet Halley. *Geophys. Res. Lett.* **17**, 1813-1816.

Bond, G.C. 1962. *Catalysis by Metals*, Academic Press, London.

Boss, A.P., G.E. Morfill, and W.M. Tscharnutter 1989. Models of the formation and evolution of the solar nebula. in *The Origin and Evolution of Planetary and Satellite Atmospheres*, ed. S.K. Atreya, J.B. Pollack, and M.S. Matthews, University of Arizona Press, Tucson, pp. 35-77.

Boström, K. and K. Fredriksson 1966. Surface conditions of the Orgueil parent meteorite body as indicated by mineral associations. *Smithson. Misc. Coll.* **151**, No. 3, 39pp.

Bradley, J.P. and D.E. Brownlee 1986. Analytical electron microscopy of thin-sectioned interplanetary dust particles. *Science* **231**, 1542-1544.

Bradley, J.P., D.E. Brownlee, and P. Fraundorf 1984. Carbon compounds in interplanetary dust: Evidence for formation by heterogeneous catalysis. *Science* **223**, 56-58.

Bradley, J.P., S.A. Sandford, and R.M. Walker 1988. Interplanetary dust particles. in *Meteorites and the Early Solar System,* ed. J.F. Kerridge and M.S. Matthews, University of Arizona Press, Tucson, AZ, pp. 861-895.

Bratton, R.J. and G.W. Brindley 1965. Kinetics of vapor phase hydration of magnesium oxide. Part 2. Dependence on temperature and water vapor pressure. *Trans. Faraday Soc.* **61**, 1017-1025.

Briggs, M.H. 1963. Evidence for an extraterrestrial origin for some organic constituents of meteorites. *Nature* **197**, 1290.

Brown, R.D. and E. Rice 1981. Interstellar deuterium chemistry. *Phil. Trans. Roy. Soc. London* **A303**, 523-533.

Bunch T.E. and S. Chang 1980. Carbonaceous chondrites-II. Carbonaceous chondrite phyllosilicates and light element geochemistry as indicators of parent body processes and surface conditions. *Geochim. Cosmochim. Acta* **44**, 1543-1577.

Burbidge, E.M., G.R. Burbidge, W.A. Fowler, and F. Hoyle 1957. Synthesis of the elements in stars. *Rev. Mod. Phys.* **29**, 547-650.

Cameron, A.G.W. 1966. The accumulation of chondritic material. *Earth Planet. Sci. Lett.* **1**, 93-96.

Cameron, A.G.W. 1973. Abundances of the elements in the solar system. *Space Sci. Rev.* **15**, 121-146.

Cameron, A.G.W. 1978. Physics of the primitive solar accretion disk. *Moon and Planets* **18**, 5-40.

Cameron, A.G.W. 1982. Elemental and nuclidic abundances in the solar system. in *Essays in Nuclear Astrophysics,* eds., C.A. Barnes, D.D. Clayton, and D.N. Schramm, Cambridge University Press, Cambridge, pp. 23-43.

Cameron, A.G.W. 1985. Formation and evolution of the primitive solar nebula. in *Protostars and Planets II,* eds., D.C. Black and M.S. Matthews, University of Arizona Press, Tucson, pp. 1073-1099.

Cameron, A.G.W. and M.B. Fegley 1982. Nucleation and condensation in the primitive solar nebula. *Icarus* **52**, 1-13.

Cameron, A.G.W., S.A. Colgate, and L. Grossman 1973. Cosmic abundance of boron. *Nature* **243**, 204-207.

Christoffersen, R. and P.R. Buseck 1983. Epsilon carbide: A low temperature component of interplanetary dust particles. *Science* **222**, 1327-1329.

Clarke, F.W. 1889. The relative abundances of the chemical elements. *Bull. Phil. Soc. Washington* **11**, 131.

Clayton, R.N. and T.K. Mayeda 1984. The oxygen isotope record in Murchison and other carbonaceous chondrites. *Earth Planet. Sci. Lett.* **67**, 151-161.

Clayton, R.N., L. Grossman, and T.K. Mayeda 1973. A component of primitive nuclear composition in carbonaceous meteorites. *Science* **182**, 485-488.

Combes, M., V.I. Moroz, J. Crovisier, T. Encrenaz, J.P. Bibiring, A.V. Grigoriev, N.F. Sanko, N. Coron, J.F. Crifo, R. Gispert, D. Bockelée-Morvan, Y.U. Nikolsky, V.A. Krasnopolsky, T. Owen, C. Emerich, J.M. Lamarre, and F. Rocard 1988. The 2.5 - 12 μm spectrum of comet Halley from the IKS-Vega experiment. *Icarus* **76**, 404-436.

Coustenis, A., B. Bézard, and D. Gautier 1989. Titan's atmosphere from Voyager infrared observations II. The CH_3D abundance and D/H ratio from the 900-1200 cm^{-1} spectral region. *Icarus* **82**, 67-80.

Crovisier, J., D. Despois, D. Bockelée-Morvan, P. Colom, and G. Paubert 1991. Microwave observations of hydrogen sulfide and searches for other sulfur compounds in comets Austin (1989c1) and Levy (1990c). *Icarus* **93**, 246-258.

Danks, A.C., D.L. Lambert, and C. Arpigny 1974. The $^{12}C/^{13}C$ ratio in comet Kohoutek (1973f). *Astrophys. J.* **194**, 745-751.

Davidson, D.W., M.A. Desando, S.R. Gough, Y.P. Handa, C.I. Ratcliffe, J.A. Ripmeester, and J.S. Tse 1987. A clathrate hydrate of carbon monoxide. *Nature* **328**, 418-419.

DeBergh, C., B.L. Lutz, T. Owen, J. Brault, and J. Chauville 1986. Monodeuterated methane in the outer solar system. II. Its detection on Uranus at 1.6 microns. *Astrophys. J.* **311**, 501-510.

DeBergh, C., B.L. Lutz, T. Owen,, and J.P. Maillard 1990. Monodeuterated methane in the outer solar system. IV. Its detection and abundance on Neptune. *Astrophys. J.* **355**, 661-666.

DeBergh, C., B. Bézard, T. Owen, D. Crisp, J.P. Maillard, and B.L. Lutz 1991. Deuterium on Venus:Observations from Earth. *Science* **251**, 547-549.

Delsemme, A.H. 1976. Chemical nature of the cometary snows. *Mém. Soc. Roy. Sci. Liège* **IX**, 135-145.

Delsemme, A.H. 1982. Chemical composition of cometary nuclei. in *Comets*, ed. L.L. Wilkening, University of Arizona Press, Tucson, pp. 85-130.

Delsemme, A.H. 1988. The chemistry of comets. *Phil. Trans. Roy. Soc. London* **A325**, 509-523.

Delsemme, A.H. and D.C. Miller 1970. Physico-chemical phenomena in comets--II. Gas adsorption in the snows of the nucleus. *Planet. Space Sci.* **18**, 717-730.

Delsemme, A.H. and P. Swings 1952. Hydrates de gaz dans les Noyaux Cométaires et les Grains Interstellaires. *Annales d'Astrophys.* **15**, 1-6.

Delsemme, A.H. and A. Wenger, Physico-chemical phenomena in comets--I. Experimental study of snows in a cometary environment. *Planet. Space Sci.* **18**, 709-715.

dePater, I., P. Palmer, and L.E. Snyder 1991. A review of radio interferometric imaging of comets. in Comets in the Post-Halley Era, ed. R. Newburn and J. Rahe, Kluwer Academic Publishers, Dordrecht, Netherlands.

Despois, D., J. Crovisier, D. Bockelée-Morvan, J. Schraml, T. Forveille, and E. Gerard 1986. Observations of hydrogen cyanide in comet Halley. *Astron. Astrophys.* **160**, L11-L12.

Dictor, R.A. and A.T. Bell 1986. Fischer-Tropsch synthesis over reduced and unreduced iron oxide catalysts. *J. Catalysis* **97**, 121-136.

Donahue, T.M., J.H. Hoffman, R.R. Hodges, Jr., and A.J. Watson 1982. Venus was wet: A measurement of the ratio of D to H. *Science* **216**, 630-633.

Drapatz, S., H.P. Larson, and D.S. Davis 1987. Search for methane in comet P/Halley. *Astron. Astrophys.* **187**, 497-501.

Dry, M.E. 1981. The Fischer-Tropsch synthesis. in *Catalysis Science and Technology,* vol. 1, eds., J.R. Anderson and M. Boudart, Springer-Verlag, Berlin, pp. 159-255.

Dry, M.E., T. Shingles, and L.J. Boshoff 1972. Rate of the Fischer-Tropsch reaction over iron catalysts. *J. Catalysis* **25**, 99-104.

DuFresne, E.R. and E. Anders 1962. On the chemical evolution of the carbonaceous chondrites. *Geochim. Cosmochim. Acta* **26**, 1085-1114.

Eberhardt, P., D. Krankowsky, W. Schulte, U. Dolder, P. Lämmerzahl, J.J. Berthelier, J. Woweries, U. Stubbemann, R.R. Hodges, J.H. Hoffman, and J.M. Illiano 1987a. The CO and N_2 abundance in comet P/Halley. *Astron. Astrophys.* **187**, 481-484.

Eberhardt, P., U. Dolder,, W. Schulte, D. Krankowsky, P. Lämmerzahl, J.H. Hoffman, R.R. Hodges, J.J. Berthelier, and J.M. Illiano 1987b. The D/H ratio in water from comet P/Halley. *Astron. Astrophys.* **187**, 435-437.

Eberhardt, P., R. Meir, D. Krankowsky, and R.R. Hodges 1991. Methanol abundance in comet P/Halley from in-situ measurements. *Bull. Amer. Astron. Soc.* **23**, 1161.

El Goresy, A., K. Nagel, and P. Ramdohr 1978. Fremdlinge and their noble relatives. *Proc. Lunar Planet. Sci. Conf.* **9**, 1249-1266.

Engel, S., J.I. Lunine, and J.S. Lewis 1990. Solar nebula origin for volatile gases in Halley's comet. *Icarus* **85**, 380-393.

Fegley, B., Jr. 1983. Primordial retention of nitrogen by terrestrial planets and meteorites. *Proc. 13th Lunar Planet. Sci. Conf. J. Geophys. Res.* **88**, A853-A868.

Fegley, B., Jr., 1988. Cosmochemical trends of volatile elements in the solar system. in *Workshop on the Origins of Solar Systems*, ed. J.A. Nuth and P. Sylvester, LPI Technical Report No. 88-04, pp. 51-60.

Fegley, B., Jr. 1990. Disequilibrium chemistry in the solar nebula and early solar system: Implications for the chemistry of comets. in *Proc. of the Comet Nucleus Sample Return Workshop*, ed. S. Chang, NASA CP, in press.

Fegley, B., Jr. and T.R. Ireland 1991. Chemistry of the rare earth elements in the solar nebula. *European J. Solid State Inorg. Chem.* **28**, 335-346.

Fegley, B., Jr., and A.S. Kornacki 1984. The geochemical behavior of refractory noble metals and lithophile trace elements in refractory inclusions in carbonaceous chondrites. *Earth Planet. Sci. Lett.* **68**, 181-197.

Fegley, B. Jr., and J.S. Lewis 1980. Volatile element chemistry in the solar nebula: Na, K, F, Cl, Br, and P. *Icarus* **41**, 439-455.

Fegley, B., Jr., and H. Palme 1985. Evidence for oxidizing conditions in the solar nebula from Mo and W depletions in refractory inclusions in carbonaceous chondrites. *Earth Planet Sci. Lett.* **72**, 311-326.

Fegley, B., Jr. and R.G. Prinn 1988. The predicted abundances of deuterium-bearing gases in the atmospheres of Jupiter and Saturn. *Astrophys. J.* **326,** 490-508.

Fegley, B., Jr., and R.G. Prinn 1989. Solar nebula chemistry: Implications for volatiles in the solar system. in *The Formation and Evolution of Planetary Systems*, eds. H.A. Weaver and L. Danly, Cambridge University Press, Cambridge, pp. 171-211.

Feldman, P.D. 1991. Ultraviolet spectroscopy of cometary comae. in *Comets in the Post-Halley Era*, ed. R. Newburn and J. Rahe, Kluwer Academic Publishers, Dordrecht, Netherlands, pp. 139-148.

Festou, M.C., P.D. Feldman, M.F. A'Hearn, C. Arpigny, C.B. Cosmovici, A.C. Danks, L.A. McFadden, R. Gilmozzi, P. Patriarchi, G.P. Tozzi, M.K. Wallis, and H.A. Weaver 1986. IUE observations of comet Halley during the Vega and Giotto encounters. *Nature* **321**, 361-363.

Fredriksson, K. and J.F. Kerridge 1988. Carbonates and sulfates in CI chondrites: Formation by aqueous activity on the parent body. *Meteoritics* **23**, 35-44.

Fuchs, L. and M. Blander 1980. Refractory metal particles in refractory inclusions in the Allende meteorite. *Proc. Lunar Planet. Sci. Conf.* **11**, 929-944.

Geballe, T.R., F. Bass, J.M. Greenberg, and W. Schutte 1985. New infrared absorption features due to solid phase molecules containing sulfur in W33A. *Astron. Astrophys.* **146**, L6-L8.

Geiss, J. and H. Reeves 1981. Deuterium in the solar system. *Astron. Astrophys.* **93**, 189-199.

Geiss, J., K. Altwegg, E. Anders, H. Balsiger, W.H. Ip, A. Meier, M. Neugebauer, H. Rosenbauer, and E.G. Shelley 1991. Interpretation of the ion mass spectra in the mass per charge range 25-35 amu/e^- obtained in the inner coma of Halley's comet by the HIS-sensor of the Giotto IMS experiment. *Astron. Astrophys.* **247**, 226-234.

Gerin, M., H.A. Wooten, F. Combes, F. Boulanger, W.L. Peters, T.B.H. Kuiper, P.J. Encrenaz, and M. Bogey 1987. Deuterated C_3H_2 as a clue to deuterium chemistry. *Astron. Astrophys.* **173**, L1-L4.

Goldschmidt, V.M. 1937. Geochemische Verteilungsgesetze der Elemente IX. *Skrifter Norske Videnscaps-Akademiend, Oslo I. mat. Natur. Kl.* No. 4.

Goldschmidt, V.M. 1954. *Geochemistry*, Oxford: Clarendon Press.

Greenberg, J.M. 1991. Physical, chemical, and optical interactions with interstellar dust. in *Chemistry in Space*, ed. J.M. Greenberg and V. Pirronello, Kluwer Academic Publishers, Dordrecht, Netherlands, pp. 227-261.

Grevesse, N. 1984. Abundances of the elements in the Sun. in *Frontiers of Astronomy and Astrophysics*, ed. R. Pallavicini, Ital. Astron. Soc., Florence, Italy, pp. 71-82.

Grevesse, N., D.L. Lambert, A.J. Sauval, E.F. van Dishoeck, C.B. Farmer, and R.H. Norton 1990. Identification of solar vibration-rotation lines of NH and the solar nitrogen abundance. *Astron. Astrophys.* **232**, 225-230.

Grevesse, N., D.L. Lambert, A.J. Sauval, E.F. van Dishoeck, C.B. Farmer, and R.H. Norton 1991. Vibration-rotation bands of CH in the solar infrared spectrum and the solar carbon abundance. *Astron. Astrophys.* **242**, 488-495.

Grim, R.J.A. and J.M. Greenberg 1987. Photoprocessing of H_2S in interstellar grain mantles as an explanation for S_2 in comets. *Astron. Astrophys.* **181**, 155-168.

Grinspoon, D.H. and J.S. Lewis 1987. Deuterium fractionation in the presolar nebula: Kinetic limitations on surface catalysis. *Icarus* **72**, 430-436.

Grossman, L. 1972. Condensation in the primitive solar nebula. *Geochim. Cosmochim. Acta* **36**, 597-619.

Grossman, L. and J.W. Larimer 1974. Early chemical history of the solar system. *Rev. Geophys. Space Phys.* **12**, 71-101.

Guélin, M., W.D. Langer, and R.W. Wilson 1982. The state of ionization in dense molecular clouds. *Astron. Astrophys.* **107**, 107-127.

Hagemann, R., G. Nief, and E. Roth 1970. Absolute isotopic scale for deuterium analysis of natural waters. Absolute D/H ratio for SMOW. *Tellus* **22**, 712-715.

Haltenorth, H. and J. Klinger 1969. Diffusion of hydrogen fluoride in ice. in *Physics of Ice*, ed. N. Riehl, B. Bullemer, and H. Engelhardt, Plenum Press, NY, pp. 579-584.

Hashimoto, A. and L. Grossman 1987. Alteration of Al-rich inclusions inside amoeboid olivine aggregates in the Allende meteorite. *Geochim. Cosmochim. Acta* **51**, 1685-1704.

Hayatsu, R. and E. Anders 1981. Organic compounds in meteorites and their origins. *Topics in Current Chemistry* **99**, 1-39.

Herbst, E., N.G. Adams, D. Smith, and D.J. DeFrees 1987. Ion-molecule calculation of the abundance ratio of CCD to CCH in dense interstellar clouds. *Astrophys. J.* **312**, 351-357.

Hoban, S., M. Mumma, D.C. Reuter, M. DiSanti, R.R. Joyce, and A. Storrs 1991. A tentative identification of methanol as the progenitor of the 3.52 μm emission feature in several comets. *Icarus* **93**, 122-134.

Holweger, H., C. Heise, and M. Kock 1990. The abundance of iron in the Sun derived from photospheric Fe II lines. *Astron. Astrophys.* **232**, 510–515.

Huebner, W.F. 1987. First polymer in space identified in comet Halley. *Science* **237**, 628–630.

Huebner, W.F., D.C. Boice, and C.M. Sharp 1987. Polyoxymethylene in comet Halley. *Astrophys. J.* **320**, L149–L152.

Huebner, W.F., L.E. Snyder, and D. Buhl 1974. HCN radio emission from comet Kohoutek (1973f). *Icarus* **23**, 580-584.

Huntress, W.T., M. Allen, and M. Delitsky 1991. Carbon suboxide in comet Halley? *Nature* **352**, 316-318.

Ip, W.H., H. Balsiger, J. Geiss, B.E. Goldstein, G. Kettman, A.J. Lazarus, A. Meier, H. Rosenbauer, R. Schwenn, and E. Shelley 1990. Giotto IMS measurements of the production rate of hydrogen cyanide in the coma of comet Halley. *Ann. Geophys.* **8**, 319-326.

Irvine, W.M. and R.F. Knacke 1989. The chemistry of interstellar gas and grains in *The Origin and Evolution of Planetary and Satellite Atmospheres*, ed. S.K. Atreya, J.B. Pollack, and M.S. Matthews, University of Arizona Press, Tucson, pp. 3-34.

Jeffery, P.M. and J.H. Reynolds 1961. Origin of excess ^{129}Xe in stone meteorites. *J. Geophys. Res.* **66**, 3582-3583.

Kawara, K., B. Gregory, T. Yamamoto, and H. Shibai 1988. Infrared spectroscopic observation of methane in comet P/Halley. *Astron. Astrophys.* **207**, 174–181.

Kelley, K.K. 1937. *Contributions to the Data on Theoretical Metallurgy VII. The Thermodynamic Properties of Sulphur and its Inorganic Compounds*, U.S. Bureau of Mines Bull. No. 406, U.S. GPO, Washington, D.C.

Kerridge, J.F., A.L. MacKay, and W.V. Boynton 1979. Magnetite in CI carbonaceous meteorites: Origin by aqueous activity on a planetesimal surface. *Science* **205**, 395-397.

Kim, S.J. and M.F. A'Hearn 1991. Upper limits of SO and SO_2 in comets. *Icarus* **90**, 79-95.

Kim, S.J., M.F. A'Hearn, and S.M. Larson 1990. Multi-cycle fluorescence: Application to S_2 in Comet IRAS-Araki-Alcock 1983 VII. *Icarus* **87**, 440-451.

Kleine, M., S. Wyckoff, P.A. Wehinger, and B.A. Peterson 1991. The carbon isotope abundance ratios in comets. *Bull. Amer. Astron. Soc.* **23**, 1166.

Kornacki, A.S., and B. Fegley, Jr. 1984. Origin of spinel-rich chondrules and inclusions in carbonaceous and ordinary chondrites. *Proc. 14th Lunar Planet. Sci. Conf. J. Geophys. Res.* **89**, B588-B596.

Kornacki, A.S., and B. Fegley, Jr. 1986. The abundance and relative volatility of refractory trace elements in Allende Ca, Al-rich inclusions: Implications for chemical and physical processes in the solar nebula. *Earth Planet. Sci. Lett.* **75**, 297-310.

Korth, A., A. K. Richter, A. Loidl, K.A. Anderson, C.W. Carlson, D.W. Curtis, R.P. Lin, H. Réme, J.A. Sauvaud, C. d'Uston, F. Cotin, A. Cros, and D.A. Mendis 1986. Mass spectra of heavy ions near comet Halley. *Nature* **321**, 335-336.

Kozasa, T. and H. Hasegawa 1988. Formation of iron-bearing materials in a cooling gas of solar composition. *Icarus* **73**, 180-190.

Krankowsky, D. 1991. The composition of comets. in *Comets in the Post-Halley Era*, ed. R.L. Newburn, J. Rahe, and M. Neugebauer, Kluwer Academic Publishers, Dordrecht, Netherlands, pp. 855-877.

Krankowsky, D., P. Lämmerzahl, I. Herrwerth, J. Woweries, P. Eberhardt, U. Dolder, U. Herrmann, W. Schulte, J.J. Berthelier, J.M. Iliano, R.R. Hodges, and J.H. Hoffmann 1986. *In situ* gas and ion measurements at comet Halley. *Nature* **321**, 326-330.

Krebs, H.J., H.P. Bonzel, and G. Gafner 1979. A model study of the hydrogenation of CO over polycrystalline iron. *Surface Sci.* **88**, 269-283.

Lacy, J.H., J.S. Carr, N.J. Evans II, F. Baas, J.M. Achtermann, and J.F. Arens 1991. Discovery of interstellar methane: observations of gaseous and solid CH_4 absorption toward young stars in molecular clouds. *Astrophys. J.* **376**, 556-590.

Langer, W.D., F.P. Schloerb, R.L. Snell, and J.S. Young 1980. Detection of deuterated cyanoacetylene in the interstellar cloud TMC-1. *Astrophys. J.* **239**, L125-L128.

Larimer, J.W. 1967. Chemical fractionations in meteorites--I. Condensation of the elements. *Geochim. Cosmochim. Acta* **31**, 1215-1238.

Larimer, J.W., 1973. Chemical fractionations in meteorites--VII. Cosmothermometry and cosmobarometry. *Geochim. Cosmochim. Acta* **37**, 1603-1623.

Larimer, J.W. 1975. The effect of C/O ratio on the condensation of planetary material, *Geochim. Cosmochim. Acta* **39**, 389-392.

Larimer, J.W. 1988. The cosmochemical classification of the elements. in *Meteorites and the Early Solar System*, ed. J.F. Kerridge and M.S. Matthews, pp. 375-389, Tucson: University of Arizona Press.

Larimer, J.W. and E. Anders 1967. Chemical fractionations in meteorites, 2, Abundance patterns and their interpretation. *Geochim. Cosmochim. Acta* **31**, 1239-1270.

Larimer, J.W. and M. Bartholomay 1979. The role of carbon and oxygen in cosmic gases: some applications to the chemistry and mineralogy of enstatite chondrites, *Geochim. Cosmochim. Acta* **43**, 1453-1466.

Larson, H.P., H.A. Weaver, M.J. Mumma, and S. Drapatz 1989. Airborne infrared spectroscopy of comet Wilson (1986l) and comparisons with comet Halley. *Astrophys. J.* **338**, 1106-1114.

Latimer, W.M. 1950. Astrochemical problems in the formation of the Earth. *Science* **112**, 101-104.

Layden, G.K. and G.W. Brindley 1963. Kinetics of vapor phase hydration of magnesium oxide. *J. Amer. Ceram. Soc.* **46**, 518-522.

Lebofsky, L.A. and M.B. Fegley, Jr. 1976. Laboratory reflection spectra for the determination of chemical composition of icy bodies. *Icarus* **28**, 379-387.

Lecluse, C. and F. Robert 1992. Origin of the deuterium enrichment in the solar system. *Meteoritics* **27**, 248.

Lee, T. 1988. Implications of isotopic anomalies for nucleosynthesis. in *Meteorites and the Early Solar System*, ed. J.F. Kerridge and M.S. Matthews, University of Arizona Press, Tucson, pp. 1063-1089

Lewis, J.S. 1972a. Metal/silicate fractionation in the solar system. *Earth Planet. Sci. Lett.* **15**, 286-290.

Lewis, J.S. 1972b. Low temperature condensation from the solar nebula. *Icarus* **16**, 241-252.

Lewis, J.S. 1974. The temperature gradient in the solar nebula. *Science* **136**, 440-443.

Lewis, J.S., S.S. Barshay, and B. Noyes 1979. Primordial retention of carbon by the terrestrial planets. *Icarus* **37**, 190-206.

Lewis, J.S., and R.G. Prinn 1980. Kinetic inhibition of CO and N_2 reduction in the solar nebula. *Astrophys. J.* **238**, 357-364.

Lewis, R.S., T. Ming, J.F. Wacker, E. Anders, and E. Steel 1987. Interstellar diamonds in meteorites. *Nature* **326**, 160-162.

Lin, D.N.C. and J. Papaloizou 1985. On the dynamical origin of the solar system in *Protostars and Planets II*, ed. D.C. Black and M.S. Matthews, University of Arizona Press, Tucson, pp. 981-1072.

Lohn, B. and A. El Goresy 1992. Morphologies and chemical composition of individual magnetite grains in CI and CM chondrites: A potential genetic link to their origin? *Meteoritics* **27**, 252.

Lunine, J.I., 1989. Primitive bodies: Molecular abundances in comet Halley as probes of cometary formation environments. in *The Formation and Evolution of Planetary Systems*, ed. H.A. Weaver and L. Danly, Cambridge University Press, Cambridge, pp. 213-242.

Lunine, J.I. and D.S. Stevenson 1985. Thermodynamics of clathrate hydrate at low and high pressures with application to the outer solar system. *Astrophys. J. Suppl.* **58**, 493-531.

Lunine, J.I., S. Engel, B. Rizk, and M. Horanyi 1991. Sublimation and reformation of icy grains in the primitive solar nebula. *Icarus* **94**, 333-344.

MacLeod, J.M., J.W. Avery, and N.W. Broten 1981. Detection of deuterated cyanodiacetylene (DC_5N) in Taurus Molecular Cloud 1. *Astrophys. J.* **251**, L33-L36.

MacPherson, G.J., D.A. Wark, and J.T. Armstrong 1988. Primitive material surviving in chondrites: Refractory inclusions. in *Meteorites and the Early Solar System*, ed. J.F. Kerridge and M.S. Matthews, University of Arizona Press, Tucson, pp. 746-807.

Marconi, M.L. and D.A. Mendis 1988. On the ammonia abundance in the coma of Halley's comet. *Astrophys. J.* **330**, 513-517.

Marconi, M.L., D.A. Mendis, A. Korth, R.P. Lin, D.L. Mitchell, and H. Réme 1990. The identification of H_3S^+ with the ion of mass per charge (m/q) 35 observed in the coma of comet Halley. *Astrophys. J.* **352**, L17-L20.

Mason, B. (ed.) 1971. *Handbook of Elemental Abundances in Meteorites.* Gordon & Breach, New York.

Mason, B. 1979. Cosmochemistry. Part 1. Meteorites. in *Data of Geochemistry, Sixth Edition,* ed. M. Fleischer, Geol. Surv. Prof. Paper 440-B-1, U.S. Govt. Print. Office, Washington, D.C.

Meier, R., P. Eberhardt, D. Krankowsky, and R.R. Hodges, The spatial distribution of the hydrogen sulfide and formaldehyde sources in comet P/Halley, *Bull. Amer. Astron. Soc.* **23**, 1167, 1991.

Mendybayev, R.A., A.B. Makalkin, V.A. Dorofeyeva, I.L. Khodakovsky, and A.K. Lavrukhina 1986. The role of CO and N_2 reduction kinetics in the chemical evolution of the protoplanetary cloud. *Geochem. Intl.* **8**, 105-116.

Miller, S.L. 1961. The occurrence of gas hydrates in the solar system. *Proc. Natl. Acad. Sci. USA* **47**, 1798-1808.

Miller, S.L. 1969. Clathrate hydrates of air in antarctic ice. *Science* **165**, 489-490.

Miller, S.L. and W.D. Smythe 1970. Carbon dioxide clathrate in the Martian ice cap. *Science* **170**, 531-533.

Misener, D.J. 1974. Cationic diffusion in olivine to 1400°C and 35 kbar. in *Geochemical Transport and Kinetics*, edited by A.W. Hofmann, B.J. Giletti, H.S. Yoder, Jr., and R.A. Yund, Carnegie Institution of Washington, Washington, D.C., pp. 117-129.

Mitchell, D.L., R.P. Lin, K.A. Anderson, C.W. Carlson, D.W. Curtis, A. Korth, H. Rème, J.A. Sauvaud, C. D'Uston, and D.A. Mendis 1987. Evidence for chain molecules enriched in carbon, hydrogen, and oxygen in comet Halley. *Science* **237**, 626-628.

Mitchell, D.L., R.P. Lin, K.A. Anderson, C.W. Carlson, D.W. Curtis, A. Korth, H. Rème, J.A. Sauvaud, C. d'Uston, and D.A. Mendis 1989. Complex organic ions in the atmosphere of comet Halley. *Adv. Space Res.* **9**, 35-39.

Morfill, G.E. and H.J. Volk 1984. Transport of dust and vapor and chemical fractionation in the early protosolar cloud. *Astrophys. J* **287**, 371-395.

Mumma, M.J. and D. Reuter 1989. On the identification of formaldehyde in Halley's comet. *Astrophys. J.*, **344**, 940-948.

Mumma, M.J., W.E. Blass H.A. Weaver and H.P. Larson 1988. Measurements of the ortho-para ratio and nuclear spin temperature of water vapor in comets Halley and Wilson (1986l) and implications for their origin and evolution. in *The Formation and Evolution of Planetary Systems: A Collection of Poster Papers*, ed. H.A. Weaver, F. Paresce, and L. Danly, STScI publication, pp. 157-168.

Mumma, M.J., S.A. Stern, and P.R. Weissman 1992. Comets and the origin of the solar system: Reading the Rosetta stone. in *Protostars and Planets III*, ed. E.H. Levy, J.I. Lunine, and M.S. Matthews, University of Arizona Press, Tucson, in press.

Mumma, M.J., H.A. Weaver and H.P. Larson 1987. The ortho-para ratio of water vapor in comet P/Halley. *Astron. Astrophys.* **187**, 419-424.

Mumma, M.J., H.A. Weaver, H.P. Larson, D.S. Davis, and M. Williams 1986. Detection of water vapor in Halley's comet. *Science* **232**, 1523-1528.

Niederer, F.R., D.A. Papanastassiou, and G.J. Wasserburg 1980 Endemic isotopic anomalies in titanium. *Astrophys. J.* **240**, L73-L77.

Niemeyer, S. and G.W. Lugmair 1984. Titanium isotopic anomalies in meteorites. *Geochim. Cosmochim. Acta* **48**, 1401-1416.

Olberg, M., M. Bester, G. Rau, T. Pauls, G. Winnewisser, L.E.B. Johansson, and Å. Hjalmarson 1985. A new search for and discovery of deuterated ammonia in three molecular clouds. *Astron. Astrophys.* **142**, L1-L4.

Ormont, A. 1991. Circumstellar chemistry. in *Chemistry in Space*, ed. J.M. Greenberg and V. Pirronello, Kluwer Academic Publishers, Dordrecht, Netherlands, pp. 171-196.

Oró, J. and A.P. Kimball 1961. Synthesis of purines under possible primitive Earth conditions. I. Adenine from hydrogen cyanide. *Arch. Biochem. Biophys.* **94**, 217-227.

Owen, T. 1973. The isotope ratio $^{12}C/^{13}C$ in comet Tago-Sato-Kosaka 1969g. *Astrophys. J.* **184**, 33-43.

Owen, T., B.L. Lutz, and C. DeBergh 1986. Deuterium in the outer solar system: Evidence for two distinct reservoirs. *Nature* **320**, 244-246.

Owen, T., J.P. Maillard, C. DeBergh, and B.L. Lutz 1988. Deuterium on Mars: The abundance of HDO and the value of D/H. *Science* **240**, 1767-1770.

Palme, H., and B. Fegley, Jr. 1990. High-temperature condensation of iron-rich olivine in the solar nebula. *Earth Planet. Sci. Lett.* **101**, 180-195.

Palme, H., and F. Wlotzka 1976. A metal particle from a Ca, Al-rich inclusion from the meteorite Allende, and the condensation of refractory siderophile elements. *Earth Planet. Sci. Lett.* **33**, 45-60.

Pilcher, C.B., S.T. Ridgway, and T.B. McCord 1972. Galilean satellites: Identification of water frost. *Science* **178**, 1087-1089.

Pillinger, C.T. 1984. Light element stable isotopes in meteorites – from grams to picograms. *Geochim. Cosmochim. Acta* **48**, 2739-2766.

Plescia, J.B. 1987. Cratering history of the Uranian satellites: Umbriel, Titania, and Oberon. *J. Geophys. Res.* **92**, 14918-14932.

Podosek, F.A. and T.D. Swindle 1988. Extinct radionuclides. in *Meteorites and the Early Solar System*, ed. J.F. Kerridge and M.S. Matthews, University of Arizona Press, Tucson, pp. 1093-1113.

Prinn, R.G. 1990. On neglect of non-linear momentum terms in solar nebula accretion disk models. *Astrophys. J.* **348**, 725-729.

Prinn, R.G., and M.B. Fegley, Jr. 1981. Kinetic inhibition of CO and N_2 reduction in circumplanetary nebulae: Implications for satellite composition. *Astrophys. J.* **249**, 308-317.

Prinn, R.G., and B. Fegley, Jr. 1989. Solar nebula chemistry: Origin of planetary, satellite, and cometary volatiles. in *The Origin and Evolution of Planetary and Satellite Atmospheres*, ed. S.K. Atreya, J.B. Pollack, and M.S. Matthews, University of Arizona Press, Tucson, pp. 78-136.

Rambaldi, E.R. and J.T. Wasson 1981. Metal and associated phases in Bishunpur, a highly unequilibrated ordinary chondrite. *Geochim. Cosmochim. Acta* **45**, 1001-1015.

Rambaldi, E.R. and J.T. Wasson 1984. Metal and associated phases in Krymka and Chainpur: Nebular formational processes. *Geochim. Cosmochim. Acta* **48**, 1885-1897.

Renshaw, G.D., C. Roscoe, and P.L. Walker, Jr. 1970. Disproportionation of CO I. Over iron and silicon-iron single crystals. *J. Catalysis* **18**, 164-183.

Reynolds, J.H. 1960. Determination of the age of the elements. *Phys. Rev. Lett.* **4**, 8.

Richardson, S.M. 1978. Vein formation in the CI carbonaceous chondrites. *Meteoritics* **13**, 141-159.

Richet, P., Y. Bottinga, and M. Javoy 1977. A review of hydrogen, carbon, nitrogen, oxygen, sulphur, and chlorine stable isotope fractionation among gaseous molecules. *Ann. Rev. Earth Planet. Sci.* **5**, 65-110.

Ruden, S.P. and D.N.C. Lin 1986. The global evolution of the solar nebula. *Astrophys. J.* **308**, 883-901.

Schloerb, F.P., R.L. Snell, W.D. Langer, and J.S. Young 1981. Detection of deteriocyanobutadiyne (DC_5N) in the interstellar cloud TMC-1. *Astrophys. J.* **251**, L37-L42.

Schloerb, F.P., W.M. Kinzel, D.A. Swade, and W.M. Irvine 1987. Observations of HCN in comet P/Halley. *Astron. Astrophys.* **187**, 475-480.

Sears, D.W. 1978. Condensation and the composition of iron meteorites. *Earth Planet. Sci. Lett.* **41**, 128-138.

Shock, E.L. and M.D. Schulte 1990a. Amino-acid synthesis in carbonaceous meteorites by aqueous alteration of polycyclic aromatic hydrocarbons. *Nature* **343**, 728-731.

Shock, E.L. and M.D. Schulte 1990b. Summary and implications of reported amino acid concentrations in the Murchison meteorite. *Geochim. Cosmochim. Acta* **54**, 3159-3173.

Shukolyukov, A. and G.W. Lugmair 1992. First evidence for live ^{60}Fe in the early solar system. *Lunar Planet. Sci.* **XXIII**, pp. 1295-1296.

Sill, G.T., and L.L. Wilkening 1978. Ice clathrate as a possible source of the atmospheres of the terrestrial planets. *Icarus* **33**, 13-22.

Simonelli, D.P., J.B. Pollack, C.P. McKay, R.T. Reynolds, and A.L. Summers 1989. The carbon budget in the outer solar nebula. *Icarus* **82**, 1-35.

Snyder. L.E., P. Palmer, and I. dePater 1989. Radio detection of formaldehyde emission from comet Halley. *Astron. J.* **97**, 246-253.

Stawikowski, A. and J.L. Greenstein 1964. The isotope ratio $^{12}C/^{13}C$ in a comet. *Astrophys. J.* **140**, 1280-1291.

Stevenson, D.J. 1990. Chemical heterogeneity and imperfect mixing in the solar nebula. *Astrophys. J.* **348**, 730-737.

Stevenson, D.J. and J.I. Lunine 1988. Rapid formation of Jupiter by diffusive redistribution of water vapor in the solar nebula. *Icarus* **75**, 146-155.

Stewart, A.I.F. 1987. Pioneer Venus measurements of H, O, and C production in comet P/Halley near perihelion. *Astron. Astrophys.* **187**, 369-374.

Strom, S.E., S. Edwards, and K.M. Strom 1989. Constraints on the properties and environment of primitive solar nebulae from the astrophysical record provided by young stellar objects. in *The Formation and Evolution of Planetary Systems*, ed. H.A. Weaver and L. Danly, Cambridge University Press, Cambridge, pp. 91-109.

Studier, M.H., R. Hayatsu, and E. Anders 1968. Origin of organic matter in early solar system - I. Hydrocarbons. *Geochim. Cosmochim. Acta 32* , 151-173.

Suess, H.E. 1947a. Über kosmische Kernhäufigkeiten. I. Mitteilung: Einige Häufigkeitsregeln und ihre Anwendung bei der Abschätzung der Häufigkeitswerte für die mittelschweren und schweren Elemente. *Z. Naturforsch.* **2a**, 311-321.

Suess, H.E. 1947b. Über kosmische Kernhäufigkeiten. II. Mitteilung: Einzelheiten in der Häufigkeitsverteilung der mittelschweren und schweren Kerne. *Z. Naturforsch.* **2a**, 604-608.

Suess, H.E. 1965. Chemical evidence bearing on the origin of the solar system. *Ann. Rev. Astron. Astrophys.* **3**, 217-234.

Suess, H.E., and H.C. Urey 1956. Abundances of the elements. *Rev. Mod. Phys.* **28**, 53-74.

Tang, M. and E. Anders 1988. Isotopic anomalies of Ne, Xe, and C in meteorites. II. Interstellar diamond and SiC: Carriers of exotic noble gases. *Geochim. Cosmochim. Acta* **52**, 1235-1244.

Tegler, S. and S. Wyckoff 1989. NH_2 fluorescence efficiencies and the NH_3 abundance in comet Halley. *Astrophys. J.* **343**, 445-449.

Thiemens, M.H. 1988. Heterogeneity in the nebula: Evidence from stable isotopes. in *Meteorites and the Early Solar System*, ed. J.F. Kerridge and M.S. Matthews, University of Arizona Press, Tucson, pp. 899-923.

Tielens, A.G.G.M. 1983. Surface chemistry of deuterated molecules. *Astron Astrophys.* **119**, 177-184.

Tielens, A.G.G.M. and L.J. Allamandola 1987. Evolution of interstellar dust. in *Physical Processes in Interstellar Clouds*, ed. G.E. Morfill and M. Scholer, D. Reidel, Netherlands, pp. 333-376.

Tomeoka, K. and P.R. Buseck 1985. Indicators of aqueous alteration in CM carbonaceous chondrites: Microtextures of a layered mineral containing Fe, S, O, and Ni. *Geochim. Cosmochim. Acta* **49**, 2149-2163.

Turkdogan, E.T., W.M. McKewan, and L. Zwell 1965. Rate of oxidation of iron to wüstite in water-hydrogen gas mixtures. *J. Phys. Chem.* **69**, 327-334.

Urey, H.C. 1952. *The Planets,* New Haven: Yale University Press.

Urey, H.C. 1953. Chemical evidence regarding the Earth's origin. in *XIIIth International Congress Pure and Applied Chemistry and Plenary Lecture,* Almqvist & Wiksells, Stockholm, pp. 188-217.

Vannice, M.A. 1975. The catalytic synthesis of hydrocarbons from H_2/CO mixtures over the group VIII metals. *J. Catal.* **37**, 449-461.

Vannice, M.A. 1982. Catalytic activation of carbon monoxide on metal surfaces. in *Catalysis Science and Technology*, vol. 3, eds. J.R. Anderson and M. Boudart, Springer-Verlag, Berlin, pp. 139-198.

Vanýsek, V. 1977. Carbon isotope ratio in comets and interstellar medium. in Comets, Asteroids, and Meteorites: Interrelations, Evolution, and Origins, ed. A.H. Delsemme, University of Toledo Press, Toledo, OH, pp. 499-503.

Virag, A., B. Wopenka, S. Amari, E. Zinner, E. Anders, and R.S. Lewis 1992. Isotopic, optical, and trace element properties of large single SiC grains from the Murchison meteorite. *Geochim. Cosmochim. Acta* **56**, 1715-1733.

Wai, C.M., and J.T. Wasson 1977. Nebular condensation of moderately volatile elements and their abundances in ordinary chondrites. *Earth Planet. Sci. Lett.* **36**, 1-13.

Wai, C.M., and J.T. Wasson 1979. Nebular condensation of Ga, Ge and Sb and the chemical classification of iron meteorites. *Nature* **282**, 790-793.

Walmsley, C.M., W. Hermsen, C. Henkel, R. Mauersberger, and T.L. Wilson 1987. Deuterated ammonia in the Orion hot core. *Astron. Astrophys.* **172**, 311-315.

Warnatz, J. 1984. Rate coefficients in the C/H/O system. in *Combustion Chemistry*, ed. W.C. Gardiner, Jr., Springer-Verlag, New York, pp. 197-360.

Wasserburg, G.J. 1985. Short-lived nuclei in the early solar system. in *Protostars and Planets II*, eds., D.C. Black and M.S. Matthews, University of Arizona Press, Tucson, pp. 703-754.

Wasson, J.T. 1985. *Meteorites*. New York: W.H. Freeman and Co.

Weaver, H.A. 1989. The volatile composition of comets. in *Highlights of Astronomy* **8**, 387-393.

Weaver, H.A., M.J. Mumma, and H.P. Larson 1991. Infrared spectroscopy of cometary parent molecules. in Comets in the Post-Halley Era, ed. R. Newburn and J. Rahe, Kluwer Academic Publishers, Dordrecht, Netherlands.

Whipple, F.L. 1966. Chondrules: Suggestions concerning their origin. *Science* **153**, 54-56.

Wood, J.A., and G.E. Morfill 1988. A review of solar nebula models. in *Meteorites and the Early Solar System*, ed. J.F. Kerridge and M.S. Matthews, University of Arizona Press, Tucson, pp. 329-347.

Woods, T.N., P.D. Feldman, K.F. Dymond, and D.J. Sahnow 1986. Rocket ultraviolet spectroscopy of comet Halley and abundance of carbon monoxide and carbon. *Nature* **324**, 436-438.

Woolum, D.S. 1988. Solar-system abundances and processes of nucleosynthesis. in *Meteorites and the Early Solar System*, ed. J.F. Kerridge and M.S. Matthews, University of Arizona Press, Tucson, pp. 995-1020.

Wootten, A. 1987. Deuterated molecules in interstellar clouds. in *Astrochemistry*, eds. M.S. Vardya and S.P. Tarafdar, D. Reidel, Dordrecht, Netherlands, pp. 311-320.

Worrell, W.L. and E.T. Turkdogan 1968. Iron-sulfur system, Part II: Rate of reaction of hydrogen sulfide with ferrous sulfide. *Trans. AIME* **242**, 1673-1678.

Wyckoff, S., E. Lindholm, P.A. Wehinger, B.A. Peterson, J.M. Zucconi, and M.C. Festou 1989. The $^{12}C/^{13}C$ abundance ratio in comet Halley. *Astrophys. J.* **339**, 488-500.

Wyckoff, S. and J. Theobald 1989. Molecular ions in comets. *Adv. Space Res.* **9**(3), 157-161.

Wyckoff, S., S. Tegler, and L. Engel 1989. Ammonia abundances in comets. *Adv. Space Res.* **9**(3), 169-176.

Wyckoff, S., S. Tegler, and L. Engel 1991a. Ammonia abundances in four comets. *Astrophys. J.* **368**, 279-286.

Wyckoff, S., S. Tegler, and L. Engel 1991b. Nitrogen abundance in comet Halley. *Astrophys. J.* **367**, 641-648.

Wyckoff, S., S. Tegler, P.A. Wehinger, H. Spinrad, and M.J.S. Belton 1988. Abundances in comet Halley at the time of the spacecraft encounters. *Astrophys. J.* **325**, 927-938.

Yang, J. and S. Epstein 1983. Interstellar organic matter in meteorites. *Geochim. Cosmochim. Acta* **47**, 2199-2216.

Yung, Y.L., R.R. Friedl, J.P. Pinto, K.D. Bayes, and J.S. Wen 1988. Kinetic isotopic fractionation and the origin of HDO and CH_3D in the solar system. *Icarus* **74**, 121-132.

Zinner, E. 1988. Interstellar cloud material in meteorites. in *Meteorites and the Early Solar System*, ed. J.F. Kerridge and M.S. Matthews, University of Arizona Press, Tucson, pp. 956-983.

Zinner, E., M. Tang, and E. Anders 1987. Large isotopic anomalies of Si, C, N, and noble gases in interstellar silicon carbide from the Murray meteorite. *Nature* **330**, 730-732.

EARLY EVOLUTION OF THE ATMOSPHERE AND OCEAN

James F. Kasting
Department of Geosciences
211 Deike
The Pennsylvania State University
University Park, PA 16802
USA

ABSTRACT. The early evolution of the atmosphere and oceans is discussed with particular emphasis on factors relevant to the origin of life. Both the atmosphere and ocean formed early as a consequence of impact degassing of planetesimals during accretion. The post-accretionary atmosphere was probably denser than the present atmosphere and was dominated by carbon compounds, principally CO_2 and CO. The greenhouse effect of this atmosphere could have kept the early Earth significantly warmer than today despite reduced solar luminosity at that time. The atmosphere is believed to have been weakly reducing; that is, it contained tens to hundreds of parts per million of H_2 and very little free O_2. Highly reduced gases such as methane and ammonia are generally considered to have been present in only minute quantities. However, several different lines of evidence, including Mars' climate history, theoretical factors affecting hydrogen escape, and new evidence bearing on the early mantle redox state, all indicate that reduced gases may have been more plentiful than is usually assumed.

1. INTRODUCTION

One of the principal difficulties in understanding how life originated is that we have little firm evidence concerning the conditions on the early Earth. If life originated in the near-surface environment, understanding that process requires some idea of the Earth's surface temperature and the oxidation state of its atmosphere. If life originated in

J. M. Greenberg et al. (eds.), The Chemistry of Life's Origins, 149–176.

hydrothermal vents, one needs to know something about the chemistry of the ocean and the interaction of seawater with the oceanic crust. Even if life originated in space and was brought to the Earth by asteroids or comets, it would be useful to know the density of the atmosphere (to determine whether incoming objects could be effectively decelerated) along with something about the conditions that such 'panspermic' organisms would have encountered once they arrived. Since the geologic record is extremely sparse prior to the time that life is thought to have arisen, >3.5 billion years (b.y.) ago (Schopf and Packer, 1987), most of what we can learn about the early Earth must be ascertained by indirect methods. One method that yields some useful constraints on the problem is computer modeling of atmospheric composition and climate. This chapter describes some of the results predicted by such models.

2. FORMATION OF THE ATMOSPHERE AND OCEAN

The Earth's atmosphere and oceans are composed of volatile elements and compounds, principally nitrogen, oxygen, water, and carbon dioxide. The nitrogen in the atmosphere (~0.8 bar) and the water in the oceans (~1.4×10^{21} l) probably represent the bulk of the surface inventories of these compounds (Holland, 1978). The carbon dioxide in the atmosphere (3×10^{-4} bar) represents only a small fraction of the surface carbon inventory. Most of the Earth's carbon, some 1×10^{23}g C, is locked up in the crust as carbonate rocks (ibid.). If present in the atmosphere as CO_2, this carbon would produce a CO_2 partial pressure of 60 to 80 bars. The water in the oceans, if vaporized, would produce an H_2O partial pressure of ~270 bars. Thus, water and carbon dioxide are quantitatively the most important volatile compounds on our planet. Molecular oxygen (O_2) is an important atmospheric volatile today but is unlikely to to have been so in the past since it is almost exclusively a product of photosynthesis.

All of these volatile compounds were brought to the Earth as a component of solid materials that accreted from the solar nebula. We can be fairly certain that that they were not captured gravitationally (as gases) by the young Earth because this process would have also brought in more nonradiogenic rare gases (e.g. ^{20}Ne, ^{36}Ar, ^{84}Kr) than are found in our present atmosphere (Holland, 1984). (Once incorporated into the planet, rare gases would have been difficult to remove without removing most of the other volatile elements at the same time.) Two types of solid bodies thought to be representative of primitive, volatile-rich, solar system materials are carbonaceous chondrites and comets. Carbonaceous chondrites probably originate from asteroids in the region between Mars and Jupiter (Lewis and Prinn, 1984). Comets originate now from the Oort cloud far beyond the orbit of Pluto; however, they are thought to have originally formed in the vicinity of Uranus and Neptune (Weissman, 1985). Both types of bodies contain carbon and nitrogen in the form of organic compounds. Water is bound up primarily as water ice in comets and as hydrated silicate minerals in carbonaceous chondrites. The more common ordinary chondrites also contain volatiles, but in far smaller amounts. Presumably, this is because they originated closer to the Sun where nebular temperatures were higher. An Earth composed entirely of ordinary chondritic material would nevertheless contain sufficient volatiles to account for the observed abundances (Lewis and Prinn, 1984).

Where the volatile-rich bodies originate is relevant to the origin of life because it affects the timing of formation of the atmosphere and ocean. If the Earth is composed largely of ordinary chondritic material that condensed near 1 AU, then volatiles would have been accreted simultaneously with the planet. As long as the planet was small, most of these volatiles would have remained trapped within the solid material. However, once the radius of the growing Earth reached approximately 40 percent of its present value, the shock of impact onto the surface should have caused extensive degassing, causing most of the volatiles to be released directly into the atmosphere (Lange and Ahrens, 1982). The gravitational potential energy released by the infalling bodies would have prevented any water from condensing at the Earth's surface (Matsui and Abe, 1986; Zahnle et al., 1988), so the atmosphere would have consisted of superheated steam along with smaller amounts of other volatile compounds. Once the main accretion period was over, ~100 million years (m.y.) after it had begun or 4.5 b.y. before present (Wetherill, 1991), most of the water vapor would have condensed out to form an ocean. The remaining atmosphere would have consisted of a warm, dense mixture of CO_2, CO, N_2, and some still appreciable quantity of H_2O (Holland, 1984; Walker, 1985; Kasting, 1990).

This outline of atmospheric formation is complicated, however, by the likely occurence of a giant, Moon-forming impact at some late stage in the accretion process (Benz et al., 1986). Formation of the Moon appears to require that the Earth collide obliquely with a body about the same size as Mars (~1/10 of an Earth mass). An impact of this magnitude would probably have stripped off most or all of the volatiles present at the Earth's surface. The atmosphere and ocean would then need to have been resupplied either by subsequent impacts or by outgassing from the interior. Bodies that impacted the Earth during the latter stages of the accretion process are likely to have originated from farther out in the solar nebula and ought therefore to have been more volatile rich and more oxidized. (Longer orbital periods for the more distant planetesimals would have extended their accretion time. The higher oxidation state is a result of the incorporation of oxygen in the form of water ice.) The early Earth may therefore have been 'veneered' by oxidized, volatile-rich planetesimals similar in composition to carbonaceous chondrites or comets (Dreibus and Wanke, 1989). Much of the Earth's present surface volatile inventory may have been brought in at this time. Judging from the lunar impact record this accretionary tail lasted until at least 3.8 b.y. ago, at which time the impact flux declined dramatically (Wilhelms, 1989). Some fraction of the impactors during this 'heavy bombardment' period may have been comets, perturbed from their original orbits by the gravitational influence of the giant planets. Chyba (1987) has estimated that, if even 10 percent of the late impactors were comets, they could have supplied all the water now present in the ocean.

At the same time that volatiles were being imported from space, they should also have been exchanged between the Earth's surface and its interior. The conventional view for the origin of the atmosphere (cf. Holland, 1984) is that it was mostly supplied by volcanic outgassing. The preceding discussion should make it clear that this is, at best, only part of the story. Indeed, when one considers the efficiency of impact degassing, it seems more likely that most of the volatiles were emplaced directly at the surface and that the mantle initially ingassed volatile compounds. Ingassing of water presently occurs when seafloor that has been hydrated by hydrothermal circulation at the midocean ridges is subducted back into the Earth's interior (Schubert et al., 1989). The efficiency of this recycling process is uncertain because one does not known how much of the water is driven off during the descent of the oceanic slab. However, evidence from deep-focus earthquakes (Meade and Jeanloz, 1991) indicates that some of the water makes it down to at least 300 km depth. Carbon may be ingassed by a similar process involving carbonation of seafloor at midocean ridge spreading centers (Staudigal et al., 1989; Francois and Walker, 1992) or by subduction of carbonate-rich sediments. Whether similar ingassing processes occurred on the early Earth is uncertain because plate tectonics may not have operated in the same manner as today. It nevertheless seems plausible that volatiles could have been transported from the surface to the interior until a steady state was achieved in which the ingassing rate was balanced by volcanic outgassing (McGovern and Schubert, 1989). Kasting and Holm (1992) have suggested that this exchange process constitutes a dynamical control mechanism that maintains a constant depth of water (~2.5 km) above the fast-spreading midocean ridges. If this hypothesis is correct, then the oceans may have been close to their present depth for a very long time.

To summarize then, the atmosphere and ocean were formed early in Earth's history by impact degassing of volatile-rich planetesimals. The earliest atmosphere was probably denser than today's atmosphere because the amount of volatiles delivered by impacts is related to the total volatile inventory of the planet, which is larger (especially for carbon) than the atmospheric inventory. Significant quantities of volatiles may have been added throughout the heavy bombardment period, so the Earth's volatile inventory should have continued to grow until ~3.8 b.y. ago. Exchange of water and carbon between the crust and the mantle gradually established the steady-state abundances of these compounds at the Earth's surface.

3. CLIMATE OF THE EARLY EARTH

The climate of the early Earth or, to be more specific, its mean surface temperature, would have been determined by the radiative properties of its atmosphere and the illumination provided by the young Sun. Of these two factors, the luminosity history of the Sun is the easier to estimate. Standard theoretical solar evolution models (e.g. Gough, 1981; Gilliland, 1989) predict that the Sun was ~30 percent less luminous than today at the time when it entered the main sequence, ~4.6 b.y. ago, and that it has increased in brightness more or less linearly since that time. Gough (1981) (Fig. 1) approximates the luminosity increase by the formula

$$\frac{S}{S_0} = \frac{1}{1 + 0.4\left(\frac{t}{t_0}\right)} \quad (1)$$

where S_o is the present solar flux at Earth's orbit (= 1365 W m^{-2}), t is time before present, and t_o (= 4.6 b.y. ago) is when the Sun first entered the main sequence.

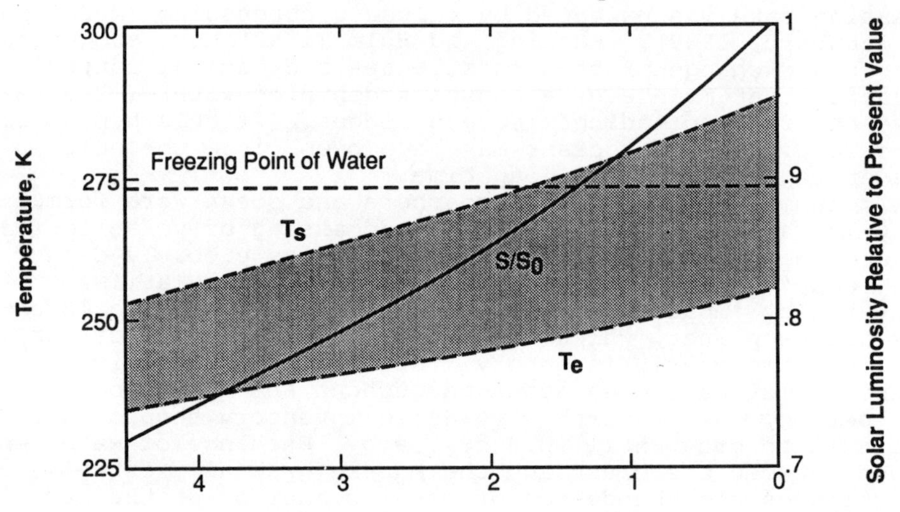

Figure 1. Diagram illustrating the faint young sun problem. The solid curve (right-hand scale) represents solar luminosity as given by equation (1). The dashed curves represent the Earth's effective radiating temperature, T_e, and its mean surface temperature, T_s. The shaded area between these two curves shows the magnitude of the greenhouse effect. The calculation was performed using a one-dimensional, radiative-convective climate model, assuming present-day CO_2 levels and fixed relative humidity. (From Kasting, 1989).

The cause for the luminosity increase in these models is the gradual conversion of hydrogen to helium by nuclear fusion. As helium accumulates, the mean molecular weight of the Sun increases, along with its density. The higher density results in a larger self-gravity; this, in turn, requires higher central pressures and temperatures to keep the core of the Sun from collapsing. Higher core temperatures increase the rate of nuclear fusion; consequently, the luminosity must increase to carry off the extra energy.

The 'faint young sun,' as it is usually termed, is a fairly robust prediction of stellar evolution theory (Newman and Rood, 1977) but it is not entirely unavoidable. One way of changing this prediction would be if the Sun has lost appreciable amounts of mass during its main sequence lifetime. (The current rate of mass loss by the solar wind, $\sim 10^{-14}$ M_o/yr, is three or four orders of magnitude too small to be significant.) A more massive early Sun would have a hotter core temperature and, hence, a higher luminosity than the standard model. Such mass-losing, nonstandard, solar evolution models have been proposed twice in the recent literature (Willson et al., 1987; Boothroyd et al., 1991). The latter authors suggest that this would be one way to explain the observed lithium depletion in the Sun. To lose mass, the young Sun would have had to either pulsate energetically (Willson et al., 1987) or had extremely high flare activity. So far, no supporting observational evidence for appreciable mass loss in other young, solar-type stars has appeared. Until it does, one can probably ignore these nonstandard solar models, bearing in mind the fact that the climate history of the planets would be quite different if they proved to be correct (Graedel et al., 1991).

The mean surface temperature of the Earth, T_s, can be formally expressed by the two equations

$$\sigma T_e^4 = \frac{S}{4}(1 - A) \tag{2}$$

$$T_s = T_e + \Delta T \tag{3}$$

Here, T_e is the effective radiating temperature of the Earth (or, equivalently, the mean surface temperature in the absence of an atmosphere); S is the solar flux (eq. 1); σ is the Stefan-Boltzmann constant (5.67×10^{-8} W m^{-2}); A is the planetary albedo, or reflectivity; and ΔT represents the magnitude of the atmospheric greenhouse effect. Plugging in the current observed values of A (~ 0.3) and T_s (288 K) yields T_e = 255 K and ΔT = 33 degrees for the modern Earth.

The mean surface temperature of an early Earth with S ζ 0.7(1365 W m^{-2}) would obviously have been less than today unless either the planetary albedo was smaller in the past or the greenhouse effect was larger. Indeed, as shown by Figure 1, T_s would have been below the freezing point of water prior to ~ 2 b.y. ago under these circumstances. The apparent conflict between this prediction and the geological evidence for liquid water as far back as 3.8 b.y. ago has been termed the 'faint young Sun paradox' (Sagan and Mullen, 1972; Kasting, 1987, 1989; Kasting and Grinspoon, 1991). Since it has been extensively discussed elsewhere, I will say little about it here except to point out that it is not a paradox if one treats atmospheric CO_2 concentration as a variable quantity that is controlled by the carbonate-silicate cycle. Such a treatment leads naturally to the prediction that atmospheric CO_2 levels would have been high enough in the past to prevent the oceans from freezing.

One can actually go a step beyond this and argue that the early Earth was probably warmer than today, despite the low solar luminosity. The reason is that most of the Earth's CO_2 was deposited directly into its atmosphere, as discussed earlier, so the greenhouse effect should originally have been very large. Holland (1984) estimates that ~1/3 of the Earth's CO_2, or 20 bars, could have been present in the atmosphere at the end of accretion. This dense CO_2 atmosphere could have persisted throughout the heavy bombardment period. It would have been particularly easy to maintain if the continents were originally small or nonexistent, as predicted by some crustal evolution models (refs. in Walker, 1985). Under these conditions, the rate of silicate weathering (the long-term loss process for CO_2) would have been small and CO_2 would have been preferentially partitioned into the atmosphere and ocean instead of being locked up in carbonate sediments. A steady-state CO_2 partial pressure of ~10 bars is theoretically possible even without considering the addition of carbon by impacts (ibid.). The mean surface temperature of such an atmosphere, given 70 percent present solar luminosity, should have been of the order of 85°C (Fig. 2). Although this is very warm compared to today (T_s = 15°C), such temperatures cannot be ruled out by either the geologic record or by considerations of atmospheric stability. (Kasting and Ackerman (1986) showed that such conditions would not lead to boiling of the oceans or to rapid loss of water to space.) Oxygen isotope ratios from cherts (Knauth and Epstein, 1976) and chert-phosphate pairs (Karhu and Epstein, 1986) actually support high ocean temperatures throughout most of the Precambrian, i.e. prior to ~600 m.y. ago. This chemical evidence is difficult to accept, however, because it

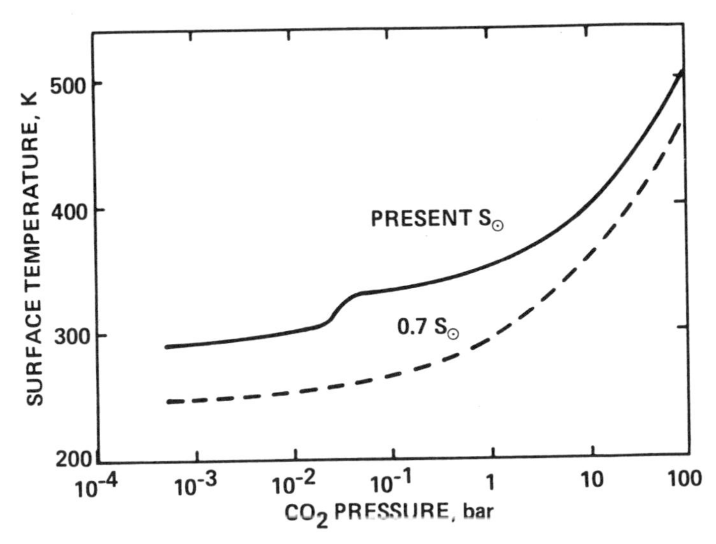

Figure 2. Mean surface temperature of the Earth as a function of atmospheric CO_2 partial pressure for two different values of solar luminosity. The dashed curve corresponds to expected conditions very early in Earth history. (From Kasting and Ackerman, 1986).

is inconsistent with physical evidence for continental glaciation around 2.3 b.y. ago and again between 800 and 650 m.y. ago (Frakes, 1979). One is tempted to conclude that the high isotopic temperatures reflect conditions within buried sediments, not in the ocean.

One recent development that may have implications for the climate of early Earth is a new simulation of the climate of early Mars (Kasting, 1991). Previous calculations (e.g. Pollack et al., 1987) had indicated that early Mars, like early Earth, could have been warmed by the greenhouse effect of a dense, CO_2-H_2O atmosphere. This model was put forward as one explanation for the presence of channels on the ancient, heavily-cratered parts of the Martian surface. (Another explanation is that the channels were formed by sapping of subsurface groundwater on a generally cold, early Mars.) The new climate calculations showed, however, that a CO_2-H_2O greenhouse was incapable of warming early Mars (Fig. 3) because the CO_2 would have condensed to form clouds. These clouds should have cooled the surface by reflecting incident

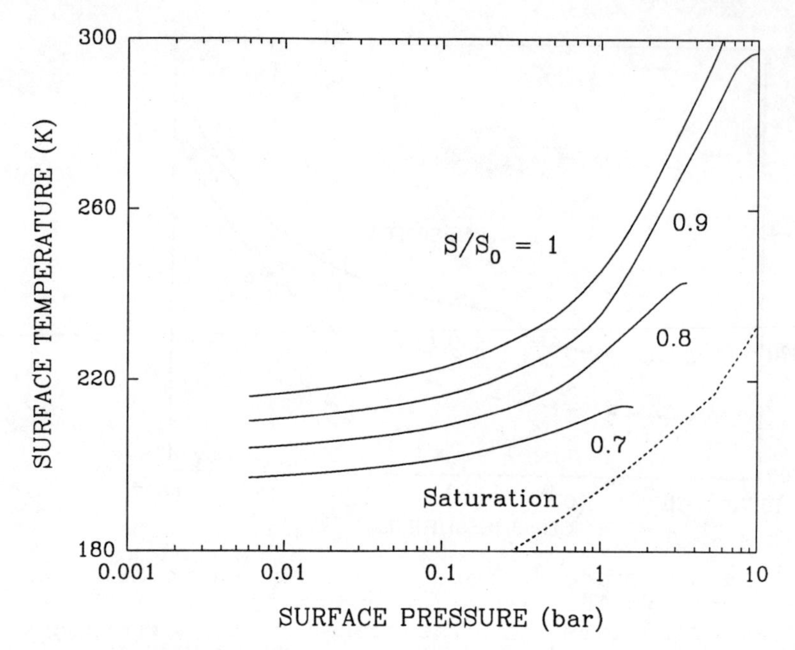

Figure 3. Mean surface temperature of the planet Mars as a function of atmospheric CO_2 partial pressure for several different values of solar luminosity. For $S/S_o < 0.85$, it is impossible to raise the mean temperature above the freezing point of water because of the formation of CO_2 clouds in the upper troposphere. (From Kasting, 1991).

sunlight and by reducing the convective lapse rate in the region where they formed. (Water clouds have a similar effect on Earth: the release of latent heat during droplet condensation causes the rate at which temperature declines with altitude to be less than predicted for a dry atmosphere.)

There are several possible solutions to this new early Mars climate problem. One, as just mentioned, is that the planet may not have ever been very warm. I personally find this difficult to accept because it requires an efficient mechanism for recharging groundwater reservoirs in the absence of precipitation. Geothermally-induced groundwater circulation may work in certain localized areas but seems implausible as the driving force for all of the widespread valley networks. A second possibility, discussed earlier, is that the young Sun was not as faint as has been previously assumed. The most likely solution, however, is that the atmosphere of early Mars contained other greenhouse gases in addition to CO_2 and H_2O. Ammonia (NH_3) was suggested originally by Sagan and Mullen (1972); methane (CH_4) is another possibility. Both of these gases are photochemically unstable: ammonia converts to N_2 and H_2 and methane to CO_2 and H_2 (Kuhn and Atreya, 1979; Kasting et al., 1983). If these gases were to have been present in concentrations of parts per million (ppm) or greater in the early Martian atmosphere, they would have required continuous sources and they would probably need to have been shielded from direct UV photolysis. Whether these two conditions could have been satisfied is a question that deserves further study. If it could be demonstrated that these highly reduced gases were present in the atmosphere of early Mars, it would require only a small logical step to suggest that they were present in Earth's early atmosphere as well. Thus, the problem of explaining the Earth's early climate may require consideration of atmospheric redox state. The next section describes another reason why redox state is important.

4. THE WEAKLY REDUCING ATMOSPHERE AND THE ORIGIN OF LIFE

Current chemical models for the early terrestrial atmosphere (e.g., Walker, 1977; Levine, 1982; Kasting, 1987, 1990) suggest that it consisted of a weakly reducing mixture of N_2, CO_2, and H_2O with lesser amounts of H_2, CO, SO_2, and H_2S. CH_4 and NH_3 are generally considered to have been absent, or else present in only minute concentrations, for reasons given above. Free O_2 is modestly abundant in the stratosphere, where it is produced from CO_2 photolysis, but is virtually nonexistent near the Earth's surface (Fig. 4).

Although this general model for the primitive atmosphere is widely accepted by photochemical modelers (those, at least, who have thought about the problem), it is not very popular with many origin of life researchers. The reason, as demonstrated by Stribling and Miller (1987), is that it is a relatively poor medium in which to form the complex organic molecules required for the origin of life. The one biological precursor molecule that can be readily formed in such an atmosphere is formaldehyde (H_2CO). The formation process begins with photolysis of H_2O and CO_2 and proceeds by way of

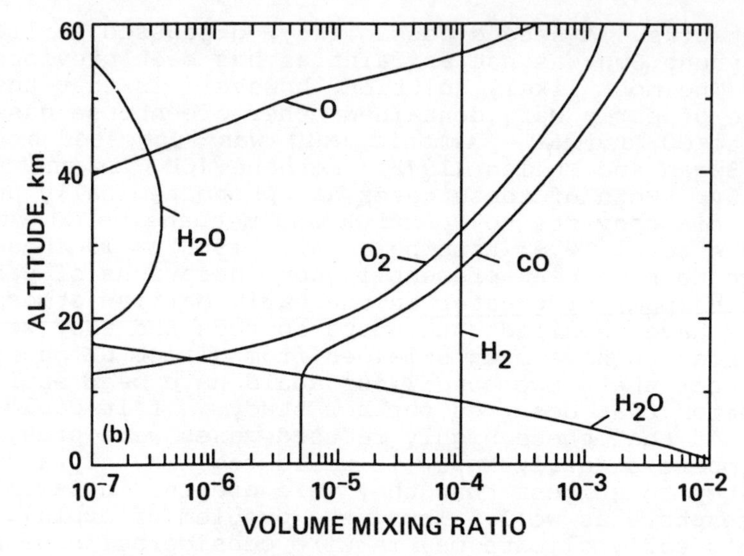

Figure 4. Vertical profiles of major atmospheric constituents in a 'standard,' weakly reducing primitive atmosphere. The atmosphere is assumed to consist of 0.8 bars of N_2 and 0.2 bars of CO_2. Other species are calculated using a one-dimensional photochemical model. (From Kasting, 1990).

reactions involving the formyl radical (HCO):

$$H_2O + h\nu \rightarrow H + OH$$
$$CO_2 + h\nu \rightarrow CO + O$$
$$H + CO + M \rightarrow HCO + M$$
$$HCO + HCO \rightarrow H_2CO + CO$$

Since formaldehyde is a basic building block for both sugars and amino acids, it is essential to have a good supply of this molecule, especially if one believes that life originated in the near-surface environment without any help from organic molecules imported from space.

Another molecule that is needed as a precursor to amino acids is hydrogen cyanide (HCN). (Alternatively, one might say that it is necessary to form C-N bonds. HCN is the simplest molecule that contains such a bond.) It is in synthesizing HCN that the weakly reducing primitive atmosphere runs into difficulty. If one starts from CO and N_2 as the initial carbon- and nitrogen-containing molecules (the CO being itself produced from photolysis of CO_2), it is necessary to break the strong C≡O and N≡N triple bonds to form HCN. This can be done with short wavelength (λ < 900 nm) ultraviolet radiation, but such photons are few in number and are absorbed high up in the atmosphere. Consequently, little HCN production is expected from photolysis of CO and N_2. Alternatively, the breaking of C≡O and N≡N bonds could occur in the high-temperature plasma surrounding lightning discharges. Formation of products in such a process can be approximated by assuming thermodynamic equilibrium at high temperatures (2000 to 3000 K) between the different molecules. For example, shock heating of CO_2 and N_2 should result in formation of at least four other molecules by way of the reversible reactions

$$CO_2 + \tfrac{1}{2} N_2 \rightleftharpoons CN + O_2 \qquad \text{(i)}$$
$$CO_2 + \tfrac{1}{2} N_2 \rightleftharpoons CO + NO \qquad \text{(ii)}$$

Both theory (Chameides and Walker, 1981) and experiment (Stribling and Miller, 1987), however, have shown that reaction (ii) above goes much farther toward the right than does reaction (i); that is, nitrogen is fixed overwhelmingly as NO rather than CN when the starting materials are N_2 and CO_2. Significant amounts of CN are formed only when the starting C/O ratio exceeds unity as it would, for example, in an N_2-CH_4 atmosphere.

A second pathway for forming HCN in the primitive atmosphere has been suggested by Zahnle (1986). His method relies on nitrogen atoms from the upper atmosphere that flow downward and react with the byproducts of methane photolysis in the lower atmosphere. This method has some advantage over the lightning synthesis in that it requires only small amounts of methane (ppm or greater), but it still fails to occur in the 'standard' weakly reducing atmosphere.

Thus, there are at least two reasons to suspect that something is wrong with the standard primitive atmosphere model: 1) it doesn't satisfy some researchers' demands for the origin of life, and 2) it's extension to early Mars leads to problems in explaining Mars' early climate. Both of these problems have alternative solutions, some of which have already been mentioned. In the spirit of open-mindedness, however, let us take a look at the assumptions that go into the standard model to see which, if any, might deserve to be reexamined.

5. CONTROLS ON ATMOSPHERIC REDOX STATE

The basic approach to calculating the redox balance of the primitive atmosphere was first formulated by Walker (1977). (Earlier attempts to do this (e.g. Berkner and Marshall, 1965; Brinkman, 1969) failed because the investigators did not have a sufficient understanding of the processes controlling hydrogen escape.) Walker realized that the problem amounted to balancing the hydrogen budget for the early atmosphere. The source of hydrogen in his model was assumed to be volcanic outgassing. Using figures from Holland (1978, p. 292), which I consider to be more reliable than Walker's original estimates, the present volcanic release rate of H_2 is 1.1 x 10^{12} moles/yr, or 4 x 10^9 H_2 molecules cm^{-2} s^{-1}. (Henceforth, I will omit the word 'molecules' and express rates in units of 'cm^{-2} s^{-1}'.) The sink for hydrogen in Walker's model was escape to space. Escape of hydrogen from the present Earth is limited by the rate at which it can diffuse upwards through the 100-km level (the homopause) (Hunten, 1973; Walker, 1977). The diffusion rate, in turn, is directly proportional to the total hydrogen mixing in the stratosphere, i.e. the sum of the mixing ratios, f_i, of all hydrogen-bearing molecules weighted by the number of hydrogen atoms they contain. Mathematically, one can express the escape rate as

$$\Phi_{esc}(H_2) \simeq 2.5 \times 10^{13} f_T(H_2) \quad cm^{-2} s^{-1} \tag{4}$$

$$f_T(H_2) \equiv f(H_2) + f(H_2O) + 2 f(CH_4) + \ldots \tag{5}$$

Here, all rates are written in terms of H_2 molecules. The same formulas apply if one deals with H atoms, except that the weighting factors in equation (5) differ by a factor of two. The term 'mixing ratio' means the same as 'mole fraction', that is, the number density of species i divided by the number density of the atmosphere. The constant in equation (4) is approximate; it is derived by assuming a specific ratio of H atoms to H_2 molecules at the 100-km level (the homopause). It's actual value could vary from ~2 x 10^{13} for pure H_2 to ~4 x 10^{13} for pure atomic H.

If one now assumes that the H_2 outgassing rate on the early Earth was the same as today, one can estimate $f_T(H_2)$ by equating the escape rate with the outgassing rate. This yields $f_T(H_2)$ ~ 1.6 x 10^{-4}, which is the same as a concentration of 160 ppm. For the standard, weakly reducing atmosphere the only hydrogen-containing species that one expects to find in the stratosphere are H_2 and H_2O. The concentration of H_2O in the stratosphere is limited by condensation at the tropopause cold trap (the cold region at the top of the convective lower atmosphere); its value for the current Earth is ~3 to 5 ppm. Since the ozone-free, CO_2-rich stratosphere of the primitive Earth was probably colder and drier than today (Kasting and Ackerman, 1986), the bulk of the 160 ppm of hydrogen-containing gases would have to have been in the form of H_2. Photochemical modelling shows that this H_2 should have a relatively long chemical lifetime and would therefore be well-mixed; thus, its surface concentration would also be close to 160 ppm. Changing this model slightly by allowing for a higher volcanic outgassing rate can produce the species mixing ratios shown in Figure 4.

In the course of creating computer models of primitive atmospheres over the past several years, I have come to realize that it is necessary to generalize Walker's approach by considering the influence of other gases on the hydrogen budget. One reason is that outgassing of another reduced gas can create hydrogen by reacting with H_2O -- not directly, but through various UV-catalyzed photochemical mechanisms. For example, outgassing of carbon monoxide leads to H_2 production by way of the reaction sequence

$$
\begin{array}{lrcl}
 & H_2O + h\nu & \rightarrow & H + OH \\
 & CO + OH & \rightarrow & CO_2 + H \\
 & \underline{H + H + M} & \underline{\rightarrow} & \underline{H_2 + M} \\
\text{Net:} & CO + H_2O & \rightarrow & CO_2 + H_2
\end{array}
$$

If the resulting CO_2 molecule dissolves in the ocean, then the atmosphere has gained one H_2 molecule. Similarly, formation of a reduced species in the atmosphere from a more oxidized one can result in a drain on atmospheric hydrogen. Thus, when formaldehyde is formed from CO_2

$$
\begin{array}{lrcll}
 & CO_2 + h\nu & \rightarrow & CO + O & (\times 2) \\
 & H_2 + O & \rightarrow & H + OH & (\times 2) \\
 & CO + OH & \rightarrow & CO_2 + H & \\
 & H + OH + M & \rightarrow & H_2O + M & \\
 & H + CO + M & \rightarrow & HCO + M & (\times 2) \\
 & \underline{HCO + HCO} & \underline{\rightarrow} & \underline{H_2CO + CO} & \\
\text{Net:} & CO_2 + 2\,H_2 & \rightarrow & H_2CO + H_2O &
\end{array}
$$

the result is the consumption of two H_2 molecules. This formaldehyde molecule must be removed from the atmosphere by rainout or surface deposition in order to constitute a net loss for atmospheric H_2. The formaldehyde formation pathway shown here is more complicated than the one shown in Section 4 because it is necessary to balance production and loss of the intermediate species. Clearly, the specific reaction pathway assumed does not affect the net change in oxidation state.

What one really needs to balance is not the hydrogen budget but the electron budget. We can cast this in terms of hydrogen molecules, though, by defining 'neutral' oxidation states for various elements and then expressing changes in redox state by reactions involving H_2 and H_2O (as in the 'net' reactions above). Convenient neutral oxidation species for the important volatile elements are H_2O, CO_2, N_2, and SO_2. With these definitions, we can write the atmospheric redox balance equation as

$$\begin{aligned} \Phi_{out}(H_2) + \Phi_{out}(CO) + 3\Phi_{out}(H_2S) + \Phi_R(Ox) \\ = \Phi_{esc}(H_2) + \Phi_R(Red) \end{aligned} \tag{6}$$

Here, $\Phi_{out}(i)$ represents the outgassing rate of species i; the terms labelled $\Phi_R(i)$ represent the removal rates of soluble oxidized and reduced gases by rainout and surface deposition weighted by their effect on the redox balance. For the species included in the Kasting (1990) photochemical model, the most important contributions are

$$\Phi_R(Ox) = \Phi_R(H_2O_2) + \Phi_R(H_2SO_4) + \frac{1}{2}\Phi_R(HNO) + \ldots \tag{7}$$

$$\Phi_R(Red) = 2\,\Phi_R(H_2CO) + 3\,\Phi_R(H_2S) + 16\,\Phi_R(S_8) + \ldots \tag{8}$$

The rainout and surface deposition rates of individual species are calculated from parameterizations of these processes in an atmospheric photochemical model. Rainout rates, along with other important photochemical parameters such as species photolysis rates, depend on the density and surface temperature of the atmosphere. As demonstrated by Kasting (1990), the differences in redox state between a 1-bar, N_2-CO_2 atmosphere and an 11-bar atmosphere can be substantial even when the assumed boundary conditions are exactly the same. It is therefore necessary to consider the climatic factors discussed in Section 3 in order to solve for the atmospheric redox balance.

What this analysis implies about the validity of the standard, weakly reducing atmosphere model can be summarized by two statements: 1) The concentration of highly reduced gases could have been large only if they were protected from photolysis and if either the surface source of reduced gases was much higher or the rate of hydrogen escape was much lower than typically assumed. Otherwise, the atmospheric redox budget would be badly out of balance. (This statement does not place limits on carbon monoxide, which does not contain hydrogen atoms and, thus, does not contribute directly to hydrogen escape. Kasting (1990) has shown that the CO partial pressure could have been quite large under some circumstances.) 2) The concentration of reduced gases in the primitive atmosphere cannot have been much lower than assumed in the standard model even if the rate of volcanic outgassing has been overestimated. The reason is the presence of the $\Phi_R(Ox)$ term in equation (6). At low atmospheric H_2 mixing ratios, more soluble oxidized gases are formed from photochemical processes, resulting in a greater effective production of H_2 (Fig. 5). Thus, the H_2 mixing ratio could not have fallen below ~10^{-4} even if volcanic outgassing had ceased entirely. A corollary to this last statement is that the high prebiotic O_2 concentrations postulated repeatedly by Towe (1981, 1983, 1990) would have been impossible either to produce or to maintain.

6. POSSIBLE PERTURBATIONS TO THE WEAKLY REDUCING MODEL

The two ways to increase the concentrations of highly reduced gases, as indicated above, would be to either increase the surface source or to decrease the hydrogen escape rate. Let us consider whether either of these changes to the standard model might be warranted.

Looking first at the escape rate, here is a problem which emminently deserves reconsideration. Hydrogen escapes from Earth at the diffusion-limited rate at present but it need not have done so in the past. The two potential bottlenecks for hydrogen escape are at the homopause (~100 km altitude) and at the exobase (~500 km altitude). The homopause is where the dominant mode of vertical transport switches from eddy diffusion (below this level) to molecular diffusion (above this level); the exobase is the height at which the atmosphere becomes collisionless. On present Earth, hydrogen escapes readily once it reaches the exobase. The reason is two-fold: First, the exosphere is hot, 1000 to 1500 K, so

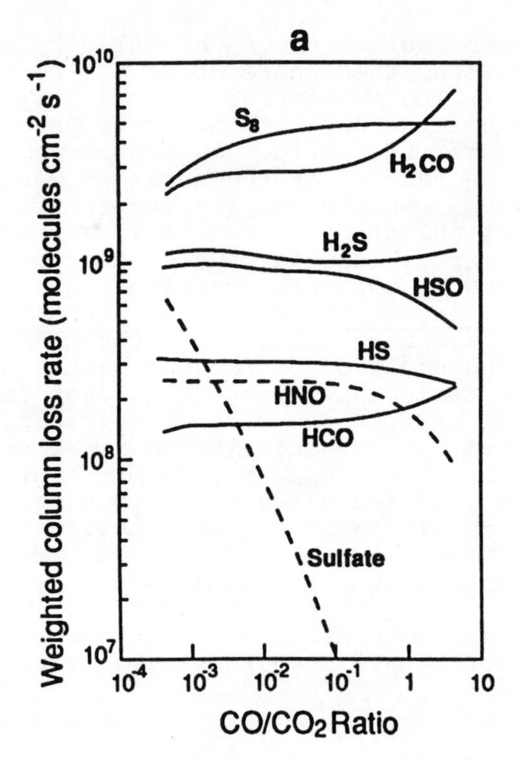

Figure 5. Calculated rainout plus surface deposition rates of reduced (solid curves) and oxidized (dashed curves) species as a function of atmospheric H_2 mixing ratio. Rates are weighted by their coefficients in the hydrogen budget (equations 7 and 8). The photochemical model used to calculate these rates is the same as in Kasting (1990).

thermal (or 'Jeans') escape is relatively rapid. It is especially rapid at times of maximum sunspot activity, when the solar EUV flux and, hence, the exospheric temperature are at their high points. And, second, because it has an internal magnetic field, the Earth is surrounded by a relatively dense plasmasphere filled with energetic protons. These protons can undergo a charge exchange reaction with neutral hydrogen atoms

$$H + H^+(\text{hot}) \rightarrow H^+ + H(\text{hot}) \ .$$

The resultant hot H atoms have more than enough energy to escape from the atmosphere.

Hydrogen escape from the early Earth could have been more difficult because a CO_2-rich, O_2 poor exosphere might have been much colder. (O_2 is a good absorber of solar ultraviolet radiation and CO_2 is an effective infrared radiator.) The present exospheric temperatures of Mars and Venus (both of which have CO_2-rich atmospheres) are of the order of 350 to 400 K (Hunten and Chamberlain, 1987). Jeans escape of H and H_2 still proceeds on Mars because of its low gravity, but is negligible on Venus. The exospheric temperature of primitive Earth might have been higher than 400 K because of the enhanced solar EUV flux (Zahnle and Walker, 1982; Canuto et al., 1982, but it would probably have still been significantly lower than today. Thus, most of the escaping hydrogen would have to have left by way of the charge exchange process. Whether or not this mechanism is capable of handling a large escape flux remains to be determined. Note that the escape rates required to balance the expected outgassing rates of reduced volcanic gases are of the order of 10^{10} H atoms cm^{-2} s^{-1}, which may be compared with the present escape rate of ~3×10^8 H atoms cm^{-2} s^{-1} (Walker, 1977). A mechanism that works adequately at the present modest escape flux might not have been able to accomodate the much larger hydrogen escape flux from the primitive atmosphere. A study of hydrogen escape from early Venus (Kumar et al., 1983) indicates that such 'nonthermal' escape mechanisms tend to saturate somewhere in the range of 10^{10} H atoms cm^{-2} s^{-1}. Thus, it is conceivable that hydrogen escape could have been 'bottled up' at the exobase and that the concentrations of H_2 and/or other reduced gases might have been considerably larger than predicted by equation (4).

Turning now to the other side of the question, one way to increase the source of reduced gases to the primitive atmosphere would be if the Earth's early mantle were more reduced than it is today. Today's mantle appears to be relatively oxidized, with most samples hovering near the QFM (quartz-fayalite-magnetite) oxygen buffer (Holland, 1978). (Reports of a wider range of present mantle oxidation states are now thought to be mostly explained by inaccuracies associated with the 'intrinsic' method for measuring oxygen fugacities (Eggler and Kasting, 1992).) To a non-mineralogist, the QFM buffer can be thought of as the equilibrium between FeO and Fe_3O_4. The volcanic gases released from a melt of present mantle material are also relatively oxidized. Both thermodynamic equilibrium calculations and measurements of actual volcanic gases from Hawaii and other locales show that the average ratio of H_2/H_2O is ~1/100 and the CO/CO_2 ratio is ~1/30 (Holland, 1978). Nitrogen is outgassed almost entirely as N_2 and sulfur mostly as SO_2. Negligible amounts of CH_4 and NH_3 are released.

It has been recognized for a long time that the volcanic gases released from a more reduced magma would themselves be more reduced. For example, Holland (1962) assumed that the primitive mantle would have had an oxygen fugacity near the IW (iron-wustite) buffer, representing equilibrium between Fe and FeO . The gases released from early volcanos would then have contained substantial amounts of H_2, CH_4, NH_3, and H_2S. Holland's atmospheric model fell out of favor, however, because it was based on Harold Urey's model of an Earth that formed cold and did not differentiate a metallic core until several hundred million years into its history. We now believe that the Earth formed hot and that the core formed at essentially the same time as did the planet (Stevenson, 1983). (The basic difference from Urey's theory is that the presumed accretion time is shorter.) Early separation of metallic iron from silicates could, in principle, have led to a high oxygen fugacity for the early mantle. This is especially true if the Earth was veneered with oxidized planetesimals during the latter stages of accretion, as discussed in Section 2.

Until recently, one of the principal arguments in favor of the veneering process has been the observed distribution of siderophile elements in the Earth's mantle (Ringwood, 1990). Siderophile elements, of which iridium (Ir) is perhaps the best-known example, are elements that dissolve preferentially in molten iron as opposed to molten silicates. As a consequence, they are presumed to have been concentrated in the Earth's core at the time of its formation, leaving the upper parts of the Earth depleted. (This is why the overabundance of Ir in the clay layer at the Cretaceous-Tertiary boundary is taken as evidence for the impact of an extraterrestrial, Ir-rich object.) Ringwood (1990) and others have pointed out that the observed concentrations of siderophile elements in the Earth's mantle are substantially higher than one would predict based on measured 'partition coefficients' that describe the partitioning of elements between silicate and iron melts (Fig. 6). The suggested explanation is that the excess siderophiles were imported after the core had already formed as part of the oxidized, late veneer.

A new calculation by Murthy (1991), however, throws this whole line of reasoning into question. Taking into consider

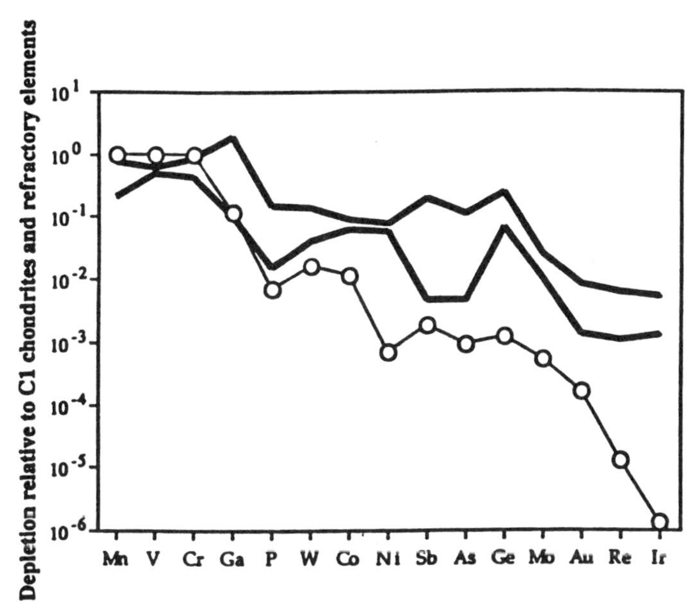

Figure 6. Depletion factors of various siderophile elements in the Earth's mantle compared to C1 carbonaceous chondrites. Solid curves show upper and lower bounds on observed mantle abundances; open circles represent values predicted using partition coefficients evaluated at 1600K. (From Murthy, 1991).

ation recent ideas about the formation of the Moon by a giant impact (Section 2), Murthy realized that the effective equilibration temperature between iron and silicates during core formation could have been in the range of 3000 to 3500 K, as opposed to the 1600 K temperature at which partition coefficients have been measured. Theoretical extrapolation to higher temperatures predicts lower values for these coefficients and brings the observed mantle abundances into accord with predicted values (Fig. 7). If this calculation is correct, no late veneer may be needed to explain the observed mantle composition.

The implications of Murthy's idea for the oxidation state of the early mantle remain to be worked out. My current understanding is that the mantle oxygen fugacity would have been at or below IW shortly after the core had formed. With no substantial late veneer to bring in more oxidized material, the mantle oxygen fugacity could only have been raised by exchange of volatiles with the surface. The process that was quantitatively most important would likely have been the

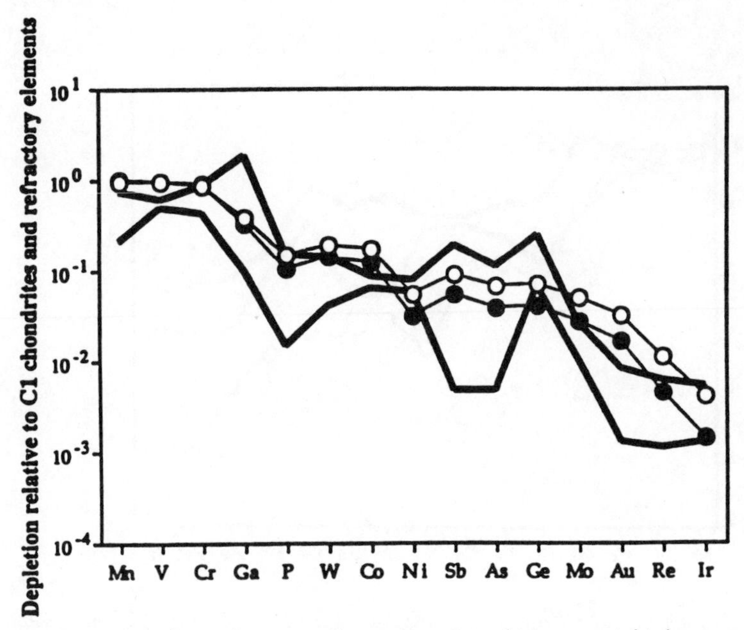

Figure 7. As in Figure 6, except with partition coefficients evaluated at 3000 K (solid circles) and 3500 K (open circles). (From Murthy, 1991).

recycling of seawater, discussed briefly in Section 2. Subduction of hydrated oceanic crust would have brought water down into the mantle; volcanism would have returned it to the surface. If the initial mantle oxygen fugacity was near IW, however, approximately half of this water would have been converted to H_2 before it was outgassed. The oxygen would have been left behind to combine with mantle minerals. The rate at which the mantle became oxidized would have depended on how rapidly water was recycled. This might have been very fast if tectonic cycling rates scaled linearly with internal heat release (McGovern and Schubert, 1989); however, it might have been slow if plate subduction was inhibited on the hot, early Earth. Theory alone is not likely to be able to answer this question; one needs measurements of internal oxygen fugacity on samples of Archean mantle material. What this argument does indicate, however, is that the uniformitarian view of constant mantle oxidation state is not necessarily correct. Thus, it is possible, even probable, that the volcanic source of reduced gases on the primitive Earth was larger than is customarily assumed.

Finally, this section would be incomplete without mention of one other possible source of highly reduced gases on the early Earth, namely, the photochemical reduction of dissolved gases in the ocean. It has been recognized for some time that water can be reduced to H_2 by UV irradiation of a solution containing dissolved ferrous iron (Braterman et al., 1983, and references therein). Recently, Borowska and Mauzerall (1988) claimed that bicarbonate (i.e. dissolved CO_2) could be reduced to formaldehyde under the same circumstances. It is even conceivable that the reduction might proceed all the way to methane. Borowska and Mauzerall's results have since been retracted (private communication, D. Mauzerall, 1990) because of problems with the experiment, but the possibility that carbon (or nitrogen) might be photochemically reduced in solution remains. The fate of dissolved carbon and sulfur gases in the oceans is important for other reasons as well (Kasting et al., 1989; Kasting, 1990). Predicted atmospheric concentrations of SO_2 and CO depend strongly on what happens to these gases once they dissolve. Aqueous chemistry experiments in conditions simulating the early oceans are needed to allow further progress in modelling primitive atmospheric composition.

7. CONCLUSIONS

Much can be learned about the composition and climate of the early atmosphere by developing theoretical models, but there are still many more uncertainties than certainties. The greatest utility of such models is in guiding our thinking so that we learn how to ask the right questions. A good example of this process is the way in which recent theoretical models for the origin of the Moon have pointed toward a new understanding of the distribution of siderophile elements in the Earth's mantle. Whether improved models for the atmosphere and oceans of the early Earth will allow us to ask the right questions about the origin of life remains to be seen. At the very least, such models help provide a conceptual framework into which one's speculations on the subject may be woven.

REFERENCES

Benz, W., Slattery, W.L., Cameron, A.G. W. (1986) 'The origin of the Moon and the single impact hypothesis', Icarus 66, 515-535.

Berkner, L.V., Marshall, L.C. (1965) 'On the origin and rise of oxygen concentration in the Earth's atmosphere', J. Atmos. Sci. 22, 225-261.

Boothroyd, A.I., Sackmann, I.-J., Fowler, W.A. (1991) 'Our Sun II. Early mass loss of 0.1 Mo and the case of the missing lithium', Ap. J. 377, 318-329.

Borowska, Z., Mauzerall, D. (1988) 'Photoreduction of carbon dioxide by aqueous ferrous iron: An alternative to the strongly reducing atmosphere for the chemical origin of life', Proc. Natl. Acad. Sci. 85, 6577-6580.

Braterman, P.S., Cairns-Smith, A.G., Sloper, R.W. (1983) 'Photooxidation of hydrated Fe+2 -- significance for banded iron formations', Nature 303, 163-164.

Brinkman, R.T. (1969) 'Dissociation of water vapor and evolution of oxygen in the terrestrial atmosphere', J. Geophys. Res. 74, 5355-5368.

Canuto, V.M., Levine, J., Augustsson, T., Imhoff, C. (1982) 'UV radiation from the young Sun and oxygen levels in the pre-biological paleoatmosphere', Nature 296, 816-820.

Chamberlain, J. W., Hunten, D. M. (1987) Theory of Planetary Atmospheres, Academic Press, Orlando, 481 pp.

Chameides, W.L., Walker, J.C. G. (1981) 'Rates of fixation by lightning of carbon and nitrogen in possible primitive terrestrial atmospheres', Origins of Life 11, 291-302.

Chyba, C.F. (1987) 'The cometary contribution to the oceans of primitive Earth', Nature 330, 632-635.

Dreibus, G., Wanke, H. (1989) 'Supply and loss of volatile constituents during the accretion of terrestrial planets', in S.K. Atreya, J.B. Pollack, and M.S. Matthews (eds.), Origin and Evolution of Planetary and Satellite Atmospheres, pp. 268-288, University of Arizona Press, Tucson.

Eggler, D., Kasting, J.F. 'Diamond sulfides and redox of Archean mantle', manuscript in preparation.

Frakes, L. A. (1979) Climates Throughout Geologic Time, Elsevier, New York, 310 pp.

Francois, L.M., Walker, J.C. G. (1992) 'Modelling the Phanerozoic carbon cycle and climate: Constraints from the 87Sr/86Sr isotopic ratio of seawater', manuscript in

preparation.
Gilliland, R.L. (1989) 'Solar evolution', Global Planet. Change 1, 35-55.
Gough, D.O. (1981) 'Solar interior structure and luminosity variations', Solar Phys. 74, 21-34.
Graedel, T.E., Sackmann, I.-J., Boothroyd, A.I. (1991) 'Early solar mass loss: A potential solution to the weak sun paradox', Geophys. Res. Lett. 18, 1881-1884.
Holland, H. D. (1984) The Chemical Evolution of the Atmosphere and Oceans, Princeton University Press, Princeton, 582 pp.
Holland, H. D. (1978) The Chemistry of the Atmosphere and Oceans, Wiley, New York, 351 pp.
Holland, H.D. (1962) 'Model for the evolution of the Earth's atmosphere', In A.E.J. Engel, H.L. James, B.F. Leonard, Petrologic Studies: A Volume to Honor A.F. Buddington 447-477, Geol. Soc. Am., New York.
Hunten, D.M. (1973) 'The escape of light gases from planetary atmospheres', J. Atmos. Sci. 30, 1481-1494.
Karhu, J., Epstein, S. (1986) 'The implication of the oxygen isotope records in coexisting cherts and phosphates', Geochim. Cosmochim. Acta 50, 1745-1756.
Kasting, J.F. (1987) 'Theoretical constraints on oxygen and carbon dioxide concentrations in the Precambrian atmosphere', Precambrian Res. 34, 205-229.
Kasting, J.F. (1989) 'Long-term stability of the Earth's climate', Palaeogeogr., Palaeoclimat., Palaeoecol. 75, 83-95.
Kasting, J.F. (1990) 'Bolide impacts and the oxidation state of carbon in the Earth's early atmosphere', Origins of Life 20, 199-231.
Kasting, J.F. (1991) 'CO2 condensation and the climate of early Mars', Icarus 94, 1-13.
Kasting, J.F., Ackerman, T.P. (1986) 'Climatic consequences of very high CO2 levels in the earth's early atmosphere', Science 234, 1383-1385.
Kasting, J.F., Grinspoon, D.H. (1991) 'The faint young sun problem', in C.P. Sonett, M.S. Giampapa, M.S. Matthews (eds.), The Sun in Time, pp. 447-462, University of Arizona Press, Tucson.
Kasting, J.F., Holm, N.G. (1992) 'What determines the volume of the oceans?', Earth Planet. Sci. Lett., in press.
Kasting, J.F., Zahnle, K.J., Walker, J.C. G. (1983) 'Photochemistry of methane in the Earth's early atmosphere', Precambrian Res. 20, 121-148.

Kasting, J.F., Zahnle, K.J., Pinto, J.P., Young, A.T. (1989) 'Sulfur, ultraviolet radiation, and the early evolution of life', Origins of Life 19, 95-108.
Knauth, L.P., Epstein, S. (1976) 'Hydrogen and oxygen isotope ratios in nodular and bedded cherts', Geochim. Cosmochim. Acta 40, 1095-1108.
Kuhn, W.R., Atreya, S.K. (1979) 'Ammonia photolysis and the greenhouse effect in the primordial atmosphere of the Earth', Icarus 37, 207-213.
Kumar, S., Hunten, D.M., Pollack, J.B. (1983) 'Nonthermal escape of hydrogen and deuterium from Venus and implications for loss of water', Icarus 55, 369-389.
Lange, M.A., Ahrens, T.J. (1982) 'The evolution of an impact generated atmosphere', Icarus 51, 96-120.
Levine, J.S. (1982) 'The photochemistry of the paleoatmosphere', J. Molec. Evol. 18, 161-172.
Lewis, J. S., Prinn, R. G. (1984) Planets and Their Atmospheres: Origin and Evolution, Academic Press, Orlando, Florida, 470 pp.
Matsui, T., Abe, Y. (1986) 'Evolution of an impact-induced atmosphere and magma ocean on the accreting Earth', Nature 319, 303-305.
McGovern, P.J., Schubert, G. (1989) 'Thermal evolution of the Earth: effects of volatile exchange between the atmosphere and interior', Earth Planet. Sci. Lett. 96, 27-37.
Meade, C., Jeanloz, R. (1991) 'Deep-focus earthquakes and recycling of water into the Earth's mantle', Science 252, 68-72.
Murthy, V.R. (1991) 'Early differentiation of the Earth and the problem of mantle siderophile elements: a new approach', Science 253, 303-306.
Newman, M.J., Rood, R.T. (1977) 'Implications of solar evolution for the earth's early atmosphere', Science 198, 1035-1037.
Pinto, J.P., Gladstone, C.R., Yung, Y.L. (1980) 'Photochemical production of formaldehyde in the earth's primitive atmosphere', Science 210, 183-185.
Pollack, J.B., Kasting, J.F., Richardson, S.M., Poliakoff, K. (1987) 'The case for a wet, warm climate on early Mars', Icarus 71, 203-224.
Ringwood, A.E. (1990) 'Earliest history of the Earth-Moon system', in H.E. Newsom, J.H. Jones (eds.), Origin of the Earth, pp. 101-134, Oxford University Press, New York.
Sagan, C., Mullen G. (1972) 'Earth and Mars: Evolution of atmospheres and surface temperatures', Science 177, 52-56.

Schopf, J.W., Packer, B.M. (1987) 'Early Archean (3.3 billion to 3.5 billion-year-old) microfossils from Warrawoona Group, Australia', Science 237, 70-73.
Schubert, G., Turcotte, D.L., Solomon, S.C., Sleep, N.H. (1989) 'Coupled evolution of the atmospheres and interiors of planets and satellites', In S.K. Atreya, J.B. Pollack, M.S. Matthews, (eds.), Origin and Evolution of Planetary and Satellite Atmospheres, pp. 450-483, University of Arizona Press, Tucson, Arizona.
Staudigal, H., Hart, S.R., Schmincke, H.U., Smith, B.M. (1989) 'Cretaceous ocean crust at DSDP site 417 and 418: Carbon uptake from weathering versus loss by magmatic outgassing', Geochim. Cosmochim. Acta. 53, 3091-3094.
Stevenson, D.J. (1983) 'The nature of the Earth prior to the oldest known rock record: the Hadean Earth', In J.W. Schopf, Earth's Earliest Biosphere: Its Origin and Evolution, pp. 32-40, Princeton University Press, Princeton, New Jersey.
Stribling, R., Miller, S.L. (1987) 'Energy yields for hydrogen cyanide and formaldehyde syntheses: the HCN and amino acid concentrations in the primitive ocean', Origins of Life 17, 261-273.
Towe, K.M. (1990) 'Aerobic respiration in the Archaean?', Nature 348, 54-56.
Towe, K.M. (1981) 'Environmental conditions surrounding the origin and early Archean evolution of life: a hypothesis', Precambrian Res. 16, 1-10.
Towe, K.M. (1983) 'Precambrian atmospheric oxygen and banded iron formations: a delayed ocean model', Precambrian Res. 20, 161-170.
Urey, H.C. (1952) The Planets: Their Origin and Development, Yale University Press, New Haven, Conn., 245 pp.
Walker, J.C.G. (1985) 'Carbon dioxide on the early Earth', Origins of Life 16, 117-127.
Walker, J.C.G. (1977) Evolution of the Atmosphere, Macmillan, New York.
Weissman, P.R. (1985) 'Dynamical evolution of the Oort cloud', in A. Carusi, G.B. Valsecchi (eds.), Dynamics of Comets: Their Origin and Evolution, pp. 87-96, D. Reidel, Dordrecht.
Wetherill, G.W. (1991) 'Occurrence of Earth-like bodies in planetary systems', Science 253, 535-538.
Wilhelms, D.E. (1984) in M.H. Carr (ed.), The Geology of the Terrestrial Planets 107-205, NASA SP-469.

Willson, L.A., Bowen, G.H., Struck-Marcell, C. (1987) 'Mass loss on the main sequence', Comments Astrophys. 12, 17--34.

Zahnle, K.J. (1986) 'Photochemistry of methane and the formation of hydrocyanic acid (HCN) in the Earth's early atmosphere', J. Geophys. Res. 91, 2819-2834.

Zahnle, K.J., Kasting, J.F., Pollack, J.B. (1988) 'Evolution of a steam atmosphere during Earth's accretion', Icarus 74, 62-97.

Zahnle, K.J., Walker, J.C. G. (1982) 'The evolution of solar ultraviolet luminosity', Rev. Geophys. Space Phys. 20, 280-292.

ORIGIN AND EVOLUTION OF MARTIAN ATMOSPHERE AND CLIMATE AND POSSIBLE EXOBIOLOGICAL EXPERIMENTS

L. M. MUKHIN
Max-Planck-Institut für Chemie
Saarstrasse 23
D-6500 Mainz
F.R.Germany
and
Space Research Institute (IKI)
Moscow
Russia

ABSTRACT. The early evolution of the atmosphere of Mars suggests the possibility that life at some level may have evolved. Scientific techniques are suggested for exploration and physical and chemical analyses of the surface of Mars to probe for possible past and present exobiological processes and events.

1. Introduction.

There are several principal facts due to which the continuation of systematic studies of Mars by space means is of particular interest.

The first fact is that the observational data available at present allow us to suggest with a great extent of probability that 3.6 - 3.8 billions of years ago the climate of Mars was much different. Present-day Mars is the cold deserted planet. The atmosphere of this planet consists mainly of CO_2, N_2, and Ar with O_2, O_3, CO, H_2O as admixture. The mean pressure near the surface is about 6 mb that corresponds to the pressure in the Earth's atmosphere at a level of 33 km; the daytime temperature at lower latitudes reaches 290 K and 145 K at poles. Nevertheless, as it will be shown below, the Martian atmosphere should have changed essentially for last billions of years; and probably at the earlier stages of its evolution Mars had the warm climate and the denser atmosphere which "permitted" the existence of free water bodies on Mars.

It is this fact that forces us to return to the discussion of the fundamental problem on the chemical evolution of early Mars and on the possible origination of life on this planet. In this connection, not going into details, I would like to say that some results of Viking Mission do not yield to absolute "negative" interpretation.

The second, not less important, problem which in many respects warms up the interest of mankind to the exploration of Mars is its possible colonization and artificial changes of its climate. It is obvious that Mars is the only planet of the solar system which in the nearest future can be considered as a possible place for human settlings.

There are naturally a few other fundamental problems of Mars studies. Among

J. M. Greenberg et al. (eds.), The Chemistry of Life's Origins, 177–184.

them are: the inner structure of the planet, the modern climate and weather on Mars, the detailed study of Martian geochemistry and some others.

In this paper I would like to review briefly the status of activity concerning the first problem, that is the problem of the evolution of the Martian atmosphere and climate, and also to consider a few possible experiments for future space missions which are reasonable to perform for studying processes of the chemical evolution on this planet.

2. Paleoclimate of Mars.

The problem of Martian paleoclimate is intimately linked with the problem of volatile budget on this planet, water and carbon dioxide in particular. Numerous traces of water-erosion activity seen on photo images of the planet taken by Marine-9, Mars-5, Viking-1 and Viking-2 evidence obviously in favor of the existence of open water bodies on Mars several billions of years ago. Interpretations of photo images of the Martian surface indicate the different nature of water streams on Mars.

Part of them which are of comparatively recent origin can be associated with warming up of ice-saturated soils by magmatic focuses and following break-out of powerful water streams. It is evident that the similar process is not at all necessarily associated with the presence of the warm climate on the planet. However, ramified river valleys in the zone of ancient impact craters could certainly be formed only if the dense atmosphere exists. The question of geomorphological evidences that the climate on Mars was warm has been discussed in detail elsewhere [1,2,3]. Naturally the question raises on parameters of the early atmosphere of Mars, reasons of its drastical change, and, as it has been mentioned before, on the volatile budget of this planet.

Let us discuss the latter question which was the subject of many publications where different approaches to this problem were considered. Obviously, Lewis [4] was the first who noted in 1972 that due to the great distance between the Sun and Mars it had been formed under conditions of volatile abundance much higher than on the Earth and Venus. Recently, Dreibus and Wänke [5, 6] have analyzed in details this problem using the latest data on SNC-meteorites composition and the two component model of accretion of Mars. The main assumption on which conclusions made in these papers are based is that Mars is indeed a parent body for SNC-meteorites. In some additional assumptions they estimated the water content on Mars that is the layer 130 m thick which cover the whole planet. Carr [7] employed the entirely different approach where he estimated the water layer on Mars based upon geomorphological features of this planet. His estimates are much higher - the layer 0.5 to 1 km thick. Even higher values, 3.1 km, were obtained in the paper of Lin-Gun Liu [8]. This value is already comparable with the Earth's hydrosphere thickness. It should be noted, however, that estimates of the volcanic activity on Mars yield the water layer 46 m [9]. On the other hand, minimum estimates obtained from the analysis of comparative velocities for hydrogen and deuterium escapes from the Martian atmosphere [10], result in the layer of degassed water equal to only 3.6 m.

McElroy et al. [11] evaluated the amount of degassed water on Mars based upon the estimate of isotope ratios for nitrogen under assumption that N_2/H_2O on Mars is the same as that on the Earth. They obtained the very wide spread in values: from 8 to 133 m.

One can take the simplest way and make the lower-bound estimate for the water content (as also other volatiles on Mars), assuming that the H2O content in units of a volatile gram by a planetary matter gram is approximately the same as for Mars as for the

Earth. Then, we can obtain a water layer of 1200 m. Table 1 shows different estimates of the water content on Mars.

Let us now discuss the estimates of other most important total high volatiles inventory on Mars - CO_2 and N_2. Here, there is also a significant spread in estimates. So, Pollack and Black [12] assessed the CO_2 abundance (reserve) on Mars equal to 1 - 3 bars, based on the data on the content of nitrogen and noble gases in the Venus atmosphere. At the same time Carr [7] obtained the estimate of 10 - 20 bars which is maximum among those available at present. As in the case with water he used the analysis of morphological features of the Martian surface. Minimum estimates of the CO_2 abundance (reserve) on Mars are within first hundreds of millibars [13, 14].

Recently, the idea has been widely discussed that there is the great amount of CO_2 adsorbed by Martian regolith [15, 16]. Table presents the data on the adsorbed CO_2 inventory. Some indirect indications that there are powerful carbonate deposites equivalent to the 1.2 bar pressure of CO_2 near the surface were received while interpreting the Viking X-ray data [17] on Mars (see Table 2). It should be noted that the problem of volatile inventory on Mars is quite undetermined; this is clearly understood from the estimates discussed above. Nevertheless, for the analysis of the Martian paleoclimate problem it is important to indicate that within most models the assumed total inventory of CO_2 on Mars is capable to provide pressure near the surface higher than 1 bar.

3. The Evolution of the Climate.

The problem of the Martian paleoclimate has been widely discussed elsewhere [18 - 21]. Evidently, in 1977 Moroz and Mukhin [22] were the first who calculated the dry and wet greenhouse effect on Mars taking into account the lower luminosity of the Sun 4 x 10^9 years ago. After that a series of publications appeared which had used some additional assumptions and another method of calculation.

Maybe the most interesting fact is here that the estimates made by different authors are approximately the same. The most detailed calculations made recently in the paper of Pollack et al. [20] show that CO_2 with 0.75 to 5 bars was sufficient near the Martian surface in order to ensure the existence of the free water body on Mars. The significant achievement of the paper of Pollack et al. [20] is that along with estimates of the greenhouse effect they made the detailed calculation of weathering velocities. This is of extremely importance for discussing the problem of the irreversible change in the early climate of Mars. It is followed from their estimates that the "lifetime" of the warm climate is only about 107 years for Mars with the CO_2 pressure of 0.75 bar, and, therefore, in order to maintain this climate during 109 years it is necessary to have the CO_2 storage equivalent to a pressure of about 50 bars. This value exceeds any estimate discussed above. hence, it is evident that the juvenile source should have acted during $\geq 10^8$ years. It is in this period of time when the epoch of the "Martian paradise" existed. This epoch was quite favorable for originating the life on this planet. Naturally, the main question whether these possibilities have been realized is still open, and, therefore the next global attack on Mars (new space missions) is so principally significant. To build any model of the evolution of the Martian climate and atmosphere the problem is to be solved why climatic conditions on present day Mars and on early Mars are drastically different. In other words, it is necessary to explain where those significant amounts of water and CO_2 are which existed on Mars billions of years ago. In order to answer these questions it is necessary to

Table 1. Estimates of H_2O Reservoir on Mars

Present Mars atmosphere	7 µm
Analysis of HDO only exchangeable reservoir	3.6 m
SNC-meteorites, accretional model with	
100% release of H_2O	130 m
Scaling based on N-isotopes	8 - 133 m
Geomorphic estimates	0.5 - 3 km
Thermal model	3 km
Volcanoes only	> 46 m
Earth Scaling	1200 m
Post-accretionary outgassing factor	625 - 3530 m
Polar caps	15 m

Table 2. Estimates of CO_2 and N_2 Reservoir on Mars

	CO_2 (mb)	N_2 (mb)
Present Mars atmosphere	7	0.2
Earth Scaling*	27000	300
N_2, noble gases on Venus*	1000-3000	6.6-66
Scaling based on N-isotopes*	1760	21.5
SNC-meteorites, accretional model*	3000	33
Volcanoes only*	>1000	>11
Viking X-ray	1200	≈12

* from McKay C.P. and Stoker, C.R. (1989) Rev. Geophys. 27, 189-214.

consider the problem on sinks for volatiles. In principle, there are two types of sinks - reversible and irreversible. Examples of reversible sinks are polar caps or the layer of regolith where sorption - desorption processes are occurring due to changes in temperature.

Among irreversible sinks are first of all the atmospheric erosion and hydrodynamical escape. Both these mechanisms are evidently efficient enough in order to remove its whole atmosphere in the early period of the Mars formation (earlier than 3.8 x 10^9 years). For the case of atmospheric cratering this fact was rather conclusively demonstrated by Melosh and Vickerey [23], whereas the hydrodynamical escape was estimated in the papers of Hunten et al. [24] and Hayashi et al. [25], where this process was proved to be highly efficient, also.

The other atmospheric sinks related to thermal and non-thermal mechanisms of escape are less efficient. However, the situation with the complete removal of the atmosphere is not so unambiguous as it might appear at first glance. The point is that processes of Mars degassing have not been understood yet properly. It is still vague when the planetary atmosphere began forming and how many volatiles were added at last stages of accretion. Hence, the conclusion that the atmosphere had been removed completely due to impact processes cannot believe to be undisputed. On the other hand, the estimates of the age of volcanic formations on Mars indicate obviously that the degassing processes after the late heavy bombardment occurred efficiently.

Thus, it is necessary to have regard to the other type of sinks - sinks onto the planetary surface. Let us consider the most significant sink which ensures the possible existence of the warm climate - the carbonate formation. The analyses made by Moroz and Mukhin [22] and Pollack et al. [20], clearly show that the Martian atmosphere is instable on a rather short scale of time ($\approx 10^7$ years). Carr [26] indicated a possibility to decompose carbonates in the zone of volcanic focuses and due to impact processes. Note also that Clark and Van Hart [27] gave attention to the instability of carbonates in the presence of sulphates, though kinetic parameters of this event are unclear. Somehow or other, there were sources for recovering the atmosphere on early Mars though their power is rather indefinite. Evidently, the most real mechanism of the CO_2 sink is the formation of carbonates. It is interesting to note that experiments on simulation of carbonate formation processes under present day conditions on Mars yielded very high velocities for the CO_2 sink to regolith [28]. It turns out if the results of these experiments are extrapolated directly to the present day atmosphere of Mars it becomes unstable and should disappear practically completely on a scale of time not longer than 10^4 - 10^5 years. This paradoxical result forces us to suggest the presence of degassing sources on Mars even at present. The more trivial conclusion is that the results of experiments are wrong. None of these possibilities can be excluded.

4. Exobiological Experiments.

Summarizing the results discussed above, one can say that the carbonate formation mechanism is the most evident mechanism of depletion of the Martian atmosphere and, accordingly, of the ceasing of the "Martian paradise" era. Assuming somewhat arbitrarily that the characteristic lifetime of the warm climate on Mars is 10^7 to 10^8 years, it is now needed to answer a question: Could the life originate on Mars during such a short period of time? If no, what could the level of the chemical evolution be at early stages of the

Mars evolution? It is clear that there is no exact answer to these questions. However, it is possible to make very important conclusion that the time scale for the life origination on the Earth was short enough (< 0.5×10^9 years) [29]. This fact is borne out by several convincing experimental data [30]. Therefore, one cannot exclude a possibility for originating of primitive living forms on Mars. As to the level of the chemical evolution, simple estimates can be obtained from the papers of Fegley et al. [31] and of Mukhin et al. [32]. The meaning of these estimates is in the following. Either the former paper or the latter one demonstrated the significant role of impact processes for the synthesis of precursors of complex organic compounds. The difference between models is: In the model of Fegley et al. [31] precursors are formed in the atmosphere during passing the impactor through it, and, hence, the yield of products depends on the composition and density of the atmosphere. In the model of Mukhin et al. [32] the yield of products is practically independent of the composition and pressure of the atmosphere since the synthesis occurs in the hot plasma cloud formed at the instant of planetesimals impact with the planetary surface. Either the former model or the latter one produces significant yields of precursors (see Figures).

Hence, in principle there are no reasons to doubt that intense processes of organic compound formation occurred in conditions of early Mars. At present organic molecules cannot survive on the Martian surface due to their fast destruction by the ultra-violet radiation and intense oxidation processes. With the denser atmosphere, however, processes of the chemical evolution could occur on Mars rather intensively.

Here, we came near the main problems arising during the development of plans for studying Mars by means of space vehicles. these problems can be formulated as follows:

1.) Search for carbonates as a main reservoir of bounded carbon dioxide which ensured the presence of the warm and wet climate on early Mars.
2.) Search for traces of organic compounds on Mars.
3.) Search for traces of extinct life on Mars.
4.) Search for living forms on Mars.

I will not discuss here the very important problem how to select landing sites for studying the surface. I will only list possible experiments which can be implemented.
It is clear, however, that to carry out exobiological experiments and, in general, a wide range of studies on the Martian surface the rover is the best technical device.
Let us discuss briefly several technical versions developed now in Russia. It should be noted that a rover form studying Mars is also under development in the USA and ESA.

I) Large scale rover. Its mass is 350 kg, the transverse size is 1400 mm, the length 220 mm. The rover is a six wheel device with a folded frame. A diameter of each wheel is 600 mm. The minimum motion velocity is 0.5 km/h. The rover can cover 500 Km over the surface. The rover manipulator is capable to take soil samples from a depth of 90 cm whereas a drill from a depth of several meters. The operation resource is about 3 years. The mass of its scientific instruments is 50 kg.

II) Small rover has a mass of 70 - 100 kg only and is capable to carry scientific instruments ≈ 15 kg in mass (Figure). this rover is also a six wheel device, however, the wheel diameter is only 250 mm, the length 1000 mm, and the width 900 mm. The velocity is also 0.5 km/h. If chemical batteries are used the rover's lifetime will be 100 days. If RTG is employed, the lifetime can be increased up to 1 - 2 years.

The main scientific objectives of small rover are:

- to make high-resolution color stereo mosaics of the surface of Mars;
- to determine the elemental composition of the surface and subsurface of Martian soil;
- to determine organic compounds, free and bound volatiles in the Martian soil;
- to study the temporal and spatial variations of minor components of the Martian atmosphere;
- to study the permafrost zone depth;
- to study meteo parameters;
- to study dust storm dynamics.

Scientific instrumentation package (preliminary):
- TV-system;
- Alpha X,p,n-spectrometers: Alpha-backscattering,
  - proton,
  X-fluorescence,
  - neutron;
- gas chromatograph, sample pyrolysis, DTA;
- optical spectrometer,
- Mössbauer spectrometer,
- electromagnetic sounding,
- impurity sensors for minor components of Mars' atmosphere;
- meteor complex;
- samplers (vibropenetrator, driller) and devices for sample distribution.

Unfortunately, the rover becomes part of the mission only in 1996. Nevertheless, it seems to be evident only the development of a Martian rover along with a system of return Martian soil samples to the Earth makes it possible to formulate better the fundamental problems on origin and evolution of the life in the solar system and the problems of evolutionary planetology.

4. References.

[1] Piery, D.C. (1980) 'Martian valleys: Morphology, distribution, age, and origin. Science **210**, 895-897.

[2] Sharp, R.P.and Malin, M.S. (1975) 'Channels on Mars.' Bull. Geol. Soc. Amer. **86**, N5, 593-609.

[3] Pollack, J.B. (1979) 'Climatic change on the terrestrial planets.' Icarus **37**, 479-553.

[4] Lewis, J.S. (1972) 'Low temperature condensation from solar nebula.' Icarus **16**, 241-252.

[5] Dreibus, G. and Wänke, H. (1987) 'Volatiles on Earth and Mars: A comparison.' Icarus **71**, 225-240.

[6] Dreibus, G. and Wänke, H. (1989) 'Supply and loss of volatile constituents during the accretion of terrestrial planets.', in S.K. Attreya, J.B. Pollack and M.S. Matthews (eds.), Origin and Evolution of Planetary and Sattelite Atmospheres, University of Arizona Press, Tucson, pp. 268-288.

[7] Carr, M.H. (1986) 'Mars: A water rich planet?' Icarus **68**, 187-216.
[8] Lin-Gun-Liu (1988) 'Water in the terrestrial planets and the Moon.' Icarus 74, 98-107.
[9] Greely, R. (1987) 'Release of juvenile water on Mars: Estimated amounts and timing associated with volcanism.' Science **236**, 1653-1654.
[10] Yuk, L. Yung et al. (1988) 'HDO in the Martian atmosphere: Implication for the abundance of crustal water. Icarus **76**, 146-159.
[11] McElroy, M.B., Yuk Ling Yung,and Nier, A.O. (1976) Isotopic composition of nitrogen: Implication for the past history of Mars' atmosphere.' Science **194**, 70-72.
[12] Pollack, J.B. and Black, D.C. (1979) 'Implication of the gas compositional measurement of Pioneer on Venus for origin of planetary atmosphere.' Science **205**, 56-59.
[13] Anders, E. and Owen, T. (1977) 'Mars and Earth. Origin and abundances of volatiles.' Science **198** N4316, 453-465.
[14] Clark, B.C. and Baird, A.K. (1979) 'Volatiles in the Martian regolith. J. Geophys. Res. Lett. **6**, 811-814.
[15] Fanale, F.P. (1976) 'Martian volatiles: Their degassing history and geochemical fate.' Icarus **28**, 179-202.
[16] Zent A.P., Fanale, F.P.and Postawko, S.E. (1987) 'Carbon dioxide: Adsorption on palagonite and partitioning in the Martian regolith.' Icarus **71**, 241-249.
[17] Toulmin, P., Baird, A.K., Clark, B.C. et al. (1977) 'Geochemical and mineralogical interpretation of the Viking inorganic chemical results.' J. Geophys. Res. **82**, 4625-4634.
[18] Pollack, J.B. and Yung, Y. L. (1980) 'Origin and Evolution of planetary atmospheres.' Ann. Rev. Earth and Planet. Sci. **8**, 425-487.
[19] Postawko, S.E. and Kuhn, W.R. (1986) 'Effect of the greenhouse gases (CO_2, H_2O, SO_2) on Martian paleoclimate.' J. Geophys. Res. **91**, D431-D438.
[20] Pollack, J.B., Kasting, J.F., Richardson, S.M. and Poliakoff, K. (1987) 'The case for a wet, warm climate on early Mars.' Icarus **71**, 203-224.
[21] Hoffert, M.I., Gallegary, A, J., Hsieh, C.I. and Ziegler, W. (1918) 'Liquid water on Mars: An energy balance climate model for CO_2/H_2O atmospheres.' Icarus **47**, 112-129.
[22] Moroz, V.I. and Mukhin, L.M. (1977) 'Early evolutionary stages of the atmosphere and climate of the terrestrial planets.' Cosmic Res. **15**, 769-791.
[23] Melosh, H.J. and Vickery, A.M. (1989) 'Impact erosion of the primordial atmosphere of Mars.' Nature **338**, 487-489.
[24] Hunten, D.M. et al. (1989) in S.K. Atreya, J. B. Pollack and M. S. Matthews (eds.). Origin and Evolution of Planetary and Satellites Atmospheres, University of Arizona Press, Tucson, pp. 386-422.
[25] Hayashi, C. et al. (1985) 'Formation of the solar system'. in D.C. Black and M. S. Matthews (eds.), Protostars and Planets II, University of Arizona Press, Tucson, pp. 1100-1153.
[26] Carr, M.H. (1989) 'Recharge of the early atmosphere of Mars by impact release of CO2.' Icarus **79**, 311-327.
[27] Clark, B.C. and van Hart, D. C. (1981) 'The salts on Mars.' Icarus **45**, 370-378.
[28] Booth, M.S. and Kiffer, H.M. (1978) 'Carbonate formation in Mars like environment.' J. Geophys. Res. B. **83**, 1809-1815.
[29] Mukhin, L.M. and Gerasimov, M.V. (1992) 'The possible pathways of the synthesis of precursors on the early Earth.' (this volume).
[30] Schidlowski, M. (1988) 'A 3.800 million year isotopic record of life from carbon in sedimentary rocks.' Nature **333**, 313-318.
[31] Fegley, B., Jr., Prinn, R. G., Hartmann, H., Hauptmann, G., Watkin (1986) Nature **319**, 305-307.
[32] Mukhin, L.M., Gerasomiv, M.V. and Safonova, E.N. (1989) 'Origin of precursors of organic molecules during evaporation of meteorites and mafic terrestrial rocks. Nature **340**, 45-48.

THE POSSIBLE PATHWAYS OF THE SYNTHESIS OF PRECURSORS ON THE EARLY EARTH

L. M. MUKHIN and M. V. GERASIMOV
Max-Planck-Institut für Chemie
Saarstrasse 23
D-6500 Mainz, F.R.Germany
and
Space Research Institute (IKI)
Moscow, Russia

ABSTRACT. A theory of the origin of life is still far from being complete at this time. The most promising part of this theory is the understanding of synthesis of the precursors of complex organic compounds on the early Earth. In fact, almost all these processes are connected with the origin and chemical evolution of the Earth's atmosphere. Therefore, the understanding of the origin and evolution of the early Earth's atmosphere is very important not only for planetary sciences but also for the origin of life.

1. Introduction.

Currently existing paleontological data offer clear evidence of the presence of life on Earth 3.5 billion years ago (Schopf and Packer (1987)). Furthermore, we can make the fundamental conclusion from the analysis of carbon isotope ratios $^{13}C/^{12}C$ that an almost contemporary biogeochemical carbon cycle (Schidlowski (1988)) existed on Earth 3.8 billion years ago. Moreover, there is some reason to believe that this last dating could be pushed further back to four billion years (Schidlowski, personal communication). There is no doubt that both the prebiological evolution processes and the global biogeochemical carbon cycle can only occur in a sufficiently dense atmosphere and hydrosphere that have already formed or are forming. Taking into account data from isotope systematics (Faure (1986)), with which one can estimate the Earth's age at 4.55 to 4.57 billion years old, we conclude that the time scale for prebiological evolution and the emergence of life on Earth was sufficiently brief: possibly less than 0.5 billion years. This fact means, however, that the Earth's protoatmosphere must have been formed prior to the appearance of the biogeochemical cycle, that is, over a period of time significantly less than 0.5 billion years. Additional indirect evidence of the early emergence of the Earth's atmosphere can be found in analyzing the isotope relationships of the noble gas Ar and Xe (Ozima and Kudo (1972); Ozima (1975); Kuroda and Crouch (1962); Kuroda and Manuel (1962); Phinney et al. (1978)).

Therefore, existing and observed data provide evidence of the very early formation of a dense atmosphere on Earth. We shall consider possible scenarios for the formation of the Earth's early atmosphere and its initial chemical composition.

J. M. Greenberg et al. (eds.), The Chemistry of Life's Origins, 185–194.

The former point is crucial for the problem of origin of life. It is useful to remind that the idea about strong reduced early atmosphere, consisting mainly of hydrogen, methane, ammonia and water vapor was supported by Miller's experiments on synthesis of precursors and simple biomolecules. These experiments for many years had influence on theories of atmospherical evolution. Actually, therefore, the models of reduced (solar composition) early atmosphere were very popular some decades after Miller's experiments. From this point of view, it will be interesting to provide a brief historical review of different models of atmospheric evolution (see also chapter by J. Kasting).

2. Sources and Sinks for the Early Atmosphere.

A model for Earth's early atmosphere, formed from solar composition gas captured gravitationally during the final stages of accretion, was explored in Hayashi's studies (1981, 1985). Hydrogen and helium are the primary components of this atmosphere. According to the estimates of several authors, the total mass of the initial atmosphere could have reached 10^{26}-10^{31} g (Hayashi et al. (1985); Cameron and Pine (1973)). However, the inference by these models that the entrapment process occurred isothermically may lead to significant error: they overlook the heating up of the gas during accretion (Lewis and Prinn (1984)). We will note that there are some additional difficulties in a model of an isothermal, initial atmosphere. They stem from the diffusive concentration of heavy gases in the initial atmosphere (Walker (1982)). The pressure of the initial atmosphere for Earth is only 0.1 bar in the more realistic models of the adiabatic, gravitational capture of gas from a protoplanetary nebula (Lewis and Prinn (1984)).

Clearly, an initial atmosphere could only have formed if the processes of gas dissipation from the protoplanetary nebula had not ceased by the time Earth's accumulation had concluded. Gas dissipation from a protoplanetary nebula is determined by a solar wind from a young T Tauri Sun and EUV (Sekiya et al. (1980); Zahnle and Walker (1982); Elmegreen (1978); Horedt (1978). Canuto et al. (1983)) have estimated the time scale for dissipation of the gaseous component of a protoplanetary nebula, using observed data on T Tauri star luminosity. Their estimates show that this time scale is not more than 10^7 years. It may only be several million years, beginning with the onset of the convective phase in the history of the Sun's development. This time scale is appreciably shorter than the approximately 10^8 years estimated for Earth's accumulation (Safronov (1969); Wetherill (1980)). Hence, current theories of stellar evolution, coupled with observed astronomical data, raise the possibility that only a very weak initial atmosphere existed at the very inception of planets' accumulation process. Moreover, using very simple estimations it is possible to show that the enormous high partial pressure of Ne should be in modern atmosphere if Hayashi type atmosphere was in the past. But the modern atmosphere does not contain a "rare gas memory" from past. This is a very serious argument against the formation of an initial atmosphere on Earth as proposed by Hayashi. Hence, various scenarios for a dense initial atmosphere appear to be highly improbable for the above reasons. Therefore, the very rapid formation of a dense Earth atmosphere was apparently a function of different physical processes.

One of the suggested mechanisms for the rapid formation of the atmosphere is the intense, continuous degassing of Earth in the conventional sense of this term, including magmatic differentiation and volcanism (Walker (1977); Fanale (1971)). There is still no answer to the question of whether continuous degassing of the Earth could have been intensive enough to allow for the formation of the atmosphere and the hydrosphere over a very short time span (< 10^8 years). Such a possibility exists where there is a strong, overall heating of the planet during its accretion. However, this entails the inclusion of a number of additional inferences regarding a very brief accretion scale: < 5 x 10^5 years (Hanks and Anderson (1969)).

3. Impacts as the Source of Prebiology.

Walker's recent analysis of the process of continuous degassing (Walker (1991)), shows that the present rate of degassing is clearly insufficient to be responsible for forming the atmosphere in a very brief time scale. At the same time, Walker's proposed numerical estimate of the accumulation rate of hot lava on primordial Earth, which could have provided for the necessary intensity of degassing (1.3 x 10^{19} g/yr), appears to be unjustifiably high. Such intensive volcanism infers that ≈ 1.3 X 10^{27} g of magma would pour out over 10^8 years of accumulation on the surface, or a quarter of the entire present mass of the Earth.

Under the currently adopted planetary accumulation models with an approximate 108 year time scale, such a hypothesis would only be justified in the case of a gigantic impact (Kipp and Melosh (1986); Wetherill (1986)). There is no question that a considerable portion of the planet would have melted as a result of a gigantic impact which would have released a huge quantity of gases. This amount would have been sufficient to have formed an atmosphere. Yet, there is no detailed, physical-chemical model of this process at the present time. Furthermore, it cannot be considered that the very fact of a giant impact in the Earth's history has been firmly established. Clearly, if such a catastrophic event did actually take place in the earliest history of our planet, its consequences would have been so great that they would have been reflected in the geochemical "records". However, this is not the case. None the less, the impact-induced degassing played a main role in the origin of the Earth's atmosphere.

Indeed, the role of impact processes in forming Earth's protoatmosphere was discussed more than 20 years ago in Florenskiy's study (1965). During the years that followed, this idea was developed in a number of studies (Arrhenius et al. (1974); Benlow and Meadows (1977); Gerasimov and Mukhin (1979); Lang and Ahrens (1982); Gerasimov et al. (1985)). The core of this idea is the loss of volatiles by planetesimals and the embryo of growing Earth in the time of accretion process. Indeed, the velocity of the collision of a random body of a preplanetary swarm with the embryo of primordial Earth is no less than the escape velocity. Therefore, the range of velocities of planetesimal collision with a growing planet embryo varies from meters per second at the initial stages of its expansion, to ≥ 12 kilometers per second at the final stages of accumulation. In reality, collision velocities could have been significantly higher in the case of a collision between Earth and bodies escaping from the asteroid belt, or with comets. In this case, when the mass of an expanding Earth reached approximately 10% of its current mass, the escape velocity became equal to roughly five kilometers per second; the melting of silicate matter began during a collision between planetesimals and an embryonic planet (Ahrens and O'Keefe (1972)). Beginning with the point where the mass of the embryo was equal to approximately 0.5 of the mass of present-day Earth, impact processes were paralleled by the partial vaporization of silicates (Ahrens and O'Keefe (1972)). The more high-speed planetesimals, reaching a speed of more than 16-20 kilometers per second, were completely vaporized. Several works (Ahrens and O'Keefe (1972); Gault and Heitowit (1963); McQueen et al. (1973); Gerasimov (1979); O'Keefe and Ahrens (1977)) analyze in detail the physics of how colliding matter is heated as a shock wave passes through that matter and estimate the efficiency of impact-induced degassing. They demonstrate that impact degassing is most efficient when melting and partial vaporization of silicate matter begin. It is worth noting that the release of the primary volatile components begins long before the point when the planet's accumulation process reaches collision velocities corresponding to the melting point of matter. Pioneering works (Lange and Ahrens (1982, 1983)) establish the beginning of volatile loss at extremely low collision velocities of approximately one kilometer per second. Using the Murchison meteorite as an example, the authors established that about 90% of the initial amount of volatiles is already lost at a collision velocity of 1.67 kilometers per second (Tuburczy et al. (1986)). This loss is due to breakdown of the meteorite's water-, carbon-, and sulphur-containing minerals at impact. These experiments are proof of the beginning of volatile release at a very early stage of the Earth's accumulation. Its mass was less than 0.01 of its final mass. Water and carbon dioxide are the main constituents involved in the processes of shock-induced dehydration and decarbonatization of minerals.

As the mass of the embryonic planet increases, escape velocity rises. Consequently, there is also a rise in the velocity at which planetesimals fall. There occurs a corresponding increase in the heating of matter of the planetesimal and the surface of the planet in the central shock zone. The nature of degassing processes undergoes qualitative changes. The shock process becomes a considerably high-temperature one. Chemical reactions occurring in the vaporized cloud become increasingly significant. High-temperature chemical processes begin to play a predominant role during the final stages of Earth's accumulation, instead of the relatively simple processes involved in the breakdown of water-containing compounds and carbonates. Colliding matter undergoes total meltdown in the central zone of impact at collision velocities of five to eight kilometers per second: volatile components are released from the melted matter and interact with it. The chemical composition of the released gases must correspond to the volcanic gases for magma of corresponding composition and temperature. CO_2, H_2O, and SO_2 will clearly be the primary components in such a gas mixture. The gases CO, H_2, H2S, CH_4, and others may be competitors to these components, depending on the extent to which planetesimal matter is reduced (Holland (1964)). Where collision velocities exceed eight to nine kilometers per second, vaporization at impact of a portion of planetesimal matter becomes significant.
The chemistry of a high-temperature, gas-vapor cloud is too complex to be able to judge it solely in terms of the thermodynamical equilibrium in its gaseous phase. Condensation of silicate particles and catalytic activity occur during expansion and cooling of the vaporized cloud. These processes can significantly alter the equilibrious gas chemical composition. This circumstance imposes certain requirements on the search for both theoretical and experimental approaches to the study of chemical processes in a cloud of impact-vaporized matter.

Gerasimov and collaborators (Gerasimov et al. (1984); Gerasimov and Mukhin (1984)) have used laser radiation to examine the chemical composition of gases which form during high-temperature, pulse-induced vaporization of the Earth's rocks and meteorites. Regardless of how reduced the initial matter is, the primary volatile elements H, C, and S are released as oxides: H_2O, CO, CO_2 ($CO/CO_2 \geq 1$), and SO_2. In addition to the oxides, a certain amount of reduced components and organic molecules is formed in the cloud: H_2, H_2S, CS_2, COS, HCN, saturated, unsaturated, and aromatic hydrocarbons from CH_4

to C_6 and CH_2CHO. Molecular nitrogen is released (see Table).

The vaporization-induced gaseous mixtures for samples belonging to both crust and mantle rock, as well as for conventional and carbonaceous chondrite, are qualitatively homogeneous. This is seen in the preponderant formation of oxides, and in both the comparable (within one order of magnitude) ratios of the gases CO/CO_2 and the correlation between the various hydrocarbons. The gas mixtures formed at high temperatures are in non-equilibrium for normal conditions. Therefore, their chemical composition will be easily transformed under the impact of various energy sources.

One should note that the passage of a planetesimal through the atmosphere exerts a substantial influence on the formation of the atmosphere's chemical composition. Fegley and collaborators (Fegley et al. (1986); Prinn and Fegley (1987), Fegley and Prinn (1989)) have analyzed this issue in the greatest detail in recent years. In the physics of the process, a body of large dimension (approximately 10 kilometers in diameter) passes through primordial Earth's atmosphere at a speed of approximately 20 kilometers per second; shock waves send a large amount of energy into the atmosphere. If we put the planetesimal asteroidal density of ≈ 3 g/cm^3 (which matches the chondrite composition), and the angle of entry into the atmosphere at 45°, as first approximation, we would have 2.2×10^{27} ergs. This figure is 0.07% of the energy of an asteroid. The energy passes directly into the atmosphere as the body flies through, and even more energy ($\approx 3.2 \times 10^{29}$ ergs) is "pumped" into the atmosphere by a supersonic discharge of matter from the impact crater (Fegley and Prinn (1989)). The shock wave front (formed in the atmosphere during this process) compresses and heats the atmospheric gas to several thousand degrees Kelvin. It is clear that various thermochemical and plasmochemical reactions are occurring in this region. Due to recombination processes, new compounds are formed as cooling occurs. This substantially alters the initial composition of the atmosphere.

We can estimate the chemical composition of the gas mixtures at high temperatures, using the standard methods of thermodynamical equilibrium (given the presupposition that the time scale for the breakdown of a given molecule is less than the time required to cool the elementary gas exchange). The most detailed computations of these processes were made in the studies of Fegley et al. (1986) and Fegley and Prinn (1989). It seems obvious that these findings must be critically dependent on the initial chemical composition of an unperturbed atmosphere. The atomic ratio C/O is an important parameter here: it determines the "oxidized" and "reduced" state of the atmosphere. Fegley and Prinn (1989) demonstrated that if $C/O > 1$ (reduced atmosphere), precursors of biomolecules, such as HCN and H_2CO, are formed as the asteroid passes through this atmosphere. Nitrogen oxides appear in the oxidized atmosphere instead of prussic acid, supporting the formation of nitric acid if rain falls. Fegley and Prinn (1989) considered several possible chemical compositions for unperturbed atmospheres, and they calculated the commensurate alterations in the composition owing to the passage of large bodies through the atmosphere.

We would like to make the following comment regarding their work. The authors used computation methods employed for purely gaseous reactions. Heterogeneous catalysis on mineral particles (present in the atmosphere during impact-induced discharge of matter) must play a significant role in the actual natural process. Heterophase reactions must considerably affect the evolution of the atmosphere's chemical composition during impact reprocessing. However, it is quite difficult to account for these reactions at this time. An account of the fluxes of such important components as prussic acid and formaldehyde is an unquestionable achievement in the work of Fegley and Prinn. Their efforts made it

TABLE. The results of some typical analyses of the chemical composition of residual gases

Sample	Initial concentration (wt %)		Mass loss	Gases ($x10^{-6}$g)										
	C	S	(mg)	H_2	N_2	CO	CO_2	HC^+	SO_2	H_2S	HCN	CH_3CHO	$\frac{CO}{CO_2}$	$\frac{CO+CO_2}{HC}$
Detection limit				0.2	0.1	0.1	0.02	0.02	0.1	0.08	0.06	0.02		
He atmosphere														
Augite	ND	ND	10.8	<0.2	0.4	3.3	2.7	0.7	<0.1	<0.08	<0.06	<0.02	1.2	8.6
Basalt	0.04*	<0.01*	20.1	<0.2	0.5	3.5	5.9	0.6	<0.1	<0.08	<0.06	≈0.02	0.6	16
Peridotite	0.03*	0.03*	10.2	2.6	0.1	25.4	21.4	1.3	ND	ND	ND	ND	1.2	36
Gabbro	0.05*	3.04*	14.0	19.6	0.9	36.3	28.3	2.6	240	1.8	0.7	0.2	1.3	25
L5 (Tsarev)	ND	1.92 a	7.3	6.2	1.9	36.3	19.8	2.5	17.7	ND	0.2	≈0.08	1.8	22
C3O (Kainsaz)	0.61 b	2.06 c	10.5	18.6	1.9	236	94.3	12.2	<0.1	ND	ND	≈0.04	2.5	27
H_2 atmosphere														
Augite			9.1	ND	ND	25.6	9.0	8.2	<0.1	<0.08	<0.06	<0.02	2.8	4.2
L5 (Tsarev)			7.1	ND	ND	16.9	10.8	7.0	<0.1	162	ND	≈0.1	1.6	4.0
C3O (Kainsaz)			9.7	ND	ND	266	46.4	25.6	<0.1	149	0.4	≈0.2	5.8	12

ND, not determined; * Certified by IGEM Acad. of Sci., Russia; +HC, the sum of hydrocarbons a) Barsukova, L.D., Kharitonova,, V.Ya. and Bannykh, L.N. (1982) Meteoritika 41, 41-43 (in Russian) b) Gibson, E.K., Moore, C.B. and Lewis, C.F. (1971) Geochim. Cosmochim. Acta 35, 599-604 c) Ahrens, L.H., Willis, J.P. and Erlank, A.J. Meteoritics 8, 133-139 (1973).

possible to estimate the stationary concentrations of these components in a modelled early atmosphere. We shall note that the numerical values of the strength of the source of cyanic hydrogen formation in atmospheric reprocessing and in experiments on laser modelling of the processes of shock-induced vaporization are comparable (Mukhin et al. (1989). Hydrocarbons and aldehydes output in the latter case is significantly higher.

4. Concluding Remarks.

It is clear that impact processes created a practically completely determined chemical composition of early atmosphere on the Earth. This atmosphere was very favorable for synthesis of the precursors because it contains all initial components. Moreover, this atmosphere from the very beginning contained some principal precursors. Therefore, we think that the impact processes are very realistic and powerful mechanisms for early stage of chemical evolution on the Earth. Nevertheless, we are very pessimistic in the sense of the origin of life problem. Indeed we have a lot of possible pathways and sources of energy for the creation of the precursors on the early Earth: impacts, volcanoes, electrical discharge, UV, etc. Moreover, we do not see today the serious troubles for the creation of biomolecules and its polymers: polypeptides, polynucleotides, etc. But life is much more complicated than organic chemistry and we can see three main unresolved problems for origin of life:

1. Origin of information in living systems, or origin of genetic code, and its evolution.

2. Origin and evolution of translation machine.

3. Origin and evolution of reduplication processes.

Even if one can think that the early stages of chemical evolution on the Earth are more or less clear today, the above points are far from being fully understood.

Acknowledgements:
This work was supported by Max-Planck-Institut für Chemie, Mainz.

References:

Ahrens, T.J. and O'Keefe, J.D. (1972) 'Shock melting and vaporization of lunar rocks and minerals ', The Moon 4, 214-249.

Arrhenius, G., De, B.R. and Alfven, H. (1974) 'Origin of the ocean', in E.D. Goldberg (ed.), The Sea, Vol. 5, Wiley, New York.

Benlow, A. and Meadows, A.J. (1977) 'The formation of the atmospheres of the terrestrial planets by impact', Astrophys. Space Sci. 46, 293-200.

Cameron, A.G.W. and Pine, M.R. (1973) 'Numerical models of the primitive solar nebular', Icarus 18, 377-406.

Canuto, V.M., Levine, J.S., Augustsson, T.R., Imhoff, C.L. and Giampapa, M.S. (1983) 'The young Sun and the atmosphere and photo-chemistry of the early Earth', Nature 305, 281-286.

Elmegreen, B.G. (1978) 'On the interaction between a strong stellar wind and a surrounding disk nebular', Moon and Planets 19, 261-277.

Fanale, F.P. (1971) 'A case for catastrophic early degassing of the Earth', Chemical Geology 8, 79-105.

Faure, G. (1986) Principals of Isotopes Geology. Sec. ed. Wiley, New York.

Fegley, B. Jr. and Prinn, R.G. (1989) 'Chemical reprocessing of the Earth's present and primordial atmosphere by large impacts', in: G. Visconti (ed.) Interactions of the Solid Planet with the Atmosphere and Climate, in press.

Fegley, B. Jr., Prinn, R.G., Hartmann, H. and Watkins G.H. (1986) 'Chemical effects of large impacts on the Earth's primitive atmosphere', Nature 319, 305-308.

Florenskiy, K.L. (1965) 'On the initial stage of the differentiation of Earth's matter ', Geochemiya 8, 909-917.

Gault, D.E. and Heitowit, E.D. (1963) 'The partition of energy for hypervelocity impact craters formed in rock. Proc. 6th Hypervelocity Impact Symp. 2, 419-456.

Gerasimov, M.V. (1979) 'On the mechanisms of impact degassing of planetesimals', Letters to AJ. 5 (5), 251-256.

Gerasimov, M.V. and Mukhin L.M. (1984) 'Studies of the chemical composition of gaseous phase released from laser pulse evaporated rocks and meteorites materials' (abstract), in: Lunar and Planetary Sci,-XV, Lunar and Planetary Institute, Houston, pp. 298-299.

Gerasimov, M.V. and Mukhin, L.M. (1979) 'On the mechanism forming the atmosphere of the Earth and the terrestrial planets at the stage of their accretion', Letters to AJ. 5(8), 411-414.

Gerasimov, M.V., Mukhin, L.M., Dikov, Yu.P. and Rekharskiy, V.I. (1985) 'The mechanisms of Earth's early differentiation', Vestnik of the USSR Academy of Sciences 9, 10-25.

Hanks, T.C. and Anderson, D.L. (1969) 'The early thermal history of the Earth', Phys. Earth Planet. Interiors 2, 19-29.

Hayashi, C. (1981) 'Formation of planets', in D. Sugimoto, D.Q. Lamb and D.N. Schramm (eds.), Fundamental Problems in the Theory of Stellar Evolution, IAU Symposium No. 93, D. Reidel, Dordrecht, pp. 113-128.

Hayashi, C., Nakazawa, K. and Nakagawa, Y. (1985) 'Formation of the solar system', in D.C. Black and M.S. Matthews (eds.) Protostars and Planets II, University of Arizona Press, Tucson, pp. 1100-1153.

Holland, H.D. (1964) 'On the chemical evolution of the terrestrial and cytherian atmospheres', in P.J. Brancazio and A.G.W. Cameron (eds.) The Origin and Evolution of Atmospheres and Oceans, Wiley, New York, pp. 86-101.

Kipp, M.E. and Melosh, H. (1986) 'Origin of Moon: A preliminary numerical study of colliding planets (abstract)', in Lunar and Planetary Sci.-XVII, Lunar and Planetary Institute, Houston, pp. 420-421.

Kuroda, P.K. and Crouch, W.A.Jr. (1962) On the chronology of the formation of the solar system. 2. Iodine in terrestrial rocks and the xenon 129/136 formation in the Earth', J. Geophys. Res. 67, 4863-4866.

Lange, M.A. and Ahrens, T.J. (1982) 'The evolution of an impact generated atmosphere'; Icarus 51, 96-120.

Lange, M.A. and Ahrens, T.J. (1982) 'Shock-induced dehydration of serpentine: First quantitative results and implications for a primary planetary atmosphere (abstract)', in Lunar and Planetary Sci.-XIII, Lunar and Planetary Institute, Houston, pp. 419-420.

Lewis, J.S. and Prinn, R.G. (1984) 'Planets and their atmospheres: Origin and evolution', in W.L. Donn (ed.) International Geophysics Series, Vol. 33, Academic Press, Inc. Orlando.

McQueen, R., Marsh, S. Taylor, J., Fritz, G and Carter, W. (1973) 'A formula for solid state based on the results of investigation into shock waves, in B.N. Nikolaevskiy (ed.) High Velocity Impact Phenomena, Mir, Moscow, pp. 299-427.

Mukhin, L.M., Gerasimov, M.V. and Safonova, E.N. (1989) 'Origin of precursors of organic molecules during evaporation of meteorites and mafic terrestrial rocks', Nature 340, 45-48.

O'Keefe, J.D. and Ahrens, T.J. (1977) 'Impact induced energy partitioning, melting, and vaporization on terrestrial planets ', Proc. Lunar Sci. Conf. 8, 3357-3374.

Ozima, M. (1975) 'Ar isotopes and earth-atmosphere evolution models ', Geochim. Cosmochim. Acta 39, 1127-1134.

Ozima, M. and Kudo (1972) 'Excess argon in submarine basalts and the Earth-atmosphere evolution model ', Nature Phys. Sci. 239, 23.

Phinney, D., Tennyson, J. and Frick, V. (1978) 'Xenon in CO2 well gas revisited',

Prinn, R.G. and Fegley, B. Jr. (1987) 'Bolide impacts, acid rain, and biospheric traumas at the Cretaceous-Tertiary boundary', Earth Planet. Sci. Lett 83, 1-15.

Safronov, V.S. (1969) Evolution of the Preplanetary Cloud and the Formation of the Earth and the Planets, Nauk, Moscow.

Schidlowski, M. 81988) 'A 3,800-million-year isotopic record of life from carbon in sedimentary rocks ', Nature 333 (6171), 313-318.

Schidlowski, M., private communication.

Sekiya, M.K., Nakazawa, K. and Hayashi, C. (1980) 'Dissipation of the primordial terrestrial atmosphere due to irradiation of the solar EUV ', Prog. of Theor. Phys. 64, 1968-1985.

Tyburczy, J.A., Frisch, B. and Ahrens, T.J. (1986) 'Shock-induced volatile loss from a carbonaceous chondrite: Implications for planetary accretion', Earth Planet. Sci. Lett. 80, 201-207.

Walker, J.C.G. (1977) Evolution of the Atmosphere, Macmillan, New York.

Walker, J.C.G. (1982) 'The earliest atmosphere of the Earth', Precambrian Research 17, 147-171.

Walker, J.C.G. (1991) Degassing in Planetary Science, Proc. from the US-USSR Workshop on Planetary Science, N.A. Press, Washington, D.C. pp. 191-202.

Wetherill, G.W. (1980) 'Formation of the terrestrial planets', Ann. Rev. Astron. Astrophys. 18, 77-113.

Wetherill, G.W. (1986) 'Accumulation of the terrestrial planets and implications concerning lunar origin', in W.K. Hartmann, R. J. Phillips, and G.J. Taylor (eds.) Origin of the Moon, Lunar and Planetary Institute, Houston, pp. 519-550.

Zahnle, K.J. and Walker, J.C.G. (1982) Evolution of solar ultraviolet luminosity. Rev. Geophys. Space Phys. 20, 280-292.

PHYSICAL AND CHEMICAL COMPOSITION OF COMETS - FROM INTERSTELLAR SPACE TO THE EARTH

J. MAYO GREENBERG
Huygens Laboratory
Niels Bohrweg 2
P.O. Box 9504
2300 RA Leiden
The Netherlands

ABSTRACT. The chemical and morphological structure of comets is based on a fluffy aggregate of interstellar dust with an average porosity of P = 0.8. Observations of the volatile and dust (refractory) components of comets are seen to provide evidence of the preservation of a major fraction of the interstellar dust composition in periodic comets - represented mostly by P/Halley - even after 4½ billion years in the Oort cloud. The effects of thermal evolution do not appear to introduce any gross chemical modification. The fluffy character of the nucleus in combination with the submicron units of its structure indicate a possible mechanism for delivery of a substantial fraction of the prebiotic interstellar dust molecules by nucleus fragmentation and ablation so long as comet impacts were cushioned by a dense atmosphere on the early earth.

1. Introduction

That comets are the most primitive objects in the solar system is indisputable. What is questionable is how closely comets may be related to the protosolar nebula material out of which the solar system was born. At this time there are observations of the dust and gas in a wide variety of molecular clouds and there are theories for how to explain in good part the chemical composition of the gas and dust in moderately dense clouds, but not in clouds which have just evolved to the protosolar stage representative of the initial cometary environment. The closest we come to this is in the study of that stage which appears after star formation. There thus remains a missing link between interstellar chemistry and comets. Nevertheless there is considerable evidence that comets retain many of the signatures of interstellar chemistry and it may, even in the absence of our ultimate knowledege of the protosolar composition, be shown that a quite good theory of comets can be built simply on their being aggregated interstellar dust as we may project it to be in the first stage of cloud contraction before the birth of the solar system.

In this chapter interstellar dust will be considered as the basic building blocks of comets. On the one hand the relative amount and kinds of volatiles of interstellar dust will be related to the volatiles of comets. Along with the refractory components of the dust they will be used to derive the chemical and physical properties of comets and comet dust. The initial morphological structure of comet nuclei will be assumed to derive directly from the assumption that comet formation occurs in a relatively cool environment - probably as low a temperature as 30K

J. M. Greenberg et al. (eds.), The Chemistry of Life's Origins, 195–207.

(Yamamoto, 1991a, 1991b; Mumma et al 1993) and that the aggregation of the dust is gentle enough not to heat or disrupt the dust grains. The aggregation occurs as a result first of grain-grain collisions and then by collisions between aggregates in a hierarchy of sizes leading to intermediate cometesimals and ultimately to comets (Weidenschilling et al, 1989). See Figure 1 for a schematic of such aggregation.

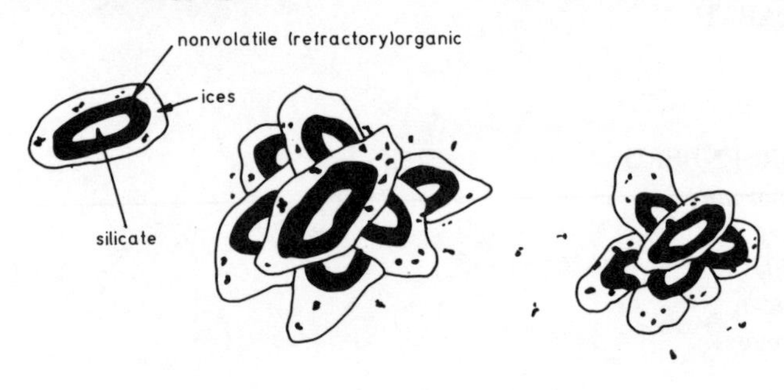

Fig. 1

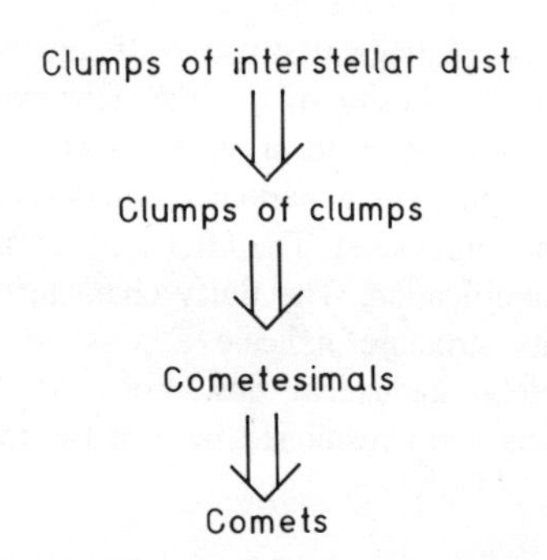

The time scale for this process of growth by aggregation is variously estimated as a million years or less. This kind of aggregation leads to comets being low density, precisely how low is still a matter of debate but, in any case, it is low enough to significantly affect the thermal conductivity of comets. This bears directly on the degree to which comets are self-preserving.

It is completely accepted that comets which formed at whatever distance from the sun - at least those formed in the Uranus/Neptune region and also at larger distances - are subsequently ejected to much greater distances. It is suggested that there are two basic populations of comets - those which reside in a spherical cloud between 50000 to 100000 AU from the sun, called the outer or classical Oort cloud (those which formed in the Uranus/Neptune region) and the Kuiper belt comets (Kuiper, 1951) which circulate in a flattened distribution been 35 to 60 AU from the sun (see the review by Weissman and Campins, 1992). All these comets are presumed to have been formed 4½ billion years ago and have been subject to physical processes which have modified them in varying degrees and to various depths. How well preserved comets are before they enter the inner solar system and how much they are affected by solar heating after they become comets as we "see" them in the inner solar system, plays an important role in our discussion of their relevance to the origins of life. The most critical factors influencing comet nucleus preservation appear to be heating by residual radioactive species and thermal evolution resulting from solar radiation. Other factors, such as cosmic ray penetration while in the Oort cloud and comet collisions are also important but will be seen to be not critically so in our considerations of nucleus interior properties of average comets.

We shall start with an initial comet nucleus whose chemical and morphological compositions are derived from a model of interstellar dust presumed to exist in the latest stages of protosolar nebula condensation. Based on this we shall deduce possible evolutionary tracks of comet nuclei in their 4½ billion year lifetimes and their subsequent thermal evolution after becoming "new" comets. The relationship and relative abundances between observed comet volatiles and refractories will be compared with the presumed interstellar dust composition. The thermal emission properties of comet dust will be used to confirm the low density of comets as well as the idea of interstellar dust particles as the building blocks.

The final discussion will be on the grand design picture of the preservation of presolar system organics in interstellar dust from their time of aggregation into comets to their landing on the earth.

2. The Chemical Composition of Comets

A quantitative estimate of the chemical composition of comets deduced in 1982 directly from interstellar modelling (Greenberg, 1982) has subsequently undergone some modification based both on our increased knowledge of the dust as well as that of comets. In Table 2.1 we present a comet chemical composition, which represents also an updated version of the relative abundances of the common condensable elements with the exception of sulfur. We shall here briefly give the justifications as well as the qualifications on the various components.

TABLE 2.1 Protostellar Dust → Comet Model

	Component	Mass Fraction
D		
U	Silicate (cores)	0.2
S	Carbonaceous	0.05 ($a<0.01\mu m$)
T	Organic residue	0.19
G	H_2O	0.39
	CO	0.03
A	CO_2	0.02
S	Other molecules ($CH_3OH=0.02$, $NH_3<0.01$, $H_2CO=0.03$, $CH_4\leq0.01$, $OCN^-=0.005+....$)	0.12

dust/gas = refractory/volatile = 0.44/0.56 ≈ 0.8

Firstly, the total composition is constrained by the relative abundances of the principal condensable species (C, O, N, Mg, Si, Fe) and secondly by the distribution in molecular species. In Table 2.1 all the condensable species are fully represented. The underabundance of hydrogen relative to its cosmic abundance reflects the composition of the interstellar dust molecules and not the interstellar gas. The 20% for silicates implies that all silicon, magnesium and iron are bound up in rocky materials. The 19% for the organic refractory dust grains is based on the dust model in which the combination of cosmic abundance of the rockies, the total extinction relative to hydrogen, and the wavelength dependence of the extinction lead to a volume ratio of organic mantles to silicate cores of about 2 : 1 (Greenberg, 1989a). Using these and a density ratio of ≈1 : 2 for the mantle to core material gives an approximately equal mass of each. The 5% estimate for very small carbonaceous particles is based on the strength of the extinction (absorption) hump at 216 nm and on the far ultraviolet extinction assuming it is produced by large molecules like polycyclic aromatic hydrocarbons (PAH's). Since neither of these contributions is yet well quantified in terms of absorption per unit mass it may be uncertain by a factor as high as two although a significantly higher carbon content would violate our cosmic abundance constraint.

For the volatile components the H_2O abundance is based on an extrapolation of theory and observation of dust/gas chemistry in fairly dense clouds. A relative error of 20% is not excluded but any increase in H_2O would require a decrease in the CO and CO_2 within cosmic abundance limitations for oxygen. The CO and CO_2 abundances are similarly based on theory plus observation as are the H_2CO and CH_3OH abundances. The CO and CH_3OH abundances will be seen to play important roles in the interpretation of the comet dust properties even though they are obviously volatile components. The rest of the components included under "other" in the volatiles, such as CH_4, NH_3, H_2S, OCS, are known to exist in interstellar dust but not necessarily in all environments. We summarize here constraints on the sources of various comet chemical constitutents as detected or inferred from comet observations. Table 2.2 is reproduced with some modification from Mumma et al (1993)

TABLE 2.2 Some Key Indicators of Cometary Origins: UID ≡ Unmodified Interstellar Dust (Mumma, Stern,Weissman. Protostars and Planets III, 1993)

CO:	Variable abundance supports condensation near 25-50K.
H_2O:	Enriched D/H relative to solar (i.e. H_2) supports interstellar origin Consistent with unmodified interstellar dust (UID) Low spin temperature in Halley supports interstellar origin (UID). High spin temperature in Wilson consistent with radiation damage.
HCN:	Abundance much higher than expected for solar nebula or giant planet subnebulae; supports interstellar origin (UID).
H_2CO:	Abundance much higher than expected for solar nebula or giant planet subnebulae; but about right for interstellar origin. Polymer not expected for solar nebula origin (UID).
$^{12}C/^{13}C$:	Value in CN consistent with solar abundance ratio, but higher than local interstellar medium. Inconclusive, since both grains and volatiles contribute to observed CN.
NH_3:	Low abundance (0.2%) consistent with interstellar origin (UID).
CH_4:	Low abundance (0.2%) consistent with interstellar origin (UID).
CH_3OH:	Abundance consistent with interstellar origin, but not with solar nebula origin (UID).
N_2/NH_3:	Ratio consistent with interstellar origin (UID).
CO/CH_4:	Ratio consistent with interstellar origin (UID).
S_2:	Inconclusive. Telling us something important, but what?
CHON:	Easily destroyed, unique chemistry favors interstellar origin. UID in form of organic refractory (O.R.) grain mantles.
Outbursts:	Significance not yet clear. Perhaps a result of thermal processing.

The question of the volatile composition of comets being attributable to chemical processes in the protosolar or solar nebula has been extensively treated elsewhere (see Fegley this volume and references therein).On the other hand, as can be see from Table 2.2,there is strong evidenced that the low abundance of CH_4 and the relatively high ratio of CO/CH_4 are more consistent with an interstellar dust origin than with a solar system origin which required evaporation of the interstellar dust ices. According to Larson et al. (1989) th CH_4 abundance in the dynamically new comet Wilson and in P/Halley is too low for equilibrium compositional models and much too high for disequilibrium condensation in the solar nebula (Huebner, 1985; Prinn and Fegley, 1988). Insofar as the organic refractory component is concerned the requirement that this be produced by Fischer-Tropsch type reactions catalyzed by hot Fe grains presumes a high abundance of iron particles which were produced by homogeneous condensation of the iron subsequent to complete evaporation of the interstellar dust including its silicate component. This appears to be a very stringent requirement because it demands: (1) reconstituted silicate particles which have grown to the same size as in interstellar dust, (2) the FTT created organics form mantles on the silicate particles, (3) these particles then acquire additional ice mantles and are loosely aggregated in order for the comet dust to reproduce the observed infrared emission features at 3.4 μm and 9.7 μm as shown by Greenberg and Hage (1990). The evidence for tar balls in IDP's is then consistent with these being the residue of such fluffy comet dust out of which the more volatile organics have been evaporated during the time spent in the solar system. In any case whether the FTT reactions lead to too much CH_4 or to organics remains quite uncertain and, with reference to the solar nebula, the theory of grain catalyzed chemistry is not a very robust one.

The general conclusion has to be that most observations favor comets being made of "unmodified interstellar dust material". The question is: "Even if the comet nucleus consists initially of UID, how does it maintain its pristine composition after 4½ billion years in the Oort cloud and subsequent heating by the sun after it becomes a periodic comet". This question will be addressed in the following section.

3. The Morphology and Density of the Comet Nuclei

The density of comet nuclei can only be deduced from indirect observations. One of these methods is based on following the effect of non-gravitational forces on the comet orbits. Various authors have derived densities ranging from as much as 0.6 g cm^{-3} to as little as ≈0.25 g cm^{-3}. The latter (actually 0.28 g cm^{-3}) is preferred by Rickman (1986, 1989, 1991) who presents arguments against such higher values as ~0.7 g cm^{-3} as derived by Sagdeev et al (1987). If one assumes homogeneous aggregation of the nucleus from interstellar material it is possible to derive a reliable upper limit on nucleus densities based on comet dust properties. This was first done by Greenberg (1982) and subsequently updated by Greenberg and Hage (1990) using the chemical composition given in Table 2.1. The upper limit on comet nucleus density of ϱ_{nuc}=1.54 g cm^{-3} can only occur if the dust grains are fully packed in the aggregation process. This appears to be impossible if only because of the elongated core-mantle structure of the tenth micron dust grains. A more likely appearance would be that of a tangled bird's nest as suggested by Greenberg and Gustafson (1981) in which case there will be many voids in the structure. The degree of porosity (P = 1 - filling factor) must be substantial but how to quantify this? An upper limit on comet nucleus density deduced by Sekanina and Yeomans (1985) from the gravitational force splitting of a comet which passed near Jupiter was $\rho_{nuc} \leq 0.3$ g cm^{-3}. Another way to arrive at nucleus densities is to work backwards from meteor densities and from interplanetary particle densities (IDP's) "measured" for particles captured in the earth's upper atmosphere assuming these are both of cometary origin. It is often stated that the IDP's have a density of $\approx$1 g cm^{-3}. Noting that the major composition of IDP's is in the form of silicates with no more than about a 10% contribution of organics (see Greenberg and Hage, 1990, for references) leads to a porosity of P= 0.7 and, with the presumption that these IDP's have become compacted during their ~10,000 years in the system (Mukai and Fechtig 1983) leads to a rather low initial comet density.

The most complete quantitative method designed to deduce comet densities from comet dust was developed by Greenberg and Hage (1990). It used simultaneously: (1) the size distribution of comet dust (McDonnell et al, 1989), (2) the 3.4 μm and 9.7 μm emission strengths (Hanner et al, 1987, Danks et al, 1987, Gertz and Ney, 1986), (3) the shape of the 9.7 μm silicate emission feature(Hanner et al, 1987), (4) mass absorption efficiencies of organics and silicates (Schutte, 1988). At the time it was assumed that the 3.4 μm emission was totally a result of thermal emission by solid particles and, using an organic mass absorption coefficient of ~400 cm^3 gm^{-1} one derived required porosities of P > 0.95 for the comet dust. From this we deduced comet nucleus densities $\rho_{nuc} \approx 0.3$ g cm^{-3}. Recently the question of the contribution to the 3.4 μm emission by fluorescence of the methanol molecule in the gas phase (CH_3OH emission) has been considered. If the extra emission at 3.52 μ in the wing of the 3.4 μm emission is attributed to CH_3HO it has been suggested (Reuter, 1992) that as much as 40% of the 3.4 μm strength is due to this molecule rather than to thermal emission by solid organics. Recent modelling of the ratio of dust extinction to the 3.4 μm absorption to the galactic center and new laboratory measurements (Jenniskens et al, 1993) suggest that a $\kappa_{OR} \approx 200$ may be more realistic than the previously used value of ~400. This compensates in the theoretical results because keeping the amount of organics (needed to heat the dust) as high as before, it emits less at 3.4 μm so that the predicted value of comet dust densities should be essentially as given earlier. Consequently the derived comet density remains little changed from our initial estimate of $\rho_{nuc} \approx 0.3$ gm cm^{-3}. It is not just the nucleus density which is derived but also that the morphological structure of the nucleus required by the comet dust observation must be that of tenth micron units of silicate core-organic refgractory-icy outer mantle particles, i.e., interstellar dust grains. This is important in trying to estimate the thermal properties of the comet nucleus.

4. Thermal Evolution

The major factor modifying the interior of comets is that of heating. The sources of this heating are residual radioactive nuclides, accumulated along with the rest of the interstellar material, and the heating by the sun. Should the comets aggregate during an active phase of stellar evolution they would then also be effected by energetic winds. It is reasonable to assume that such a phase occurs after most of the comets have formed if the rate of comet aggregation is well under 10^6 years.

The two properties of comets which determine their thermal conductivity are: (1) thermal conductivity of the material constitutents and (2) porosity or morphological structure. The observations of H_2O imply a low temperature of comet formation and the fact that the H_2O in interstellar dust is amorphous leads to quite low comet nucleus thermal conductivity. Experimental measurements of the thermal conductivity of ice have resulted in values of the order of 10^6 times less than that of crystalline ice (Kouchi et al, 1992). We note here that unpublished results by Kouchi on H_2O:CO mixtures may lead to an increase in thermal conductivity by a factor of 10. The low density structure of the comet nucleus leads to a further decrease in thermal conductivity because of the reduction of contact surface to conduct the heat. There are many uncertainties in the estimates of the porosity effect. While it has been often assumed by some authors that the reduction, α, is proportional to the volume reduction factor, $\alpha = 1\text{-}P$, this is undoubtedly too weak an assumption. Considered as an aggregate of particles with limited contact surfaces between them already suggests smaller values extending to as little as zero corresponding to hard particle interactions. A model of a comet consisting of equal size spheres in a fully packed cubic lattice where the fractional contact area between adjacent spheres is f leads to a reduction factor of $\alpha \approx 8f(1 - P_0)$ where P_0 is the minimum porosity ($P_0 \rightarrow 0$) of the configuration. A significant squashing of the spheres by 1/10 of their radius is already required for f = 1/20. Thus even for fully packed particles (P=0) one may easily have a reduction by <0.5 corresponding to P = 0.5 in the formula, $\alpha = 1\text{-}P$.

The thermal conductivity first plays a critical role in determining how hot the nucleus becomes as a result of internal heating by radioactive nuclides (Haruyama et al, 1993). Unless we assume that a supernova exploded in the solar neigborhood just when comets were forming - as inferred from ^{26}Al evidence in certain meteorites - the only important nuclide is ^{40}K. This has a very long lifetime, $\sim 10^9$ years so that its effect is long-lived. Haruyama et al. (1993) have examined its effects - particularly with regard to heating the comet to a point of making the ice crystalline - and have established that for a 5 km comet the degree of crystallinity achieved is probably less than ~ 3% if the overall nucleus thermal conductivity (including the porosity effect) is no greater than $\sim 0.3\kappa_{c\text{-}m}$ where $\kappa_{c\text{-}m}$ is the thermal conductivity of the typical silicate core - "ice" mantle grain. If it is significantly lower than this value, the entire nucleus will be heated enough to crystallize the ice; i.e., a low thermal conductivity traps the heat to the point that crystallinity is induced while a high thermal conductivity allows the heat to escape. The upper limit on the effective thermal conductivity which preserves the amorphicity of ice is still quite low by classical standards. It may be shown that it is indeed so low that during the entire time from when a comet is born its maximum temperature during the radiogenic heating is ≤80K. When it is transported out to the Oort cloud where it remains for 4½ billion years it goes down to ~3K and when it is finally brought back to the inner solar system, its temperature over the major volume of the nucleus remains low. Such a comet, when it returns from the Oort cloud to the inner solar system, is a "new" comet, and its temperature will be, over most of its volume, much lower than 80 K, probably ≈50 K (Greenberg et al, 1993). Note that the radionuclide (^{40}K) heating effect is less the smaller the nucleus so that for nuclei with R≈1km, the thermal conductivity remains close to that which it had initially.

5. Comet Impact Delivery of Organics to the Earth - The Primary Source of the Prebiotic Organic Soup?

As early as 1961 Oró (1961) suggested that comets could have supplied part or all of the initial inventory of organic matter for chemical evolution. Of course, at that time, the Whipple ice model of comets was thought of as containing the simplest molecules so that basically the idea was that this inventory was contained in volatile species which provided the "atomic" and simple molecular ingredients needed for further chemical evolution.

Our current understanding of the chemical composition of comet nuclei suggests the possibility that comets could have delivered advanced stages of prebiotic molecules along with the water so that subsequent chemical evolution would have had a head-start. Given in terms of mass alone, the *complex* organics present in one comet with the size of Halley is

$$M_{org} = f_{org}\, M_{com} = 0.19(4\pi/3)\ \rho_c R_c^3 = 3 \times 10^{16} g$$

where f_{org} is the organic refractory fraction, $\rho_c = 0.26 gcm^{-3}$, R_c = 5km.

This is of the order of 1/10 of the entire current biomass of the Earth. Even if only a small fraction of the organics are biologically relevant they would - if delivered intact - have had a really substantial impact on the prebiotic chemistry of the earth. We should note that within the volatile fraction of the comet nucleus as well as in the organic refractory component (see Greenberg and Mendoza-Gomez in this volume) there are also substantial quantities of biologically significant molecules like formaldehyde and various CN containing moleculues including HCN as deduced from precometary interstellar dust evolution. It is well known that, even with a substantial atmosphere, a comet impacting the earth will be strongly shocked. The heat generated by the shock with a comet hitting at 50 km s^{-1} is enormous and at the high temperatures all molecules would be pyrolized so that the initial chemistry would be lost; i.e, the interstellar organics would be substantially destroyed. Is there a way to alleviate the effect of the energetic shocks produced when the comet enters the earth's atmosphere? Such shocks could substantially destroy a major fraction of the *initial* comet organics. This problem has led Clark (1988) to suggest that a soft landing, either a glancing collision or a lower than average impact velocity, might have provided the right condition. But the requirement of a very special condition like this does not give one a completely comfortable feeling of inevitibility. The requirement of soft impact may possibly be already partially satisfied by the nature of comets themselves. So far most calculations of the effect of impact on comets have considered them to be relatively compact objects. We now know that this is not the case. Their low density along with their submicron morphological stucture must be incorporated into the impact physics. Aerobraking of comets has been calculated for a variety of cometary sizes (Chyba, 1990, Chyba et al, 1990) but only for a (*high*) density of 1 g cm^{-3} compared with the densities in 0.3 g cm range derived from comet dust and from non gravitational forces. What has not been considered is the ablation of such a fragile structure as we believe is characteristic of the fluffy comet described in the preceding section. Chyba has, however, shown that the effect of reducing the comet density is roughly equivalent to making the atmosphere thicker ($\rho_c P_{atm}$= constant). The acceleration is reduced the lower the comet density. We expect that the survivability of comet organics depends on both the size of the comet and the scale height of the atmosphere as well as the comet density *and* morphological structure. With a 20 bar CO_2 atmosphere suggested as possible for the early earth (Kasting, 1988, and this volume) the scale height would be large enough to accomodate the deceleration of comets of the order of 1 km. How hot does the comet *material* become if, upon impacting the atmosphere, the comet breaks up into small fragments? In the case of comet fragments made up of aggregated interstellar dust, the heating would be concentrated at the surface because of the exceedingly low thermal conductivity. This would lead to rapid water (ice) evaporation and further fragmentation. This sequence leads to smaller and smaller fragments whose deceleration in the atmosphere is progressively less and whose heating is consequently smaller.

The general result is then that a comet, rather than impacting as a single body, breaks up into smaller and smaller components, each of which is subjected to lower degrees of heating. As described earlier by Greenberg (1986) the strong heating by shock compression of very porous bodies leads to a rarefaction shock (Zel'dovitch and Raizer, 1967). The heat is concentrated in the shock by virtue of the very low thermal conductivity of the *bulk* comet nucleus. The shock dissipates more energy in the low sound velocity (amorphous ice) components which therefore provide an energy sink so that they are quickly evaporated with less heat going into the refractories. In a manner of speaking each grain acts like a free surface unloading material in the direction of the initial motion of the shock. The impulsive evaporation of the volatiles provides the basis for fragmentation and thus can lead to the survival of the core-organic refractory mantle particles. In general, boundaries between the hierarchy of sizes from cometesimals down to individual submicron units would provide further fragile dividing layers splitting the comet into many fragments following a similar hierarchy. The well observed splitting of comets subjected to gravitational forces when passing near massive bodies like Jupiter is a good example of such levels of fragility. Thus rather than the shock leading to complete evaporation of the comet ingredients it leads to *selective* evaporation of the more volatile components (in the amorphous ice) leaving the complex organics relatively cool.

Getting back to the question of *when* comet organics delivery would have been effective in providing the basis for living organism evolution we consider the competition with the destructive effects of high comet flux impacts on the *very* early earth. The impact evidence on extraterrestrial bodies is that up to about the first 5 x 10^8 year the comet and asteroid bombardment would have been so intense that even with a high delivery of organics, no organism is likely to have survived the frequent high temperatures at the earth's surface; i.e. global sterilization was the prevailing mode (Maher and Stevenson, 1988). Therefore in order for life to have *succeeded* it had to have started during the tail-off of the bombardment period. At this time the mass input by comets is greatly reduced. But what is important? Is it the *average* mass input or the concentrated effect of one comet bringing with it all the required ingredients - a *substantial local* region of organics and water? From a survival point of view it appears that the latter seems a reasonable basis for saying that comet's (read interstellar) organics were the basic precursors of life's origins.

6. Concluding Remarks

I would like to conclude by quoting from the paper by Clark (1988) which seems very appropriate to the previous arguments. "This model [the interstellar dust model of comets] cannot yet address all of the known problems of the concept that arose from a natural organic soup, but it does offer a richness of chemical ingredients, physicochemical microenvironments, and succession of processes in a way which is natural and inevitable. Ultimately, much of the theory is testable. We will probably never find a sample of Earth's early atmosphere and apparently not even the primordial oceanic soup. We can, in fact, obtain a sample of bulk cometary matter."

7. Acknowledgements

I would like to thank Dr. Mendoza-Gomez and Dr. Menno de Groot for manny helpful discussions. This work was supported in part by NASA Grant # NGR-33018148 and by a grant from the Netherlands Space Research Organization (SRON).

8. References

Agarwal, V.K., Schutte, W., Greenberg, J.M., Ferris, J.P., Briggs, R., Connor, S., van de Bult, C.P.E.M. and Baas, F. (1985) 'Photochemical reactions in interstellar grains, photolysis of CO, NH_3 and H_2O', Origins of Life 16, 21-40.

Briggs, R., Ertem, G., Ferris, J.P., Greenberg, J.M., McCain, P.J., Mendoza-Gómez, C.X., and Schutte, W. (1992) 'Comet Halley as an aggregate of interstellar dust and further evidence for the photochemical formation of organics in the interstellar medium', Origins of Life and Evolution of the Biosphere 22, 287-307.

Chyba, C.F. (1991), Icarus 92, 217-233.

Chyba, C.F., Thomas, P.J., Brookshaw, L, and Sagan, C. (1990) 'Cometary delivery of organic molecules to the early Earth', Science 249, 366-373.

Clark, B.C. (1988) 'Primeval procreative comet pond', Origins of Life 18, 209-238.

Cronin, J., and Pizarrello, S. (1990) 'Aliphatic hydrocarbons of the Murchison meteorite', Geochim. Cosmochim. Acta 54, 2859-2868.

Danks, A.C., Encrenaz, T.,Bouchet, P., LeBertre, T. and Chalabaev, A. (1987) 'The spectrum of comet P/Halley from 3.0 to 4.0 μm', Astron. Astrophys. 184, 329-332.

Ehrenfreund, P., Robert, F., d'Hendecourt, L. and Behar, F. (1991) 'Comparison of interstellar and meteoritic organic matter at 3.4 μm', Astron. Astrophys. 252, 712-717.

Gehrz, R.D. and Ney, E.P. (1986) in Exploration of Halley's Comet. ESA SP-250, p. 1010.

Greenberg, J.M. (1977) 'From dust to comets', in A.H. Delsemme (ed.), Comets, Asteroids and Meteorites, Univ. of Toledo, pp. 491-497.

Greenberg, J.M. (1986) 'The chemical composition of comets and possible contribution to planet composition and evolution', in R. Smoluchowski, J.N. Bahcall and M.S. Matthews (eds.), The Galaxy and the Solar System, Un. of Arizona Press, pp. 103-115.

Greenberg, J.M. (1989) 'Interstellar dust as the source of organic molecules in comet Halley', Adv. Space Res. 9, nr. 2, 13-22.

Greenberg, J.M. (1989a) 'Interstellar dust, an overview of physical and chemical evolution', in A. Bonetti, J.M. Greenberg and S. Aiello (eds.), Evolution of Intersstellar Dust and Related Topics, North Holland, pp. 7-51.

Greenberg, J.M. (1989b) 'Comet nuclei as aggregated interstellar dust. Comet Halley results', in A. Bonetti, J.M. Greenberg, S. Aiello (eds.), Evolution of interstellar dust and related topics,

Amsterdam, North Holland, pp. 383-395.

Greenberg, J.M. (1990) 'The evidence that comets are made of interstellar dust', in J.W. Mason (ed.),Comet Halley: investigations, results, interpretations, vol. 2: Dust, nucleus, evolution, New York, Ellis Horwood, pp. 99-120.

Greenberg, J.M. (1991b) 'The interplanetary medium is thriving', in A.C. Levasseur-Regourd, H. Hasegawa (eds.), Origin and evolution of interplanetary dust, Kluwer, pp. 443-451.

Greenberg, J.M. (1991a) 'Physical, chemical and optical interactions with interstellar dust', in J.M. Greenberg, V. Pirronello (eds.), Chemistry in Space Kluwer, Dordrecht, pp. 227-261.

Greenberg, J.M. and Gustafson, B. (1981) 'A comet fragment model for zodiacal light particles', Astron. Astrophys. 93, 35-42.

Greenberg, J.M. and Hage, J.I. (1990) 'From interstellar dust to comets: a unification of observational constraints', Astrophys. J. 361, 260-274.

Greenberg, J.M. and Hage, J.I. (1991a) 'From interstellar dust to comets to comet dust: a test of the interstellar dust model of comets', in J.M. Greenberg, V. Pirronello (eds.), Chemistry in Space , Dordrecht, Kluwer, pp. 363-382

Greenberg, J.M. and Hage, J.I. (1991b) 'The interstellar connection to solar system bodies', Space Sci. Rev. 56, 75-82.

Greenberg, J.M. and Mendoza-Gómez, C.X. (1992) 'The seeding of life by comets'. Adv. Space Res. 12, nr. 4, 169-180.

Greenberg, J.M., Mendoza-Gómez, C.X. and Hage, J.I. (1990) 'Interstellar dust in solar system bodies', in J. Krelowski and J. Papaj (eds.), Physics and composition of interstellar matter, Torun, Poland.

Hanner, M.S., Tokunaga, A.T., Golisch, W.E., Gripe, D.M. and Kaminski, C.D. (1987), Astron.Astrophys. 187, 653.

Haruyama, J., Yamamoto, T., Mizutani, H., Greenberg, J.M. (1993) 'Thermal history of comets during residence in the Oort cloud: effect of radiogenic heating in combination with the very low thermal conductivity of amorphous ice', J. Geophys. Res., in press.

Kasting, J.F., Ackerman, T.P. (1986) 'Climatic consequences of very high CO_2 levels in the earth's early atmosphere', Science 234, 1383-1385.

Kissel, J. and Krueger, F.R. (1987) 'The organic component in dust from comet Halley as measured by the PUMA mass spectrometer on board Vega 1', Nature 326, 755-760.

Kouchi, A., Greenberg, J.M., Yamamoto, T. and Mukai, T (1992) 'Extremely low thermal conductivity of amorphous ice: relevance to comet evolution. Astrophys. J. 388, L73-L76.

Kuiper, G.P. (1951) 'On the origin of the solar system', in J.A. Hynek (ed.), Astrophyiscs. McGraw Hill, New York, pp. 357-424.

Léger, A., d'Hendecourt, L., Verstraete, L and Ehrenfreund, P. (1991) 'Small grains and large aromatic molecules' in J.M. Greenberg, V. Pirronello (eds.), Chemistry in Space Kluwer, Dordrecht, p. 221-225.

Maher, K.A. and Stevenson, D.J. (1988) 'Impact frustration of the origin of life', Nature 331, 612-614.

McDonnell, J.A.M., Pankiewicz, G.S., Birchley, P.N.W., Green, S.F. and Perry, C.H. (1989) 'The in-situ cometary particulate size distribution measured for one comet: P/Halley' in Proc. workshop on Analysis of Returned Comet Nucleus Samples, Milpitas, CA, 16-18 Jan. 1987.

Mendoza Gómez, C,X. (1992) 'Complex irradiation products in the interstellar medium', Thesis Leiden.

Mendoza-Gómez, C.X., Greenberg, J.M., (1991) 'Laboratory studies of grain mantles in interstellar space', in A.C. Levasseur-Regourd, H. Hasegawa (eds.), Origin and evolution of interplanetary dust, Kluwer, pp. 437-440.

Mendoza-Gómez, C.X., Greenberg, J.M., Eijkel, G.B. and J.J. Boon (1991) 'Astromacromolecules: formation of very large molecules in interstellar space' in J.M. Greenberg and V. Pirronello (eds.), Chemistry in Space, Kluwer Academic Publishers, Dordrecht, pp. 455-457.

Mukai, T., and Fechtig, H. (1983) 'Packing effect of fluffy particles', Planet Space Sci. 31, 655.

Mumma, M.J., Stern, S.A., and Weissman, P.R. (1993) 'Comets and the origin of the solar system: Reading the Rosetta stone', in E.H. Levy, J.I. Lunine and M.S. Matthews (eds), Planets and Protostars III, Univ. of Arizona Press, Tucson, 1177-1252.

Oberbeck, V.R. and Fogleman, G. (1989) 'Estimates of the maximum time required for origination of life' Paper , Origins of life and Evolution of the Biosphere.

Oró, J. (1961) 'Comets and the formation of biochemical compounds on the primitive earth', Nature 190, 389-390.

Pirronello, V. (1991) 'Physical and chemical effects induced by fast ions in ices of astrophysical interest', in J.M. Greenberg, V. Pirronello (eds.), Chemistry in Space Kluwer, Dordrecht, pp. 263-303.

Reuter, D.C. (1992), 'The contribution of methanol to the 3.4 micron emission feature in comets', Astrophys. J. 386, 330-335

Rickman, H. (1986) 'Masses and densities of comets Halley and Kopff', in O. Melitta (ed.), Comet Nucleus Sample Return Mission, ESA SP-249, pp. 195-205.

Rickman, H. (1989) 'The nucleus of comet Halley: Surface structure, mean density, gas and dust production', Adv. Space Res. 9 nr. 3, 59-71.

Rickman, H. (1991) 'The thermal history and structure of cometary nuclei' in R.L. Newburn, M. Neugebauer and J. Rahe (eds.), Comets in the Post-Halley Era, pp. 733-760.

Sagdeev, R.Z., Elyasberg, P.E., Moroz, V.I. (1987) 'Is the nucleus of comet Halley a low density body?', Nature 331, 240-242.

Schutte, W. (188) 'The evolution of interstellar organic grain mantles', Ph.D. thesis, Leiden.

Sekanina, Z and Yeomans, D.K. (1985) 'Orbital motion, nucleus precession and splitting of periodic Comet Brooks 2', Astron. J. 90, 2335-2352.

Smoluchowski, R. (1985) 'Amorphous and porous ices in cometary nuclei' in J. Klinger, D. Benest, A. Dollfus and R. Smoluchowski (eds.), Ices in the Solar System, Reidel, Dordrecht, p. 397.

Tielens, A.G.G.M. (1991) 'Characteristics of interstellar and circumstellar dust', in A.C. Levasseur-Regourd, H. Hasegawa (eds.), Origin and evolution of interplanetary dust, Kluwer, pp. 405-412.

Weidenschilling, S.J., Donn, B. and Meakin, P. (1989), in H.A. Weaver, F. Paresce and L. Danly (eds.), Formation and Evolution of Planetary Systems, Cambridge University Press.

Weissman, P.R. and Campins, H. (1992) 'Short period comets' in Near Earth Resources, Un. Arizona press.

Yamamoto, T. (1991a) 'Chemical composition of dust expected from condensation models', in A.C. Levasseur-Regourd and H. Hasegawa (eds.), Origins and Evolution of Interplanetary Dust, Kluwer Academic Publishers, Dordrecht, pp. 413-420.

Yamamoto, T. (1991b) 'Chemical theories on the origin of comets' in R.L. Newburn, M. Neugebauer and J. Rahe (eds.), Comets in the Post-Halley Era, pp. 361-376.

Zel'dovitch, Ya.B., and Raizer, Yu. P. (1967) 'Physics of shock waves and high-temperature hydrodynamic phenomena', Academic Press, New York.

ORGANIC MATTER IN METEORITES: MOLECULAR AND ISOTOPIC ANALYSES OF THE MURCHISON METEORITE

JOHN R. CRONIN
Department of Chemistry and Biochemistry
Arizona State University
Tempe, AZ 85287-1604
USA

SHERWOOD CHANG
Planetary Biology Branch
NASA-Ames Research Center
Moffett Field, CA 94035
USA

ABSTRACT. Carbonaceous chondrites comprise a unique subset of meteorites. Two classes of carbonaceous chondrites, the so-called CI1 and CM2 chondrites, are particularly interesting, in part because of their relatively high carbon content and the fact that most of this carbon is present as organic matter. This material is largely macromolecular but also contains a complex mixture of organic compounds that include carboxylic acids, dicarboxylic acids, amino acids, hydroxy acids, sulfonic acids, phosphonic acids, amines, amides, nitrogen heterocycles including purines and a pyrimidine, alcohols, carbonyl compounds, and aliphatic, aromatic, and polar hydrocarbons. The organic-rich CI1 and CM2 chondrites also contain an extensive clay mineralogy and other minerals that are believed to be indicative of an early episode of hydrous activity in the meteorite parent body. Recent stable isotope measurements have shown the organic matter in general, to be substantially enriched in deuterium and the discrete organic compounds to be enriched in ^{15}N and somewhat enriched in ^{13}C relative to terrestrial matter. These findings suggest that the organic matter is comprised of, or is closely related to, interstellar organic compounds. The organic chemistry of these meteorites is consistent with a formation scheme in which (1) a parent body was formed from volatile-rich icy planetesimals containing interstellar organic matter, (2) warming of the parent body led to an extensive aqueous phase in which the interstellar organics underwent various reactions, and (3) residual volatiles were largely lost leaving behind the suite of nonvolatile compounds that now characterize these meteorites.

J. M. Greenberg et al. (eds.), The Chemistry of Life's Origins, 209–258.

1. Introduction

Over the last 30 years, we have learned that organic chemistry, i.e., the chemistry of carbon-containing compounds, plays or has played an important role in a variety of space environments, e.g., gas phase and grain processes in the interstellar medium (Turner 1989), the atmospheric chemistry of several planets and planetary satellites (Lunine, 1989), as well as the chemical evolution of asteroids (Gaffey *et al.*, 1989) and comets (Krueger *et al.* 1991). The occurrence of organic chemistry in space environments, many of which were once viewed as inimical to the formation or survival of organic compounds, reminds us of the ubiquity of the major biogenic elements, H, C, N, and O, and the fact that they comprise four of the five most abundant elements in the universe. It would not be an exaggeration to view the chemistry of the cosmos as largely organic chemistry. This perception underscores the potential for chemical evolution of organic matter in the solar system and beyond. Such potential holds important implications for the origin and distribution of life and has added a new dimension to theory and experiment in this area.

Prior to this period, astronomical observations had revealed simple species like CH_4, C_2, and CN in Jupiter's atmosphere, comets and circumstellar regions, however, meteorites, particularly the carbonaceous chondrites, provided the only samples for direct study of extraterrestrial organic chemistry. Analyses of meteoritic organic matter and speculation about its origin date back more than 150 years (Nagy, 1975). These investigations, particularly those carried out over the last 25 years, now provide a detailed record of extraterrestrial organic chemistry which illuminates not only chemical processes in the early solar system, but , as we now believe, presolar events as well.

The early attempts to determine the organic composition of carbonaceous chondrites were, in many cases, compromised by the use of samples that were significantly contaminated by terrestrial organic matter (Hayes, 1967). In most instances, these have been superseded by more recent work employing better samples and more powerful analytical techniques. Therefore, we will not review this area comprehensively , but rather will concentrate on the work that most influences our current view of the nature and origin of meteorite organic compounds. Much of this will focus on work done with the Murchison meteorite since 1969, the year of its fall. The suite of compounds found in Murchison appears to be representative, in a qualitative way at least, of meteorites of its type (CM2), however, it should be noted that these meteorites are not identical with respect to their organic content and that Murchison is not even homogeneous in this respect (Cronin and Pizzarello, 1983).

Most of the work done on meteorite organic chemistry prior to the fall of Murchison has been critically reviewed by Hayes (1967) and a book by Nagy (1975) contains additional material of considerable historical interest. The subject has been reviewed more recently by Hayatsu and Anders (1981), Mullie and Reisse (1987), and Cronin *et al.*, (1988). Pillinger (1984) has reviewed light element stable isotope abundances in meteorites.

1.1. METEORITES

This section provides a brief and selective introduction to meteorites and to some relevant terminology unique to meteoritics. It focusses on the carbonaceous chondrites and attempts to describe their special characteritics and their place among meteorites, in general. A recent, comprehensive treatment of meteoritics can be found in *Meteorites and the Early Solar System,* edited by Kerridge and Matthews (1988).

Meteorites are classified as belonging to one of three general categories: *stones*, *irons*, and *stony-irons*. Most numerous are the stones, which account for more than 90 percent of meteorite falls. Stony meteorites are subdivided into two classes, *chondrites*, which account for greater than 90 percent of the stony meteorites, and *achondrites*. The defining characteristic of chondrites is the presence of *chondrules*, mm-sized spherical bodies distributed throughout the stone, which appear to have solidified from melt droplets suspended in a gas phase prior to their incorporation into the chondrite. Chondrules evidence a pervasive, although poorly understood, thermal event(s) that occurred prior to agglomeration of the body from which a chondritic meteorite was derived, i.e., the so-called *parent body*. The various parent bodies required to account for the properties of chondritic meteorites are thought to have originated in the asteroid belt, some 2 to 4 AU from the sun. (One AU, i.e., astronomical unit, is defined as the mean distance from the Earth to the Sun.) The chondrules are imbedded in a *matrix* composed of fine-grained minerals and larger mineral fragments. Together, matrix and chondrules represent a regolith environment on an asteroidal body where components were accreted, comminuted by impacts, and modified by thermal metamorphism and/or aqueous alteration.

Chondrites are considered to be *primitive* solar system materials on the basis of their chemical composition. With the exception of the volatile elements, hydrogen, helium, carbon, nitrogen, and oxygen, these meteorites have an elemental composition equivalent to that of the Sun and, since more than 99 percent of the solar system mass resides in the Sun, to that of the solar nebula from which the Sun and planets formed. Thus these meteorites have not experienced processes leading to extensive chemical fractionation such as partial evaporation or gravitational separation from a melt.

Chondrites are subclassified using a two-parameter system in which they are assigned to one of nine *chemical classes*: CI, CM, CV, CO, H, L, LL, EH, and EL. These classes are distinguished by small, systematic compositional differences. The first four classes comprise the *carbonaceous (C) chondrites*, so named because of their relatively high content of carbon and the other volatile elements, H, N, O, and S. These are subdivided on the basis of their similarity to four, prototypical meteorites, Ivuna, Mighei, Vigarano, and Ornans, thus the designations, I, M, V, and O. The next three classes, H, L, and LL, comprise the *ordinary chondrites*, the most abundant groups, and the last two classes, EH and EL, comprise the *enstatite (E) chondrites*.

The second parameter by which the chondrites are subclassified is by *petrographic type*, a numerical measure, scaling from 1 to 6, of the extent to which a chondrite has been altered by aqueous processing, thermal effects, or shock.

1.1.1. *Carbonaceous Chondrites*. Carbonaceous chondrites are unique among the chondrites in several respects. (1) They are the most primitive in terms of elemental composition, i.e., the least depleted in volatile elements. For example, their carbon contents range from 0.3 to >3% weight. (2) They are relatively oxidized with respect to the ratio Fe(II):Fe(III). (3) They provide the only examples among the chondrites of petrographic types 1 (CI chondrites) and 2 (CM chondrites). On the basis of elemental composition, it was once believed that these chondrite classes had suffered minimal alteration relative to all other chondrites. They are now known to be composed of mineral assemblages dominated by hydrous, layer lattice silicates (clay minerals) and it is widely accepted that the CI1 and CM2 chondrites are unique in having experienced *aqueous alteration* of their original anhydrous silicate matrix. (4) The CI1 and CM2 chondrites are rich in organic matter. It should be emphasized that abundant organic compounds are observed only in those chondrites containing abundant clays, i.e., only in those chondrites in which originally anhydrous minerals were converted to hydrous assemblages by the action of liquid water.

The unusual properties of the CI1 and CM2 chondrites have led to the suggestion that they are not asteroidal, but rather fragments of devolatilized cometary nuclei, a proposition against which Anders (1975,1978) has argued convincingly. On the other hand, a role for cometary material cannot be excluded. Bunch and Chang (1980) pointed out that accretion of icy cometesimals in the parent bodies followed by melting would have provided the liquid water and organic compounds for these objects.

2. Carbon Distribution in Carbonaceous Chondrites

The carbon content of carbonaceous chondrites varies systematically with petrographic type. Mason (1963) reported mean values for 16 specimens as follows: C1(I), 3.54%; C2(M), 2.46%; and C3, 0.41%. Gibson *et al.*, (1971) analyzed 19 carbonaceous chondrites and found the carbon content of C1(I) to be > 3%, C2(M) to range from 1.8 to 2.6% and C3 to range from 0.22% to 1.15%. The carbon content of C1(I) and C2(M) chondrites is dominated by an organic component, but also includes carbonate minerals and so-called "exotic" carbon phases.

2.1. SOLUBLE CARBON

The carbonaceous matter of CM2 chondrites can be fractionated on the basis of solubility. A *soluble fraction* is obtained by extracting a powdered meteorite sample sequentially with a series of solvents varying in polarity. These extracts contain a complex mixture of discrete organic compounds such as amino acids, aliphatic hydrocarbons, etc. Hayes (1967) has summarized the results of several experiments in which the mass extracted by each solvent was determined. These results indicate that about 30 to 40% of the carbonaceous matter of CM2 chondrites is soluble, however, these figures are an overestimate since the amounts were not corrected for inorganic salts coextracted by the more polar solvents.

2.2. INSOLUBLE CARBON

The bulk of the organic matter is retained in the residual *insoluble fraction* along with the exotic carbon phases and carbonate minerals. It is composed primarily of a poorly characterized, structurally heterogeneous, macromolecular material, which has been variously named, but commonly referred to as either meteorite "polymer" or "kerogen-like" material. This macromolecular carbon, along with relatively minor amounts of the exotic carbon phases, is obtained by digesting the insoluble fraction with HF-HCl mixtures, a process that dissolves most inorganic minerals including the carbonates. The exotic carbon species, which were first recognized by their content of isotopically anomalous Ne and Xe, are believed to have formed in the outflows of carbon stars. They have recently been isolated by sedimentation after HF-HCl digestion of the insoluble fraction followed by progressive oxidation of the macromolecular carbon, and characterized as diamond, ~100 ppm (Lewis *et al.*, 1987; Blake *et al.*, 1988); silicon carbide, ~6 ppm (Tang *et al.*, 1989); and graphite, ≤ 2 ppm (Amari *et al.*, 1990).

Carbonate minerals are found in samples of the Murchison meteorite in amounts that account for between 2 and 10% of the total carbon (Grady *et al.*, 1988) and in similar amounts in other CM2 chondrites (Smith and Kaplan, 1970). Several Ca, Mg, and Fe-containing carbonate minerals have been identified (Grady *et al.*, 1988).

2.2.1. *Macromolecular Carbon.* The major organic component of CM2 meteorites is the macromolecular carbon. Smith and Kaplan (1970) reported values as high as 54% of total carbon for the macromolecular carbon of four CM2 chondrites. This is very likely a minimal value since a significant fraction of the total carbon was unaccounted for in their attempt to obtain a carbon balance for these meteorites.

Preparations of this material obtained after exhaustive HF-HCl digestion of a Murchison powder (approximately 4% residual ash) gave the empirical formula, $C_{100}H_{71}N_3O_{12}S_2$, calculated on an ash-free basis (Hayatsu *et al.*, 1977; Hayatsu *et al.*, 1980). Zinner (1988) calculated a formula of $C_{100}H_{48}N_{1.8}O_{12}S_2$ for this material based on pyrolytic release studies. The ^{13}C NMR spectrum of a partially demineralized (6.65 wt. %C) Murchison sample is relatively simple with two broad features characteristic, respectively, of sp^2-(olefinic/aromatic) and sp^3-hybridized (aliphatic) carbon atoms in a variety of structural environments (Cronin *et al.*, 1987). Both the H/C ratio and simplicity of the ^{13}C-NMR spectrum are reminiscent of the more aromatic (type III) terrestrial kerogens (Miknis *et al.*, 1984).

Hayatsu *et al.*, (1977,1980) investigated the structure of the Murchison macromolecular carbon using a variety of degradative techniques. They concluded that the material is comprised of condensed aromatic, heteroaromatic and hydroaromatic ring systems in up to four-ring clusters, cross linked by short methylene chains, ethers, sulfides and biphenyl groups. Hayatsu *et al.* (1983) also noted the similarity between their results with the Murchison macromolecular carbon and those obtained with vitrinite macerals of low-volatile bituminous coal or type III mature kerogen. Despite a general similarity to these terrestrial materials, significant differences in the detailed structure have been brought out by comparison of CuO-oxidation products (Hayatsu *et al.*, 1980).

Isotopic analyses accompanying stepwise pyrolysis or oxidation suggest the presence of three or four isotopically and structurally distinct components within the macromolecular carbon which are inferred to be predominantly aliphatic or aromatic in nature (Kerridge *et al.*, 1987).

The preceding findings point toward a hybrid aliphatic-aromatic character for the macromolecular component, however, none indicates whether these are characteristics of a single carbon phase or properties of a mixture with discrete aliphatic and aromatic components. High resolution electron microscopy (HREM) of the HF-HCl digested insoluble fraction has been unable to distinguish between these alternatives. Lumpkin (1986) found this material to occur predominantly in irregular clumps having both amorphous and poorly developed turbostratic structures. The clumps are composed of particles in the range 0.01 to <0.1 μm (Reynolds et al., 1978). Not surprisingly, the insoluble carbon was observed to be heterogeneous and small amounts of other carbon phases were apparent, e.g., circular grains with poorly developed turbostratic structure, carbon black-like particles, and polycrystalline sheets of partially ordered graphite.

Some portion of the macromolecular carbon may occur in fluorescent, μm-size carbonaceous inclusions of various morphologies. Rossignol-Strick and Barghoorn (1971) thoroughly surveyed the particles liberated by HF-digestion of the Orgueil meteorite and identified numerous hollow spheres as well as less abundant irregular objects having membranous and spiraled structures. They suggested that these objects are composed of macromolecular carbon and proposed abiotic schemes for their formation. The hollow spheres may have originated as coatings on mineral grains. Alpern and Benkheiri (1973) studied sections of the Orgueil meteorite by fluorescence microscopy and observed both small spherical bodies and larger, irregular objects formed around a central grain. These structures are reminiscent of the organic refractory mantles in Greenberg's (1973, 1977) interstellar dust model. Similar fluorescent particles have been observed in Murchison sections (Deamer 1985). Fluorescent material can also be extracted with a chloroform-methanol mixture suggesting that the fluorescence is not entirely attributable to macromolecular carbon. The chloroform-methanol extracts of Murchison were shown to form fluorescent droplets when dispersed in aqueous solution, a finding that has led Deamer (1985) to speculate on a possible role in prebiotic membrane formation.

3. Organic Compounds

3.1. AMINO ACIDS

Amino acids are essential for all terrestrial life, and it has been widely assumed that they were essential for the origin of life. Consequently, their presence in meteorites has been a matter of considerable interest. Recent reports that RNA can function as both catalyst and genetic material weaken this assumption (Joyce, 1989). Nonetheless, meteoritic amino acids constitute unequivocal evidence of abiotic synthesis of biologically important compounds, thus providing a firm foundation for the central concept that natural processes gave rise to the building blocks of life and ultimately to life itself.

The first positive results from amino acid analyses of meteorites were reported by Degens and Bajor (1962). Although these were followed by several confirming reports the recognition by Hamilton (1965) and Oró and Skewes (1965) that the amino acids reported closely resembled those found in dust or human skin caused the early results to be discredited. It was not until the benchmark analyses of Kvenvolden *et al.* (1970) that amino acids were firmly established as indigenous to a carbonaceous chondrite (Murchison). The occurrence of chiral amino acids as a 1:1 mixture of D- and L-enantiomers and the discovery of several amino acids not found in proteins, and thus unlikely to be terrestrial contaminants, provided compelling evidence of extraterrestrial origin. By 1975, 22 amino acids had been positively identified in Murchison hot-water extracts using ion exchange chromatography (IEC), gas chromatography (GC), and combined gas chromatography-mass spectrometry (GC-MS) (Kvenvolden *et al.*, 1970); Oró *et al.*, 1971; Kvenvolden *et al.*, 1971; Lawless, 1973; Buhl, 1975). These included eight of the protein amino acids, ten of more restricted biological occurrence, and four that were considered to be very rare, if not nonexistent in the biosphere. Since then, more than 50 additional amino acids have been identified, bringing the total to more than 70 (Cronin *et al.*, 1981; Cronin *et al.*, 1985; Cronin and Pizzarello, 1986). Most of the more recently identified amino acids are biologically unknown and thus may be unique to carbonaceous chondrites.

Structurally, these amino acids are of two types: monoamino alkanoic acids [I] and monoamino alkandioic acids [II]. Two variations on these basic structures have been observed: N-alkyl derivatives [III] and cyclic amino acids in which an alkyl side chain is bonded to the α-amino group [IV]. Within the limits posed by these classes, complete structural diversity prevails. For example, there are 14 possible chain isomers for seven-carbon α-amino acids. Four of these have two chiral centers, i.e., exist as two diastereomeric forms. Thus, counting diastereomers but excluding enantiomers, 18 isomeric forms exist which are, in principle, separable by ordinary chromatographic methods. All 18 isomers have been identified in a Murchison extract (Cronin and Pizzarello, 1986). Other structural characteristics of the meteoritic amino acids are:

1. At each carbon number the abundance order is $\alpha > \gamma > \beta$.
2. Within isomeric sets, branched carbon chain isomers are more abundant than the straight chain isomers.
3. Within homologous series, e.g., the straight chain α-amino acids, there is a smooth, exponential decline in amount with increasing carbon number (slope ≅ -0.7).

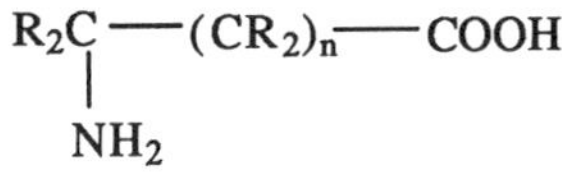

$R_2C(NH_2)$—$(CR_2)_n$—COOH

I

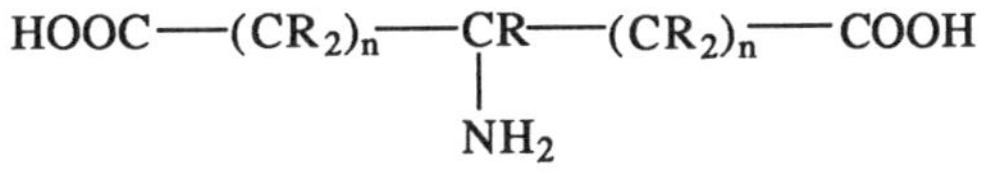

HOOC—$(CR_2)_n$—$CR(NH_2)$—$(CR_2)_n$—COOH

II

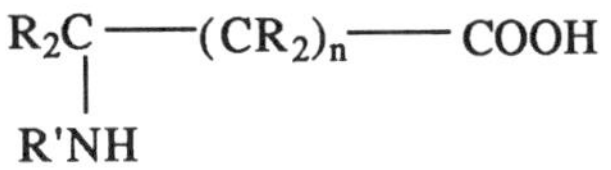

$R_2C(R'NH)$—$(CR_2)_n$—COOH

III

(pyrrolidine ring, NH) CH—COOH

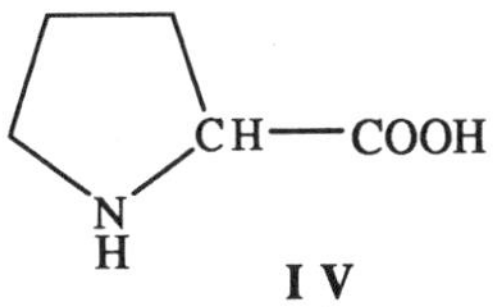

IV

$R = H$ or C_nH_{2n+1}

$R' = C_nH_{2n+1}$

The amino acids found in greatest abundance are usually glycine, the lowest carbon number α-amino acid, and α-aminoisobutyric acid, the lowest carbon number branched chain α-amino acid. The abundance of this four-carbon amino acid clearly illustrates the apparent preference for branched chain structures noted above. Amino acids may occur individually at concentrations up to about 100 nmol g^{-1}. The total amino acids may reach a concentration approaching 700 nmol g^{-1}. In discussing amino acid composition, it is important to note that there is significant quantitative heterogeneity among Murchison specimens. For example, in some samples the composition is dominated by α-methyl-α-amino alkanoic acids, i.e., α-aminoisobutyric acid, isovaline, etc. (Cronin and Pizzarello, 1983).

The hydroxy amino acids, serine and threonine, are commonly found in meteorite extracts, however, since these amino acids, serine in particular, are easily acquired contaminants, their status as meteorite constituents is questionable. Recently, Murchison samples were obtained by drilling from the surface toward the center of a stone in increments of a few millimeters. These were analyzed for serine and threonine and their enantiomers by HPLC. Although small amounts of both amino acids were found in the interior samples, the enantiomeric analyses were consistent with a biological origin (Pizzarello and Cronin, unpublished).

3.1.1. *Amino Acid Precursors.* Amino acids occur in a Murchison hot-water extract as both free amino acids and as derivatives or precursors that can be converted to amino acids by acid hydrolysis (Cronin and Moore, 1971). The total amino acid content of an extract is approximately doubled by hydrolysis, with increases in individual amino acids ranging from about 40% to 800% (Cronin, 1976a). The acid-labile compounds that account for these increases have been partially characterized by ion exchange chromatography. About 70 percent were found to be acidic compounds (Cronin, 1976a) suggesting that they may be derivatives in which the basicity of the amino group has been lost, for example, as in N-acyl amino acids.

Recently, Cooper and Cronin (unpublished) investigated the composition of the neutral and acidic fraction of a Murchison hot-water extract by GC-MS analysis of silylated derivatives. They found an extensive series of 2-carboxy-γ-lactams (pyroglutamic acid and alkyl substituted pyroglutamic acids), 2-carboxy-δ-lactams, as well as alkyl substituted and unsubstituted γ-and δ-lactams, compounds which on hydrolysis would yield, respectively, glutamic acid and higher γ-carboxy amino acid homologues, α-amino adipic acid and higher δ-carboxy-α-amino adipic acid homologues, γ-amino butyric acid and higher γ-amino acid homologues, and δ-amino valeric acid and higher δ-amino acid homologues. In addition, N-acetyl derivatives of glycine, alanine, α-amino-isobutyric acid, and aspartic acid were found in small amounts. The carboxy-γ- and δ-lactams and N-acetyl amino acids have the properties required of the acidic amino acid precursors. It remains to be determined whether the amounts of these compounds are sufficient to account for the amino acid increases seen on acid hydrolysis.

3.1.2. *Amino Acid Chirality*. The enantiomeric ratios of the Murchison chiral amino acids have been of considerable interest. Early GC analyses of volatile diastereomeric derivatives (e.g., N-TFA-(+)-2-butyl esters) showed them to be racemic, within experimental error, when extracted from uncontaminated interior samples (Kvenvolden *et al.*, 1970; Oró *et al.*, 1971; Pollock *et al.*, 1975). These results provided a powerful argument for an extraterrestrial origin for those Murchison amino acids which also occur biologically. Conflicting results were reported by Engel and Nagy (1982) who found an excess of the L-enantiomer in five Murchison amino acids (all protein amino acids). This work was criticized by other workers in the field who found reasons to attribute the result to terrestrial contamination of the sample (Bada *et al.*, 1983). Recently, Engel *et al.* (1990a) have claimed an excess of the L-enantiomer in alanine from the Murchison meteorite. They found, in addition, that the $\delta^{13}C$ values of both L and D-enantiomers showed enrichments in ^{13}C relative to terrestrial organic matter (see below) and that the difference in isotopic composition between them was too small ($\Delta\delta^{13}C$ = 3‰) to be consistent with the terrestrial contamination necessary for the L-enantiomer excess. Although the isotopic data are consistent with Engel's earlier contention, the observation that the enantiomeric excess occurred only in the biological enantiomer (L) of a common protein amino acid, the lack of a plausible abiotic explanation consistent with the formation conditions for these amino acids, the uncertainties associated with the isotopic measurements, and the contradiction of earlier results obtained with a fresher and presumably more pristine meteorite sample (Kvenvolden *et al.*, 1970) all give reason to reserve judgement regarding indigenous enantiomeric excesses.

3.1.3. *Stable Isotopic Analyses*. Carbon isotopic analyses of various preparations of the Murchison amino acids are summarized in Table 1. Chang *et al.* (1978) found a preparation of the total amino acids to have a $\delta^{13}C$ of +24.6 ‰, a value in good agreement with those obtained later for similar preparations (Epstein *et al.*, 1987; Pizzarello *et al.*, 1991). Fractionation of the amino acids gave subsets having $\delta^{13}C$ values between +23 and +44‰ (paper chromatography) and +30 to +41 ‰ (ion-exchange chromatography). Engel *et al.* (1990a) obtained $\delta^{13}C$ values between +5 and +30 for N-TFA-isopropyl esters of individual amino acids or amino acid enantiomers using a gas chromatography-isotope-ratio mass spectrometer (GC-IRMS) and correcting for the carbon added in derivatization.

TABLE 1. Carbon and hydrogen isotopic composition of Murchison amino acids

Preparation	$\delta^{13}C$ (‰)	δD (‰)	Reference
Bases and Amino Acids	+24.6	—	(1)
Amino Acids (four fractions)	+23 to +44[a]	—	(1)
Amino Acids (total)	+23.1[a]	+891[a] (+1370)[b]	(2)
Amino Acids (total)	+26[a]	+1137[a] (+1751)[b]	(3)
Glycine-Alanine fraction	+44[a]	+1072[a] (+2448)[b]	(3)
Valine-Isovaline fraction	+30[a]	+713[a] (+1014)[b]	(3)
α-Aminoisobutyric acid	+5	—	(4)
L-Glutamic acid	+6	—	(4)
Isovaline	+17	—	(4)
Glycine	+22	—	(4)
D-Alanine	+30	—	(4)
L-Alanine	+27	—	(4)

(1) Chang *et al.*, 1978; (2) Epstein *et al.*, 1987; (3) Pizzarello *et al.*, 1991; (4) Engel *et al.*, 1990a.

(a) Corrected for procedural blank. (b) corrected for exchangable hydrogen.

Thus meteoritic amino acids are clearly enriched in ^{13}C relative to their terrestrial counterparts, an additional criterion by which to confirm their extraterrestrial origin. Furthermore, individual amino acids seem to vary significantly with respect to ^{13}C content, although it is not clear from the existing data whether there is a systematic variation with carbon chain length as has been observed for the monocarboxylic acids (Section 3.3.). It is noteworthy that similarly strong ^{13}C enrichments occur in the amino acids, CO_2 and carbonates of Murchison, suggesting derivation from the same isotopic reservoir(s) of carbon.

The Murchison amino acids are also enriched in ^{15}N (Epstein *et al.*, 1987). The $\delta^{15}N$ of +90‰ measured for a preparation of the total amino acids lies well outside the -20 to +20‰ range of terrestrial nitrogen values. For comparison, the results of N-isotopic measurements made on several other Murchison phases are collected in Table 2. Ammonia is the source of the α-amino group nitrogen of the amino acids formed in a Strecker-cyanohydrin synthesis (Section 5.4.1.1.). Thus the substantial ^{15}N enrichment observed in amino acid nitrogen (Epstein *et al.*, 1987) implies at least as great an enrichment in the precursor ammonia. Kung and Clayton (1978) found a $\delta^{15}N$ value of +70 for "exchangable" ammonium ion from Murchison, a large enrichment in ^{15}N by terrestrial standards although not quite to the degree observed for the amino acids.

TABLE 2. Nitrogen isotopic composition of the Murchison meteorite

Preparation	N (ppm)	$\delta^{15}N$ (‰)	Reference
Whole rock[a]	845	+43	(1)
Whole rock[b]	828	+43.9	(2)
Whole rock[b]	950	+42	(2)
Mainly inorg. N[c]	170	+72	(2)
"Exchangable" NH_4^+[d]	24	+70	(2)
Macromolecular matter[e]	—	+18	(3)
Methanol extract		+77,+98,+90	(4)
Total Amino acids		+90	(5)

(1) Injerd & Kaplan, 1974; (2) Kung & Clayton, 1978; (3) Robert & Epstein, 1982; (4) Becker & Epstein, 1982; (5) Epstein *et al.*, 1987
(a) sealed tube Kjeldahl; (b) vacuum pyrolysis; (c) bomb Kjeldahl;
(d) MgO distillation; (e) HF-HCL insoluble

Epstein *et al.* (1987) also measured the D:H ratio of the Murchison total amino acids and found them to be substantially enriched in deuterium (δD = +1370 ‰) relative to terrestrial matter. This result was confirmed by Pizzarello *et al.* (1991) who found even greater D-enrichment in the glycine-alanine fraction obtained by ion exchange chromatography (δD = +2448 ‰). These results indicate D-enrichments in the Murchison amino acids of up to 3.5 times with respect to the terrestrial bulk value and of about 28 times with respect to the protosolar nebula D/H of 2×10^{-5} (Geiss and Reeves, 1981). Deuterium enrichments of this order have also been observed in the insoluble carbon of carbonaceous chondrites (Robert and Epstein, 1982). They are attributed (Kolodny *et al.*, 1980; Geiss and Reeves, 1981; Robert and Epstein, 1982) to isotopic fractionation effects in the formation of the precursor organic compounds by low-temperature (<50K), zero activation energy, gas phase ion-molecule reactions (Watson, 1976) and, possibly, grain-mediated (Tielens, 1983) reactions in the interstellar cloud from which the solar system formed. Even greater D-enrichments are observed in the organic molecules of dark interstellar clouds (Millar *et al.*, 1989). Whether such ion-molecule and grain processes could also have occurred in the solar nebula remains to be evaluated. Ordinary neutral-neutral chemical reactions occurring in the cooling solar nebula seem to be ruled out for the formation of these D-rich species. Reactions at the low temperatures (<200K) required to achieve comparable deuterium enrichments through thermodynamic equilibration with nebular H_2 would be too sluggish to be effective over the theoretical lifetime of the solar nebula (Geiss and Reeves, 1981). The more modest enrichments observed in ^{13}C and ^{15}N are consistent with an interstellar cloud origin for the amino acid precursors. The lower ΔE values for ion molecule ^{12}C-^{13}C and ^{14}N-^{15}N exchange reactions dictate lower enrichments for ^{13}C and ^{15}N than for deuterium at a common temperature (Wannier,

1980). The implications of these isotopic enrichments for the formation of the amino acids are discussed in Section 5.

3.2. HYDROXYCARBOXYLIC ACIDS

Hydroxycarboxylic acids were first identified in the Murchison meteorite by Peltzer and Bada (1978). They obtained a suite of seven α-hydroxycarboxylic acids [V] by extracting the powdered meteorite with hot-water, followed by ion exchange and silica-gel chromatography. The individual hydroxycarboxylic acids were identified by GC-MS of the methyl esters and found to comprise the complete set of α-isomers having five or fewer carbon atoms. Volatile diastereomeric derivatives were prepared for four of the five chiral hydroxy acids and enantiomer ratios (D/L) ranging from 0.82±0.1 to 0.93±0.3 were obtained by GC. Peltzer and Bada (1978) considered the hydroxy acids to be essentially racemic, a finding consistent with an abiotic origin.

$$R_2C(OH)—(CR_2)_n—COOH$$

V

$R = H$ or C_nH_{2n+1}

More recently, Cronin *et al.* (1993) have extended these analyses in the course of preparing the Murchison hydroxycarboxylic acids for isotopic analysis. In this work the hydroxy acids were purified by ion exchange chromatography, ether extraction and preparative HPLC, and then analyzed as t-butyl dimethylsilyl (tBDMS) derivatives by GC-MS. The results revealed a much more extensive suite of hydroxycarboxylic acids than had been observed previously. A total of 51 hydroxy acids were found, including β- and γ-substituted isomers, with carbon chains through C_8. The distribution of isomers appears to be similar to that observed previously for the amino acids, i.e., the amino- and hydroxy acids appear to comprise structurally analogous sets of compounds. The correlation between meteoritic α-hydroxy and α-amino acids supports the earlier suggestion of Peltzer *et al.* (1984) that the Murchison α-amino- and α-hydroxy acids result from Strecker-cyanohydrin syntheses. They also found that the molar ratios of three α-hydroxy- and α-amino acid pairs are consistent with their equilibration at the same ammonia concentration, as would have been the case if a common Strecker-cyanohydrin synthesis formed both sets of compounds.

TABLE 3. Carbon and hydrogen isotopic composition of Murchison hydroxy- and dicarboxylic acids

Preparation	$\delta^{13}C$ (‰)	δD (‰)	Reference
Total Hydroxy acids (Na salts)	+4	+573[a]	(1)
Total Dicarboxylic acids (Na salts) (includes hydroxy dicarboxylic acids)	-6	+357[a]	(1)

(1) Cronin *et al.*, 1993.

(a) Corrected for exchangeable hydroxyl proton.

Isotopic analyses of the total hydroxy acids gave the results shown in Table 3. A large, positive δD value was obtained, suggesting the formation of C-H bonds under cold interstellar cloud conditions. The fact that the δD and $\delta^{13}C$ values are both lower than those of the amino acids (Section 3.1.3.) is discussed in relation to the possibility of a Strecker synthesis of meteoritic amino and hydroxy acids in Section 5.4.1.1.

3.3. CARBOXYLIC ACIDS

The carboxylic acids [VI] of the Murchison meteorite have been the subject of a series of detailed analyses, carried out by Yuen and coworkers (Yuen and Kvenvolden, 1973; Lawless and Yuen, 1979; Yuen *et al.*, 1984; Epstein *et al.*, 1987). This work has provided several significant insights to the origin of these compounds and, by inference, to the origin of meteorite organic compounds in general.

$$R{-}COOH$$

VI

$$R = H \text{ or } C_nH_{2n+1}$$

The results are summarized as follows. Carboxylic acids were extracted from meteorite powders with either hot water or a KOH-methanol solution in μmol g^{-1} amounts, i.e., amounts 10-100 times greater than the amino acids of corresponding carbon number. Yuen positively identified 17 saturated aliphatic carboxylic acids with chain lengths from one (Kimball, 1988; Yuen, unpublished results) to eight (Yuen and Kvenvolden, 1973). All isomers through C_5 were identified as were at least six of the eight possible C_6 carboxylic acids. Shimoyama (1989) has obtained similar results (21 carboxylic acids identified through C_{12}) with a CM2 chondrite from the Japanese Antarctic collection. The amounts decline in the straight chain series by about 60% with each additional carbon atom, decreases similar to those seen in the amino acids (Section 3.1.). The branched chain isomers are abundant, at least through C_6. The acids presumably exist in the meteorite largely as carboxylate salts. The slightly alkaline pH value of the water extract of a Murchison powder suggests similar pH conditions for the final aqueous phase experienced by the parent body. Furthermore, if the acids had been present in the protonated carboxyl form, the lower members of the series would, in all likelihood, have been lost by evaporation unless sequestered in a special environment, for example, as is methane. Whether a portion of the carboxylic acids exist as acid-labile precursors, as appears to be the case with the amino acids, has not been investigated. However, preliminary results indicate the presence of carboxamides and acetyl amino acids, which would yield carboxylic acids and acetic acid, respectively, upon acid hydrolysis (Cooper and Cronin, unpublished). The carboxamides may have been derived in turn from partial hydrolysis of nitriles (Section 5.4.2.).

Yuen *et al.* (1983) obtained the first stable isotope ratios for individual gases and organic compounds from a meteorite, i.e., $\delta^{13}C$ values for CO, CO_2, and the C_2-C_5 branched and straight chain carboxylic acids (see Table 4). These results, which have now been confirmed (Yuen *et al.*, unpublished) show a decline in $\delta^{13}C$(+22.7 to +4.5 ‰) from acetic acid through valeric acid. The branched chain isomers are slightly heavier than their straight chain counterparts at C_4 and C_5. The data suggest a kinetically-controlled chain elongation mechanism in which the higher homologues are built up by the addition of one-carbon species. Furthermore, the acids are ^{13}C-enriched relative to the macromoleclar carbon, a finding inconsistent with their formation by degradation of the macromolecular carbon. The identification of CO as the one-carbon precursor, as in FTT processes, seems to be ruled out by the relative depletion of ^{13}C in the CO obtained from the meteorite (Table 4) and the known ^{13}C-fractionation pattern for CO in the FTT synthesis (Lancet and Anders, 1970; Yuen *et al.*, 1991). The ^{13}C-enrichment of the CO_2 (and possibly the carbonate carbon) coexisting with the carboxylic acids makes it a viable candidate for a one-carbon source, although probably not in FTT processes (Yuen *et al.*, 1991).

Yuen *et al.* (1983) noted that CO_2 was more ^{13}C-enriched than the isotopically heaviest acid, acetic acid. In fact, taken as C_1, its $\delta^{13}C$ value plots smoothly, as a function of carbon number, with the carboxylic acid values. This observation, along with the finding that the aliphatic hydrocarbons are isotopically lighter than the corresponding carboxylic acids (compare Tables 4 and 5), raises the possibility that the ^{13}C content of the carboxylic acid carboxyl group is greater than that of the aliphatic group. This relative enrichment has recently been experimentally demonstrated by isotopic analyses of the decarboxylation products of Murchison acetic acid (Yuen *et al.*, unpublished). This has led to speculation that formation of the carboxylic acids involved a multi-step process in which the carboxyl and aliphatic carbon atoms were derived from isotopically distinct precursors. For example, one can imagine the addition of HCN to an alkene and hydrolysis of the resulting nitrile during aqueous processing of the meteoritic parent body, the combination of alkyl and cyanide radicals followed by hydrolysis, or the combination of carboxyl with alkyl radicals as possible alternative pathways. At least one carboxylic acid (formic acid) and both saturated and unsaturated nitriles are known to be interstellar molecules (Turner, 1989).

TABLE 4. Carbon and hydrogen isotopic composition of Murchison monocarboxylic acids

Preparation	$\delta^{13}C$ (‰)	δD (‰)	Reference
Neutral and acidic compounds	+13.1	n.d.	(1)
Individual acids (C_2-C_5)	+4.5 to +22.7	n.d.	(2)
CO_2	+29.1	—	(2)
CO	-32	—	(2)
Total acids (Na salts)	+6.7	+377	(3)
Total acids (Na salts), triplicate analyses.	-3, 0, -1	+581, +697, +679	(4)

(1) Chang *et al.*, 1978; (2) Yuen *et al.*, 1984; (3) Epstein *et al.*, 1987; (4) Krishnamurthy *et al.*, 1992.

Recently, δD values have been obtained for the Murchison monocarboxylic acids (Epstein *et al.*, 1987; Krishnamurthy *et al.*, 1992) (Table 4). They are D-rich, although not to the degree of the amino acids. Nevertheless, the enrichment may be indicative of an interstellar origin as discussed previously for the amino acids (Section 3.1.3.). Kerridge

(1991) has discussed the formation of various organic acids in the Murchison meteorite and taken the position that all of them were derived from interstellar precursors during aqueous processing of the parent body. If we are correct in the supposition that D-enrichment is indicative of an interstellar origin for meteoritic compounds or their precursors, then the intermolecular and intramolecular differences in carbon isotopic composition among the carboxylic acids discovered by Yuen and his coworkers may represent heretofore unrecognized characteristics of carbon chain formation under interstellar cloud conditions.

3.4. DICARBOXYLIC ACIDS

Dicarboxylic acids [VII] were discovered in hot-water extracts of the Murchison meteorite by Lawless *et al.* (1974). They were separated from other components of the extracts by silica-gel chromatography and extraction into ether and their analysis was carried out by GC-MS of various esters. Seventeen compounds in this series were identified having carbon numbers through C_9. Three of these have chiral centers and in the case of methylsuccinic acid, both enantiomers were identified and shown to be present in about equal amounts. Fifteen of the 17 dicarboxylic acids had saturated aliphatic carbon chains.

$$HOOC—(CR_2)_n—COOH$$

VII

$$R = H \text{ or } C_nH_{2n+1}$$

Recently, Cronin *et al.* (1993) analyzed a preparation of the Murchison dicarboxylic acids that had been prepared for isotopic analysis. In this case, the dicarboxylic acids were obtained by hot-water extraction, hydrolysis, extraction into ether and group separation from other acids by HPLC. They were converted to disilyl derivatives and analyzed by GC-MS. Forty-two aliphatic dicarboxylic acids through C_8 were recognized. All chain isomers through C_5 were recognized (8) and most (25 of 33) at C_6 and C_7. These compounds, like the amino acids and hydroxy acids, constitute a suite of great, if not complete, structural diversity.

Isotopic analysis of this preparation, which included hydroxydicarboxylic acids (see Section 3.4.1), gave the results shown in Table 3. The δD value is clearly outside the terrestrial range, although lower than the values observed for the monocarboxylic acids, the hydroxy acids, and the amino acids. The $\delta^{13}C$ value is also the lowest observed within this group although a value of -3 ‰ was obtained for one of the triplicate samples of the monocarboxylic acids. In comparing $\delta^{13}C$ and δD values between classes of compounds it is important to note that there appears to be significant isotopic heterogeneity within the Murchison soluble organics. For example, Yuen *et al.* (upublished results) obtained a range of values for individual straight chain monocarboxylic acids (C_2 through C_5) of -17.9 to +13.8 when the experiment documented in Table 4 was repeated. The δD and $\delta^{13}C$ values obtained for two preparations of the total monocarboxylic acids are also different (see Table 4). Consequently it is not possible to say with certainty that the dicarboxylic and monocarboxylic acids are isotopically distinguishable.

We assume that the dicarboxylic acids exist in the meteorite as the carboxylate dianions for the reasons discussed previously. Oxalic acid has been shown to occur as whewellite, its Ca salt (Fuchs *et al.*, 1973).

It is interesting to note that series of pyrrolidiones and piperidiones have recently been recognized in meteorite extracts (Cooper and Cronin, unpublished). These compounds are converted respectively to various β- and γ-dicarboxylic acids upon hydrolysis. It is an interesting question whether such compounds might be precursors of all meteoritic dicarboxylic acids or whether they represent an alternative pathway to their formation. Again the question of multiple pathways of formation arises.

3.4.1 *Hydroxydicarboxylic Acids.* Analysis of the Murchison dicarboxylic acids (Cronin *et al.*, 1992) disclosed an extensive suite of hydroxydicarboxylic acids (VIII), compounds which had previously escaped detection in meteorites. More than 50 hydroxydicarboxylic acids were recognized, of which five α-substituted compounds were positively identified, including malic acid, α-hydroxy glutaric acid, and 2-hydroxy-2-methyl succinic acid, which correspond to three meteoritic α-amino acids, aspartic acid, glutamic acid, and 2-methyl aspartic acid, respectively. Many β-, γ-, etc., hydroxydicarboxylic acids were found for which corresponding amino dicarboxylic acids have not been identified. The finding of the additional sets of α-hydroxy and α-amino acid analogues is expected if Strecker-cyanohydrin reactions were responsible for their formation. The finding of a more numerous set of hydroxydicarboxylic acids than dicarboxylic amino acids suggests that a search for additional members of the latter class of compounds might be productive. The hydroxydicarboxylic acids were included with the dicarboxylic acids in the preparation for which isotopic values were obtained (Table 3), thus a value for either class of compound alone is not available.

$$HOOC—(CR_2)_n—\underset{\displaystyle OH}{\underset{|}{C}R}—(CR_2)_n—COOH$$

VIII

$R = H$ or C_nH_{2n+1}

3.5. SULFONIC AND PHOSPHONIC ACIDS

Recent analyses of water extracts of Murchison have disclosed analogous suites of alkyl sulfonates [IX] and alkyl phosphonates [X] (Cooper *et al.*, 1992). Both series were identified by GC-MS after silylation of an acidic/neutral fraction obtained by cation exchange chromatography of the meteorite aqueous extract. They were also identified by ion chromatography without derivatization. The alkyl phosphonates represent the first organophosphorous compounds recognized in meteorites. In the case of the alkyl sulfonates, seven of the eight possible compounds through C4 were recognized; five of the corresponding alkyl phosphonates were observed. Both series appear to show the decline in amount with increasing carbon number that has been seen in other aliphatic acid series from Murchison.

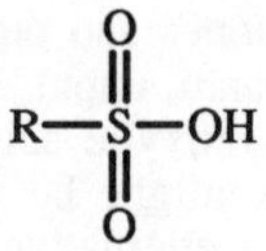

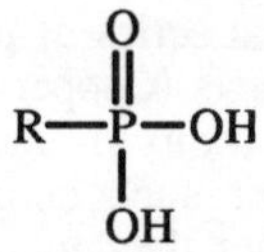

$$R-\overset{O}{\overset{\|}{\underset{\underset{O}{\|}}{S}}}-OH \qquad R-\overset{O}{\overset{\|}{\underset{\underset{OH}{|}}{P}}}-OH$$

IX **X**

$$R = C_nH_{2n+1}$$

Although isotopic analyses have not been done with these compounds, they are thought to be derived from interstellar precursors. An abundant sulfur-containing carbon chain series, C_nS, has been identified in molecular clouds (Wilson *et al.*, 1971; Saito *et al.*, 1987; Yamamoto *et al*, 1987) and the radical of the first member of a possible phosphorous-containing series, CP, has also been observed (Guélin *et al.*, 1990). Deuterium analyses of the meteoritic compounds would be particularly interesting since they could be indicative of the site of hydrogenation, i.e., interstellar or parent body.

3.6. HYDROCARBONS

The hydrocarbons of CM2 meteorites, and the Murchison meteorite in particular, have been the subject of considerable interest and analytical effort. Two distinct suites of hydrocarbons differing in molecular weight and volatility have been found.

3.6.1. *Higher (nonvolatile) hydrocarbons.* The higher molecular weight, non-volatile hydrocarbons can be extracted from meteorite powders using various organic solvents or solvent mixtures, e.g., benzene-methanol. The extract can be separated into aliphatic, aromatic, and polar hydrocarbon fractions by adsorption onto silica gel and elution with, respectively, hexane, benzene, and methanol (Meinschein *et al.*, 1963). These fractions are suitable for direct analysis by GC or GC-MS.

3.6.1.1. *Aliphatic hydrocarbons.* The nature of the higher aliphatic hydrocarbons from Murchison has been controversial. Three different laboratories analyzed these compounds soon after the meteorite fell and obtained somewhat discordant results ranging from predominance of aliphatic polycyclic compounds e.g., [XI] (Kvenvolden *et al.*, 1970) to predominance of straight-chain alkanes [XII] (Studier *et al.*, 1972). Nevertheless, it came

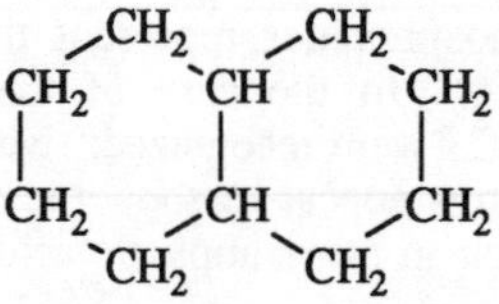

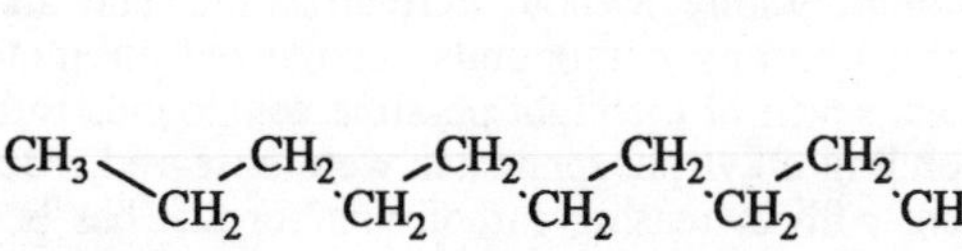

CH_2, CH_2, CH_2, CH, CH_2, CH_2, CH, CH_2, CH_2, CH_2 (bicyclic ring structure)

XI

$$CH_3-CH_2-CH_2-CH_2-CH_2-CH_2-CH_2-CH_2-CH_2-CH_3$$

XII

to be widely believed that straight-chain alkanes (n-alkanes) were the dominant components of this fraction and the specificity of the synthetic process inferred from this result has often been cited in support of a Fischer-Tropsch type process as the formation mechanism (see, for example, Hayatsu and Anders, 1981). Even so, the dominance of n-alkanes remained suspect in view of several other findings.

1. Kvenvolden *et al.* (1970) did not find abundant n-alkanes in the earliest analysis carried out with fresh interior samples of Murchison.
2. Han *et al.* (1969) analyzed the Allende meteorite within weeks of its fall and found that it had already *acquired* superficial hydrocarbons dominated by n-alkanes.
3. Evaluation of a large amount of early meteorite hydrocarbon data showed that n-alkane content correlated with the amount of isoprenoid hydrocarbons, well known terrestrial biomolecules (Hayes, 1967; Wasson, 1974).
4. The finding of straight chain dominance in the alkanes was at odds with the finding of abundant chain branching in the aliphatic portion of other Murchison organic compounds, e.g., amino acids and monocarboxylic acids.

TABLE 5. Carbon and hydrogen isotopic composition of Murchison hydrocarbons

Preparation	$\delta^{13}C$ (‰)	δD (‰)	Reference
Benzene-methanol extract	+4.4, +4.8, +5.9	—	(1)
Methanol extract	+7.2	+406	(2)
Sequential extraction			
1. CCL_4 extract	-8.5	+112	(2)
2. Methanol extract	+9.6	+486	(2)
3. Methanol extract	+8.5	+461	(2)
Methanol extract	+4	+957	(3)
Fractionated benzene-methanol extract			
Aliphatic fraction	-10, -13	+249, +280	(3)
Aromatic fraction	-5, -6	+346, +468	(3)
Polar fraction	+6, +6	+945, +947	(3)
Aliphatic fraction	-5	+103	(3)
Aromatic fraction	-5	+244	(3)
Polar fraction	+5	+751	(3)
Individual C_1-C_4 alkanes	+2.4 to +9.2	—	(4)
Total C_2-C_5 alkanes			
freeze-thaw disagg.	—	-92	(3)
hot water	+6	+217	(3)
H_2SO_4	+17	+410	(3)

(1) Kvenvolden *et al.* (1970); (2) Becker and Epstein (1982); (3) Krishnamurthy *et al.* (1992); (4) Yuen *et al.* (1984).

Recently, Cronin and Pizzarello (1990) carried out analyses of the Murchison aliphatic hydrocarbons prepared for isotopic analysis (Krishnamurthy *et al.*, 1992). When interior samples were obtained and the analyses were carried out under conditions that minimized environmental contaminants, they found the principal aliphatic components of the meteorite to be structurally diverse C_{15} to C_{30} branched alkyl-substituted mono-, di-, and tricyclic alkanes. They concluded that the n-alkanes, methyl alkanes, and isoprenoid alkanes found in some extracts are terrestrial contaminants and that the argument for an FTT synthesis based on selectivity for straight chain compounds therefore was not valid.

The origin of the polycyclic aliphatic compounds is unknown. Measurements of their isotopic composition (Table 5) show $\delta^{13}C$ values (-5 to -13 ‰) at the heavy end of the terrestrial range accompanied by significant deuterium enrichments (+103 to +280 ‰).

3.6.1.2. *Aromatic hydrocarbons.* Several analyses of the Murchison aromatic hydrocarbons have given results that are in general agreement. Recently, Krishnamurthy *et al.* (1992) analyzed a preparation that had been obtained for isotopic analysis. They found the most abundant compounds in the silica gel benzene fraction to be pyrene [XIII], fluoranthene [XIV], phenanthrene [XV], and acenaphthene [XVI] in abundance ratios of about 10:10:5:1, respectively. Other unsubstituted aromatic compounds found in lower amounts included naphthalene, biphenyl, acenaphthylene, 9H-fluorene or another $C_{13}H_{10}$ isomer, anthracene, benzo[ghi]fluoranthene, cyclopenta[cd]pyrene, four $C_{18}H_{12}$ isomers, i.e., chrysene, etc., and five $C_{20}H_{12}$ isomers, i.e., benzo[e]pyrene, etc. The remainder of the compounds are apparently alkyl substituted and partially hydrogenated forms of these polycyclic aromatic compounds. Deuterium enrichments in the aromatic hydrocarbon fraction ranged from +244 to +468 ‰ (Table 5) signifying an interstellar input. The corresponding $\delta^{13}C$ values were about -5 ‰. Gilmour *et al.* (1992) have obtained $\delta^{13}C$ values by GC-IRMS for several individual, unsubstituted aromatic compounds from a similar preparation. The values obtained for eight compounds ranged from -5.9 to -22.3 ‰. Although no systematic trend in $\delta^{13}C$ with molecular weight was observed, the higher molecular weight compounds tended to be isotopically lighter.

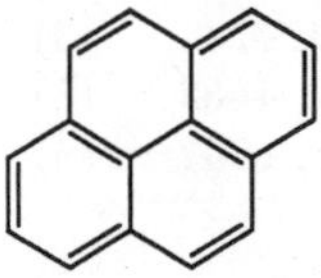

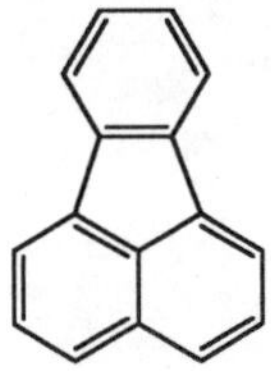

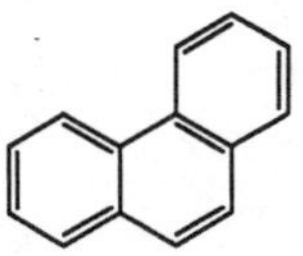

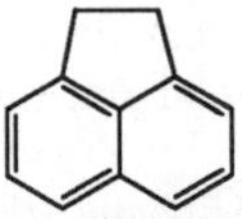

XIII **XIV** **XV** **XVI**

Results of the molecular analysis by Krishnamurthy *et al.* (1992) compare well with the findings of earlier analyses of the Murchison non-volatile aromatic compounds (Pering and Ponnamperuma, 1971; Studier *et al.*, 1972; Basile *et al.*, 1984). The minor differences among all these data sets seem to derive mainly from the extent to which the more volatile components, e.g., naphthalene, methyl naphthalenes, biphenyl, acenaphthene, and acenaphthylene, were lost by evaporation during preparation of the samples. These latter compounds are apparently present in the meteorite, but may be completely lost if the extract is rigorously dried before analysis (Basile *et al.*, 1984). In effect, the solvent extraction-GC analytical method gives qualitatively and quantitatively valid results for only a fairly narrow range of aromatic compounds. Compounds with boiling points less than about 300°C may be partially or completely lost by evaporation whereas those with boiling points greater than about 500°C are not eluted from the GC column. Among aromatic compounds, the 300-500°C boiling point range corresponds to a molecular weight range of about 178 to 252 daltons , i.e., to tri- through pentacyclic compounds and their alkyl derivatives.

Recently, several investigators have carried out analyses of meteorite aromatic hydrocarbons using the laser desorption/ionization-mass spectrometry (L^2MS) technique. Hahn *et al.* (1988) analyzed powdered Murchison samples and have reported naphthalene, phenanthrene/anthracene, and fluoranthene/pyrene and various C_1, C_2, and C_3 alkyl derivatives. Zenobi *et al.* (1989) examined powders of the Allende (CV3) meteorite using this technique and reported naphthalene and phenanthrene, their alkyl derivatives through, respectively, C_5 and C_6, alkylbenzenes, and alkyl biphenyl/acenaphthenes. Wing and Bada (1991) investigated the aromatic hydrocarbons of Ivuna (CI1) and found abundant phenathrene/anthracene (3 ring) but no naphthalene (2 ring) or pyrene/fluoranthene (4 ring). They interpret these unexpected results in terms of a geochromatographic separation of aromatic hydrocarbons during aqueous alteration. Tingle *et al.* (1991) used the SALI technique (surface analysis by laser ionization) to analyze the aromatic hydrocarbons in intact, freshly exposed interior fragments from both the Murchison and Allende meteorites. They found that unsubstituted PAHs are the dominant aromatic compounds in Murchison, whereas extensive substitution of PAHs is indicated in the Allende compounds. Kovalenko *et al.* (1992) have recently found that the aromatic compounds desorbed from acid-insoluble residues of ordinary chondrites by L^2MS also show an increase in alkylation with increasing petrographic type.

De Vries *et al.* (1991) extended the molecular weight range of chondritic aromatic hydrocarbons to encompass coronene and its alkyl-substituted derivatives. Their results were obtained by applying the L^2MS technique to sublimates obtained under high vacuum at 300, 450 and 600°C from residues of samples of Murchison and Allende which had been demineralized by HCl and HF treatments. The sublimation process favors retention of the higher molecular weight compounds and loss of the more volatile, lower molecular weight analogs. Unidentified PAHs up to 750 daltons were revealed when Fourier transform ion cyclotron resonance mass spectrometry was used to analyze a dried toluene extract of an interior Murchison sample. Another notable feature of these analyses is that an upper limit of 2 ppb was set for the presence of Buckminsterfullerene, C_{60}, the lack of which contradicts the prediction that fullerenes are abundant interstellar molecules (Kroto, 1988).

3.6.1.3. *Polar hydrocarbons*. The bulk of the material extractable from Murchison with benzene-methanol is found after silica-gel chromatography in the methanol fraction. In two preparations analyzed by Krishnamurthy *et al.* (1992), this fraction contained 69 and 74 percent of the total eluted carbon. Similar percentages were found when benzene-methanol extracts of the Orgueil and Murray meteorites were fractionated on silica gel (Meinschein *et al.*, 1963). The components of the Murchison methanol fraction, that is the more polar, heteroatom-containing hydrocarbons, have been analyzed in detail only recently (Krishnamurthy *et al.*, 1992). Earlier, Basile *et al.* (1984) found several compounds with heteroatoms in a silica-gel benzene fraction, and four of these, fluoren-9-one [XVII], anthracenone [XVIII], anthracenedione [XIX], and dibenzothiophene [XX] were also found in the methanol fraction by Krishnamurthy *et al.* (1992). Using different extraction methods, others have found several water-soluble N-heterocyclic compounds such as substituted pyridines, quinolines, isoquinolines, purines, and pyrimidines in Murchison extracts (Hayatsu *et al.*, 1975; Stoks and Schwartz, 1982) (Section 3.7).

Krishnamurthy *et al.* (1992) tentatively identified four classes of compounds in the Murchison methanol fraction on the basis of mass spectra. These were (1) several series of alkyl aryl ketones [XXI], e.g., alkyl naphthyl-, alkyl biphenyl-, and alkyl anthracyl/phenanthryl ketones; (2) aromatic ketones and diketones which, in addition to those found by Basile *et al.* (1984), include $C_{15}H_8O$ (possibly oxocyclopentaphenanthrene), $C_{17}H_{10}O$ (benzofluorenone isomers), and $C_{18}H_{10}O_2$ (benzoanthracenedione); (3) nitrogen heterocycles which include $C_{13}H_9N$ (benzoquinoline isomers) [XXII] and $C_{15}H_9N$ (azapyrene/azafluoranthene); and (4) sulfur heterocycles of which dibenzothiophene [XX] was identified. The methanol fraction contains a mixture of enormous complexity and the series identified represent only a fraction of the total suite of compounds. In most cases, homologous series could be identified only by a characteristic single ion plot. Individual compounds were generally imbedded in a mixture of structurally similar and chromatographically unresolved compounds and in many cases were not apparent as peaks in the total ion plot. Although the identified compounds were mainly aromatic, mass spectra suggested the presence, in addition, of aliphatic compounds with polar substitutents, e.g., heteroalicyclic compounds,

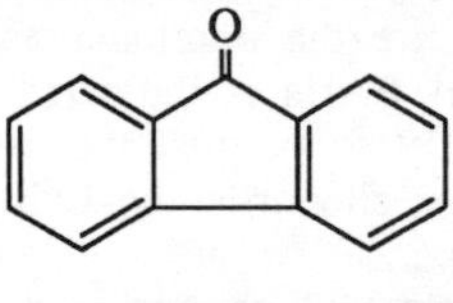

XVII 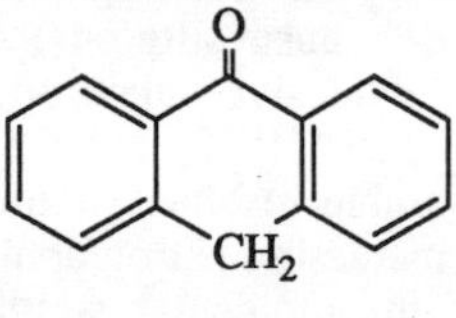 **XVIII** 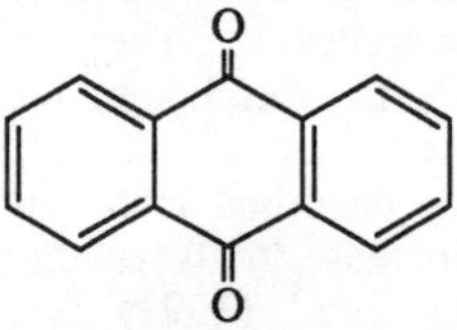 **XIX**

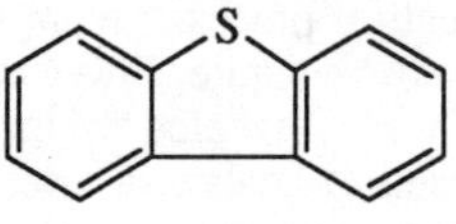

XX 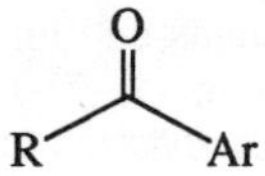 **XXI** 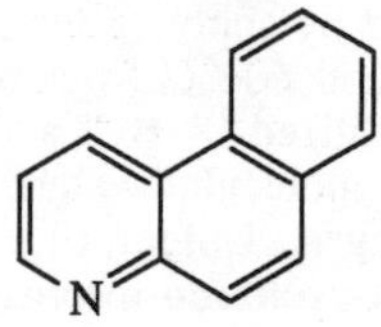 **XXII**

R= C_nH_{2n+1}

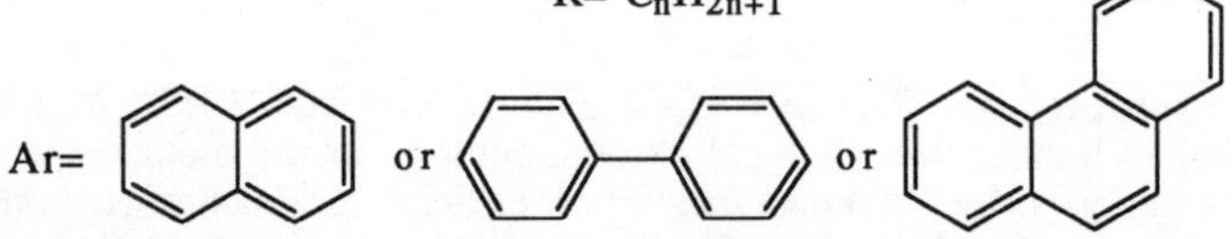

Ar= or or

and aliphatic-substituted heteroaromatic compounds. Further analytical work will be necessary to fully characterize the components of this fraction.

Among hydrocarbons extracted by solvents, the polar component exhibits the strongest enrichments in ^{13}C and deuterium relative to terrestrial values (Table 5). The bulk deuterium content of these compounds (+751 to +947 ‰) is among the highest observed and indicative of interstellar origin. How these enrichments are distributed among the four types of polar species is unknown.

Interestingly, isotopic measurements on the bulk hydrocarbon sublimates of de Vries et al. (1992) by secondary ion mass spectrometry yielded δD values of +990, +1400 and

+650 ‰, indicating deuterium contents overlapping with but extending beyond that of either bulk or specific hydrocarbon fractions extracted with solvents (Table 5). Which type of hydrocarbon is responsible for such strong enrichments is unclear because the distribution of deuterium among compound types in the sublimates was not determined. Infrared spectra of these sublimates indicated the presence of highly branched polycyclic aliphatic hydrocarbons as well as PAHs and possibly polar hydrocarbons. The overall aliphatic-H to aromatic-H ratio was about four. The strong deuterium enrichment may be associated with higher molecular weight compounds since the proportion of compounds heavier than 250 daltons is higher in the sublimates than in the solvent extracts. Krishnamurthy et al. (1992) found D-enrichment increasing in the order aliphatic < aromatic < polar hydrocarbons within solvent extractable compounds (Table 5). In contrast, Kerridge et al. (1987) suggested that in the macromolecular carbon the deuterium was more strongly concentrated in the aliphatic than the aromatic moieties. Clearly, additional measurements need to be made in order to understand these differences.

3.6.2. *Lower (volatile) hydrocarbons.* Light hydrocarbons presumably remain in CM chondrites only because they are immobilized, e.g., within crystals or between crystal boundaries, firmly adsorbed to grain surfaces, or dissolved in the macromolecular carbon phase (Belsky and Kaplan, 1970). In releasing these compounds, the analyst is faced with the dilemma of choosing conditions for disruption of the meteorite that will effect the release of these compounds quantitatively and free of contamination, but not lead to their production by degradation of other organic matter present. As might be expected, there is a good deal of variation in the analytical results reported. With respect to aliphatic hydrocarbons in the C_3 to C_8 range, Studier *et al.* (1972) found branched-chain alkanes but no straight chain isomers. Yuen *et al.* (1984) reported the presence of propane, *n*-butane and *n*-pentane along with isobutane. Belsky and Kaplan (1970) also found the lower *n*-alkanes, including heptane, as well as the branched-chain isomers and the lower alkenes.

The work of Yuen *et al.* (1984) is particularly significant in that they isolated individual members of the lower hydrocarbon series for $^{13}C/^{12}C$ measurements. They found methane and its homologues through the butanes to be isotopically heavier than their terrestrial counterparts, clearly indicating an extraterrestrial origin. More interestingly, they discovered a trend in which $^{13}C/^{12}C$ decreased with increasing C-chain length, as was observed for the monocarboxylic acids. As previously mentioned, these results have important implications in regard to the production mechanism (see Section 3.3.).

The lower aliphatic hydrocarbons trapped in carbonaceous chondrites may represent only a fraction of the meteorite's original content of such compounds. Given the amounts of nonvolatile hydrocarbons in Murchison, it would follow that the meteorite originally contained much larger quantities of the volatile lower hydrocarbons if the exponential decline in amount with increasing carbon number that characterizes several other classes of compounds is also applicable to the hydrocarbons. The loss of a significant amount of less firmly held volatile hydrocarbons is suggested by the common observation that carbonaceous chondrites have a “bitumen-like” odor immediately after falling (Vdovykin 1967). In the case of Murchison, Lovering *et al.* (1971) refer in their initial description of the meteorite to an "obvious volatile hydrocarbon component." The opportunities for preterrestrial loss might also have been great. Isotopic fractionation may have accompanied such losses, a possibility that gains support from observation of increasing enrichments of ^{13}C and deuterium in the light hydrocarbons obtained by increasingly vigorous extraction conditions (Table 5). Whether mass fractionation due to evaporative losses may have contributed to the deuterium enrichments is unknown.

3.7. NITROGEN HETEROCYCLES

The central role of purines and pyrimidines in biological information storage and transfer has evoked a strong interest in whether these compounds occur in Murchison and other carbonaceous chondrites. The early literature in this area is contradictory, however, more recent investigations by Schwartz and his colleagues (Van der Velden and Schwartz, 1977; Stoks and Schwartz, 1979, 1981a, 1981b, and 1982) have clarified the situation considerably. A summary of the work done in this area is given in the review by Cronin *et al.* (1988) and references to the earlier work are given there and in the papers by Stoks and Schwartz. The current view is that Murchison contains the purines, xanthine [XXIII], hypoxanthine [XXIV], guanine [XXV] and adenine [XXVI], and the pyrimidine, uracil [XXVII], at a total concentration of about 1.3 ppm.

Stoks and Schwartz (1982) also identified a suite of basic N-heterocyclic compounds in water and formic-acid extracts of the Murchison meteorite. Positive identifications were made of 2,4,6-trimethylpyridine, quinoline, isoquinoline, 2-methyl- and 4-methylquinoline. In addition, N-methyl-aniline, a suite of 12 additional alkyl pyridines,

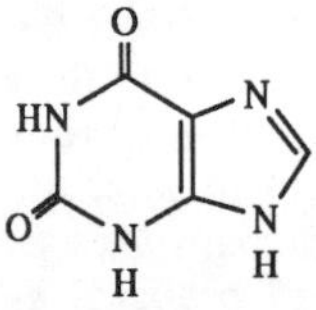

O HN N O N H N H **XXIII**

O HN N N N H **XXIV**

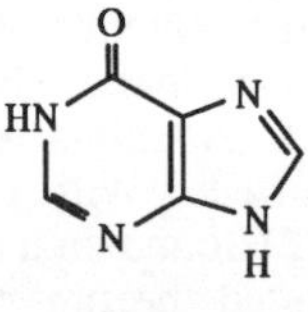

O HN N H_2N N N H **XXV**

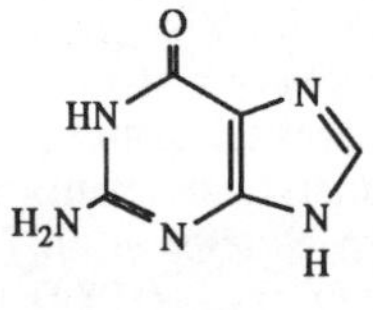

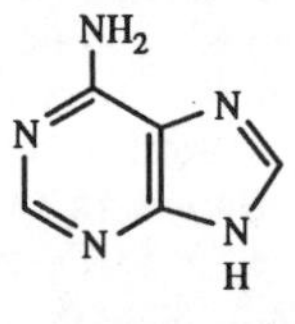

NH_2 N N N N H **XXVI**

O HN O N H **XXVII**

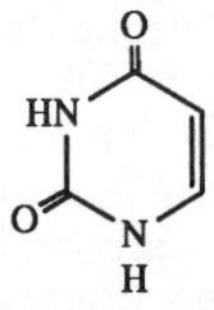

and a suite of 14 methyl-quinolines and/or isoquinolines was identified. Krishnamurthy *et al.* (1992) have recently found higher members of the quinoline/isoquinoline series among the polar hydrocarbons (Section. 3.6.1.3.).

In summary, Murchison appears to contain several classes of basic and neutral N-heterocycles, including purines, pyrimidines, quinolines/isoquinolines and pyridines. The last two groups are structurally diverse and contain a large number of isomeric alkyl derivatives. Taken together they total about 7 ppm. In contrast, only one pyrimidine and four purines are found and only in smaller amounts. All of the purines and the pyrimidines found are common biological constituents and no biologically unknown or unusual analogues seem to accompany them. Isotopic measurements have not been reported and stereoisomeric criteria are not applicable since the compounds are achiral. Thus, although blank runs testify to contamination-free procedures, the possibility that these compounds originated in terrestrial microorganisms should be kept in mind.

3.8. AMIDES AND AMINES

Hayatsu *et al.* (1975) identified guanylurea [XXVIII] (30-45 ppm) and tentatively identified urea [XXIX] (25 ppm) and some substituted ureas, perhaps including phenylureas, in a 6M HCl extract of Murchison. Recently, Cooper and Cronin (unpublished results) tentatively identified series of carboxamides [XXX] and dicarboxylic monoamides [XXXI] in Murchison water extracts. In addition they found extensive suites of cyclic amides, i.e., substituted and unsubstituted γ- [XXXII] and δ-lactams [XXXIII], pyrrolidiones [XXXIV] and piperidiones [XXXV]. Hydrolysis of these compounds would yield, respectively, γ- and δ-amino acids, dicarboxylic acids, and, in the case of the carboxamides, monocarboxylic acids. As mentioned previously (Sections 3.3. and 3.4.), these results show that Murchison contains acid-labile precursors of the monocarboxylic acids and dicarboxylic acids as well as the amino acids. These findings raise the question whether these amides are true precursors of all meteorite carboxylic acids and whether the free carboxylate form of these acids found in Murchison is a result of partial hydrolysis of such precursors on the parent body. Furthermore, one should consider whether the amides, themselves, were products of even more primitive precursors, possibly nitriles.

Jungclaus *et al.* (1979) tentatively identified 10 aliphatic amines [XXXVI] in aqueous extracts of Murchison using GC as well as ion-exchange chromatography. This suite of amines shows several of the characteristics of the amino acids: their content is increased about two-fold by acid hydrolysis; their concentrations decline sharply with increasing carbon number within the straight-chain homologous series; the branched-chain isomers are more abundant than the straight-chain isomers; and all possible isomers are present, at least for the primary amines through C_4. The concentrations of individual amines are comparable to those of the corresponding amino acids and their total content is in excess of 0.2 μmol/g (8 ppm). Under hydrolytic conditions secondary amides could have acted as amine precursors, but nitriles could not.

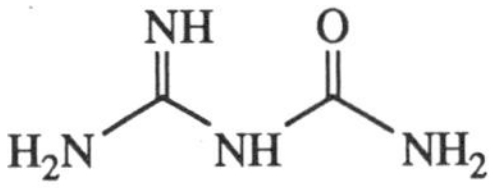

XXVIII 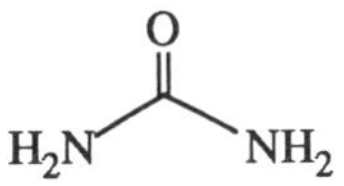 XXIX 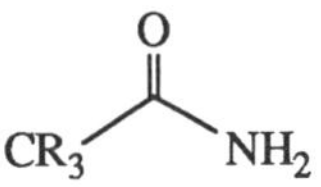 XXX 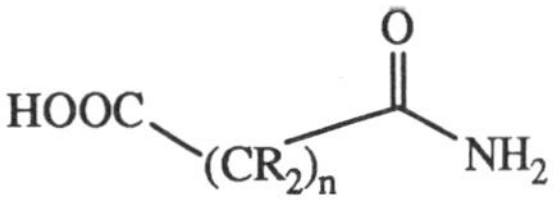 XXXI

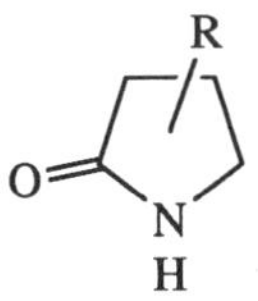

XXXII 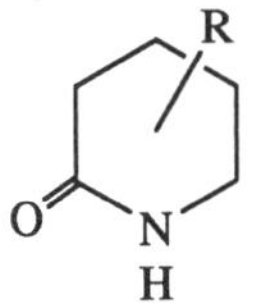 XXXIII 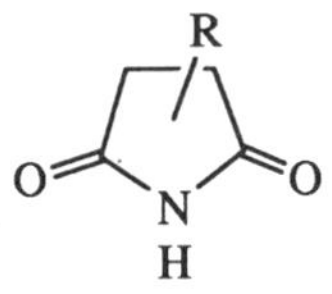 XXXIV 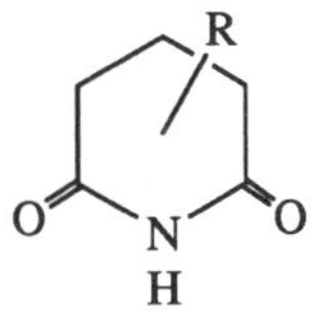 XXXV

R= H or C_nH_{2n+1}

3.9. ALCOHOLS AND CARBONYL COMPOUNDS

Jungclaus *et al.* (1976) have identified series of lower alcohols [XXXVII], aldehydes [XXXVIII] and ketones [XXXIX] in aqueous extracts of the Murchison and Murray meteorites. Alcohols and aldehydes through C_4 and ketones through C_5 were found. Declining concentrations with increasing carbon number were observed in the straight-chain homologous series of each class. All possible isomers appear to be present through C_4. As in the case of the amines, the aldehydes and ketones survived the parent body Strecker cyanohydrin syntheses that appear to have formed the amino and hydroxy acids. This suggests that either the Strecker synthesis was relatively inefficient in its consumption of these reactants or cyanide was the limiting reactant.

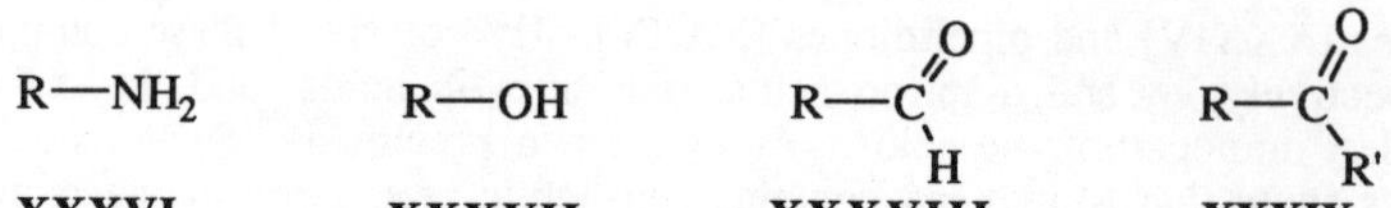

$$R{-}NH_2 \qquad R{-}OH \qquad R{-}C(=O)H \qquad R{-}C(=O)R'$$

XXXVI XXXVII XXXVIII XXXIX

$R,R' = C_nH_{2n+1}$

4. SUMMARY OF MURCHISON ORGANIC COMPOUNDS

4.1. MOLECULAR CHARACTERISTICS

The organic carbon of the Murchison meteorite has been found, in large part, to consist of macromolecular material that has not been amenable to detailed structural characterization. A smaller fraction is accounted for by the complex mixture of discrete, soluble molecules described in the immediately preceding sections. These compounds viewed as reaction products can be, in principle, a useful source of information in regard to the process(es) leading to their formation. Their comparison allows the recognition of several significant common charactersitics.

1. *Complete structural diversity.* At a given carbon number, all stable isomeric forms are present for most classes of compounds, at least among the lower homologues. This suggests synthesis by random processes involving carbon-containing free radicals and ions, independent of directing influences, for example, by catalysts. The purines and pyrimidine appear to be exceptions to this generalization.
2. *No enantiomeric preference.* With the exception of Engel and co-workers' reports of excess L-amino acids (sec. 3.1.2), all investigators have found chiral compounds to be racemic, i.e., to occur as equimolar mixtures of enantiomers.
3. *Concentrations decline with increasing carbon number in homologous series.* This observation is quite general and suggests a progressive build-up of carbon chains by one-carbon additions.
4. *Branched-chain isomers are predominant.* This observation suggests a carbon chain formation process in which radical or ion stability was influential .

4.2. ISOTOPIC CHARACTERISTICS

The results of isotopic measurements made on the major Murchison phases are summarized in Table 6. Several generalizations can be drawn from these data and those for molecular species collected in the preceding tables and discussion.

1. The indigenous organic matter is generally enriched in deuterium and found to have δD values > +100 ‰, the heavy end of the terrestrial range (Pillinger, 1984).
2. The indigenous soluble organic compounds are enriched in ^{13}C relative to terrestrial matter or, at least, lie at the heavy end of the terrestrial range.
3. The indigenous soluble organic compounds are significantly enriched in ^{15}N relative to terrestrial matter.
4. Five isotopically distinct components, exclusive of the "exotic" phases, can be recognized in Murchison:
   a. amino acids: $\delta D = +1000$ to $+2500$; $\delta^{13}C = +5$ to $+40$; $\delta^{15}N = +90$
   b. other organic compounds: $\delta D = +100$ to $+1400$; $\delta^{13}C = -10$ to $+10$; $\delta^{15}N \leq +98$
   c. insoluble carbon: δD +500 to +1000; $\delta^{13}C \cong -15$; $\delta^{15}N = +18$

TABLE 6. Carbon and hydrogen isotopic composition of major Murchison phases

Preparation	$\delta^{13}C$ (‰)	δD (‰)	Reference
Bulk powder	-7.2	—	(1)
	-0.5		(2)
	-2.4	-51	(3)
	-4.5	-6	(4)
		-65	(5)
	-10.6	—	(6)
Insoluble carbon	-10.6	—	(1)
	-15.8, -16.2, -16.8	—	(2)
	-13.0	+830	(4)
	-15.8	—	(6)
	-15.5	+664	(7)
	-14.5	+483	(3)
	—	+607	(8)
	-17.5, -17.6	+483, +945	(9)
Phyllosilicates	—	-100[a]	(5)
	—	-100[a]	(4)
		-110[a]	(7)
Carbonate + CO_2	37.3±3.1	—	(10)
	44.3		(10)
	45.4		(1)
	44.4		(1)
CO_2[b]	+29.1±.2	—	(11)
CO[b]	-32±2	—	(11)

(1) Kvenvolden *et al.* (1970); (2) Chang *et al.* (1978); (3) Yang and Epstein (1984); (4) Robert and Epstein (1982); (5) Kolodny *et al.* (1980); (6) Swart *et al.* (1983); (7) Yang and Epstein (1983); (8) Yang and Epstein (1985); (9) Kerridge *et al.* (1987); (10) Grady *et al.* (1988); (11) Yuen *et al.* (1984).

(a) Not determined directly. (b) Released by freeze-thaw disaggregation of Murchison stones.

d. carbonate: $\delta^{13}C$ = +35 to +45
e. phyllosilicates $\delta D \cong -100$

5. The monocarboxylic acids and light alkanes show a decline in $\delta^{13}C$ as carbon number increases.
6. The hydrocarbons differ systematically in δD with aliphatic < aromatic « polar hydrocarbons. Higher molecular weight hydrocarbons may be more deuterium enriched than lower molecular weight homologs.

5. Formation of Murchison Organic Matter

Twenty-five years ago, in reviewing the question of origin, Hayes (1967) stated, "This highly controversial subject is, at its simplest level, a question of whether the compounds observed are biogenic or abiogenic." This question has been clearly answered by analyses of the Murchison meteorite in favor of an abiogenic origin, i.e., meteorite organic compounds are not products of either terrestrial or extraterrestrial life. Consequently, the origin question was recast in terms of nebular and/or parent body formation processes with Fischer-Tropsch type (FTT) synthesis and Miller-Urey synthesis as prototypes, respectively, for the organic chemistry occurring in these two regimes. Each process proved to have attractive features as well as shortcomings, although a recent theoretical evaluation of potential solar nebular processes for synthesis of organic matter suggests only the FTT process as a significant contributor (Prinn and Fegley, 1989).

5.1. THE FISCHER-TROPSCH TYPE PROCESS

The FTT process fit well with the concept of a cooling, gaseous solar nebula in which CO, the thermodynamically stable form of carbon in a hot gas of solar composition, persisted to the relatively low temperatures at which suitable catalysts would condense. At these lower temperatures the nebular gases, CO, H_2 and NH_3 were presumed to react on catalytic dust grains to form the organic compounds observed in meteorites. However, it now appears that the purported catalysts, such as magnetite and clay minerals found in carbonaceous chondrites, are not nebular products, but formed later within or on the CI (Kerridge *et al.*, 1979) and CM (Bunch and Chang, 1980) carbonaceous chondrite parent bodies. Furthermore, the straight chain hydrocarbons believed to be indicative of FTT chemistry (Hayatsu and Anders, 1981) have been found to be terrestrial contaminants (Cronin and Pizzarello, 1990) and the carbon isotopic fractionation accompanying the FTT process (Lancet and Anders, 1970) is not consistent with the isotopic fractionation pattern observed for the carbonate, CO, and reduced carbon of Murchison (Chang *et al.*, 1978; Yuen *et al.*, 1984; Yuen *et al.*, 1991; Yuen *et al.*, unpublished).

5.2. THE MILLER-UREY SYNTHESIS

The Miller-Urey (MU) electric discharge synthesis garnered experimental support mainly from matches of the product mixture with early analytical data from Murchison amino acid and dicarboxylic acid analyses (Wolman *et al.*, 1972; Zeitman *et al.*, 1974). The goodness of these matches in relation to more recent analyses has been questioned (Cronin *et al.*, 1985). The Miller-Urey synthesis, in the context of meteorite organic chemistry, is assumed to have taken place at the surface of the parent body, perhaps the precursor of an asteroid (Peltzer *et al.*, 1984), however, there are difficulties with this concept relating to the ability of such a small body to retain the necessary atmosphere. More importantly, the

Miller-Urey process does not fractionate carbon (Chang *et al.*, 1983) as extensively as observed in meteorites, neither the FTT synthesis nor the MU synthesis is capable of yielding the ^{15}N enrichments (Kung *et al.*, 1979) observed in the Murchison organic compounds, nor is either process capable of achieving the deuterium enrichments observed in the Murchison organic compounds at the temperatures they require. Likewise, other ordinary solar system processes involving photochemistry and solution chemistry are incapable of yielding the range of isotopic fractionation of C, N and H observed in chondritic organic matter (Chang et al., 1983).

5.3. THE INTERSTELLAR-PARENT BODY HYPOTHESIS

As difficulties with the preceding models have become progressively more compelling, a third proposal has received serious consideration, that is, the formation of chondritic organic matter from surviving interstellar organic compounds. Such a possibility seemed at first to be precluded by the extreme temperatures thought to have been attained in the solar nebula and the consequent conversion of pre-existing molecules and minerals to a well-mixed gas of simple molecular composition. Under such circumstances the nebula would retain no structural or isotopic "memory" of its initial organic raw materials. This view has gradually given way to models of the solar nebula in which temperatures where meteorites formed, about 3 AU from the nascent sun, never exceeded more than a few hundred K (Wood and Morfill, 1988). Cameron (1973) was the first to suggest that at these distances interstellar grains might have retained some integrity and that CI1 chondrites and the matrices of CM2 and C3 chondrites were originally composed of aggregates of these remnants. The discovery of many isotopic anomalies, that is materials bearing isotopic ratios indicative of presolar nucleosynthesis, gave clear evidence for nebular inhomogeneity (Clayton, 1978). The first indication that the organic matter may also have had a presolar origin came from the discovery that the deuterium enrichment observed in bulk carbonaceous chondrite samples is concentrated in the organic fraction (Kolodny *et al.*, 1980; Robert and Epstein, 1982). This suggested the incorporation of interstellar molecules (Kolodny *et al.*, 1980; Geiss and Reeves, 1981; Robert and Epstein, 1982), an interpretation that has subsequently been widely adopted as high D/H ratios have been discovered for various classes of soluble organic compounds as well as the macromolecular carbon.

In Figure 1 are plotted all the simultaneously determined δD and $\delta^{13}C$ values for classes of soluble organic compounds obtained in collaboration over the last few years by the Arizona State-Caltech groups (Epstein *et al.*, 1987; Pizzarello *et al.*, 1991; Krishnamurthy *et al.*, 1992; Cronin *et al.*, 1993). Among these compounds there appears to be a roughly linear correlation of δD with $\delta^{13}C$. Such a correlation might arise from a homogeneous isotopic reservoir if isotopic fractionation of both elements resulted from the same process(es) operating at different efficiencies according to variations in "local" conditions. Kerridge (1980) has suggested on the basis of a bulk meteorite δD-$\delta^{15}N$ correlation a common fractionation process for these two sets of isotopes as well. The amino acid data are consistent with this suggestion, although additional data are required to establish this relationship as Kerridge noted earlier (1980).

The range of the isotopic variations is too large to be attributable to ordinary solar system processes, but it is consistent with interstellar processes. In this case, variations in "local" conditions such as temperature or fractional ionization would influence the extent of isotopic fractionation associated with gas phase ion-molecule and grain surface reactions. Thus, all the organic compounds may have originated from parental interstellar matter. The fact that some classes of compounds are much more deuterium enriched than others suggests that local conditions also influence the types of compounds that are formed as well as their characteristic isotopic enrichments. Notably, organic chemical and isotopic heterogeneities do exist in interstellar clouds, and cloud conditions could be correlated with such heterogeneities. Additional abundance and isotopic data from both astronomical observations and meteorite analyses should lead to a better understanding of the basis for the apparent correlation.

The isotopic values might also reflect mixing of small amounts of isotopically anomalous interstellar material with a much larger reservoir of isotopically normal material formed in the solar nebula. This idea presupposes that little interstellar matter survived processing in the nebula and that organic synthesis occurred readily in the solar nebula, for which evidence is lacking. Application of Occam's Principle to theories of the origin of meteoritic organic matter leads to the working conclusion that it was all of interstellar origin.

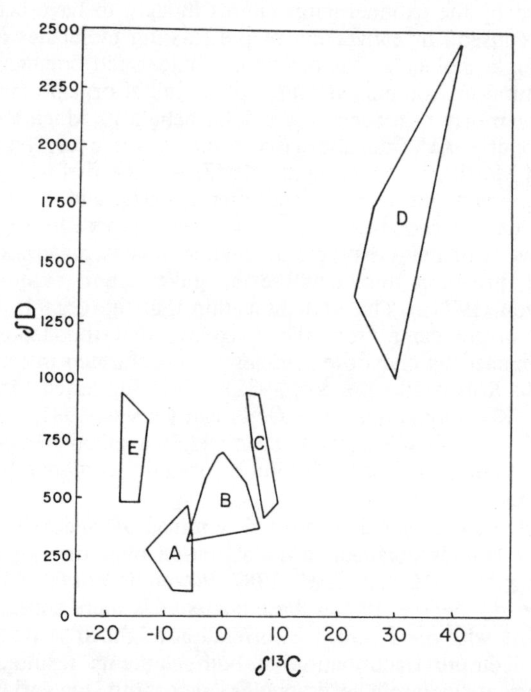

Fig. 1 δD versus δ^{13}C plot for Murchison molecular compounds and insoluble carbonaceous matter. A. *Nonpolar hydrocarbons*. Includes aliphatic and aromatic hydrocarbons and extracts made with nonpolar solvents, e.g., CCl_4; B. *Carboxylic, dicarboxylic, and hydroxy acids;* C. *Polar hydrocarbons*. Includes polar hydrocarbons from silica gel fractionation and extracts made with polar solvents, e.g., methanol; D. *Amino acids*. Includes total and fractionated amino acids; E. *Insoluble carbonaceous matter*.

5.3.1. *Nebular and Parent Body Processing.* As interstellar gas and dust entered the solar nebula it would have been processed both at the nebula accretion shock front and within the nebula itself. As a result of alteration by thermochemical processes, particle and photon irradiation, and electric discharges, the organic matter would have been partially destroyed or converted to other species, and their interstellar isotopic signatures partially diluted by exchange with bulk nebular reservoirs of ordinary composition. The severity of such alteration would have decreased with increasing heliocentric distance. Therefore, material accreted into solid objects in the outer regions of the solar nebula where comets and ice rich objects are thought to have condensed would have been most likely to preserve their interstellar characteristics (Engel *et al.*, 1990b). More refractory organic matter such as aromatic hydrocarbons or macromolecular carbon would have persisted in regions closer to the sun. In this context, accretion of cometesimal ices into the parent bodies of meteorites has been proposed as the source of the water and organic matter for these objects (Bunch and Chang, 1980). In addition to the volatile species usually observed in cometary comae, non-volatile species would also have been supplied, for example, as contained in the organic, so-called CHON particles, that make up a major fraction of the dust grains found in the coma of Comet Halley (Clark *et al.*, 1988; Fomenkova *et al.*, 1992).

Presumably, the ices melted due to warming of the parent bodies by radioactive decay of short-lived isotopes like ^{26}Al, low energy collisions or other processes. The resulting solutions converted previously anhydrous mineral assemblages to clays, carbonates, sulfates and other hydrous mineral phases (Zolensky and McSween, 1988). In analogous manner, the organic components susceptible to reactions in aqueous solution also underwent conversion to secondary products.

Peltzer *et al.* (1984) suggested that the Strecker-cyanohydrin reactions occured during the aqueous stage of a Miller-Urey-type process in which the Strecker reactants were produced by electrical discharge and/or UV irradiation in a short-lived asteroidal parent body atmosphere. The physical requirements of forming and maintaining the required atmosphere for electric discharges in contact with a reservoir of liquid water on a small body in the asteroid belt would appear daunting, but they have not yet been assessed. Maintenance of an internal reservoir of liquid water over a long period of time under a thick permafrost layer has been discussed, however, and would appear to be feasible (Dufresne and Anders, 1962) .

Ultimately the parent body lost its internal water through evaporation to space and reaction with minerals prior to lithification. Loss of volatiles was incomplete, at least in CM parent bodies, since CO_2, light hydrocarbons and low molecular weight aldehydes, ketones, amines and alcohols have been isolated from the Murchison meteorite. These species are apparently trapped in special environments like the interlayers of hydrous minerals or the interiors of the macromolecular carbon component. Significant loss of volatiles must have occurred, however, since the organic matter consists predominantly of material less volatile than naphthalene or decane. In this context, the degree to which the biogenic elements have been retained in carbonaceous chondrites may reflect the extent of conversion of the volatile interstellar component to the non-volatile organic products that characterize the CM2 and CI1 chondrites today.

5.4. FORMATION OF CHARACTERISTIC CLASSES OF ORGANIC COMPOUNDS

The formation of chondritic α-amino acids and α-hydroxy acids from interstellar precursors by Strecker reactions during aqueous alteration of the parent body comprises a satisfactory hypothesis for their formation (Section 3.2). In this section, we shall examine this process in more detail and also consider to what extent aqueous phase reactions of interstellar molecules can account for the full range of meteorite organic compounds. It is generally believed that the catalogue of known interstellar molecules represents only a fraction of the total species extant (Snyder, 1986). Consequently, we shall take some liberty, and use as reactants not only well-established interstellar molecules, but also plausible interstellar species. By this we mean (i) homologues, (ii) more or less saturated analogues, and (iii) difunctional derivatives of the known interstellar molecules. Examples of (i) and (ii) exist among the known interstellar compounds, e.g., HCN, CH_3CN, CH_3CH_2CN and CHCCN, CH_2CHCN, and CH_3CH_2CN. Examples of (iii) are unknown, but would include compounds like $NCCH_2CH_2CN$ (succinonitrile) and OCHCHO (glyoxal, van Ijzendoorn, *et al.*, 1986).

A large uncertainty in an exercise of this sort results from our ignorance of the composition of interstellar and cometary grains, which are essentially uncharacterized. The organic compounds of grain mantles may reach a higher level of chemical complexity than is achieved in the gas phase of interstellar clouds. Consequently they may provide a broader range of reactants for parent body aqueous phase reactions than we allow here. Furthermore, it is conceivable that they provide some fraction of meteoritic compounds, *per se*, that is, without the necessity of additional parent body processing. Until more is known about the full potential of organic chemistry in interstellar clouds and comets (Greenberg and d'Hendecourt, 1985), it will be difficult to define with certainty the extent to which the organic compounds of CM and CI carbonaceous chondrites represent primary products of interstellar chemistry, a second generation of grain mantle reaction products, or a third generation that retains significant interstellar structure and isotopic composition but has been substantially altered by parent body aqueous processing.

5.4.1. *Amino Acids and Hydroxy Acids*

5.4.1.1. *Strecker-cyanohydrin Synthesis.* The synthesis of α-amino acids from aldehydes, HCN, and NH_3 has been known for almost 150 years, having been reported (Strecker, 1850) only 22 years after Wöhler's historic synthesis of urea. In water, an equilibrium is established among the reactants, a cyanohydrin, and an aminonitrile. Irreversible hydrolyses of the cyanohydrin and aminonitrile give respectively, an α-hydroxy acid and α-amino acid. The prebiotic significance of these reactions was first suggested by Miller (1957) who found that aqueous phase Strecker reactions give rise to the amino acids formed in the Miller-Urey electrical discharge experiment.

The factual case for a Strecker-cyanohydrin synthesis of the Murchison α-hydroxycarboxylic acids and α-amino acids can be summarized as follows.

1. The meteorite contains both classes of compounds in comparable amounts.
2. They comprise structurally analogous suites of compounds in which all (α-amino acids) or nearly all (α-hydroxy acids) chain isomeric forms have been identified through C_7.
3. In three instances, molar ratios of structurally analogous α-hydroxy and α-amino acid pairs are consistent with equilibration of the precursor α-hydroxy- and α-aminonitriles at a common ammonia concentration.

Apparently, the period of aqueous alteration of the Murchison parent body provided the necessary reaction conditions for Strecker-cyanohydrin reactions. DuFresne and Anders (1962) suggested a pH in the 6 to 8 range for the aqueous phase of CM chondrites on the basis of the observed mineral assemblage and the Eh-pH relationship. Whereas pH values in the acid range ($pH < 5$) lead to large excesses of hydroxy acids in a Strecker-cyanohydrin synthesis, at pH 8 comparable amounts of α-hydroxy- and α-amino acids are produced (Miller and Van Trump, 1981) as observed in the meteorite. Peltzer *et al.* (1984) calculated the NH_4^+ concentration of the parent body aqueous phase to be $2x10^{-3}$ M based on a Strecker origin for the observed meteoritic ratios of three α-hydroxy acid:α-amino acid pairs. Assuming synthesis of the meteoritic β-alanine and succinic acid from acrylonitrile, the HCN concentration was estimated to be 10^{-3} M to 10^{-2} M. If NH_4^+ and HCN provided the dominant buffer components for the parent-body aqueous phase, a pH of 9±1 would be expected.

$$RR'CO + HCN \rightleftharpoons R\text{-}CR'(OH)\text{-}CN \longrightarrow R\text{-}CR'(OH)\text{-}C(=O)\text{-}NH_2 \longrightarrow R\text{-}CR'(OH)\text{-}COOH$$

$$RR'CO + HCN \xrightleftharpoons{+NH_3} R\text{-}CR'(NH_2)\text{-}CN \longrightarrow R\text{-}CR'(NH_2)\text{-}C(=O)\text{-}NH_2 \longrightarrow R\text{-}CR'(NH_2)\text{-}COOH$$

$$R,R' = H \text{ or } C_nH_{2n+1}$$

Strecker-Cyanohydrin Synthesis

When carried out under plausible parent body conditions, however, Strecker-cyanohydrin syntheses also yield imino acids (e.g, iminodiacetic acid) as prominent by-products. For example, in the synthesis of glycine the ratio of iminodiacetic acid to glycine was >0.2 under all conditions examined (Lerner et al., 1992). To date, imino acids have not been reported in meteorites, let alone at such relative abundances, even though they should have been readily identified by GC-MS methods. If amino acids arose primarily from the Strecker-cyanohydrin syntheses, the apparent absence of imino acids remains to be understood.

As discussed previously the anomalous enrichments of D and ^{15}N in amino acids and of D in hydroxy acids suggest inheritance by the parent body of interstellar organic matter formed by gas-phase ion-molecule and possibly grain surface reactions at low temperatures. The suggestive evidence for Strecker-cyanohydrin reactions and these isotopic observations can be reconciled if the aldehydes and ketones required for synthesis of the amino- and hydroxy acids were interstellar in origin. Among the reactants required for the Strecker-cyanohydrin synthesis, only the hydrogen atoms of aldehydes and ketones would resist rapid exchange with water. Therefore, amino acids formed by such synthesis must inherit their deuterium enrichments from aldehyde and ketone reactants. Strecker-cyanohydrin syntheses carried out under plausible meteorite parent body conditions using deuterated aldehydes and ketones yielded amino acids (and imino acids) which typically retained 50-98% of the original isotopic label (Lerner et al., 1991). Therefore, the requirement that the deuterium signature of interstellar aldehydes and ketones be retained in the resulting amino acid is fulfilled.

Preliminary results from carbon and hydrogen isotopic analysis of the total Murchison hydroxy acids, however, appear to be inconsistent with the simple hypothesis that they and the amino acids were produced simultaneously by a Strecker-cyanohydrin synthesis. If the α-hydroxy and α-amino acids were formed by a common pathway from common precursors, their isotopic compositions should be very similar. Contrary to these expectations, the hydroxy acids were found to exhibit a positive δD value 800-1200 ‰ lighter than those obtained from the total amino acids (Table 4), although well beyond the terrestrial range (Cronin *et al.*, 1993). Likewise, the $\delta^{13}C$ value was about 20 ‰ less positive than that of the total amino acids. Several factors may contribute to these results. First, the molecular composition of the isolated hydroxy acids was not identical to that of the amino acid suite analyzed. Hydroxy dicarboxylic acids were absent from the hydroxy acids whereas amino dicarboxylic acids were present in the amino acids analyzed. Similarly, the content of β-, γ-, and δ-substituted isomers appeared to be greater in the amino acids than in the hydroxy acids. Secondly, the occurrence of β–, γ–, and δ-substituted compounds which cannot be formed by Strecker-cyanohydrin reactions demands syntheses by alternative pathways. If these additional pathways contributed to the formation of the α-substituted compounds also, different isotopic compositions might well be the consequence. Third, although the Strecker-cyanohydrin synthesis may have been entirely responsible for formation of both classes of α-compounds, time-dependent variations in the reaction conditions, especially pH, and in the isotopic composition of the available reactants, could give rise to variation in the molecular and isotopic compositions of the products. Finally, although the measured isotopic ratios were corrected for the contributions of procedural blanks, there is always a possibility of random and unrecognized "contamination" affecting one class of compound more than the other and differentially decreasing the δ-value. To establish the nature of the isotopic distinctions between these two classes of acids and to understand the cause(s), further detailed analyses relating isotopic composition to molecular structure among both hydroxy- and amino acids will be necessary.

5.4.1.2. *Alternative synthetic processes.* The other possibilities, apart from the Strecker-cyanohydrin synthesis, alluded to above for the formation of α-hydroxy and α-amino acids include parent body hydrolysis of α-amino or α-hydroxy nitriles originally formed in the interstellar medium. In the presence of ammonia and HCN, however, such hydrolysis would have been equivalent to a Strecker-cyanohydrin synthesis, provided conditions permitted attainment of equilibrium between the hydroxy and amino nitrile pairs prior to hydrolysis to form the corresponding acids. Absent such equilibration, the observed hydroxy and amino acid abundances would reflect precursor abundances determined by pre-parent body processes and kinetically controlled hydrolysis.

Another possibility is the hydrolysis of products of cyanide polymerization which may have occurred under conditions of aqueous alteration. Cyanide polymerization could have dominated Strecker-cyanohydrin syntheses if ammonia and amines or aldehydes and ketones were underabundant relative to cyanide. Such polymerization, however, might be expected to yield deuterium depleted products unless D-rich carbonyl or nitrile compounds were also incorporated in the products. It is noteworthy that deuterium-free as well as deuterated glycine was formed when deuterated formaldehyde was used to simulate Strecker-cyanohydrin syntheses under plausible parent body conditions (Lerner et al., 1991). This observation underscores the possibility that meteoritic amino acids were formed by a number of processes of which cyanide polymerization may have been one.

Direct formation of amino acids, hydroxy acids and other complex, non-volatile organic compounds by interstellar gas phase and/or grain surface reactions remains a possibility. The detection of organic compounds in interstellar clouds is clearly biased toward abundant small molecules with large dipole moments. There are undoubtedly less abundant, more complex organic species that have gone undetected in the gas phase, however, searches for glycine in interstellar clouds have been unsuccessful (Hollis et al.,

1980. Infrared observations of organic matter in interstellar dust (summarized by Allamandola et al., 1988) and laboratory simulations of organic photochemical synthesis on interstellar ice grains (Agarwal et al., 1988) emphasize the potential importance of grain chemistry in providing an interstellar reservoir of both simple and complex organic material. Indeed glycine and hydroxy acids are among the products formed in these simulations (Briggs et al., 1992)

Thermochemical calculations by Shock and Schulte (1990) suggest that reactions of polycyclic aromatic hydrocarbons in the aqueous solutions responsible for mineral alterations of parent bodies could have afforded amino acids and other compounds as a result of metastable equilibria. Preservation of observed deuterium enrichments under such conditions is problematic, however.

5.4.1.3. *Formation of β-, γ-, etc., Isomers.* The Murchison meteorite contains abundant amino- and hydroxy-position isomers of the α-amino and α-hydroxy acids, e.g., β-alanine, γ-aminobutyric acid, and β-lactic acid. The β-analogues have also been found in spark discharge experiments and are believed to form under these conditions by Michael addition of ammonia to acrylonitrile (Miller, 1957) followed by hydrolysis.

$$CH_2=CH\text{–}CN \xrightarrow{NH_3(H_2O)} H_2N\text{–}CH_2CH_2\text{–}CN\ (HO\text{–}CH_2CH_2\text{–}CN) \xrightarrow[-NH_3]{2H_2O}$$

$$H_2N\text{–}CH_2CH_2\text{–}COOH\ (HO\text{–}CH_2CH_2\text{–}COOH)$$

In meteorites, this reaction could proceed from interstellar acrylonitrile as shown, or perhaps by hydrolysis of pre-existing β-amino- and β-hydroxy propionitrile. δ-Amino acids could arise similarly from the homologous cyanodiene by 1,6 addition of ammonia although a subsequent reduction of the 2,3 double bond would be necessary. γ-Amino acids are problematical, but could be formed by decarboxylation of pyroglutamic acid/alkyl pyroglutamic acids, which are abundant in Murchison extracts. Pyroglutamic acids could be cyclization products of glutamic acid/alkyl glutamic acids formed, in turn, from acrolein by sequential Michael additions of cyanide and Strecker reactions. One can imagine a similar route to the cyclic amino acid, pipecolic acid, by 1,6 amination of the acrolein homologue, 2,4-pentadienal, although a reductive step would again be required to eliminate the 2,3-double bond.

$$CH_2=CH\text{–}CHO \xrightarrow{HCN} NC\text{–}CH_2CH_2\text{–}CHO \xrightarrow{HCN,\ NH_3,\ H_2O}$$

$$HOOC\text{–}CH_2CH_2\text{–}CH(NH_2)\text{–}COOH \xrightarrow{-H_2O} \text{(pyroglutamic acid: 5-oxo ring, N–H, COOH)} \xrightarrow{-CO_2}$$

$$\text{(2-pyrrolidone: O=, N–H)} \xrightarrow{H_2O} H_2N\text{-}CH_2\text{-}CH_2\text{-}CH_2\text{-}COOH$$

A route to proline is not obvious, although the tentative identification of pyrrolidine in the interstellar medium (Turner, 1980) suggests a possibility.

$$\text{pyrrole} \xrightarrow{HCN} \text{2-cyano-3-pyrroline} \xrightarrow{Red'n} \text{2-cyanopyrrolidine} \xrightarrow{H_2O} \text{proline (2-COOH pyrrolidine)}$$

N-Alkyl amino acids could have formed by Strecker reactions in which an amine replaced ammonia, and the acetyl amino acids could be products of the reaction of interstellar ketene with amino acids or their precursors, e.g., amino nitriles. This latter possibility may provide an explanation for the apparent absence of imino acids in meteorites. The acetylation reaction competes favorably with the reaction of ketene with water to form acetic acid (Bergman and Stern, 1930).

The efficacy of these processes in aqueous solution remains to be demonstrated, since many alternative competing reactions can be imagined. The Michael-type and other additions to double bonds suggested may have been assisted by mineral catalysts on the parent body.

5.4.2. *Mono- and Dicarboxylic Acids*

The abundant monocarboxylic acids may be unaltered interstellar molecules, since the first member in this series, formic acid, has been identified in interstellar clouds (Turner, 1989). Another possibility for their formation is the parent-body hydrolysis of alkyl nitriles such as methyl cyanide (acetonitrile) and ethyl cyanide, also known interstellar molecules. An extensive series of unsaturated nitriles, the cyanopolyynes, extending through C_{11} have been identified in interstellar sources, however, the reduction and hydrolysis of these compounds is clearly not the major route to carboxylic acids as this pathway would yield predominantly straight-chain, odd carbon number acids. One cannot rule out the possibility that dinitriles or even dicarboxylic acids themselves are formed by ion-molecule reactions in interstellar clouds, although neither have been reported.

Free radical reactions on interstellar grains or in parent body environments could also have yielded nitriles as well as dinitriles and a host of other precursor species which would have been subject to aqueous alteration. Synthesis of parent molecules on dust grains by recombination of cyano, amino, hydroxyl, and alkyl free radicals could readily account for the observation of virtually complete structural diversity among the amino- and hydroxy-, as well as the mono- and dicarboxylic acids.

Dicarboxylic acids might also arise from addition reactions between unsaturated nitriles and HCN, followed by hydrolysis. It has been suggested that succinic acid, for example, was formed during aqueous alteration by the addition of HCN to acrylonitrile (Peltzer *et al.*, 1984).

$$CH_2{=}CH{-}CN \xrightarrow{HCN} NC{-}CH_2CH_2{-}CN \xrightarrow[-2NH_3]{4H_2O} HOOC{-}CH_2CH_2{-}COOH$$

The recent finding of 2,5-pyrrolidiones, 2,6-piperidiones, and dicarboxylic monoamides in the meteorite raises the question whether the hydrolysis of these compounds may have given rise to dicarboxylic acids as shown below, specifically β- and γ-dicarboxylic acids in the first two cases, and, if so, whether these 5- and 6-membered cyclic imides are interstellar molecules.

$$\text{succinimide} \xrightarrow{H_2O} HOOC{-}CH_2CH_2{-}\overset{O}{\overset{\|}{C}}{-}NH_2 \xrightarrow[-NH_3]{H_2O} HOOC{-}CH_2CH_2{-}COOH$$

5.4.3. *Amines, Amides, and N-Heterocycles.* The amines, like the carboxylic acids, may be unaltered products of interstellar chemistry. The first member of the series, methyl amine, has been identified in space. Other possible routes include decarboxylation of α-amino acids, the reduction of imines, either of interstellar origin or formed during aqueous alteraction, and the reduction of nitriles. Where and how such reductions could have occurred remain open issues.

$$HOOC{-}CHR{-}NH_2 \xrightarrow[-CO_2]{} RCH_2{-}NH_2$$

$$RCHO \xrightarrow[-H_2O]{NH_3} RCH{=}NH \xrightarrow{Red'n} R{-}CH_2NH_2$$

$$RC{\equiv}N \xrightarrow{Red'n} RCH_2{-}NH_2$$

The amides, urea and guanylurea, are easily made from the interstellar molecules, NH_3, isocyanic acid, and cyanamide.

$$HN{=}C{=}O \xrightarrow{NH_3} NH_2{-}\overset{O}{\overset{\|}{C}}{-}NH_2 \xrightarrow{NC{-}NH_2} H_2N{-}\overset{NH}{\overset{\|}{C}}{-}NH{-}\overset{O}{\overset{\|}{C}}{-}NH_2$$

The purines and the pyrimidine found in Murchison could have been formed by several plausible routes (Ferris and Hagan, 1984). Oligomerization of cyanide in NH_3 solutions followed by hydrolysis gives adenine, hypoxanthine, and uracil. Guanine and xanthine also have been obtained from the HCN tetramer, diaminomaleonitrile, after reaction with formamidine and urea.

5.4.4. *Alcohols, Aldehydes, and Ketones.* The low molecular weight alcohols, aldehydes, and ketones found in Murchison may have originated in the interstellar medium and persisted unchanged through aqueous processing. The first members of each series have been identified in interstellar clouds, i.e., methanol, ethanol, formaldehyde, acetaldehyde, and acetone. Hydration of unsaturated interstellar precursors could, in principle, have also provided these compounds as shown below. Since hydrations of alkenes and alkynes typically occur under strongly acidic conditions, Lewis acid catalysts may have been required.

$$H_2C{=}CH_2 \xrightarrow{H_2O} CH_3CH_2OH$$

$$CH{\equiv}CH \xrightarrow{H_2O} CH_3{-}\overset{O}{\overset{\|}{C}}{-}H$$

$$CH{\equiv}C{-}CH_3 \xrightarrow{H_2O} CH_3{-}\overset{O}{\overset{\|}{C}}{-}CH_3$$

5.4.5. *Sulfonic and Phosphonic Acids*. The sulfonic acids may be derived from various interstellar organosulfur compounds. The C=S series through C_3S, has been found as have thioformaldehyde and methanethiol (Section 3.5).

$$C{=}S \xrightarrow{\text{Red'n}} CH_2{=}S \xrightarrow{\text{Red'n}} H_3C{-}SH \xrightarrow{3/2\ O_2} CH_3{-}SO_3H$$

The CP radical has recently been identified in an interstellar source. CP and higher homologues could conceivably have given rise to the phosphonic acids.

$$CP \xrightarrow{H} HCP \xrightarrow{2H_2O,1/2\ O_2} CH_3{-}PO_3H_2$$

5.4.6. *Hydrocarbons*. The molecular and isotopic composition of hydrocarbons found in the C2M chondrites is compatible with origins in a variety of environments. Saturated aliphatic hydrocarbons higher than methane have not been observed in interstellar regimes, however, their observation poses serious difficulties (Herbst, 1985). Even though unobserved, their existence is inferred on the bases of homology with methane and infrared observations of aliphatic C-H bonds in interstellar dust (Willner *et al*, 1979; Butchard *et al*., 1986; Adamson *et al*., 1990) and Comet Halley dust.

Aromatic hydrocarbons appear to be abundant in interstellar space based on the observation of aromatic C-H bonds in the infrared (Duley and Williams, 1981; Allamandola *et al*., 1985) and should have survived the transit from interstellar cloud into meteorite parent bodies (Allamandola *et al*., 1987). Thus both classes of hydrocarbons may represent relatively unaltered interstellar matter. It has been suggested that the aromatic hydrocarbons were formed by carbon condensation in the outflows of carbon stars (Frenklach and Feigelson, 1989; Jura, 1990; Beiqing, 1990).

It has also been been suggested (Basile *et al*., 1984; Pering and Ponnamperuma, 1971; Olson *et al*., 1967) that the aromatic hydrocarbons of carbonaceous chondrites are pyrolysis products of pre-existing aliphatic hydrocarbons. Such pyrolysis could have occurred as interstellar dust entered the shock front of the solar nebula. Morgan et al. (1991) suggested nebular pyrolysis-condensation of light hydrocarbons inherited from the parent cloud as a production mechanism for the aromatic moieties of the macromolecular carbon.

It is not at all clear how hydrocarbons, i.e., cycloalkanes, polycylic aromatic compounds, and heterocycles, might have formed from simple precursors by aqueous phase chemistry in a chondrite parent body. One possibility, analogous to the formation of terrestrial petroleum, is that they are derived from the macromolecular carbon, the similarity of which to terrestrial kerogen was noted previously (Sec. 2.2.1). If this were the case, and the conversion was pyrolytic as occurs with terrestrial kerogens, it would be expected that the macromolecular carbon would be isotopically heavier with respect to ^{13}C after thermal maturation. This has been found to be the case in experimental studies carried out with terrestrial kerogens (Hoefs, 1987), however, isotopic analyses of the relevant Murchison materials indicate the opposite relationship, i.e., the hydrocarbons are isotopically heavier than the macromolecular carbon.

Thermal aromatization of aliphatic hydrocarbons (Wing and Bada, 1991) has been suggested as a consequence of secondary processing on a parent body. On the one hand, the hydrogen isotopic data do not rule out this possibility, since pyrolysis should lead to preferential loss of H from the aliphatic source material and formation of relatively D-enriched aromatic products as observed. On the other hand, the temperatures required for such aromatization are incompatible with the survival in the same environment of either the interstellar organic precursors needed for aqueous phase synthesis, or the products of these syntheses, i.e., most of the other compounds now found in carbonaceous chondrites. Therefore, if some of the aromatic compounds had a pyrolytic origin, they must have experienced a different thermal history, in the nebula or at another location within the parent body, than the observed thermally labile compounds. This conclusion is consistent with the views that refractory minerals and organics coaccreted with volatile rich ices in the parent bodies and that the CI and CM meteorites have sampled only parts of the original bodies.

The isotopic data and molecular analyses are also consistent with another possibility, that is, the derivation of polycyclic aliphatic hydrocarbons from the aromatics. To the extent to which secondary reactions, e.g., hydrogenation or hydration, can add additional carbon-bound hydrogen atoms, they can also lead to differences in D/H ratios. If, for example, the aliphatic hydrocarbons were derived from interstellar aromatic species by nebular or parent body hydrogenation reactions, they would be isotopically lighter because the content of D-rich hydrogen originating in the precursor aromatic compounds would have been diluted by the isotopically lighter hydrogen added. Other interstellar species with higher resistance to hydrogenation would yield relatively D-rich products.

In summary, the hydrocarbon isotopic data appear to be compatible with several possibilities, i.e., (1) that both the aliphatic and aromatic compounds were formed by presolar processes, (2) that aliphatic precursors underwent aromatization, or (3) that aromatic precursors experienced partial reduction. Based on the limited available data, the δD vs $\delta^{13}C$ correlation, which implies *de novo* interstellar synthesis, holds for hydrocarbons as well as other soluble organic compounds (Fig. 1). Until further observations and experiments yield compelling reasons to consider other genetic relationships, the simplest working hypothesis is that the hydrocarbons, per se, were also of interstellar origin..

6. Meteoritic Organic Compounds and the Origin of Life

Origin of life studies comprise perhaps the ultimate interdisciplinary field. Subdisciplines of the earth sciences, biology, chemistry, astronomy, and the space sciences have all contributed to the contemporary understanding of this profound problem. The paleontology of microorganisms (Schopf and Packer, 1987), isotope geochemistry (Schidlowski, 1988), and inferences drawn from the lunar impact record (Maher and Stevenson, 1988) combine to constrain the time frame for the origin of life to a period between 3.5 and 3.9 billion years ago. RNA based studies of bacterial phylogeny identify contemporary organisms of primitive lineage, and studies of their biochemistry and physiology suggest the nature of the progenote, the ancestral cell-type for extant life (Woese, 1987). Geological approaches to understanding the problem of the origin of life are limited in their ability to probe the period of interest by the fragility of organic compounds and the paucity of the ancient sedimentary record. While a better preserved record of early planetary history may exist on Mars, it will remain inaccessible until samples can be returned to Earth (Klein, 1992). Biological/biochemical approaches deal only with existing organisms and what might be inferred or extrapolated backward in time from their biochemical structures and mechanisms to more primitive possibilities. For example, the suggestion of an "RNA-world," which offers an escape from the "which came first: nucleic acid or protein?" paradox, exemplifies a useful conjecture of this type (Orgel, 1986).

But can our understanding be extended to a more fundamental level, i.e., to the elucidation of molecular structures and mechanisms that are at once conceivable as arising spontaneously, robust or probable enough to persist, and complex enough to be self-maintaining and self-replicating? Organic chemical studies of synthetic processes occurring spontaneously under presumed primitive conditions have been successful in identifying routes to many of the biomonomers (Ferris, 1992), e.g., amino acids, and even to macromolecules (Fox, 1965); but the plausibility of many such processes is uncertain when viewed within the bounds of modern models of the prebiotic environment. Moreover, attempts to model chemical evolution at the level of complexity of the RNA world have met with little success (Joyce, 1991). Sophisticated chemistry mediated by laboratory researchers is required to form the reactive monomers, i.e., ribonucleotides or ribonucleotide analogues. A significant gap remains between the RNA world and even the best outcomes of the chemical modeling approach, suggesting that if this gap is to be closed, life at a level of complexity less than that of the RNA world remains to be discovered, i.e., "low tech" life, to use the terminology of Cairns-Smith (1982).

The extrapolation to low-tech life from the biological-geological side of the gap has an appeal that derives from the seeming logic of reaching insights by retracing the evolutionary process backward through time to arrive at simpler systems. Chemical evolution studies, i.e., the laboratory modeling of possible prebiotic synthetic processes, approach the possibility of low-tech life from the opposite, i.e., time-forward, direction, beginning with conditions and processes inherent in models of the prebiotic Earth. The space sciences also provide an approach from this direction that has a logical appeal equivalent to that of the biological extrapolation just described. It is possible to reconstruct a cosmic history of organic matter starting with interstellar matter (Wood and Chang, 1984). Interstellar clouds, the birthplace of stars and planetary systems, are rich in simple organic compounds and radicals in the gas phase and may contain more complex molecules and even macromolecular organic matter on grains. Further insight into these possibilities may be gained through astronomical observations and study of cometary samples returned to Earth. As we have attempted to show here, the carbonaceous chondrites already provide evidence that this interstellar organic matter has survived the rigors of the solar nebula and achieved a higher level of complexity as a result of aqueous processing in the parent body. Meteorite studies have made the outline of this cosmic history readable, and they provide the basis for extrapolating the occurrence of organic chemical evolution and the potentiality for life, itself, beyond our solar system.

It is now clear that the primitive earth accumulated meteoritic matter ranging in size from dust particles to giant impactors. Indeed, the earth may have received the major fraction of its volatile elements from meteorites and comets, and some part of this total undoubtedly arrived under conditions that allowed the survival of organic compounds (Anders, 1989; Chyba *et al.*, 1990; Chyba and Sagan, 1992). What new level of chemical complexity was reached by further reactions on the primitive earth is unknown, however, the outlines of a progression in chemical complexity may become apparent as the potential of extraterrestrial organic compounds for further chemical evolution is explored. One such exploration is the work by Deamer *et al.* (1985) who showed that organic matter in Murchison chondrite was capable of forming membranous vesicles.

Thus a major and unique contribution of meteorite organic chemistry to origin of life studies is to provide a natural inventory of at least a part of the organic milieu available for the origin of life. This inventory is clearly not a recipe for constructing the progenote, nor is it a complete list of ingredients. In addition to the vesicle-forming compounds, only eight of the protein amino acids are available, only short chain carboxylic acids are abundant, and simple sugars have not been found. Moreover, the inventory poses additional questions: If a meteorite-like assemblage of organic compounds were available, either through impacts or by endogenous synthetic processes, how was the limited suite of biochemical building blocks selected or produced from the inventory? And how was the inventory itself preserved against dilution and destruction in the oceans? In the face of such large uncertainties, it is unrealistic to make the extrapolation from meteoritic organic chemistry to life. Rather, the challenge in origin of life studies is to characterize the structure and function of chemical systems at that point on the continuum of chemical complexity where time-forward extrapolations from chemical evolution studies and meteorite studies coincide with those inferred by time-backward extrapolation from biological evolution, the point where necessity encounters chance (Monod, 1971), where physics meets history (Smith and Morowitz, 1982), and where the diversity that characterizes chemical evolution as exemplified in carbonaceous chondrites gives way to the selectivity of life.

References

Adamson, A. J., Whittet, D. C. B. and Duley, W. W. (1990) The 3.4-μm interstellar absorption feature in Cyg OB2 no. 12*. *Mon. Not. R. astr. Soc.* **243**, 400-404.

Agarwal, V.K., Schutte, W., Greenberg, J.M., Ferris, J.P., Briggs, R., Connor, S., Van de Bult, C.P.E.M. and Baas, F. (1985) Photochemical reactions in interstellar grains. Photolysis of CO, NH_3 and H_2O. *Origins of Life*, **16**, 21-40.

Allamandola, L.J., Tielens, A. G. G. M. and Barker, J. R. (1985) Polycyclic aromatic hydrocarbons and the unidentified infrared emission bands: Auto exhaust along the Milky Way? *Astrophys. J.* **290**, L25-L28.

Allamandola, L.J., Sandford, S.A. and Wopenka, B. (1987) Interstellar polycyclic aromatic hydrocarbons and carbon in interplanetary dust particles and meteorites, *Science* **237**, 56-59.

Allamandola, L.J., Sandford, S.A. and Valero, G.J. (1988) Photochemical and thermal evolution of interstellar/precometary ice analogs, *Icarus*, **76**, 225-252.

Alpern, B. and Benkheiri, Y. (1973) Distribution de la matiére organique dans la météorite d'Orgueil par microscopie en fluorescence. *Earth Planet. Sci. Lett.* **19**, 422-428.

Amari, S., Anders, E., Virag, A. and Zinner, E. (1990) Interstellar graphite in meteorites. *Nature* **345**, 238-240.

Anders, E. (1975) Do stony meteorites come from comets? *Icarus* **24**, 363-371.

Anders, E. (1978) Most stony meteorites come from the asteroid belt, in D. Morrison and W. C. Wells (eds.) Asteroids: An Exploration Assessment, NASA CP-2053, U.S. Govt. Printing Office, Washington, D.C., pp. 57-75.

Anders, E. (1989) Pre-biotic organic matter from comets and asteroids. *Nature* **342**, 255-256.

Bada, J. L., Cronin, J. R., Ho, M.-S., Kvenvolden, K. A., Lawless, J. G., Miller, S. L., Oró, J. and Steinberg, S. (1983) On the reported optical activity of amino acids in the Murchison meteorite, *Nature* **301**, 494-497.

Basile, B. P., Middleditch, B. S. and Oró, J. (1984) Polycyclic aromatic hydrocarbons in the Murchison meteorite. *Org. Geochem.* **5**, 211-216.

Becker, R. H. and Epstein, S. (1982) Carbon, hydrogen and nitrogen isotopes in solvent-extractable organic matter from carbonaceous chondrites. *Geochim. Cosmochim. Acta* **46**, 97-103.

Beiging, J. H. (1990) Carbon chemistry of circumstellar envelopes, in J. C. Tarter, S. Chang, and D. J. DeFrees (eds.), Carbon in the Galaxy: studies from Earth and Space, NASA Conf. Publ. 3081, pp. 147-158.

Belsky, T. and Kaplan, I. R. (1970) Light hydrocarbon gases, ^{13}C, and origin of organic matter in carbonaceous chondrites. *Geochim. Cosmochim. Acta* **34**, 257-278.

Bergmann, M. and Stern, F. (1930) Notiz über acetylierung von amino-säuren mittels ketens. *Ber* **63B**, 437-439.

Blake, D.F., Freund, F., Krishnan, K.F.M., Echer, C.J., Shipp, R., Bunch, T.E., Tielens, A.G., Lipari, R.J., Hetherington, C. J.D. and Chang, S. (1988) The nature and origin of interstellar diamond, *Nature*, **332**, 611-613.

Briggs, R., Erten, G., Ferris, J.P., Greenberg, J.M., McCain, P.J., Mendoza-Gomez, C.X. and Schutte, W. (1992) Comet Halley as an aggregate of interstellar dust and further evidence for the photochemical formation of organics in the interstellar medium. *Origins of Life*, **22**, 287-307.

Buhl, P. (1975) An Investigation of organic compounds in the Mighei meteorite. Ph.D. Thesis, Univ. of Maryland, College Park.

Bunch, T. E. and Chang, S. (1980) Carbonaceous chondrites: II. Carbonaceous chondrite phyllosilicates and light element geochemistry as indicators of parent body processes and surface conditions. *Geochim. Cosmoshim. Acta* **44**, 1543-1577.

Butchard, I., McFadzean, A. D., Whittet, D. C. B., Geballe, T. R. and Greenberg, J. M. (1986) Three micron spectroscopy of the galactic center source IRS 7. *Astron. Astrophys. Lett.* **154**, L5-L7.

Cairns-Smith, A. G. (1982) Genetic takeover and the mineral origins of life, Cambridge Univ. Press, Cambridge.

Cameron, A. G. W. (1973) Interstellar grains in museums? in J. M. Greenberg and H. C. Van de Hulst (eds.), Interstellar dust and related topics, D. Reidel Publishing Co., Dordrecht, pp. 545-547.

Chang, S., Mack, R. and Lennon, K. (1978) Carbon chemistry of separated phases of Murchison and Allende meteorites. *Lunar Planet. Sci.* **IX**, 157-158.

Chang, S., Des Marais, D., Mack, R., Miller, S.L. and Strathearn, G.E. (1983) Prebiotic organic synthesis and the origin of life. in Earth's Earliest Biosphere (J.W. Schopf, ed.), Princeton University Press, pp. 53-92.

Chyba, C. F., Thomas, P. J., Brookshaw, L. and Sagan, C. (1990) Cometary delivery of organic molecules to the early earth. *Science*, **249**, 366-373.

Chyba, C. F. and Sagan, C. (1992) Endogenous production, exogenous delivery and impact-shock synthesis of organic molecules: an inventory for the origins of life. *Nature* **355**, 125-132.

Clark, B., Mason, L.W., and Kissel, J. (1987) Systematics of the "CHON" and other light-element particle populations in Comet Halley. *Astron. Astrophys.* **187**, 779-784.

Clayton, R. N. (1978) Isotopic anomalies in the early solar system. *Ann. Rev. Nucl. Part. Sci.* **28**, 501-522.

Cooper, G. W., Onwo, W. M. and Cronin, J. R. (1992) Alkyl phosphonic acids and sulfonic acids in the Murchison meteorite, *Geochim. Cosmochim. Acta* **56**, 4109-4115.

Cronin, J. R. and Moore, C. B. (1971) Amino acid analyses of the Murchison, Murray, and Allende carbonaceous chondrites. *Science* **172**, 1327-1329.

Cronin, J. R. (1976) Acid-labile amino acid precursors in the Murchison meteorite. I. Chromatographic fractionation. *Origins of Life* **7**, 337-342.

Cronin, J. R., Gandy, W. E. and Pizarello, S. (1981) Amino acids of the Murchison meteorite: I. Six carbon acyclic primary α-amino alkanoic acids. *J. Molec. Evol.* **17**, 265-272.

Cronin, J. R. and Pizzarello, S. (1983) Amino acids in meteorites. *Adv. Space Res.* **3**, 3-18.

Cronin, J. R., Pizarello, S. and Yuen, G. U. (1985) Amino acids of the Murchison meteorite: II. Five carbon acyclic primary β-, γ-, and δ-amino alkanoic acids. *Geochim. Cosmochim. Acta* **49**, 2259-2265.

Cronin, J. R. and Pizzarello, S. (1986) Amino acids of the Murchison meteorite. III. Seven carbon acyclic primary α-amino alkanoic acids. *Geochim. Cosmochim. Acta* **50**, 2419-2427.

Cronin, J. R., Pizzarello, S. and Frye, J. S. (1987) ^{13}C NMR spectroscopy of the insoluble carbon of carbonaceous chondrites. *Geochim. Cosmochim. Acta* **51**, 299-303.

Cronin, J. R., Pizzarello, S. and Cruikshank, D. P. (1988) Organic matter in carbonaceous chondrites, planetary satellites, asteroids, and comets. in J. F. Kerridge and M. S. Matthews (eds.) Meteorites and the Early Solar System, Chap. 10.5, pp. 819-857, Univ. Arizona Press.

Cronin, J. R. and Pizzarello, S. (1990) Aliphatic hydrocarbons of the Murchison meteorite. *Geochim. Cosmochim. Acta* **54**, 2859-2868.

Cronin, J. R., Pizzarello, S., Epstein, S. and Krishnamurthy, R. V. (1993) Molecular and isotopic analyses of the hydroxy acids, dicarboxylic acids, and hydroxydicarboxylic acids of the Murchison meteorite. *Geochim. Cosmochim. Acta*, in press.

Deamer, D. W. (1985) Boundary structures are formed by organic components of the Murchison carbonaceous chondrite. *Nature* **317**, 792-794.

Degens, E. T. and Bajor, M. (1962) Amino acids and sugars in the Brudesheim and Murray meteorites. *Naturwiss.* **49**, 605-606.

Dufresne, E. R. and Anders, E. (1962) On the chemical evolution of the carbonaceous chondrites. *Geochim. Cosmochim. Acta* **26**, 1085-1114.

Duley, W. W. and Williams, D. A. (1981) The infrared spectrum of interstellar dust: Surface functional groups on carbon. *Mon. Not. R. Astron. Soc.* **196**, 269.

Engel, M. H. and Nagy, B. (1982) Distribution and enantiomeric composition of amino acids in the Murchison meteorite. *Nature* **296**, 837-840.

Engel, M. H., Macko, S. A. and Silfer, J. A. (1990a) Carbon isotope composition of individual amino acids in the Murchison meteorite. *Nature* **348**, 47-49.

Engel, S., Lunine, J.I. and Lewis, J.S. (1990b) Solar nebula origin for volatile gases in Halley's Comet. *Icarus* **85**, 380-393.

Epstein, S., Krishnamurthy, R. V., Cronin, J. R., Pizzarello, S. and Yuen, G. U. (1987) Unusual stable isotope ratios in amino acid and carboxylic acid extracts from the Murchison meteorite. *Nature* **326**, 477-479.

Ferris, J. P. and Hagen, W. J. (1984) HCN and chemical evolution: the possible role of cyano compounds in prebiotic synthesis. *Tetrahedron* **40**, 1093-1120.

Ferris, J. P. (1992) Chemical markers of prebioltic chemistry in hydrothermal systems. *Origins of Life* **22**, 109-134.

Fomenkova, M., Chang, S. and Mukhin, L. (1992) Classification of carbonaceous components in Comet Halley "CHON" particles. *Lunar and Planet Sci.* **XXIII**, 379-380.

Fox, S. W. (1965) A theory of macromolecular and cellular origins. *Nature* **205**, 328-340.

Frenklach, M. and Feigelson, E. D. (1989) Formation of polycyclic aromatic hydrocarbons in circumstellar envelopes. *Astrophys. J.* **341**, 372-384.

Fuchs, L. H., Olsen, E. and Jensen, K. J. (1973) Mineralogy, crystal chemistry and composition of the Murchison (C2) meteorite. *Smithsonian Contrib. Earth Sci.* **10**, 1-39.

Gaffey, M. J., Bell, J. F. and Cruikshank, D. P. (1989) Reflectance spectroscopy and asteroid surface mineralogy. in R. P. Binzel, T. Gehrels, and M. S. Mathews, (eds.), Asteroids II, Univ. Ariz. Press, Tucson, pp. 98-127.

Geiss, J. and Reeves, H. (1981) Deuterium in the early solar system. *Astron. Astrophys.* **93**, 189-199.

Gibson, E. K., Moore, C. B. and Lewis, C. F. (1971) Total nitrogen and carbon abundances in carbonaceous chondrites. *Geochim. Cosmochim. Acta* **35**, 599-604.

Gilmour, I. and Pillinger, C. (1992) Isotopic differences between PAH isomers in Murchison. *Meteoritics* **27**, 224-225.

Grady, M. M., Wright, I. P., Swart, P. K. and Pillinger, C. T. (1988) The carbon and oxygen isotopic composition of meteoritic carbonates. *Geochim. Cosmochim. Acta* **52**, 2855-2866.

Greenberg, J. M. (1973) Chemical and physical properties of interstellar dust, in M. A. Gordon and L. E. Snyder (eds.), Molecules in the Galactic Environment, Wiley, N.Y., pp. 94-124.

Greenberg, J.M. (1977) From dust to comets. in A. H. Delsemme (ed.) Comets, Asteroids, and Meteorites, The University of Toledo, pp. 491-497.

Greenberg, J. M. and d'Hendecourt (1985) Evolution of ices from interstellar space to the solar system, in J. Klinger, D. Benest, A. Dollfus and R. Smouluchowski (eds.), Ices in the Solar System, Kluwer, Dordrecht, pp. 185-204.

Guélin, M., Cernicharo, J., Paubert, G. and Turner, B. E. (1990) Free CP in IRC^{+} 10216. *Astron. Astrophys.* **230**, L9-L11.

Hahn, J. H., Zenobi, R., Bada, J. L. and Zare, R. N. (1988) Application of two-step laser mass spectrometry to cosmogeochemistry: direct analysis of meteorites. *Science* **239**, 1523-1525.

Hamilton, P. B. (1965) Amino acids on hands. *Nature* **205**, 284-285.

Han, J., Simoneit, B. R., Burlingame, A. L. and Calvin, M. (1969) Organic analysis on the Pueblito de Allende meteorite. *Nature* **222**, 364-365.

Hayatsu, R., Studier, M. H., Moore, L. P. and Anders, E. (1975) Purines and triazines in the Murchison meteorite. *Geochim. Cosmochim. Acta* **39**, 471-488.

Hayatsu, R., Matsuoka, S., Scott, R. G., Studier, M. and Anders, E. (1977) Origin of organic matter in the early solar system–VII. The organic polymer in carbonaceous chondrites. *Geochim. Cosmochim. Acta* **41**, 1325-1339.

Hayatsu, R., Winans, R. E., Scott, R. G., McBeth, R. L., Moore, L. P. and Studier, M. H. (1980) Phenolic ethers in the organic polymer of the Murchison meteorite. *Science* **207**, 1202-1204.

Hayatsu, R and Anders, E. (1981) Organic compounds in meteorites and their origins. *Topics Curr. Chem.* **99**, 1-37.

Hayatsu, R., Scott, R. G. and Winans, R. E., (1983) Comparative structural study of meteoritic polymer with terrestrial geopolymers coal and kerogen. *Meteoritics* **18**, 310 (abstract).

Hayes, J. M. (1967) Organic constituents of meteorites–A review. *Geochim. Cosmochim. Acta* **31**, 1395-1440.

Herbst, E. (1985) On the formation and observation of complex interstellar molecules. *Origins of Life* **16**, 3-19.

Hoefs, J. (1987) *Stable Isotope Geochemistry*, 3rd ed. Springer-Verlag.

Hollis, J. M., Snyder, L. E., Suenram, R. D. and Lovas, F. J. (1980) Search for the lowest-energy conformer of interstellar glycine. *Astrophys. J.* **241**, 1001-1006.

Injerd, W. G. and Kaplan, I. R. (1974) Nitrogen isotope distribution in meteorites. *Meteoritics* **9**, 352-353.

van Ijzendoorn, L. J., Allamandola, L. J., Baas, F., Körnig, S. and Greenberg, J. M. (1986) Laser-induced fluorescence and phosphorescence of matrix-isolated glyoxal: evidence for exciplex formation in the $\underline{\tilde{A}}\,^1A_u$ and $\tilde{a}\,^3A_u$ states. *J. Chem. Phys.* **85**, 1812-1825.

Joyce, G.F. (1986) RNA evolution and the origins of life. *Nature* **338**, 217-225.

Joyce, G. F. (1991) The rise and fall of the RNA world. *New Biologist* **3**, 399-407.

Jungclaus, G., Cronin, J. R., Moore, C. B. and Yuen, G. U. (1979) Aliphatic amines in the Murchison meteorite. *Nature* **261**, 126-128.

Jura, M. (1990) Astronomical observations of solid phase carbon, in J. C. Tarter, S. Chang, and D. J. DeFrees (eds.), Carbon in the Galaxy: studies from Earth and Space, NASA Conf. Publ. 3061, pp. 39-46.

Kerridge, J. F., Mackay, A. L. and Boynton, W. V. (1979) Magnetite in CI carbonaceous meteorites: origin by aqueous activity on a planetesimal surface. *Science* **205**, 395-397.

Kerridge, J. F., Chang, S. and Shipp, R. (1987) Isotopic characterization of kerogen-like material in the Murchison carbonaceous chondrite. *Geochim. Cosmochim. Acta* **51**, 2527-2540.

Kerridge, J. F. and Mathews, M. S. (1988) Meteorites and the Early Solar System, Univ. of Arizona Press, Tucson.

Kerridge, J. F. (1991) A note on the prebiotic synthesis of organic acids in carbonaceous meteorites, *Origins of Life* **21**, 19-30.

Kimball, B. A. (1988) Determination of formic acid in chondritic meteorites. M. S. Thesis, Arizona State Univ., Tempe.

Klein, H. P. (1992) The Viking biology experiments: epilogue and prologue. *Origins of Life* **21**, 255-261.

Kolodny, Y., Kerridge, J. F. and Kaplan, I. R. (1980) Deuterium in carbonaceous chondrites. *Earth Planet. Sci. Lett.* **46**, 149-158.

Kovalenko, L. J., Maechling, C. R., Clemett, S. J., Philippoz, J.-M. and Zare, R. N. (1992) Microscopic organic analysis using two-step laser mass spectrometery: application to meteoritic acid residues. *Anal. Chem.* **64**, 682-690.

Krishnamurthy, R. V., Epstein, S., Cronin, J. R., Pizzarello, S. and Yuen, G. U. (1992) Isotopic and molecular analyses of hydrocarbons and monocarboxylic acids of the Murchison meteorite. *Geochim. Cosmochim. Acta* **56**, 4045-4058.

Kroto, H. W. (1988) The chemistry of the interstellar medium. *Phil. Trans. Roy. Soc. A* **325**, 405-421.

Krueger, F. R., Korth, A. and Kissel, J. (1991) The organic matter of comet Halley as inferred by joint gas phase and solid phase analyses. *Space Sci. Rev.* **56**, 167-175.

Kung, C.C. and Clayton, R.N. (1978) Nitrogen abundances and isotopic composition in stony meteorites. *Earth Planet. Sci. Lett.* **38**, 421-435.

Kung, C.C., Hayatsu, R., Studier, M.H. and Clayton, R.N. (1979) Nitrogen isotope fractionations in the Fischer-Tropsch synthesis and in the Miller-Urey reaction. *Earth Planet. Sci. Lett.* **46**, 141-146.

Kvenvolden, K., Lawless, J., Pering, K., Peterson, E., Flores, J., Ponnamperuma, C., Kaplan, I. R. and Moore, C. (1970) Evidence for extraterrestrial amino acids and hydrocarbons in the Murchison meteorite. *Nature* **228**, 923-926.

Kvenvolden, K., Lawless, J. G. and Ponnamperuma, C. (1971) Nonprotein amino acids in the Murchison meteorite. *Proc. Natl. Acad. Sci. USA* **68**, 486-490.

Lancet, M. S. and Anders, E. (1970) Carbon isotope fractionation in the Fischer-Tropsch synthesis and in meteorites. *Science* **170**, 980-982.

Lawless, J. G. (1973) Amino acids in the Murchison meteorite. *Geochim. Cosmochim. Acta* **37**, 2207-2212.

Lawless, J. G., Zeitman, B., Pereira, W. E., Summons, R. E. and Duffield, A. M. (1974) Dicarboxylic acids in the Murchison meteorite. *Nature* **251**, 40-41.

Lawless, J. G. and Yuen, G. U. (1979) Quantification of monocarboxylic acids in Murchison carbonaceous meteorite. *Nature* **282**, 396-398.

Lerner, N.R., Peterson, E. and Chang, S. (1993) The Strecker synthesis as a source of amino acids in carbonaceous chondrites: deuterium retention during synthesis. *Geochim. Cosmochim. Acta*, in press.

Lewis, R. S., Tang, M., Wacker, J. F., Anders, E. and Steel, E. (1987) Interstellar diamonds in meteorites. *Nature* **326**, 160-162.

Lovering, J. F., LeMaitre, R. W. and Chappell, B. W. (1971) Murchison C2 carbonaceous chondrite and its inorganic composition. *Nature* **230**, 18-20.

Lumpkin, G. R. (1986) Electron microscopy of carbonaceous matter in Allende acid residues. *Proc. Lunar Planet. Sci. Conf.* **12B**, 1153-1166.

Lunine, J. I. (1989) The Urey prize lecture: Volatile processes in the outer solar system. *Icarus* **81**, 1-13.

Maher, K. A. and Stevenson, D. J. (1988) Impact frustration of the origin of life. *Nature* **331**, 612-614.

Mason, B. (1963) The carbonaceous chondrites. *Space Sci. Rev.* **1**, 621-646.

Meinschein, W. G. (1963) Hydrocarbons in terrestrial samples and the Orgueil meteorite. *Space Sci. Rev.* **2**, 653-679.

Miknis, F. P., Lindner, A. W., Gannon, J., Davis, M. F. and Maciel, G. E. (1984) Solid state ^{13}C NMR studies of selected oil shales from Queensland, Australia. *Org. Geochem.* **7**, 239-248.

Millar, T. J., Bennett, A. and Herbst, E. (1989) Deuterium fractionation in dense interstellar clouds. *Ap. J.* **340**, 906-920.

Miller, S. L. (1957) The mechanism of synthesis of amino acids by electric discharges. *Biochim. Biophys. Acta* **23**, 480-489.

Miller, S. L. and Van Trump, J. E. (1981) The Strecker synthesis in the primitive ocean. in Y. Wolman (ed.), Origin of Life, D. Reidel Publishing Co., Dordrecht, pp. 135-141.

Monod, J. (1971) Chance and necessity: an essay on the natural philosophy of modern biology, Knopf, New York.

Morgan, W. A., Jr., Feigelson, E. D., Want, H. and Frenklach, M. (1991) A new mechanism for the formation of meteoritic kerogen-like material. *Science* **252**, 109-112.

Mullie, F. and Reisse, J. (1987) Organic matter in carbonaceous chondrites. *Topics in Current Chemistry* **139**, 85-117.

Olson, R. J., Oró, J. and Zlatkis, A. (1967) Organic compounds in meteorites: II. Aromatic hydrocarbons. *Geochim. Cosmochim. Acta* **31**, 1935-1948.

Orgel, L. E. (1986) RNA catalysis and the origins of life. *J. theor. Biol.* **123**, 127-149.

Oró, J. and Skewes, H. B. (1965) Free amino acids on human fingers: the question of contamination in microanalysis. *Nature* **207**, 1042-1045.

Oró, J., Gibert, J., Lichtenstein, H., Wikstrom, S. and Flory, D. A. (1971) Amino acids, aliphatic, and aromatic hydrocarbons in the Murchison meteorite. *Nature* **230**, 105-106.

Peltzer, E. T. and Bada, J. L. (1978) α-Hydroxycarboxylic acids in the Murchison meteorite. *Nature* **272**, 443-444.

Peltzer, E. T., Bada, J. L., Schlesinger, G. and Miller, S. L. (1984) The chemical conditions on the parent body of the Murchison meteorite: Some conclusions based on amino, hydroxy, and dicarboxylic acids. *Adv. Space Res.* **4**, 69-74.

Pering, K. L. and Ponnamperuma, C. (1971) Aromatic hydrocarbons in the Murchison meteorite. *Science* **173**, 237-239.

Pillinger, C. T. (1984) Light element stable isotopes in meteorites From grams to picograms. *Geochim. Cosmochim. Acta* **48**, 2739-2766.

Pizzarello, S., Krishnamurthy, R. V., Epstein, S. and Cronin, J. R. (1991) Isotopic analyses of amino acids from the Murchison meteorite. *Geochim. Cosmochim. Acta* **55**, 905-910.

Pollock, G. E., Chang, C.-N., Cronin, S. E. and Kvenvolden, K. E. (1975) Stereoisomers of isovaline in the Murchison meteorite. *Geochim. Cosmochim. Acta* **39**, 1571-1573.

Prinn, R.G. and Fegley, B., Jr. (1989) Solar nebula chemistry: origin of planetary, satellite, and cometary volatiles. in S.K. Atreya, J.B. Pollack and M.S. Matthews (eds.) Origin and Evolution of Planetary and Satellite Atmospheres, University of Arizona Press, pp. 78-136.

Reynolds, J.H., Frick, U., Neil, J.M. and Phinney, D.L. (1978) Rare-gas-rich separates from carbonaceous chondrites. *Geochim. Cosmochim. Acta* **42**, 1775-1797.

Robert, F. and Epstein, S. (1982) The concentration and isotopic composition of hydrogen, carbon, and nitrogen in carbonaceous meteorites. *Geochim. Cosmochim. Acta* **46**, 81-95.

Rossignol-Strick, M. and Barghoorn, E. (1971) Extraterrestrial abiogenic organization of organic matter: The hollow spheres of the Orgueil meteorite. *Space Life Sci.* **3**, 89-107.

Saito, S., Kawaguchi, K., Yamamoto, S., Ohishi, M., Suzuki, H. and Kaifu, N. (1987) Laboratory detection and astronomical identification of a new free radical, CCS ($^3\Sigma^-$). *Astrophys. J.* **317**, L115-L119.

Schidlowski, M. (1988) A 3800-million year isotopic record of life from carbon in sedimentary rocks. *Nature* **333**, 313-318.

Schopf, J. W. and Packer, B. M. (1987) Early archaen (3.3-billion to 3.5-billion-year-old) microfossils from Warrawoona group, Australia. *Science* **237**, 70-73.

Shimoyama, A., Naroka, H., Komiya, M. and Harada, K. (1989) Analyses of carboxylic acids and hydrocarbons in Antarctic carbonaceous chondrites, Yamato-74662 and Yamato-793321, *Geochem. J.* **23**, 181-193.

Shock, B. L. and Schulte, M. D. (1990) Amino-acid synthesis in carbonaceous meteorites by aqueous alteration of polycyclic aromatic hydrocarbons. *Nature* **343**, 728-731.

Smith, J. W. and Kaplan, I. R. (1970) Endogenous carbon in carbonaceous meteorites. *Science* **167**, 1367-1370.

Smith, T. F. and Morowitz, H. J. (1982) Between history and physics. *J. Molec. Evol.* **18**, 265-282.

Snyder, L. E. (1986) The search for biomolecules in space, in K. I. Kellerman and G. A. Seielstad (eds.), The Search for Extraterrestrial Intelligence, pp. 39-50.

Stoks, P. G. and Schwartz, A. W. (1979) Uracil in carbonaceous meteorites. *Nature* **282**, 709-710.

Stoks, P. G. and Schwartz, A. W. (1981a) Nitrogen-heterocyclic componds in meteorites: Significance and mechansim of formation. *Geochim. Cosmochim. Acta* **45**, 563-569.

Stoks, P. G. and Schwartz, A. W. (1981b) Nitrogen componds in meteorites: A reassessment. in Y. Wolman (ed.), Proc. 6th Internatl. Conf. on the Origin of Life, D. Reidel, Dordrecht, pp. 59-64.

Stoks, P. G. and Schwartz, A. W. (1982) Basic nitrogen-heterocyclic compounds in the Murchison meteorite. *Geochim. Cosmochim. Acta* **46**, 309-315.

Strecker, A. (1850) Über die künstliche bildung der milchsauer und einem neuen dem glycocoll homologen körper. *Ann. Chem.* **75**, 27.

Studier, M. H., Hayatsu, R. and Anders, E. (1972) Origin of organic matter in the early solar system–V. Further studies of meteoritic hydrocarbons and a discussion of their origin. *Geochim. Cosmochim. Acta* **36**, 189-215.

Swart, P. K., Grady, M. M., Pillinger, C. T., Lewis, R. S. and Anders, E. Interstellar carbon in meteorites. *Science* **220**, 406-410.

Tang, M., Anders, E., Hoppe, P. and Zinner, E. (1989) *Nature* **339**, 351-354.

Tielens, A. G. G. M. (1983) Surface chemistry of deuterated molecules. *Astron. Astrophys.* **119**, 177-184.

Tingle, T. N., Becker, C. H. and Malhotra, R. (1991) Organic compounds in the Murchison and Allende carbonaceous chondrites studied by photoionization mass spectrometry. *Meteoritics* **26**, 117-127.

Turner, B. E. (1980) On the identification of MM-wavelength U-lines, in B. H. Andrew (ed.), Interstellar Molecules, D. Reidel, Dordrecht, pp. 45-46.

Turner, B. E. (1989) Recent progress in astrochemistry. *Space Sci. Rev.* **51**, 235-337.

Van der Velden, W. and Schwartz, A. W. (1977) Search for purines and pyrimidines in the Murchison meteorite. *Geochim. Cosmochim. Acta* **41**, 961-968.

Vdovykin, G. P. (1967) *Carbonaceous Matter in Meteorites (Organic Compounds, Diamonds, Graphite)* (Moscow: Nauka Press). In Russian. Trans. NASA-TT-F-582.

de Vries, M.S., Reihs, K., Wendt, H.R., Golden, W.G., Hunziker, H.E., Fleming, R., Peterson, E. and Chang, S. (1993) Search for C60 in the Murchison meteorite. *Geochim. Cosmochim. Acta,* in press.

Wasson, J. T. (1974) *Meteorites, Classification and Properties.* Springer-Verlag.

Watson, W. D. (1976) Interstellar molecule reactions. *Rev. Mod. Phys.* **48**, 513-552.

Willner, S. R., Russell, R. W., Puetter, R. C., Soifer, B. T. and Harvey, P. M. (1979) The 4-8μ spectrum of the galactic center. *Astrophys. J. Lett.* **229**, L65-L68.

Wilson, R. W., Solomon, P. M., Penzias, A. A. and Jefferts, K. B. (1971) Millimeter observations of CO, CN, and CS emission from IRC^{+} 10216. *Astrophys. J.* **169**, L35-L37.

Wing, M. R. and Bada, J. L. (1991) Geochromatography on the parent body of the carbonaceous chondrite Ivuna. *Geochim. Cosmochim. Acta* **55**, 2937-2942.

Woese, C. R. (1987) Bacterial Evolution. *Microbiol. Rev.* **51**, 221-271.

Wolman, Y., Haverland, W. J. and Miller, S. L. (1972) Nonprotein amino acids from spark discharges and their comparison with the Murchison meteorite amino acids. *Proc. Natl. Acad. Sci. USA* **69**, 809-811.

Wood, J. A. and Chang, S. Eds. (1985) The Cosmic History of the Biogenic Elements and Compounds. NASA-SP-476, U.S. Government Printing Office, Washington, D.C.

Wood, J.A. and Morfill, G.E. (1988) Solar nebula models. in J.F. Kerridge and M.S. Matthews (eds.), Meteorites and the Early Solar System, University of Arizona Press, pp. 329-347.

Yamamoto, S., Saito, S., Kawaguchi, K., Kaifu, K., Suzuki, H. and Ohishi, M. (1987) Laboratory detection of a new carbon-chain molecule C_3S and its astronomical identification. *Astrophys. J.* **317**, L119-L121.

Yang, J. and Epstein, S. (1983) Interstellar organic matter in meteorites. *Geochim. Cosmochim. Acta* **47**, 2199-2216.

Yang, J. and Epstein, S. (1984) Relic interstellar grains in Murchison meteorite, *Nature* **311**, 544-547.

Yang, J. and Epstein (1985) A search for presolar organic matter in meteorites. *Geophys. Res. Lett.* **12**, 73-76.

Yuen, G. and Kvenvolden, K. A. (1973) Monocarboxylic acids in Murray and Murchison carbonaceous meteorites. *Nature* **246**, 301-302.

Yuen, G., Blair, N., Des Marais, D. J. and Chang, S. (1984) Carbon isotope composition of low molecular weight hydrocarbons and monocarboxylic acids from Murchison meteorite. *Nature* **307**, 252-254.

Yuen, G. U., Pecore, J. A., Kerridge, J. F., Pinnavaia, T. J., Rightor, E. G., Flores, J., Wedeking, K. M. Mariner, R., Des Marais, D. J. and Change, S. (1991) Carbon isotopic fractionation in Fischer-Tropsch type reactions. *Lunar Planet. Sci.* **21**, 1367-1368.

Zeitman, B., Chang, S. and Lawless, J. G. (1974) Dicarboxylic acids from electrical discharge. *Nature* **251**, 42-43.

Zenobi, R., Philippoz, J.-M., Buseck, P. R. and Zare, R. N. (1989) Spatially resolved organic analysis of the Allende meteorite. *Science* **246**, 1026-1029.

Zinner, E. (1988) Interstellar cloud material in meteorites. in J. F. Kerridge and M. S. Mathews (eds.) Meteorites and the Early Solar System, Univ. of Arizona Press, Tucson, pp. 956-983.

Zolensky, M. and McSween, H. Y. (1988) Aqueous alteration. in J. F. Kerridge and M. S. Matthews (eds.), Meteorites and the Early Solar System, Univ. of Arizona Press, Tucson, pp. 114-143.

Acknowledgment

This work was supported by a grant from National Aeronautics and Space Administration, Life Sciences Division, Exobiology Program (NAGW-1899)

PREBIOTIC SYNTHESIS IN PLANETARY ENVIRONMENTS

SHERWOOD CHANG
Planetary Biology Branch
NASA Ames Research Center
Moffett Field, CA 94035
USA

ABSTRACT. Except for major short-term perturbations in surface environments caused by a declining flux of impactors, equable conditions for prebiotic evolution could have existed as early as 4.4 billion years ago. Giant impacts undoubtedly constrained the timing of life's origin, but quantitative statements about when the clock was set awaits stronger consensus on impactor fluxes and more refined theoretical models. Organic matter surviving impacts or synthesized in impacts would have augmented the inventory of compounds produced endogenously in surface environments. The oxidation state of the prebiotic atmosphere remains controversial, but little question exists about the reduced state of the early ocean, which may have provided a more productive medium for prebiotic synthesis than the atmosphere. If the atmosphere has been only mildly reducing since the end of the major epoch of accretion at ~4.4 Ga, the apparent lack of strong atmospheric sources for hydrogen cyanide, formaldehyde and ammonia poses a serious challenge for theories of prebiotic evolution which require these key chemical intermediates. Submarine hydrothermal systems and the wind mixed layer of the ocean are specific settings which may have favored prebiotic evolution. Especially interesting is the ocean-atmosphere interface where a complex set of physical and chemical processes operated continuously: collection of gas, aerosols and dust from the atmosphere; recycling of organic and inorganic solutes between the ocean and atmosphere through bubble formation and bursting; organic synthesis by UV radiation, cavitation and other energy sources; and formation, dissipation and reformation of surface active monolayers and bilayer vesicles. The intersection of the wind mixed ocean layer with shorelines of volcanic platforms and shallow marine hydrothermal systems may have been key sites for prebiotic evolution.

1. Introduction

1.1. UNCERTAINTIES IN PREBIOTIC SYNTHESIS

The abiotic synthesis of organic matter is required for all theories of the origin of cellular life (Oparin, 1938; Miller and Orgel, 1974; Morowitz, 1992). Even theories which hypothesize inorganic crystals for the first life forms require organic compounds to enable the transition from minerals to cells (Cairns-Smith, 1982). Yet, despite its central role and the extensive literature that has developed around it since the pioneering work of Miller (1953), a critical examination

J. M. Greenberg et al. (eds.), The Chemistry of Life's Origins, 259–299.

of progress so far shows that little certainty exists in the answers to key questions about prebiotic organic synthesis: How and in what environments did it occur? What kinds of compounds were produced and in what abundances? When did it occur and for how long? Some uncertainties have developed as improved theoretical understanding of planetary origins and extensions of the geological record backward in time raised doubts about the validity of earlier models of prebiotic conditions on Earth (Chang et al., 1983). Other dubieties have emerged as the prebiotic relevance of some geophysical environments (Corliss et al., 1981; Lerman, 1986; Oberbeck et al., 1991) and astrophysical processes (Maher and Stevenson, 1988; Oberbeck and Fogelman, 1989) have become apparent, which previously had not been taken into account. In large part, these uncertainties reflect the greater knowledge and awareness of the physical-chemical complexity of the prebiotic Earth gained as by-products of the on-going deveopment of planetary science and its interfaces with other disciplines. Consequently, the perspectives of researchers on prebiotic environments are broadening beyond the "little warm lagoon" and "prebiotic soup" of earlier and simpler conceptions.

Underlying these questions is another fundamental one: What organic compounds were required for the origin of life? As model dependent as they must be, answers to this question have and will continue to frame much thought and study of prebiotic organic chemistry. Historically, some or all the types of compounds depicted in Figure 1 have been assumed to be essential in various schemes for prebiotic evolution. Supporters of "metabolism first" (Hartmann, 1975; Weber, 1987; Wachterhauser, 1990; Morowitz, 1992) might rally behind carbohydrates and lipids. Partisans of the "protein first" view assign a seminal role to amino acids and peptides (Fox and Dose, 1972; see Brack, this volume), while advocates of the RNA world (see Schwartz, this volume; Ferris, this volume) argue for the primacy of nucleic acid components. Proponents of "cooperating proteins and nucleic acid" stipulate both sets of compounds (Lahav, 1991). These viewpoints stem from a biological perspective; that is, they arise from specific models of the first life form rather than from concepts of the environments in which such life forms might have arisen.

The ensuing discussion will reflect a planetary environmental perspective. An initial set of model conditions resulting from planetary accretion will be briefly summarized. Several selected environments will be described, and the capabilities and limitation for synthesis of primary products within these environments will be considered. The productivity of such synthesis in a given environment may be taken as a possible gauge of the potential of that environment for the origin of life. For present purposes, primary products are defined as the precursor compounds (e.g., hydrogen cyanide, formaledhyde, etc.) necessary for synthesis of the simplest biomolecules shown in Figure 1. With few exceptions, relatively little attention will be given to how these products might have been transformed into more complex organic compounds and structures (see chapters by Brack, Ferris, and Schwartz, this volume). However, it will become apparent that existing ideas about these further steps in prebiotic evolution may be fraught with even greater uncertainties than are models for synthesis of primary products. What are commonly described in the literature as "plausible" prebiotic conditions will appear less credible. The purpose of this discussion is not to provide answers, but rather to pursuade the reader that the questions raised truly lack satisfactory answers and that they pose continuing challenges to researchers seeking clearer understanding of even the first and, perhaps, simplest step on the pathway to the origin of life.

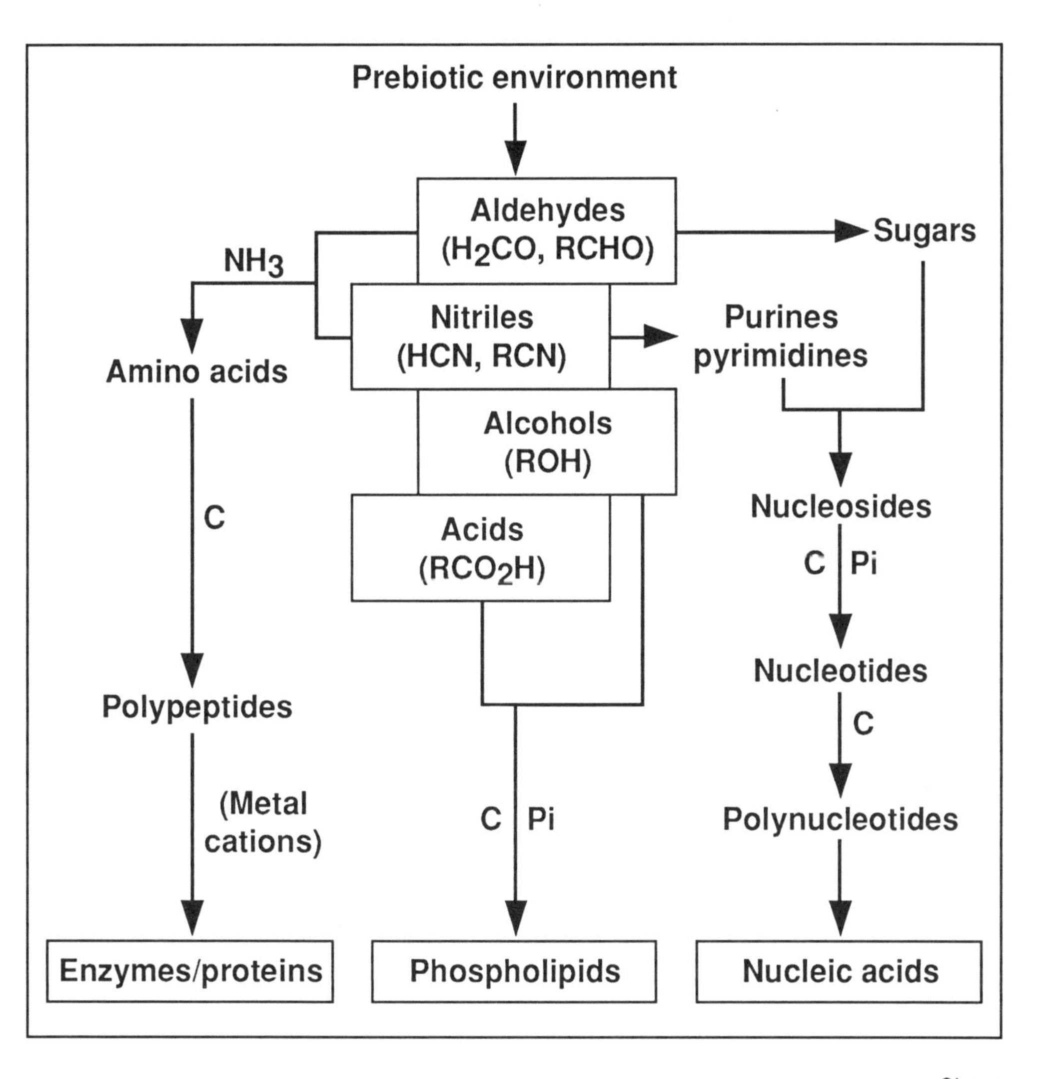

Figure 1. Typical scheme for prebiotic evolution depicting key roles of HCN and CH_2O in construction of biomolecules. The symbol, c, signifies that condensing agents are necessary in certain steps to link compounds together through loss of water (dehydration-condensation). The requirement for inorganic phosphate is indicated by Pi.

2. Initial Planetary Conditions

2.1 CONSTRAINTS IMPOSED BY ACCRETIONARY IMPACTS

In modern theories of Earth formation, accretion occurs over a period of about a hundred mllion years and produces an essentially fully formed and differentiated planet with metallic core, highly convective mantle, molten surface and massive steam atmosphere (Stevenson, 1983; Matsui and Abe, 1986; Zahnle et al., 1988; Abe and Matsui, 1988). As the accretionary energy input declined, surface temperatures dropped, a surface scum of rock solidified over the hot mantle and water rained out to form an ocean. Surface temperatures at or below100°C could have prevailed as early as 4.4 billion years ago. The thermal history of Earth following an impact with a ~450 km diameter object (Sleep et al., 1989) is depicted in Figure 2. The hot rock vapor produced in the impact encircles the globe, and its radiation downward evaporates the ocean to form a dense steam atmosphere of possibly several hundred bars. With continued heating the dry surface increases in temperature to melt the exposed rock. Radiation to space cools the steam atmosphere and surface. When the critical temperature of water is reached precipitation begins in the atmosphere, and a new ocean is rained out. Even if a giant impact capable of vaporizing an ocean of present size occurred after the main epoch of accretion, the time scale for recovery from such a catastrophic event is calculated to be on the order of a few thousand years.

Clearly, giant impacts capable of destroying Earth's ecosystem would have imposed constraints on when life originated and in what type of environment. Assuming the origin of life required 10^5 to 10^7 years, and using impactor size and frequency distributions gleaned from the lunar cratering record as a basis for their calculations, Maher and Stevenson (1988) showed that a time interval of this length could have occurred between ocean vaporizing impacts as early as 4.2 to 4.0 Ga. Calculations by others suggest that the last such catastrophic impact could have occurred as early as 4.4 Ga (Sleep et al., 1989) or later than 3.7 Ga ago (Oberbeck and Fogelman, 1989; 1990). The frequency and size of impacts increase backward in time; therefore, ecosystems at or below the ocean-sediment interface (i.e., deep sea hydrothermal systems) had the best prospect for survival early in Earth history, regardess of where they originated. Apparently, life on land and in the ocean's photic zone would have been chancy until late in this interval. As late as 3.8 Ga, the Moon suffered a large impact forming the Imbrium basin; and barring any statistical fluke, it seems improbable that Earth avoided impacts with similarly large objects (~260 km) because Earth has a much larger effective cross section than the Moon (Sleep et al., 1989).

If life arose at 3.8 Ga, the maximum time available for prebiotic evolution since the previous giant impact is estimated to be 6 million years; if life began at 3.5 Ga the maximum interval is165 million years (Oberbeck and Fogelman, 1990). These calculations suggest that shorter time intervals may have been available for prebiotic evolution than previously thought. Moreover, if the origin of life required only geologically short time scales, life might have arisen several times before establishing a permanent foothold on the planet. In this case, its spawning ground could have been the ocean photic zone as well as hydrothermal systems.

Undoubtedly, impacts constrained the timing of the origin of life. While these impact calculations are highly interesting and relevant, definitive statements about when the clock was set must await either more ancient geological evidence of surface environments than is now available or stronger consensus on impactor fluxes and more refined theoretical models. In any case, just as environmental purturbations caused by impacts are thought to have influenced biological evolution (Alvarez et al., 1980; Prinn and Fegley, 1987) impacts of varying severity throughout Earth's early history would have influenced prebiotic evolution. The nature of these effects have only begun

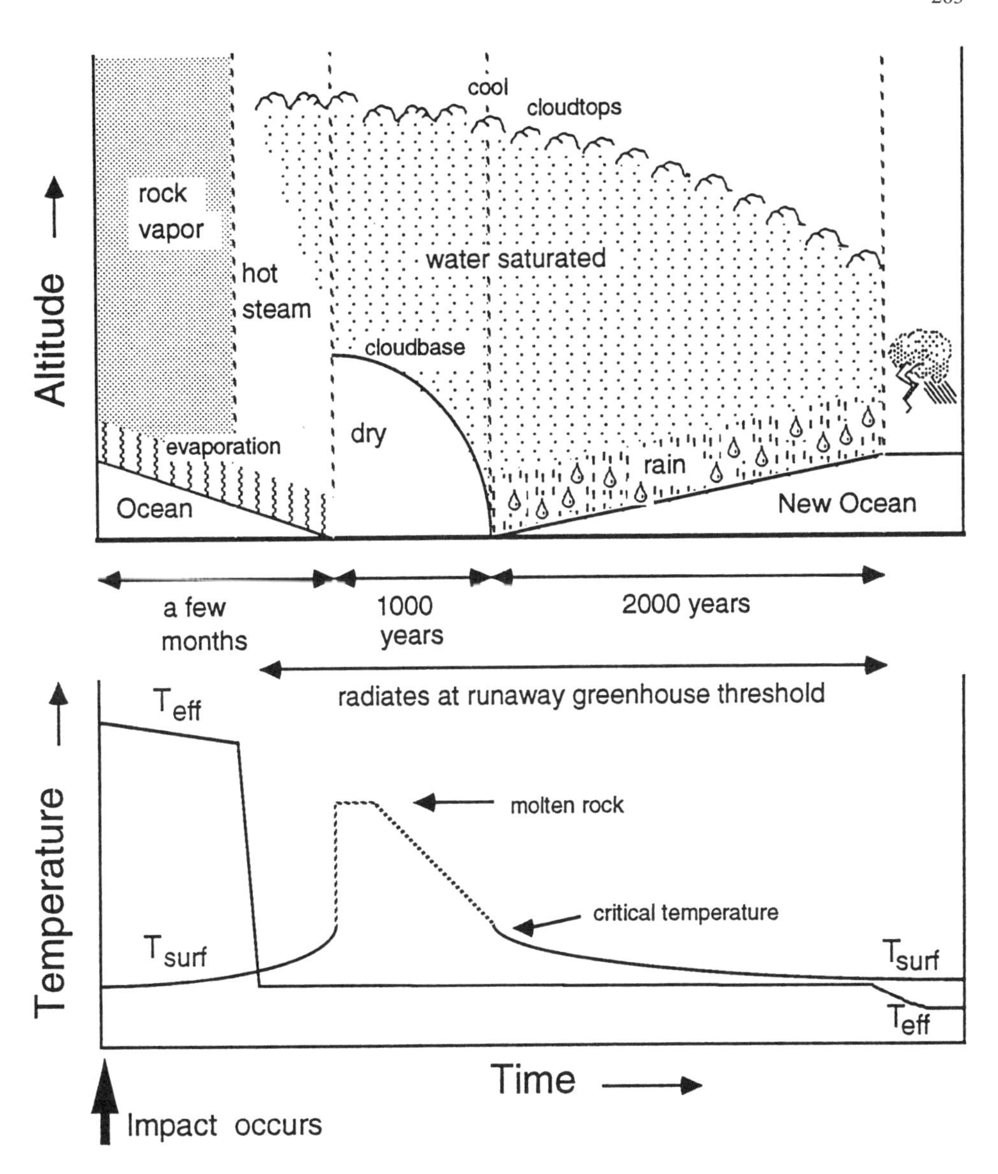

Figure 2. Thermal evolution of Earth's surface following an ocean vaporizing impact. A collision with a 500 km object results in production of 100 atmospheres of rock vapor at several thousands of degrees, radiation from which eventually evaporates the entire ocean forming a steam atmosphere. Cooling by radiation to space occurs, and a new ocean is rained out after several thousand years. (Courtesy of K. Zahnle)

to be explored (e.g., Fegley et al., 1986).

2.2 EARLY GEOPHYSICAL EVOLUTION

Although unrepresented in the geological record, the period between 4.4 and 3.9 Ga, the Hadean, must have been a time of rapid geophysical evolution in Earth history. In this interval, dissipation of internal thermal energy left over from accretion required heat flows to diminish from very high levels early on to levels later in the Archean that were about 4 times the present (Stevenson, 1983). The intensity of volcanism and hydrothermal activity diminished correspondingly. The frequency of impacts decreased in analogous fashion. The crust began to sort itself into continental and oceanic realms. Overall solar luminosity increased while the enhanced UV component emitted by the young Sun decreased from enhanced levels of up to 32 times present flux (Canuto, et al., 1982). The ocean-atmosphere system formed; land surface grew; and hydrologic and geochemical cycles developed.

The enviromental consequences of this early geophysical evolution appear first in the ~3.8 Ga metasediments of Isua, Greenland. They show the occurrence of sediments rich in volcanogenic debris laid in water, carbonate minerals, reduced graphitic carbon and basalts with melting temperatures approaching 1600°C. These in turn signify surface temperatures below 100°C, extensive bodies of liquid water; CO_2, and presumably N_2 in the atmosphere; higher heat flow and more active volcanism than today; incipient continental growth, weathering, a hydrologic cycle and a carbon geochemical cycle, perhaps driven by a form of tectonism ancestral to modern plate tectonics (Ernst, 1983). Generally, the environment recorded at the beginning of the Archean resembles some habitable environments that exist today absent atmospheric oxygen and related oxidants.

The 3.5 Ga old sediments of Western Australia contain the earliest compelling evidence of a biosphere. Sedimentological studies of this record reveal a shallow, marine environment dominated by episodic island volcanism and hydrothermal activity (Barley et al., 1979; Groves et al.,1981). The stromatolites which grew in this setting were subject to periodic agitation by waves or currents and occasional rapid burial by lava flows. Localization of microbial growth may have been influenced by hot springs or hypersaline conditions (Byerly et al., 1986). Such settings probably existed earlier than 3.8 Ga, and their occurrence as early as 4.4 ga is not contradicted by any theories or observations. Whether this type of environment spawned the first biota remains unknown, although many of its characteristics are certainly favorable (see below).

2.3 FORMATION OF THE ATMOSPHERE-OCEAN SYSTEM

Outgassing of the bulk of the atmosphere by 4.4 Ga is indicated by the isotopic systematics of radiogenic and primordial noble gases in terrestrial materials (Thomsen, 1980; Allegre et al., 1987). A similarity in chemical inertness among N_2 and noble gases, suggests that the bulk of the N_2 now in the atmosphere also outgassed at the same time and has remained at close to the same steady state abundance over time. As partially lithophilic elements, hydrogen (Khitarov and Kadik, 1973) and carbon (Berg, 1986; Tingle et al., 1988; Katsura and Ito, 1990) would have partitioned between the mantle and the atmosphere-ocean system early on as they have since then (Javoy et al., 1982; Walker, 1983). In the limiting case where all the carbon and hydrogen now in surface reservoirs were ejected into the gas phase at the end of the accretionary period, the resulting massive steam atmosphere could have contained up to ~50 atm of carbon. The time at

which an ocean of the present size ($1.4x10^{24}$ g) formed remains uncertain, although the very early existence of a substantial body of liquid water seems highly probable (Cogley and Henderson-Sellers, 1984). As a result of a declining impactor flux the bulk of the ocean is expected to have accumulated early in the period 4.4 to 3.9 Ga rather than later. At any point, however, the size of the atmosphere-ocean system would have depended on the net balance between input by volcanism and both additions and losses due to impacts.
To keep the early ocean from freezing, a global greenhouse may have been necessary to offset the effects of a faint young Sun dimmer than today by 25-30% (See Kasting, this volume. An alternative view suggests that once a near-global ocean formed, it does not freeze despite a 30% lower luminosity [Henderson-Sellers and Henderson-Sellers, 1988]). Methane and ammonia were suggested as prebiotic greenhouse gases (Sagan and Mullen, 1972). In the absence of long term sources, however, the short lifetimes of these gases against photochemical destruction (Walker, 1977; Ferris and Nicodem, 1972) would have restricted their role to the earliest 10^6 yr of the atmosphere's exposure to the Sun. Climate models (Kasting et al., 1984) confirm that 100-1000 times the present atmospheric level (PAL) of CO_2 would also have sufficed to keep oceans from freezing (Owen et al., 1979). Offsetting the lower luminosity of a young Sun, was a UV flux up to 10^4 higher than today (Canuto, et al., 1982). These large fluxes could have enhanced prebiotic photochemical synthesis and posed a hazard for early life. If an ocean and equable temperatures were the only limiting requirements, life could have arisen at any time between 4.4 Ga and the earliest record of a biosphere at 3.5 Ga. The availability of organic compounds may have been a more stringent requirement.

2.4 COMPOSITION OF THE PREBIOTIC ATMOSPHERE

In most prebiotic evolution scenarios, beginning with that of Oparin (1938), a highly reducing atmosphere has been assumed to be the primary source of chemical intermediates for synthesis of organic compounds. Although the success of Miller's (1953) pioneering experiments with electric discharges demonstrated the ease of such synthesis, the very existence of a highly reduced atmosphere remains a controversial issue with a history (see Chang et al., 1983) predating Miller's work (Rubey, 1951).
Both volcanism and impacts fueled the formation of the prebiotic atmosphere. The gases typically released today by volcanoes at magmatic temperatures, ~1200°C, are weakly reducing (Table 1). Among these gases the ratio CO/CO_2 ~0.03, H_2/H_2O ~0.01; and H_2/CO_2 ~0.06. Methane is absent. The predominant sulfur gas is SO_2. Gases exhaled at 760°C from fumaroles in Showa-shinzan, Japan, exhibit CO/CO_2 ~0.03, CH_4/CO_2 ~$2x10^{-4}$, H_2/CO_2 ~0.5 and H_2S/SO_2 ~0.01. In gases released at ~350°C from mid-ocean ridge hydrothermal systems today, $CH_{4/}CO_2$ ~0.01 and H_2/CO_2 ~ 0.3 (Welhan, 1988). These compositions are determined by the redox state of the final mineral assemblages with which volcanic gases chemically equilibrated before expulsion to the atmosphere or ocean. Provided kinetic barriers are surmountable (Sackett and Chung, 1979), equilibration at lower temperatures generally leads to more chemically reduced gases. The prevailing redox state reflects the oxygen fugacity of the mineral assemblage, which in turn is buffered by its contents of ferrous (Fe^{2+}) and ferric (Fe^{3+}) iron. Thus, measurements of the oxygen fugacities of ancient igneous rocks may provide clues to past atmospheric compositions.

Table 1. Volume percent of gases in Hawaiian volcanoes (from Walker, 1977).

H2O	79.31
H2	0.58
CO2	11.61
CO	0.37
SO2	6.48
S2	0.24
H2S	------
N2	1.29

In modern magmatic rocks, the minerals governing redox state are the fayalite-magnetite-quartz (FMQ) system. The recent ascendancy of a weakly reducing model for the prebiotic atmosphere (N_2, CO_2 >> CO >> CH_4, H_2O>>H_2, SO_2 > H_2S) is based on theoretical arguments and extrapolation from rocks younger than 3.8 Ga, whose Fe^{2+}/Fe^{3+} ratios are generally consistent with the FMQ buffer (Chang, 1983; and Holland, 1984). Thermodynamically, this system permits significant CH_4 abundances only at temperatures below ~800°C, as might be achieved in hydrothermal systems. Kasting (this volume) suggests that the ratio of Fe^{2+}/Fe^{3+} in some early Archean mantle-derived rocks points to a more highly reducing source region than manifest in modern samples. If the mantle has increased in oxidation state over time, as he suggests, the possibility is left open for a highly reduced volcanogenic atmosphere during prebiotic evolution. Clearly a more comprehensive assessment of the oxygen fugacity of magmatic rocks throughout the geological record is desirable. Extension of the redox record deeper into antiquity awaits discovery of older primary igneous rocks.

To generate a highly reducing atmosphere (N_2, CO > CH_4 > CO_2, H_2O~H_2, H_2S > SO_2) of volcanic origin, a mineral assemblage (iron-wustite system) containing metallic iron is required. For lack of a geological record to the contrary, a highly reducing atmosphere or one intermediate in redox state cannot be excluded, but its continued existence beyond the accretionary epoch would have required on-going chemical equilibration with metallic iron. If core formation occurred as supposed in accretion models, metallic iron would have been removed from the mantle and crust by 4.4 ga. Whether and for how long chemical equilibration could have persisted between core and mantle are unknown. Measurements of mantle xenolith oxygen fugacities indicating a highly reduced environment (Arculus and Delano, 1980, 1984), now appear to have been in error (Mathez et al., 1984; Virgo et al., 1988). Recent measurements by a more reliable method (Wood and Virgo, 1989) point to a modern mantle redox state bufferred close to the quartz_fayalite-magnetite system generally characteristic of basalts throughout the geological record and consistent with a weakly reducing atmosphere.

In analogous fashion, the gases produced in an impact reflect the composition and oxygen fugacity of the gas plume generated by shock-driven partial melting and vaporization of the projectile and excavated target (e.g., Zahnle, 1990). Atmospheric CO/CO_2 ratios could have been enhanced over that in a weakly reducing atmosphere by high temperature reactions of carbon with other elements in impact derived gas clouds. Impacts of metallic iron-containing asteroids would have exerted the strongest reducing effect. Kasting (1990) incorporated impact generation of CO in a photochemical model and showed that $CO/CO_2 \geq 1$ could have occurred during the first several 0.1 Ga of earth history provided that CO was not rapidly dissolved in oceans and converted to bicarbonate. Later impacts would have created isolated pulses of reducing gases in the atmosphere provided that the chemical lifetimes of the gases were greater than the global atmospheric mixing time of 1-3 yr.

Sulfur containing gases, H_2S and SO_2, would have been released into the early atmospere by impact outgassing, as well as by volcanism, in proportions governed by the prevailing redox state. In marine hydrothermal systems, precipitation of highly insoluble iron sulfides by abundant aqueous ferrous ion would have served as a highly effective sink for H_2S (Walker and Brimblecombe, 1985). In a photochemical model, SO_2 persists as the major S-containing gas in the prebiotic atmosphere. It is accompanied by minor amounts of sulfuric acid and elemental sulfur, S_8, which may have provided a protective screen against harmful ultraviolet light (Kasting et al., 1989).

Ammonia is notably absent in the foregoing discussion. Volcanic and impact processes appear incapable of producing this gas. Despite the critical role attributed to ammonia in prebiotic evolution (Bada and Miller, 1968), no primary atmospheric process has been identified as a probable source, except electrical discharges in a CH_4-H_2-N_2-containing atmosphere. Nor can NH_3 in the atmosphere survive rapid photochemical destruction (Ferris and Nicodem, 1972). Photochemical reduction of N_2 by water catalyzed by titanium oxide-rich desert sand has been demonstrated (Schrauzer et al., 1979) but the amount of surface area represented by the photo-active mineral rutile, TiO_2, was probably insignificant on the prebiotic Earth. Today, this source of NH_3 amounts to one third of the N_2 fixed by lightning discharges (Schrauzer et al., 1983); it would have been much smaller with less land surface early on. Apparently, an oceanic source is required for ammonia production (see Sections 3.2. and 4.1.1.).

Until recently, phosphorus has not been discussed in the context of volcanic gases. Yamagata, et al. (1991) have now reported the volatilization, collection and characterization of polyphosphates and trimetaphosphate produced by heating mixtures of basalt and $Ca_3(PO_4)_2$ at1200°C. The natural occurrence of these phosphate oligomers was confirmed by their identification in vapors collected by condensation at volcanic vents. Although their reactions in the atmosphere have not been explored, the polyphosphates could have served as prebiotic phosphorylating agents (Figure 1) once they rained onto the ocean. Measurements of the modern volcanogenic flux of condensed phosphates would provide a valuable basis for estimating the prebiotic flux and assessing its contribution to prebiotic chemistry.

2.5. SEAWATER COMPOSITION

In the absence of a sedimentary record, the composition of Hadean and Early Archean seawater must remain a matter of conjecture. One viewpoint takes the position that seawater composition has been been relatively conservative over time (e.g., Holland, 1984); the other contends that seawater in the Archean (and earlier, by implication) differed radically from that in today's oceans (e.g., Kempe and Degens 1985). Implicit in the material that follows is the view that the earliest seawater must have been very different, that change in its composition may have been either rapid or gradual during the Hadean or early Archaean, and that by 3.5 Ga, it was not radically different than its counterpart today, absent the influence of atmospheric oxgen.

If an ocean formed rapidly during accretion, and weathering could not keep up, a highly acidic solution could have resulted from dissolution of outgassed HCl (Holland, 1984). While an acidic ocean might have been beneficial for some prebiotic reactions, ignorance of the rate of weathering on the primordial Earth places the lifetime of a hypothetical acidic ocean and its impact on prebiotic evolution in the realm of speculation. If the dominant carbon gas was CO_2, bicarbonate concentrations would have depended on temperature, weathering rates and atmospheric CO_2 levels. Walker (1985) suggested the possibility that levels could have been as high as tens of bars immediately after accretion and up to several bars thereafter until possibly 3.8 Ga. The persistence of initially high and slowly declining CO_2 partial pressures was attributed to slow weathering rates in the absence of much continental surface. Greenhouse calculations (e.g., Kasting et al., 1984) and the sedimentological record (Walker et al., 1983; Holland, 1984; Lowe and Byerly, 1986; Buick and Dunlop, 1990; Buick, 1993, personal communication) suggest a warm (up to 40°C), bicarbonate-rich ocean with pH between 6 and 8 at 3.5 Ga, preceded possibly by a warmer and more acidic (pH >4) ocean prior to 3.8 Ga.

Today, continental weathering and river run-off exert the major influence on seawater composition. In the Hadean, higher heat flow from the interior meant higher levels of tectonic activity, including more extensive and intensive circulation of seawater through hydrothermal systems. If during prebiotic evolution continental weathering were also much less intense than today due to dearth of land surface, submarine hydrothermal fluids would have been the major influence on seawater composition (Veizer, 1983; Derry and Jacobsen, 1990). In the limit of no land surface, seawater composition would have been completely controlled by solutes extracted from hot rocks in these systems Useful analogies can be drawn, therefore, between the compositions of early seawater and modern hydrothermal fluids (corrected for an anaerobic Earth), which relative to bulk ocean are greatly enriched in silica, H_2S, Mn, Fe, Co, Cu, and Zn among other elements (Edmond et al., 1982; Von Damm et al., 1986). Interestingly, the 350°C water emerging from hydrothermal systems is acidic at pH ~3. Before the discovery of submarine hydrothermal systems, Egami (1973) proposed a model prebiotic seawater composition based on essential elements required by living organisms. His composition is remarkable for its enhanced concentrations of the trace metals listed above.

In the presence of excess ferrous iron, H_2S and sulfide ion concentrations would have been kept at negligibly low levels in seawater by precipitation of FeS. Rain out of sulfate produced by oxidation of SO_2 in the atmosphere could have yielded concentrations as high as 10^{-3} moles/liter (Walker and Brimblecombe, 1985). Just as they do today, hydrothermal systems would have served as sinks for sulfate by reducing it to sulfide.

Most importantly, the redox capacity of the prebiotic ocean far surpassed that of the atmosphere, and its reducing power would have been replenished in the Hadean on timescales much shorter than 10^7 years (Wolery and Sleep, 1976) by seawater circulation through hydrothermal systems (Veizer, 1983; Derry and Jacobsen, 1990). Aqueous Fe^{2+}, extracted from hot igneous rocks during circulation would have been the major reductant (Cloud, 1973), followed in importance by Mn^{2+}; and the deficiency of oxidants in the environment would have inhibited conversion of soluble ferrous iron to insoluble ferric iron (see Section 3.2., however). Estimates of dissolved iron exceed10^{-4} molar for the seawater from which banded iron formations precipitated in the Precambrian (Holland, 1973). In comparison, modern submarine hydrothermal solutions contain $>10^{-3}$ molar Fe^{2+} (Von Damm et al., 1985). If such a concentration existed in a primitive ocean of present surface area (5.1×10^{18} cm^2) and volume (1.4×10^{24} cm^3), the Fe^{2+} abundance would have been 1.7×10^{23} cm^{-2}. Derry and Jacobsen (1990) estimated Archean hydrothermal fluxes of Fe^{2+} as high as 3.4×10^{12} moles per yr, or 1.2×10^{10} cm^{-2} s^{-1}. Higher fluxes would have accompanied higher heat flows earlier in the Hadean. With its enormous capacity, the ocean provided an alternative to the atmosphere as a source of reducing power for prebiotic syntheses.

3. Prebiotic Syntheses of Primary Products

3.1. REACTIONS IN THE ATMOSPHERE

Although thousands of experiments have been conducted purporting to simulate prebiotic organic syntheses in the atmosphere, relatively few have yielded results suitable for quantitative comparisons. Nonetheless, a qualitative generalization has emerged: The efficiency of synthesis and the structural diversity of products decrease as the reducing power of the atmosphere decreases. Thus, reactions initiated in a CH_4 atmosphere provide an embarassment of riches (e.g., Urey and Miller, 1959; Sagan and Khare, 1971), while those carried out in CO_2 yield a dearth of products (e.g., Stribling and Miller, 1987; Pinto et al., 1980; see also Chittenden and Schwartz, 1981). Experiments involving CO gave results in-between (e.g., Bar-Nun and Chang, 1981; Hirose et al., 1990-1991).

Listed in Tables 2 and 3 are rainout rates for HCN and formaldehyde (CH_2O) produced in laboratory and computer simulations of atmospheric shocks, electric discharges and photochemistry. These two compounds are key intermediates for the prebiotic syntheses of organic compounds (Figure 1). Rainout rates were calculated from energy yields (i.e., moles per Joule) obtained in the simulations, and the revised values for lightning and corona energy fluxes provided by Chyba and Sagan (1991). After formation in the atmosphere, products must be transported by rain to the ocean surface where they can undergo further reactions. Depending on a compound's solubilitiy in water and susceptability to photochemical destruction in the atmosphere, its atmospheric residence time will vary and its rainout rate will be less than its production rate. Some workers take these factors into account and calculate a rainout rate based on the atmospheric production rate. Other workers do not, in which cases, the actual rainout rates for HCN and CH_2O may be as much or more than 99% lower than the listed values.

Impact shocks are caused by two mechanisms. In the first, hypervelocity passage of the impactor through the atmosphere produces a high temperature plasma. The plasma emits radiation and generates shock waves capable of converting atmospheric species to more complex compounds. This is the major mechanism associated with small impactors, whose kinetic energy is more or less dissipated in the atmosphere (meteor impacts). In addition to producing a shock by the first mechanism, a large impactor stiking the surface at $\geq$15 km s^{-1} generates a second shock effect as well. Impact melting and vaporization of target and projectile (Prinn and Fegley, 1989) produce a post-impact plume of hot gas. Expansion of this plasma plume imparts much more kinetic energy to and affects a greater volume of the atmosphere than does mere passage of the impactor through it. Therefore, shocks from post-impact plumes provide the largest source of impact derived compounds.

Lightning discharges produce audible thunder shocks, and the chemistry that results is thought to be the same as that associated with impacts. A shock event heats a volume of the atmosphere to >1500°C, at which temperatures the gas composition is presumed to be in thermodynamic equilibrium. As the gas cools the composition continually readjusts thermochemically, but at the so-called quench temperature, the equilibrium composition is "frozen in" by kinetic barriers to further reactions. Scattergood et al. (1989) have shown, however, that UV radiation from shock-generated plasmas also contributes to the chemistry of the gas. Therefore, thermodynamic equilibrium theory provides an incomplete model for understanding the chemistry of impact and lightning shocks.

Because of its thermochemical instability at high temperatures, the equilibrium abundance of CH_2O is orders of magnitude lower than that of HCN. For any atmospheric composition,

Table 2. Rainout rates for HCN based on laboratory and computer simulations of energetic processes in the prebiotic atmosphere.

Gas composition	Energetic process	HCN cm^{-2} s^{-1}	Authors
$CO/N_2/H_2O$	impact shocks on atmosphere	1-5 (9)[a,b]	Fegley et al. (1986)
$CO_2/N_2/H_2O$	impact shocks on atmosphere	2 (7)[c]	McKenzie & Arnold (1967)[d]
$CH_4/CO_2/N_2/H_2O$	post-impact plume shocks	2 (9)[c,e]	Chyba and Sagan (1992)
$CO_2/N_2/H_2/H_2O$	"	2 (2)[c,e]	"
n/a	vaporization/recombination of impactor	5 (6)[f]	Mukhin et al. (1989)
$CH_4/CO_2/N_2/H_2O$	solar ultraviolet radiation	1 (11)[g]	Zahnle (1986)
"	"	1 (8)[h]	"
"	"	2 (5)[i]	"
$CH_4/N_2/H_2O$l	ightning/thunder shocks	7 (7)[c,j]	Bar-Nun & Shaviv (1975)
$CH_4/N_2/H_2O$	lightning/thunder shocks	1 (8)[c,j]	Chameides &Walker (1981)
$CO/N_2/H_2O$	"	3 (4)[c,j]	"
$CO_2/N_2/H_2O$	"	9 (2)[c,j]	"
$CH_4/N_2/H_2$	Lightning/thunder shocks	6 (6)[c]	Scattergood et al. (1989)
$CO/N_2/H_2$	"	0.04-11 (3)[c,k]	"
$CO_2/N_2/H_2$	"	6-12 (2)[c,k]	"
$CH_4/N_2/H_2/H_2O$	corona discharges	0.7-2 (5)[c,j]	Stribling & Miller (1989)
$CO/N_2/H_2/H_2O$	"	2-9 (5)[c,j]	"
$CO_2/N_2/H_2/H_2O$	"	0.7-2 (4)[c,j]	"

[a] 5 (12) signifies $5x10^{12}$ molecules cm^{-2} s^{-1}.

[b] Calculated using HCN energy yields taken from Chameides and Walker (1981).

[c] These values not adjusted for up to 99% loss of HCN prior to rainout (Zahnle, 1986).

[d] Based on production rate for CN cited in Bar-Nun and Shaviv (1975).

[e] For purposes of comparison with other entries, the production rate of HCN was based on the assumption that, in the authors' calculation of shock effects over geologic time, total organic carbon occurred entirely as HCN. The total organic carbon synthesized at 4.0 Ga was used as a representative value for this table.

[f] Rainout rate estimated according to Fegley et al. (1986) from total HCN production. Entrainment of atmosphere in the plume is assumed to be negligible

[g] Rainout rate corresponding to a CH_4 flux equal to the modern biogenic CH_4 flux of $4x10^{11}$ cm^{-2} s^{-1}. Synthesis limited only by production of N in stratosphere by enhanced early solar euv flux of ~$4x10^{11}$ photons cm^{-2} s^{-1} (Kasting et al., 1983). CO_2 = 100 PAL.

[h] Rainout rate corresponding to a CH_4 flux equal to the modern CO_2 flux of $1.1x10^{10}$. CO_2 = 100 PAL.

[i] HCN production rate corresponding to a CH_4 flux equal to modern abiotic CH_4 hydrothermal flux. CO_2 = 100 PAL.

[j] Calculated using authors' energy yields and recently revised estimates of energies available in lightning or coronal discharges, $3.6x10^{-10}$ J cm^{-2} s^{-1} or $3.0x10^{-10}$ J cm^{-2} s^{-1}, respectively (Chyba and Sagan, 1991).

[k] Calculated using energy yields reported in Stribling and Miller (1989) and recently revised estimates of energies available in lightning discharges, $3.6x10^{-10}$ J cm^{-2} s^{-1} (Chyba and Sagan, 1991).

Table 3. Rainout rates for formaldehyde (CH_2O) based on laboratory and computer simulations of energetic processes in the prebiotic atmosphere.

Gas composition	Energetic process	CH_2O cm^{-2} s^{-1}	Authors
$CH_4/CO_2/N_2/H_2O$	post-impact plume shocks	9(9)[a,b,c]	Chyba and Sagan (1992)
n/a	vaporization/recombination of impactor	2(2)[a,b,c]	"
n/a	vaporization/recombination of impactor	5 (5)[b,d]	Mukhin et al. (1989)
$CH_4/CO_2/N_2/H_2O$	solar ultraviolet radiation	2 (9)[e]	Zahnle (1986)
"	"	2 (9)[f]	"
"	"	2 (9)[g]	"
CO/CO_2	solar ultraviolet radiation	1-10 (8)[h]	Kasting (1990)
$CO/N_2/H_2O$	solar ultraviolet radiation	8 (8)[i]	Bar-Nun and Chang (1983)
$CO_2/CO/N_2/H_2O$	"	5 (6)[j]	"
$CO_2/N_2/H_2O/H_2$	solar ultraviolet radiation	1.5 (10)[k]	Wen et al. (1989)
"	"	7.4 (7)[l]	"
"	"	3.5 (4)[m]	"
$CH_4/N_2/H_2/H_2O$	corona discharges	.7-7 (4)[b,n]	Stribling & Miller (1989)
$CO/N_2/H_2/H_2O$	"	7-25 (4)[b,n]	"
$CO_2/N_2/H_2/H_2O$	"	2-6 (3)[b,n]	"

[a] 3 (10) signifies 3×10^{10} molecules cm^{-2} s^{-1}.

[b] These values are not adjusted for up to 99.8% photochemical destruction of CH_2O prior to rainout (Wen et al.,1989).

[c] For purposes of comparison with other entries, the production rate of CH_2O was based on the assumption that, in the authors' calculation of shock effects over geologic time, total organic carbon was produced entirely as CH_2O. The total organic carbon synthesized at 4.0 Ga was used as a representative value in this table.

[d] Rainout rate estimated according to Fegley et al. (1986) from total experimental CH_2O production. Entrainment of atmosphere in the plume is assumed to be negligible.

[e] Rainout rate corresponding to a CH_4 flux equal to the modern biogenic CH_4 flux of 4×10^{11} cm^{-2} s^{1}. Synthesis limited only by production of N in stratosphere by enhanced early solar euv flux of ~4×10^{11} photons cm^{-2} s^{-1} (Kasting et al., 1983). CO_2 = 100 PAL.

[f] Rainout rate corresponding to a CH_4 flux equal to the modern CO_2 flux of 1.1×10^{10}. CO_2 = 100 PAL.

[g] HCN production rate corresponding to a CH_4 flux equal to modern abiotic CH_4 hydrothermal flux. CO_2 = 100 PAL.

[h] Range of rainout rates include model atmosphere with CO/CO_2 from 10^{-3} to 1 and total pressures of $CO + CO_2$ from 0.2 to 2 bar.

[i] Carbon released as CO at $6x10^{12}$ cm^{-2} s^{-1} into atmosphere from metal-equilibrated source rocks.

[j] Carbon released as CO at $5x10^{9}$ cm^{-2} s^{-1} into atmosphere from QFM-equilibrated source rocks.

[k] Mixing ratios of CO_2 and H_2 taken as 10^{-2} and 10^{-3}, respectively.

[l] Mixing ratios of CO_2 and H_2 taken as 10^{-2} and 10^{-5}, respectively.

[m] Mixing ratios of CO_2 and H_2 taken as $3x10^{-4}$ and 10^{-5}, respectively.

[n] Calculated using energy yields reported in Stribling and Miller (1989) and recently revised estimates of energies available in lightning discharges, $3.6x10^{-10}$ J cm^{-2} s^{-1} (Chyba and Sagan, 1991).

therefore, CH_2O is expected to be strongly depleted among products of shock processes. On the other hand, photochemistry in atmospheres lacking CH_4 is incapable of yielding HCN or any other nitrogen containing organic compound.

3.1.1. *HCN Synthesis.* Data in Table 2 corroborate quantitatively the general observation noted above. Production and rainout rates decrease as the oxidation state of the carbon source increases. These effects are steepest in the shock-related syntheses. The quenched equilibrium model provides a simple explanation for these observations (Chameides and Walker, 1981). Due to the greater stability of CO relative to CH_4 and H_2O at the quench temperature, C binds preferentially with O to form CO; and if there is any C left, it will react with N to form HCN. Otherwise N reacts with O to produce NO. In a CO_2 rich atmosphere, C is bound mostly to O leaving little C available for reaction with N. Consesquently, N reacts with the remaining O to yield mostly NO. In a CH_4 atmosphere containing little H_2O, C still binds with O, but now there is an excess of C rather than O, and the excess can react with N to form mostly HCN. Accordingly, $C/O > 1$ favors HCN, $C/O < 1$ favors NO.

The closest experimental simulations of the chemistry in a post-impact gas plume are the laser vaporization experiments of Mukhin et al., (1989). The gas phase produced by vaporization of samples of a meteorite contained abundant O due to vaporization of silicates in the target, yet formaldehyde, as well as HCN and hydrocarbons, were produced along with mostly CO and CO_2. Probably, in these as well as other analogous laboratory experiments the scale of the phenomenon is small, so that the surface to volume ratio of affected gases is much larger than that in natural phenomen, thus allowing non-equilibrium edge effects and inhomogeneities to play a more important role.

It is noteworthy that with CH_4 as the carbon source, experimental (Bar-Nun and Shaviv, 1975) and predicted (Chameides and Walker, 1981) production rates for HCN in shock syntheses do show good agreement. Oberbeck and Aggarwal (1992) point out, however, that production of amino acids (by way of HCN as an intermediate) in similar shock experiments (Barak and Bar-Nun, 1975) showed less than a four-fold variation when C/O ratios varied from 1 to 4 in the initial hydrocarbon/N_2/H_2O gas mixtures. Why amino acid yields should be less sensitive to C/O than HCN production rates is not known. Possibly, inhomogeneities in the shocked gas may influence amino acid and HCN syntheses in different ways. Typically, the abundance of hydrocarbons newly synthesized in these shock experiments is several time greater than that of HCN.

Perhaps the most intriguing entry in Table 2 is the result of shock experiments by McKenzie and Arnold (1967). These workers observed an unexpectedly high production rate of HCN in gas mixtures of CO_2 and N_2. Synthesis of HCN occurred with $>10^3$ higher efficiency than in any other experiment. The reason for this result is not apparent, but it may be related to the absence of water in the shock tubes and non-equilibrium effects. Repeating these experiments would be desirable.

The corona discharge experiments show less sensitivity to the C/O ratio than might have been predicted. In assigning a minimal role for shock waves in electrical discharges, Stribling and Miller (1987) suggested that lower temperatures prevail in his corona discharge than in lightning discharges and that radical and ion-molecule reactions play a more important role than quenched thermodynamic equilibrium. Were this the case, the mixing ratio of H_2 to CO_2 employed in their experiments may also have been an important factor. Their ratios were 10^3 to $4x10^3$ higher than the upper end of the range (10^{-3} to 10^{-5}) used in many prebiotic photochemical models. Kasting (1990) estimated a ratio of 10^{-3} based on a probable upper limit to the volcanic outgassing rate. The quantitative effect on the HCN production rate of decreasing the H_2/CO_2 to a more plausible model atmosphere is uncertain, but qualitatively, it can only give a lower yield than shown in Table 2.

In the photochemical model for HCN production (Zahnle, 1986), the key reaction is formation of atomic N by solar Lyman alpha irradiation of N_2 in the middle atmosphere. Reaction of N with species generated by CH_4 and H_2O photochemistry leads ultimately to synthesis of HCN. Under the most favorable conditions, HCN rainout to the early ocean occurs at the rate of 10^{11} cm^{-2} s^{-1}. This rate results from a large atmospheric mixing ratio for CH_4, $\tilde{}10^{-2}$, maintained by a high prebiotic CH_4 flux set equivalent to the modern biogenic flux of $4x10^{11}$ cm^{-2} s^{-1}. The plausibility of this mixing ratio is problematic as is the magnitude of the presumed abiotic CH_4 flux (Zahnle, 1986). For thermodynamic reason such a flux had to have been hydrothermal in origin (section 2.4). Lower mixing ratios can be maintained by lower CH_4 fluxes, but HCN production decreases. correspondingly. With a modern abiotic methane flux $\tilde{}3x10^{10}$ cm^{-2} s^{-1} (Welhan, 1988), HCN rainout is reduced to $2x10^5$ cm^{-2} s^{-1}.

Among the sets of entries in Table 2, the experimental parameters - mixing ratio of gas components, total pressure, temperature, impact velocity, size of impactor etc. - vary according to the model atmosphere or process intended for simulation by the authors. Consequently, rainout rates or production rates can be compared within sets, but, strictly speaking, not between sets. Nonetheless, the implications of a rate for a given set of conditions can still be instructive and useful for setting bounds on the efficacy of the process in the model atmosphere.

Among the processes represented in Table 2 photochemical synthesis of HCN appears to be the most productive. With a 10^{-5} CH_4 mixing ratio and a CH_4 flux equivalent to the modern total CO_2 flux of $1.1x10^{10}$ cm^{-2} s^{-1} (Gerlach, 1991), the HCN rainout rate amounts to 10^8 cm^{-2} s^{-1}. This result lies within the realm of prebiotic plausibility even for FMQ controlled gas compositions in hydrothermal systems. If the ratio of CH_4/CO_2 $\tilde{}0.01$ released today (Welhan, 1988) was also characteristic of Hadean hydrothermal fluxes, then the required ancient CO_2 flux was 10^{12} cm^{-2} s^{-1}, which is 10^2 times larger than today's. Such an increase in carbon flux is consistent with the notion of much higher heat flows in the Hadean.

In a highly reducing atmosphere, impact shocks, particularly those associated with post-impact plumes Chyba and Sagan (1992), yield more HCN than corona discharges. If the experiments of Mukhin et al. (1989) involving laser vaporization of impactors are taken at face value, that is, if the chemistry in the plasma plume thus generated is assumed to approximate that occurring in the natural post-impact plume, then in a mildly reducing atmosphere vaporization and recombination of impactors may be $\tilde{}100$ times more productive than corona discharges. Except for the singular result of McKenzie and Arnold (1967), electric discharges appear to be more productive under mildly reducing conditions than meteor shocks.

Since impacts occur as single events often separated by long periods of time, products are generated as pulses in the atmosphere. According to Fegley et al. (1986), a few large impacts every 1-5 years would maintain the rainout of products. What effects such a pulsed source of compounds might have in prebiotic evolution has received little study (see Section 3.1.3.). Certainly, discontinuities in the flux of products to the oceans would become increasingly frequent and lengthy as the frequency of impacts decreased over time. If prebiotic evolution required a continuous supply of organics, at some point in time impacts would no longer contribute.

If today's ocean were spread over the entire globe, its average depth would be $2.7x10^5$ cm. In the most favorable, if implausible, case listed in Table 2 (Zahnle, 1986), the HCN rainout rate is 10^{11} cm^{-2} s^{-1}. Suppose the ocean circulated through hydrothermal systems on a timescale of 10^7 yr equivalent to that operative today, and no loss mechanisms existed for HCN. Under these optimum conditions, the maximum steady state HCN concentration ([HCN]) could attain 0.02

molar. In the case of a more plausible photochemical rainout rate, 10^8 cm^{-2} s^{-1}, the [HCN] approaches an upper limit of $2x10^{-5}$ molar. If either oceanic recycling were faster or HCN sinks were operating, the concentration would decrease accordingly. Taking the most favorable HCN production rate from corona discharges in a methane atmosphere, and ignoring atmospheric losses prior to rainout (which could amount to 99%), [HCN] $\tilde{}4x10^{-8}$ molar. The worst case for corona discharges in a CO_2 atmosphere takes into account 99% loss of HCN in the atmosphere and yields [HCN] $\tilde{}1.4x10^{-11}$ molar. These estimates suggest a minor role for electric discharges in HCN production.

3.1.2. *Formaldehyde Synthesis..* Some generalizations about formaldehyde synthesis can be drawn from comparisons of data in Table 3, subject to the same qualifications discussed in 3.1.1. If destruction of formaldehyde prior to rainout is taken into account, impact-related syntheses turn out to be much less productive than photochemistry. Rainout rates listed for the calculations of Chyba and Sagan (1992) assume all organic carbon is synthesized as formaldehyde (as was done for HCN in section 3.1.1.). Least productive of all are electric discharges (Stribling and Miller, 1989). Photochemical yields in CH_4 atmospheres are relatively insensitive to CH_4 mixing ratios and fluxes (Zahnle, 1986). In atmospheres dominated by more oxidized carbon gases, formaldehyde rainout rates appear to be higher in CO than in CO_2 (Bar-Nun and Chang, 1983), but increases in CO/CO_2 ratio from 10^{-4} to 1 are accompanied by only a hundred-fold increase in rainout rate. (Kasting, 1990). This relative insensitivy to the nature of the carbon source is also reflected in the data for corona discharges. The results obtained by (Wen et al., 1989) for a CO_2 atmosphere, in which CO is produced *in situ* by CO_2 photodissociation, indicate that formaldehyde synthesis is very sensitive to the mixing ratios of CO_2 and H_2, Raising these ratios from $3x10^{-4}$ and 10^{-5}, respectively, to 10^{-2} and 10^{-3}, increases the rainout rates from $3.5x10^4$ cm^{-2} s^{-1} to $1.5x10^{10}$ cm^{-2} s^{-1}.

Apparently, high photochemical rainouts rates can occur regardless of whether the carbon source is CO_2 or CH_4. This interesting result suggests that formaldehyde synthesis, probably accompanied by smaller amounts of other compounds - formic acid and other acids, other aldehydes, alcohols, and even hydrocarbons (Bar-Nun and Chang, 1983; Wen et al., 1989) - was a highly probable process. Subject only to diurnal effects and occasional interruptions by impact-generated dust clouds, a more or less steady rain of CHO-containing organic compounds would have fallen on the ocean surface. As was apparent for HCN, electric discharges appear to make a relatively minor contribution to the the synthesis of CH_2O. Following the approach used to calculate [HCN] in section 3.1.1., the analogous maximum [CH_2O] can be calculated for comparison. From the data of Wen et al. (1989) are obtained upper and lower limits for [CH_2O] amounting to $3x10^{-3}$ molar and $7x10^{-9}$ molar, respectively.

3.1.3. *Comparison with Delivery of Extraterrestrial Organic Matter.* Since organic compounds (Chang et al., 1983; Stribling and Miller, 1987) and their precursors appear to have been difficult to synthesize in the mildly reducing model of the prebiotic atmosphere favored by geochemists, investigators have evaluated extraterrestrial sources of prebiotic organic matter as an alternative. Today, interplanetary dust particles (IDP, Mackinnon and Rietmeijer, 1987) and meteorites enter Earth's atmosphere and many are collected essentially intact (Anders,1989). Stony meteorites a few meters in diameter can be decelerated by the atmosphere and survive, but larger asteroidal and cometary objects are expected to suffer catastrophic destruction (Chyba et al., 1990).

Early in Earth history, the flux of IDP and material injected by asteroidal and cometary impactors would have been much higher than now, and intact extraterrestrial organic compounds would have augmented those produced by terrestrial processes (Oro, 1961; Chang, 1979; Chyba et al., 1990). From estimates of organic carbon in the infalling objects and scaling to the lunar impact flux, the flux of extraterrestrial organic carbon delivered to Earth over the period 4.4 to 3.0 Ga has been estimated (Chyba and Sagan, 1992). The IDP contribution (Anders, 1989) dominated that from comets and meteorites by orders of magnitude with flux decreasing from 10^9 to $4x10^5$ kg organic carbon yr^{-1} (10^9 kg carbon yr^{-1} = $3x10^8$ cm^{-2} s^{-1}). On an vastly different size scale, Clark (1988) suggested atmospheric deceleration during a grazing cometary impact as a possible mechanism for delivery of an intact comet to Earth's surface, the slow melting of which was speculated to provide an organic rich pond for chemical evolution.

The time dependence of the IDP flux is shown in Figure 3 along with that of production rates estimated for total organic carbon resulting from impactors (after Chyba and Sagan, 1992; several entries in Tables 2 and 3 are taken from curves in this figure). The open symbols at 4.0 Ga represent estimates for total organic carbon production obtained by lightlning discharges (squares) and ultraviolet photochemistry (triangles) in a highly reducing atmosphere; the filled symbols in a mildly reducing one. The calculations suggest that IDP were an important contributor to the total organic carbon on a prebiotic Earth endowed with a mildly reducing atmosphere. Note, however, that the IDP would have been devolatilized or partially decomposed during both exposure to solar radiation in space and transient heating upon atmospheric entry. Consequently, the organic matter remaining would have been mainly involatile and relatively refractory, rather than in the forms assumed for this discussion.

In a mildly reducing atmosphere photochemical synthesis of organic carbon (as formaldehyde only) appears to be more productive than either delivery by IDP or synthesis by lightning. This conclusion is supported by the data of Wen et al. (1989) in Table 3. The data point for lightning cited by Chyba and Sagan (1992) and shown in Figure 3 overestimates the production rate by a factor $>10^3$ as compared with all but one of the values given for HCN in Table 2. The organic production rate due to lightning in a highly reduced atmosphere is also somewhat higher than most of those listed in Table 2. The data point for ultraviolet photochemical synthesis of organics is based on hot atom reactions initiated by photodecomposition of atmospheric H_2S (Sagan and Khare, 1971), but a source for H_2S is problematic for a prebiotic ocean rich in dissolved Fe^{2+}. A comparable photochemical production rate is provided by the upper limit for CH_4 photochemistry (Zahnle, 1986; Table 2).

The results in Figure 2 and Tables 2 and 3 provide a status report rather than the final word on the relative contributions of various sources of organic carbon. Mechanisms for organic synthesis remain to be discovered; and uncertainties exist in the calculations, as do unaccounted factors in the modeling of the complexities of the natural world in laboratory and computer experiments. Among such factors are processes acting as sinks for organic matter, for example, destruction of organic compounds by ultraviolet irradiation of surface waters (Dose, 1974) or sequestration in insoluble polymers (Nissenbaum et al., 1975).

Oberbeck and Aggarwal (1992) incorporated a photochemical sink (Dose, 1974) in modeling the time dependence of amino acid concentrations following the collision of a 10 km comet (Figure 3) into a shallow sea. Two amino acid sources were assumed: ordinary electric discharge synthesis and shock synthesis in the impact plume. Their calculations showed that the concentration of comet derived amino acids decreased over time, eventually falling below the level maintained by synthesis from discharges (Figure 4). This crossover point occurred after 90 or 140 years, depending on the discharge energy flux used in the calculations. Thus for ~100 years after each impact, shock synthesized amino acids would have made the predominant contribution. Since 10% of initial amino acids were photochemically destroyed in 38 years, while cycling of

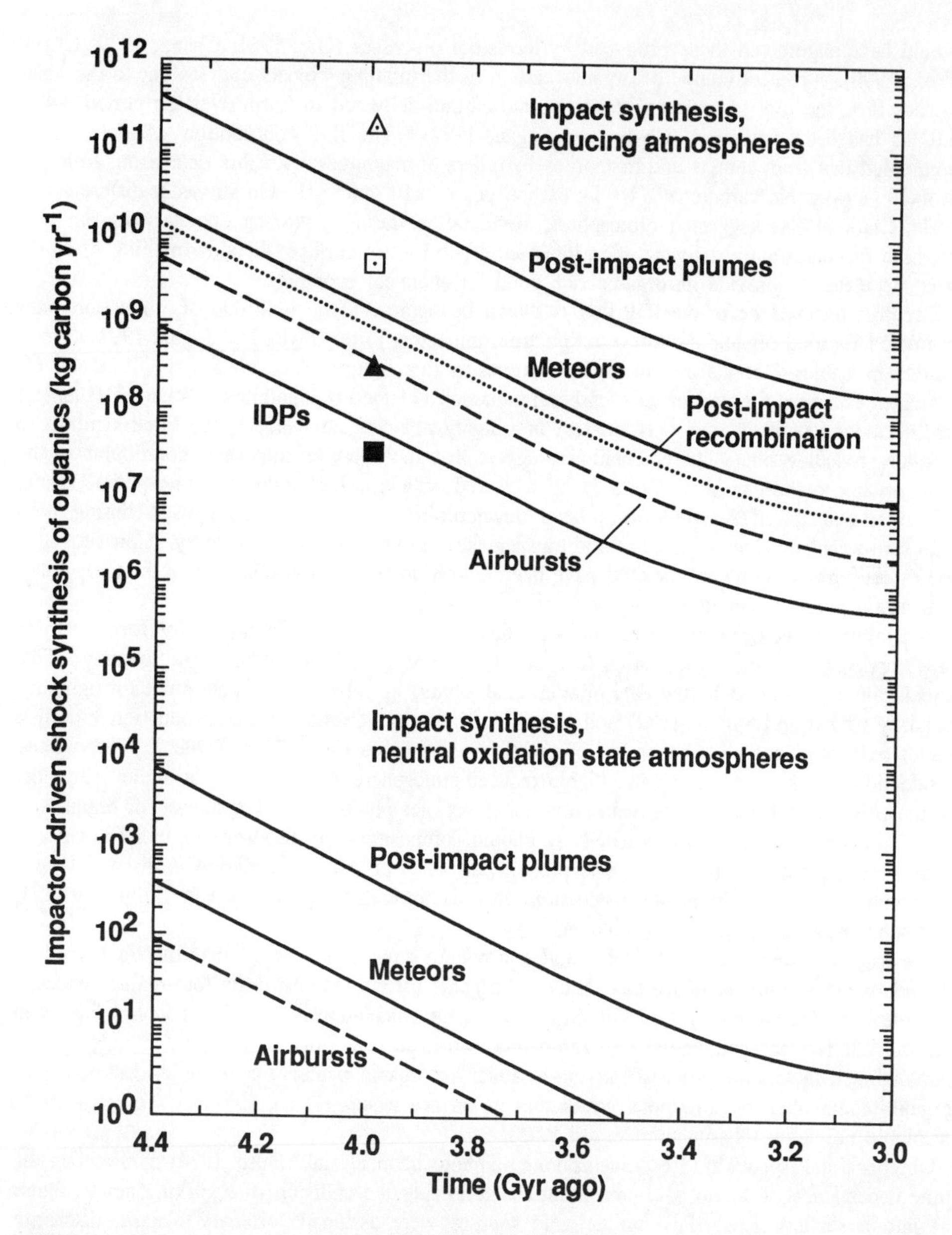

Figure 3. Rates of impact shock synthesis, exogenous delivery and endogenous production of organic carbon as a function of time. Upper set of curves for a reducing atmosphere: CH_4 + (N_2 or NH_3) + H_2O; lower set for neutral redox atmosphere: $CO_2 + N_2 + H_2O$. Dashed lines indicate upper bounds, and the dotted line represents a poorly understood estimate perhaps independent of atmospheric redox state. Open triangle and square, respectively, indicate production rates from lightning and UV light (absorption by H_2S) in highly reduced atmosphere; filled symbols apply to neutral atmosphere (absorption by CO_2). (After Chyba and Sagan, 1991)

an ocean through hydrothermal systems would have taken at least 10^5 years, the authors concluded that pyrolysis in hydrothermal systems was not a significant sink. They suggested that adsorption on clay minerals (e.g., Lahav and Chang, 1976; Hedges and Hare, 1987) might have occurred faster even than photolysis. Calculations like those of Oberbeck and Aggarwal (1992), although model dependent, take into account both sources and sinks and represent a valuable approach for assessing the contribution of any process to chemical evolution.

3.2. REACTIONS IN THE OCEAN

Perhaps the most well known reaction thought to occur in the prebiotic ocean is the Strecker synthesis of amino acids (Miller, 1957; Barak and Bar-Nun, 1975), which proceeds by way of aminonitrile intermediates. In water, the aminonitriles formed by the reaction of HCN, NH_3 (or primary and secondary amines), and aldehydes and ketones are supposed to be in equilibrium with the corresponding hydroxynitriles. At a given temperature and pH, the ratios of amino to hydroxy acids depend in a complex way on the equilibria among the aminonitriles and hydroxynitriles in the mixture, the rates of irreversible hydrolysis of these nitriles, and the rate of hydrolysis of cyanide to formate. Miller and Van Trump (1981) have analyzed this dependency for the simpler case of single amino/hydroxy acid pairs. They derived equation (1) which holds when loss of HCN to formate equals loss of HCN to amino acid synthesis. In the equation, k_{1HCN} is the rate of HCN hydrolysis, k_{1AN} is the rate of hydrolysis of the aminonitrile, and K_{AN} is the equilibrium constant for $RCHO + NH_3 + HCN \longleftrightarrow RCH(NH_2)CN$ at a given temperature and pH.

$$[RCHO]\,[NH_3] = k_{1HCN}/k_{1AN}K_{AN} \qquad (1)$$

At 0°C and pH 8, $k_{1HCN}/k_{1AN}K_{AN}$ $\sim 9\mathrm{x}10^{-12}$ (Miller and Van Trump, 1981). Therefore, if $[NH_3]$ is known, [RCHO] is readily calculated. For the upper limit of $[NH_3] = 0.01$ molar, fixed by ionic equilibria with NH_4^+ -binding marine clays (Bada and Miller, 1968), the required aldehyde concentration is $9\mathrm{x}10^{-10}$ molar, according to Van Trump and Miller (1981). The lower limit of $7\mathrm{x}10^{-9}$ molar for [CH2O] described in section 3.1.2. would require a minimum $[NH_3]$ of $1.3\mathrm{x}10^{-3}$ molar for amino acid synthesis to exceed HCN hydrolysis.

While these simple calculations suggest that the Strecker synthesis may have been viable at low reactant concentrations (Miller and Van Trump, 1981), they are unlikely to accurately reflect the complexity of reactions or possible inteferences in natural seawater. If $[NH_3]$ concentrations were much lower or if HCN concentrations were also low, as suggested at least for synthesis in a mildly reducing atmosphere, amino acid, purine or pyrimidine syntheses based on HCN would have been very problematic. In any case amino acid concentrations would have been several orders of magnitude lower than reactant aldehyde or HCN concentrations (Stribling and Miller, 1987). Such severe dilutions have led investigators to invoke evaporation in ponds and lagoons as means of concentrating prebiotic reactants. To sidestep the problem of HCN loss on evaporation, Sanchez et al. (1966) suggested freezing of prebiotic waters to concentrate HCN in a eutectic phase. The latter solution is contradicted by the apparent absence of glacial deposits in the early Archean (Walker et al., 1983) and sedimentological and climate models favoring a warm Hadean environment.

A multitude of other reactions have been studied under putative prebiotic ocean conditions. Typically the concentrations of reactants used in them far exceed the estimated values of [HCN] and $[CH_2O]$ attainable from production rates in all but the most reducing of atmospheric conditions described in Tables 2 and 3. Critical readers and workers in the field should weigh

the relevance of past and future work against the rationales offered to defend such usage.
Rarely, if ever, has synthesis of HCN in solution been the objective of a prebiotic study. Instead it is presumed to be present in the prebiotic ocean and, based on that presumption, it is used as an intermediate in many reactions. If the atmosphere has been only mildly reducing since the end of the major epoch of accretion at ~4.4 Ga, the apparent lack of strong atmospheric sources for HCN poses a serious problem for theories of prebiotic evolution. More effort needs to be directed toward finding solution pathways or novel atmospheric sources for HCN synthesis.
Interestingly, Yamagata and Mohri (1982) report the formation of cyanate (NCO^-) during electric discharges through a gas mixture containing N_2, CO_2, H_2 and water vapor. Their experiments were conducted with ratios of H_2/CO_2, from 1.5 to 0.25, and best cyanate yields were obtained at ratios <1. Cyanate reacts with phosphate to yield the phosphorylating agent, carbamyl phosphate. Other possible roles for cyanate should be explored.
Other ways in which electrical discharges may have produced organic compounds should be assessed. For instance when electrical discharges were passed through N_2 to strike the surface of a suspension of powdered $CaCO_3$ at pH 8 containing ^{14}C-$NaHCO_3$, small amounts of hydrazine (H_2NNH_2), ^{14}C-carbohydrazide ($H_2NNHCONHNH_2$) and unidentified ^{14}C-organic products were formed along with major amounts of nitrous and nitric acids (Folsome et al., 1981). Carbohydrazide yields were much higher than formic acid yields obtained in earlier experiments involving exposure of CO_2 to ionizing radiation (Garrison et al.1953). Acid hydrolysis of the unidentified material yielded ninhydrin-reactive compounds (amino acids?) and UV chromatophores. Synthesis of hydrazine and carbohydrazide also occurred in experiments in air, but not in experiments lacking $CaCO_3$. Although the relevance of carbohydrazides to prebiotic evolution is unclear, the reduction of CO_2 to organic carbon suggests that analogous experiments using geochemically reasonable reductants merit further study.
The photochemical synthesis of formaldehyde or other organic compounds from CO_2 or bicarbonate in solution has a long and largely unproductive history (see review by Chittenden and Schwartz, 1981). Unambiguous evidence for such synthesis assisted by photosensitizers began appearing in the literature in the 1960's (e.g., Getoff et al., 1960). Among the reports directly relevant to prebiotic evolution were those of Halmann et al. (1981), who showed that iron-rich clays acted as photosensitizers to produce formaldehyde and methanol, and Åkermark et al. (1980), who obtained formic acid and formaldehyde by uv irradiation of acidic solutions of Fe^{2+}. Results obtained by Halmann et al. (1981) with sunlight lead to formaldehyde and methanol production rates of $5x10^7$ cm^{-2} s^{-1} and $9x10^8$ cm^{-2} s^{-1}, respectively. These rates compare favorably with those listed in Table 3 for reactions taking place in CO_2.containing atmospheres. The work of Halmann et al. (1981) has relevance for near surface environment where iron-rich clays may occur as a result of weathering; the work of Åkermark et al. (1980) has prebiotic relevance for early acidic environments exposed to sunlight. These characteristics appear to be common among continental or shallow marine hydrothermal systems (see section 3.3.1).
Since the early 1980's, and despite the surety of a reduced primordial ocean, the prebiotic chemistry of Fe^{2+} in seawater has not been extensively explored. Irradiation of aqueous Fe^{2+} with ultraviolet light results in release of H_2 and precipitation of iron oxy-hydroxides (Braterman et al.,1983; Borowska and Mauzerall, 1987). This reaction could have provided an important source of reducing equivalents in the atmosphere and upper meters of the ocean. Photostimulated reduction of bicarbonate to formaldehyde in aqueous Fe^{2+} at neutral pH has been reported (Borowska and Mauzerall, 1988), but the results have been found irreproducible (Borowska and Mauzerall, 1991). Additional reactions mediated by iron species are described in below in section 3.3.1.

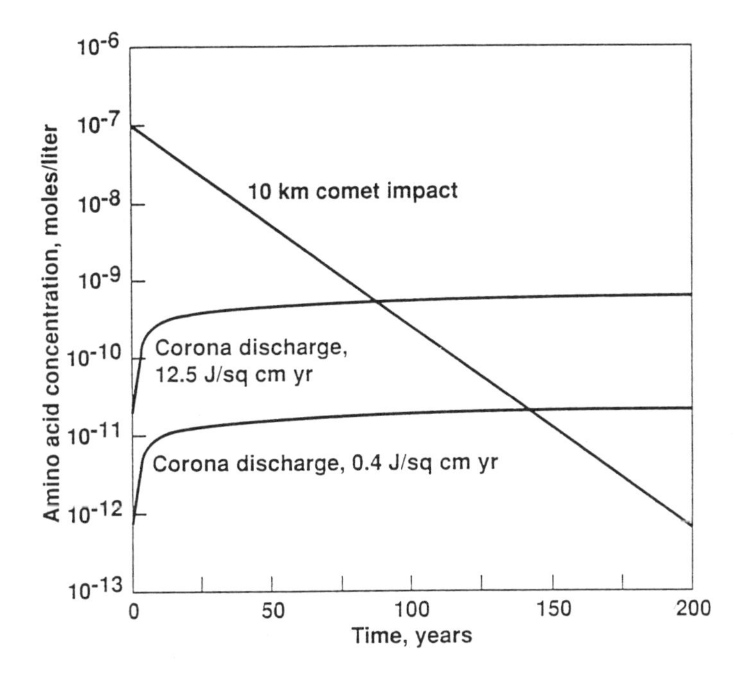

Figure 4. Time dependent concentrations of amino acids produced by comet impact and by corona discharge in a neutral redox atmosphere. Oberbeck and Aggarwal (1992) estimated an initial concentration of 10^{-7} moles/liter for amino acids produced by shock synthesis. Their calculations were based on the assumptions that the abiogenic a-amino-isobutyric acid in K/T boundary sediments (Zhou and Bada, 1989) was synthesized in the impact, and the inland Cretaceous sea at Stevns Klint, Denmark had a 100 m maximum depth (Alvarez, 1980). Production from discharges was caculated using energy fluxes of 12.5 Joule/cm^2 (Stribling and Miller, 1987) and 0.4 Joule/cm^2 (Chyba and Sagan, 1991). (Oberbeck and Aggarwal, 1992, courtesy of Kluwer Academic Press)

Nitrogen oxides fixed in the atmosphere by shocks and corona discharges and rained into the oceans as nitrite would have been reduced to NH_3 by aqueous ferrous iron (Chang et al.,1983). Model calculations based on measured rate constants for nitrite reduction at pH 7.4 and 25°C and estimated atmospheric nitric oxide production rates (Kasting, 1990) suggest maximum, steady state, ammonia concentrations between $3.6x10^{-6}$ and $70x10^{-6}$ molar, depending on whether photochemistry or hydrothermal systems was the prevailing sink for nitrite and ammonia (Summers and Chang, unpublished results). At these concentrations, the Strecker synthesis requires a minimum aldehyde concentration of $3x10^{-6}$ to 10^{-7} molar at 0°C and pH 8.
Whether the low concentrations inferred for NH_3, CH_2O and HCN were adequate for any syntheses on a prebiotic world with a mildly reducing atmosphere is unclear. The atmospheric processes included in Tables 2 and 3 represent the largest energy sources on the prebiotic Earth. Lacking additional sources of comparable strength, identification of pathways for organic synthesis at very low concentrations of these key intermediates, and possibly in the absence of HCN, poses a major challenge. Additional investigations of the chemistry of Fe^{2+}-rich seawater under photochemical and dark conditions should be undertaken to further explore its potential for chemical evolution.

4. Planetary Settings for Prebiotic Synthesis

4.1. INTERFACIAL ENVIRONMENTS

The importance of interfacial environments on the prebiotic Earth has been emphasized (Chang, 1988). Most relevant for the origin of life may have been the ubiquitous, geophysically active regions at the ocean-atmosphere interface and at the ocean-crust interface in marine hydrothermal systems. These realms contain phase boundaries between gas, liquid and solid states where disequilibrium resulting from gradients in physical and chemical properties are maintained by physical and chemical energy fluxes. Within these environments small scale interfaces are provided by aerosols, volcanic and cometary dust, hydrothermal minerals, chemical precipitates and vesicle-like structures of organic or mineral chemical composition. Inasmuch as life itself must have emerged as a phase bounded system, the formation, dissipation and reformation of small scale interfaces must have been a prerequisite for the origin of life.

4.1.1. *Hydrothermal systems.* Hydrothermal systems have been proposed as sites for organic synthesis and the origin of life by a number of authors (e.g., Ingmanson and Dowler, 1977; Corliss et al., 1981; Wachterhauser, 1988, 1988a; Shock, 1990, 1991). The main objection to these sites is the perceived difficulty of synthesizing and preserving the organic compounds necessary for the evolution of cellular life (Miller and Bada, 1988; Holm, 1992). Energy sources, metal ions and iron-bearing minerals as catalysts and mechanisms for organic synthesis have been proposed (Holm, 1985; Arrhenius, 1986; Wachterhauser, 1988; 1988a, 1990, this volume), but little research into these materials and mechanisms has been conducted under hydrothermal conditions (see, however, French, 1971; Yanagawa et al., 1988; Shock, 1990a; Bloechl et al., 1992; Hennet et al., 1992; Yanagawa and Kobayashi, 1992).
A thermodynamic basis for organic synthesis in hydrothermal systems has been advanced by Shock (1990) who proposed that abiotic synthesis could have occurred in metastable states as seawater circulated through hydrothermal systems. At stable C-H-N-O equilibrium and redox states buffered by mineral assemblages in these systems, activities for relevant organic compounds

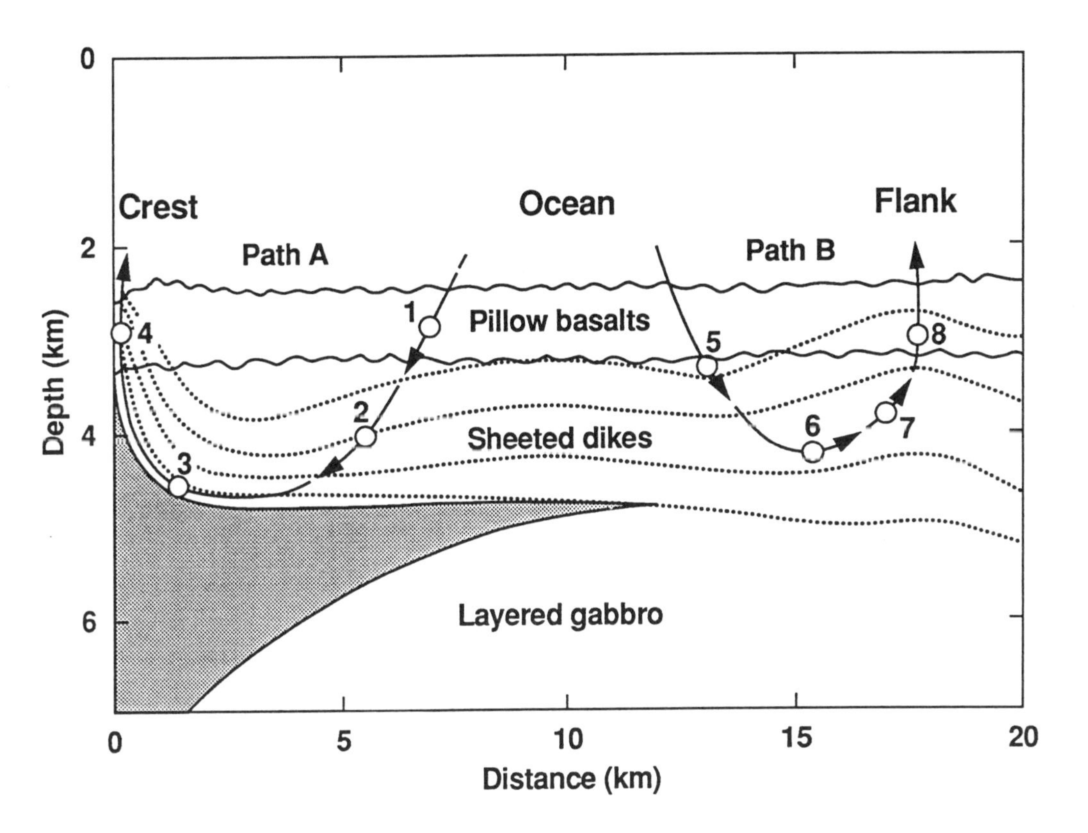

Figure 5. Water circulation paths in hydrothermal systems. Path A shows solutions taken deep into the vicinity of the magma chamber and emerging through black smokers at the ridge crest. Fluids taking Path B penetrate and emerge from the flanks of the system following a lower temperature trajectory. Dotted lines indicate isotherms. The numbered locations correspond to points for which activities of organic compounds were calculated at defined temperatures and oxidation states: 1, 100°C, hematite/magnetite buffer (HM); 2, 200°C, pyrrhotite-pyrite-magnetite buffer (PPM); 3, 400°C, fayalite-magnetite-quartz buffer (FMQ); 4, 350°C, PPM; 5, 100°C, HM; 6, 250°C, FMQ; 7, 200°C, PPM; 8, 150°C, PPM. (Shock, 1992, courtesy of Kluwer Academic Press)

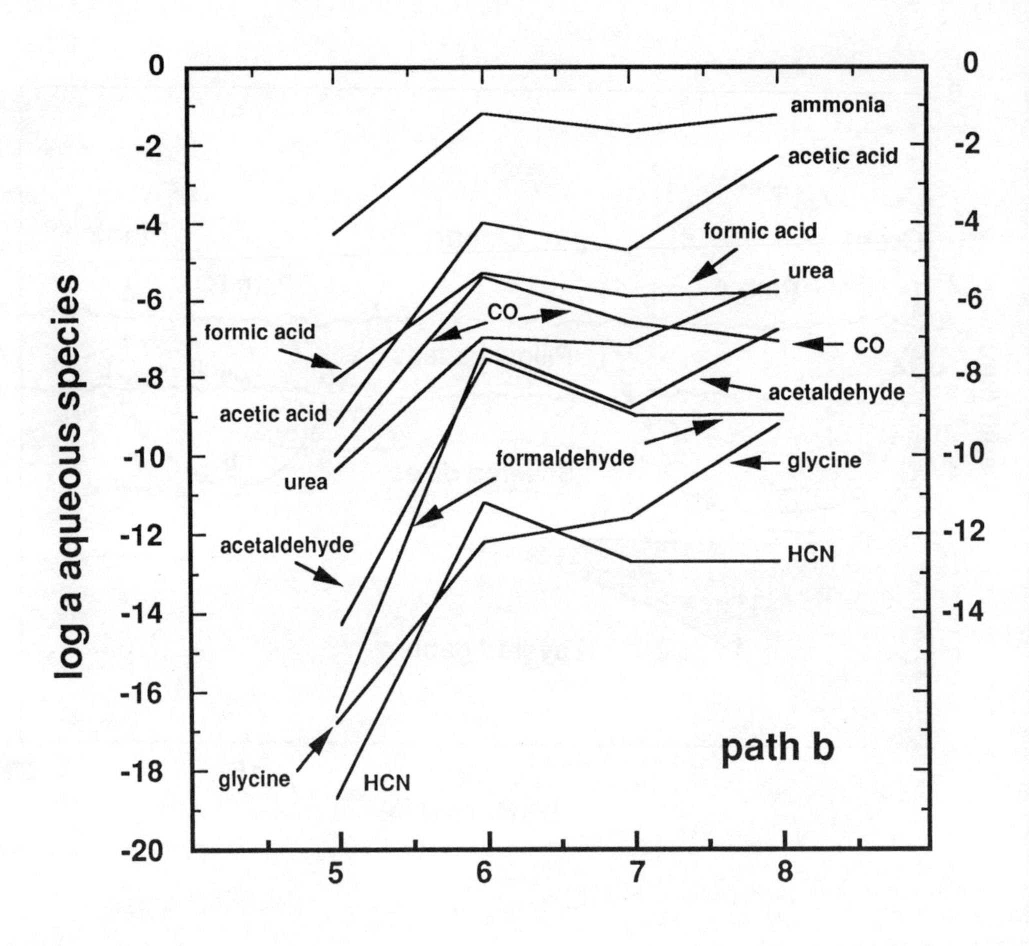

Figure 6. Plots of the log activities of selected aqueous species at points indicated along Path B in figure 5. (Shock, 1992, courtesy of Kluwer Academic Press)

are negligible. Below about 600°C, however, kinetic barriers to thermodynamic equilibrium are observed to occur allowing attainment of metastable states and much higher activities of organic compounds.

The scenario proposed by Shock (1990; 1992) for organic synthesis in hydrothermal systems starts with C and N as predominantly CO_2 and N_2 at high temperature equilibrium with mineral assemblages deep in hydrothermal systems. As fluids containing these species circulate to lower temperature regimes, they cool under the influence of mineral-bufferred hydrogen fugacities. As they cool they move from conditions under which CO_2 and N_2 predominate to those in which CH_4 and NH_3 dominate. If kinetic barriers to CO_2 and N_2 reduction are surmounted, organic compounds are produced and maintained in metastable states. Further conversion to CH_4 and NH_3 is supposed to be inhibited by other kinetic barriers.

In Figure 5 two paths are shown illustrating the circulation of water through hydrothermal systems. The calculated activities of aqueous organic compounds and ammonia were found to be significant and higher at the final point of Path B than of Path A (Figure 6). Notably, the activity of ammonia suggests a significant source of reduced nitrogen. Since flow through potentially destructive Path A is estimated to be less than 5% of that along Path B, the flanks of these systems represent sites of potentially high productivity for synthesis of organic carbon. Shock (1992) estimated this source (kinetic barriers permitting) at $2x10^8$ kg per year which is comparable to other estimates of production rates in a neutral redox atmosphere (Figure 3). If the hypothesis of hydrothermal synthesis is verified on a broad front by experiments, hydrothermal systems could emerge as sources for organic compounds and ammonia rather than sinks.

Synthesis of organic compounds is only the first step in chemical evolution, however. Along the way to more complex chemical systems, condensation reactions are necessary to form phosphate esters, lipids, peptides, and oligonucleotides from monomers (for examples, see Brack, this volume; Schwartz, this volume). These reactions require the intermolecular elimination of water, and plausible mechanisms remain to be demonstrated under hydrothermal conditions (cf., Flegmann and Tattersall, 1979; Siskin and Katritzky, 1991; Yanagawa and Kobayashi, 1992).

For any synthesis scheme to gain credibility, however, it must go beyond mere plausibility; its efficacy must be evaluated within the constraints imposed by the environment for which it is proposed. Among the general constraints applicable to hydrothermal systems is the time available for chemical evolution of organic compounds before diffusive loss to the bulk ocean or burial with minerals in sediments below levels where useful chemical transformations could take place. Extensive metalliferous sediments occur over the crests and flanks of ocean ridge hydrothermal systems today. Based on estimated fluxes of hydrothermal particulates in the northeast Pacific (Baker and Masoth, 1987; Dymond and Roth, 1988), accumulation of a 1-7 cm thick sediment layer would occur in 10^4 year. In addition the particulate phases (colloidal iron oxides and hydroxides, and iron sulfides) act as scavengers of phosphorus (and other elements) from sea water (Arrhenius, 1952; Feely et al., 1990). Co-precipitation with iron hydroxides has also been used as a means of stripping dissolved marine organic compounds from seawater (Garrett, 1967). Diffusion, burial or irreversible adsorption, and destruction by secodary reactions, were probably limiting factors for chemical evolution of organic compounds in all planetary environments.

4.1.2. *The ocean-atmosphere interface.* For this discussion the region encompassed by the ocean-atmosphere interface includes a zone from the ocean surface film to about a hundred meters below. This zone is continuously mixed by winds and waves. Residence times in the upper few hundred meters of the ocean are about 100 years. A packet of water transported to depth will have a residence time in the deep ocean water of about 10^3 years. Assuming similar residence

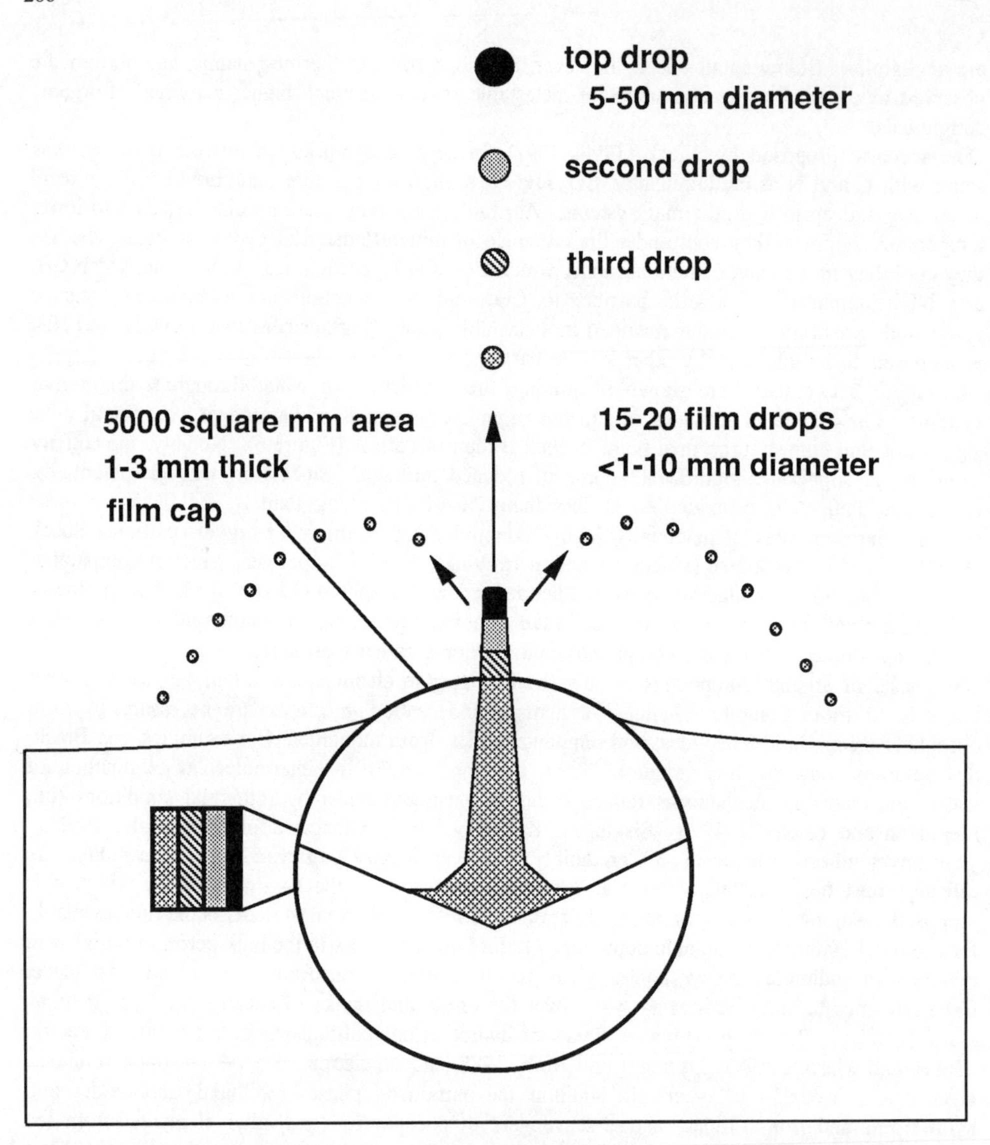

Figure 7. Composite schematic cross section of a 500 mm bubble bursting and aerosols generated by the burst at the ocean-atmosphere interface. Ejected aerosols come from the thin layer pealed off the inner wall of the bubble. Each successive drop originates from increasingly deeper onion-like shells in the bubble wall. The enlargement of the bubble wall is intended to show the origin of the jetting drops, which can be accelerated at $>10^3$ G, and the less energetically accelerated film droplets. (After MacIntyre, 1974a, and Berg and Winchester, 1978)

times early on, the deep ocean would have acted as a sink for non-buoyant and non-surface-active materials. If life arose in the wind mixed layer, and if 100 years is an inconceivably short time for the origin of life, then surface active substances must have played a critical role in sustaining chemical evolution.

Various processes acting on and within the uppermost layer of the mixed zone would have made it a complex, physically and chemically active environment perhaps well suited for chemical evolution (Lerman, 1986). Material in the atmosphere from all sources (oceanic, detrital, volcanic as well as extraterrestrial) would have fallen onto the ocean surface and mixed with organic compounds synthesized *in situ*. In turn, substances in the mixed layer would have been exposed to wave action, diurnal solar and cosmic radiation, electrical discharges and other energy sources. Concentrations of reactants and products delivered from the atmosphere would have been highest at the ocean surface and decreased with depth of mixing. Airborne polyphosphates produced by volcanism (Yamagata et al., 1991) and deposited at the ocean surface could have helped fuel phosphorylation reactions. Surface active compounds capable of forming bilayer membranes could have been supplied by extraterrestrial inputs (Deamer, 1988). Disruption of surface monolayers could have enhanced the formation, dissipation and reformation of the bilayer vesicles that were prerequisites for the origin and evolution of cellular life (Stillwell, 1980; Koch, 1985; Morowitz et al., 1988). Preferred molecular orientations and ordering at the atmosphere-monolayer, bilayer water and atmosphere-water interfaces (MacIntyre, 1974a; Pohorille and Benjamin, 1991; Wilson and Pohorille, 1991) could have promoted molecular self-assembly (Whitesides et al., 1991) and selectively enhanced some reaction pathways over others. Today the pH of rain lies in the 3-5 range resulting from dissolution of CO_2, sulfuric and nitric acid aerosols and organic acidic air pollutants (Chameides and Davis, 1983). Acid rain on the primitive Earth would have originated from atmospheric photochemistry (Kasting, et al., 1989) and aerosols of volcanic (Snetsinger et al., 1987) and impact origin (Prinn and Fegley, 1987). Possibly, a slightly acid ocean surface film could have benefited chemical evolution, for instance allowing photoreduction of bicarbonate or carbon dioxide by ferrous iron (Akermark et al., 1980).

At any given moment, bubbles cover 3-4 % of the ocean surface (MacIntyre, 1974a). Bubble formation occurs in the upper portion of the mixed zone as the result of wind and wave action; and bubble bursting at the surface ejects aerosols and particulates into the overlying atmosphere. The importance of bubble formation and bursting in accounting for differences in chemical properties between the thin surface skin of the ocean and underlying waters has been known for years (MacIntyre, 1974, 1974a; Liss, 1975; Wu, 1981; Tseng et al., 1992). Among these differences are strong enrichments in organic compounds (particularly surface active material), dissolved phosphate, Cr, Cu, Fe, Pb and Zn. Apparently, surface active and hydrophobic organic compounds complexed with other species diffuse to and concentrate at the hydrophobic inner surface of bubbles as they rise to the surface. Bursting of bubbles ejects a proportion of the inner wall into the atmosphere as aerosol droplets (Figure 7) where it can undergo other processes and be transported downwind or higher in the atmosphere.

Return of aerosols to the ocean causes chemical enrichments in the surface film relative to bulk water. Phosphate ion enrichment in drops from breaking bubbles can attain factors of 600 over the bulk solution (MacIntyre and Winchester, 1969); organic matter can be enriched by more than a factor of 10^3, with organic/salt mass ratios of 0.3-0.9 (Berg and Winchester, 1978). Airborne material today contains mineral grains, organic compounds and sea salts, all of which undoubtedly had ancient analogs. Modern estimates of the global ocean emission of sea salt alone approach 10^{12} kg per year (Chester, 1987). If 30% of this amount is organic carbon, $3x10^{10}$ kg per year or 100 times as much is processed through marine aerosols today as is estimated to have fallen to the ocean surface 4.0 Ga ago from all sources in a neutral redox atmosphere and all but the photochemical source in a highly reduced atmosphere (Figure 2). Since wind and wave action then was unlikely to have been much less intense than now, prebiotic organic matter in the ocean surface layer would have been similarly reprocessed in marine aerosols. Interestingly, total oceanic biomass amounts to ~$3x10^{12}$ kg today.

Only recently has the possible relevance of ocean bubble phenomena to the origin of life been suggested. According to Lerman (1986) material scavenged from the mixed zone by bubble formation could have taken part in multiple cycles of processing during repeated ejection into the atmosphere and return to the surface as aerosols and particulates. Supposedly, aerosols could have undergone evaporative dehydration, rehydration as condensation nuclei for rain and snow, and exposure to energy sources (sunlight, lightning and coronal discharges) before redeposition in the ocean. Could condensation reactions and synthesis of organic phosphates have occurred first in such cycles? Woese (1979) first proposed that life on Earth might have originated in clouds during the accretion epoch. More recently, Oberbeck et al. (1989; 1991) also suggested that condensation reactions of organic compounds supplied by or synthesized during injection of extraterrestrial material in the atmosphere could have been enhanced by wetting and drying cycles in cloud drops.

The enormous amount of energy in wave action would have served other purposes in addition to bubble formation. Evidence that cavitation and sonochemistry accompany wave action has been reported by Anbar (1968) who also suggested a role for these processes in prebiotic synthesis. Heating and cooling rates during cavity collapse are estimated to be billions of degrees C per second with peak temperatures above 5000°C (Suslick, 1989; Flint and Suslick, 1991). Scavenging of sonochemically produced hydroxyl radicals and other oxidants by aqueous Fe^{2+} may eliminate side reactions and favor organic synthesis. The possibilities for prebiotic synthesis in the exotic cavity microenvironment are essentially unexplored.

There are also detremental aspects to the processes acting at the ocean-atmosphere interface. Ultraviolet photochemistry can be destructive to organic compounds at the surface, and for prebiotic evolution to progress, photodecomposition cannot outpace synthesis. In addition to producing H_2, the photochemistry of aqueous ferrous iron also yields colloidal ferric oxide-hydroxide which could have scavenged organic compounds as it settled to the seafloor. Bubble phenomena also enhances particle aggregation thereby accelerating the removal of particulate bound organic matter, phosphate and trace metals from the mixed zone (Wallace and Duce, 1978; Sackett, 1978). Clearer insight into the potential of the wind mixed layer as a site for prebiotic evolution will be gained from a more comprehensive characterization and evaluation of sources and sinks for organic compounds.

Consider the implications of the fluxes of organic carbon shown in Tables 2 and 3 and Figure 3. If transport to the deep ocean on a 100 year time scale was the only sink, upper limits for the total concentration of organic carbon in the top 100 m of the ocean would have been $5x10^{-8}$ molar under a neutral redox atmosphere and $5x10^{-5}$ molar under a highly reduced atmosphere. Since typical prebiotic energy sources (e.g., lightning) produce a variety of compound types in relatively low yields from the key intermediates, concentrations would have been much lower for amino acids or any particular class of compounds deemed necessary for prebiotic evolution. The dilution of any critical single compound would have been far worse in the absence of a compound specific synthetic pathway. Because prebiotic evolution in the ocean at such high dilution is so problematic, evaporation of water and cycles of wetting and drying in shoreline environments or bodies of water with restricted access to the ocean have been proposed to overcome this dilemna (e.g., Lahav and Chang, 1976).

Mechanisms combatting dilution would have been necessary at the ocean-atmosphere interface as well. Rather than dispersing by dissolution, surface active compounds would have acccumulated at the surface film as a result of buoyancy, surface tension and hydrophobic molecular interactions. Higher steady state organic carbon concentrations could have been achieved in the upper several meters by bubble phenomena. Higher concentrations of key organic compounds in and near the surface film could have been maintained by adsorption on and transfer among surface active monolayers or within bilayer vesicles. Evidence that some of these phenomena are at work today can be found in the chemical oceanographic literature cited above. Finally, more productive and selective organic synthesis mechanisms may have existed on surface films and within vesicles than in seawater solution. Although these processes offer potential solutions to the problem of dilution, their effectiveness in a prebiotic setting remains to be evaluated.

Clearly the ocean-atmosphere interface is a dynamic environment worthy of much future study as a site for prebiotic evolution. Its intersection with the shorelines of volcanic platforms and shallow marine hydrothermal systems where possibly other sources of ammonia, organic compounds, mineral catalysts and chemical energy were available could have been extremely important for the origin of life. Perhaps, the geological setting associated with the earliest record of ecosystems was indeed a spawning ground for life.

ACKNOWLEDGEMENTS

I thank Prof. J. Mayo Greenberg and the organizers of the International School of Space Chemistry for inviting me to participate and contribute this manuscript to the proceedings. Louis Lerman recognized and brought to my attention to the roles that sea surface phenomena could play in prebiotic evolution. Support for this work was provided in part by the Exobiology Program of the National Aeronautics and Space Administration.

References

Abe, Y. and Matsui, T. (1988) Evolution of an impact-generated H_2O-CO_2 atmosphere and formation of a hot proto-ocean on Earth. J. Atmos. Sci. 45, 3081-3101.

Åkermark, B., Eklund-Westlin, U. Baecktroem, P., and Loef, R. (1980) Photochemical, metal-promoted reduction of carbon dioxide and formaldehyde in aqueous solution. Acta Chem. Scand. B34, 27-30.

Allegre, C.J., Staudacher, T. and Sarda, P. (1987) Rare gas systematics: formation of the atmosphere, evolution and structure of the Earth's mantle. Earth Planet. Sci. Lett. 81, 127-150.

Alvarez, W., Asaro, F. and Montanari, A. (1980) Iridium profile for 10 million years across the Cretaceous-Tertiary boundary at Gubbio (Italy)
Science 250, 1700-1702.

Anbar, M. (1968) Cavitation during impact of liquid water on water: Geochemical implications.

Science 161, 1343-1344.

Anders, A. (1989) Pre-biotic organic matter from comets and asteroids. Nature 342, 255-256.

Anders, E. and Owen, T. (1977) Mars and Earth: origin and abundance of volatiles. Science 198, 453-465.

Arculus, R.J. and Delano, J.W. (1980) Implications for the primitive atmosphere of the oxidation state of Earth's upper mantle. Nature 288, 72-74.

Arculus, R.J. and Delano, J.W. (1981) Intrinsic oxygen fugacity measurements: techniques and results for spinels from upper mantle peridotites and megacryst assemblages. Geochim. Cosmochim. Acta 45, 899-913,

Arrhenius, G. (1952) Sediment cores from the East Pacific. I. Properties of the sediment and their distribution. Reports of the Swedish Deep-Sea Exp[edition 1947-1948. 5, 5-91.

Arrhenius, G., De, B.R. and Alfven, H. (1974) Origin of the ocean, in E.D. Goldberg (ed.), The Sea Volume 5, John Wiley and Sons, New York, pp. 839-861.

Arrhenius, G. (1986) Dysoxic environments as models for primordial mineralisation, in A.G. Cairns-Smith, A.G. and H. Hartman, H. (eds.), Clay Minerals and the Origin of Life, Cambridge University Press, Cambridge, UK, pp. 97-104

Bada, J.L. and Miller, S.L. (1968) Ammonium ion concentration in the primitive ocean. Science 159, 423-425.

Baker, E.T. and Massoth, G.J. (1985) Hydrothermal particle plumes over the southern Juan de Fuca Ridge. Nature 316, 342-344.

Barak, I. and Bar-Nun, A. (1975) The mechanisms of amino acids synthesis by high temperature shock-waves. Origins of Life 6, 483-506.

Bar-Nun, A. and Shaviv, A. (1975) Dynamics of the chemical evolution of Earth's primitive atmosphere. Icarus 24, 197-210.

Bar-Nun, A. and Chang, S. (1983) Photochemical reactions of water and carbon monoxide in Earth's primitive atmosphere. J. Geophys. Res. 88, 6662-6672.

Barley, M.E., Dunlop, J.S.R., Glover, J.E., and Groves, D.I. (1979) Sedimentary evidence for an Archaean shallow-water volcanic-sedimentary facies, eastern Pilbara Block, Western Australia. Earth Planet. Sci. Lett. 43 74-84.

Benlow, A. and Meadows, A.J. (1977) The formation of the atmospheres of the terrestrial planets by impact. Astrophys. Space Sci. 46, 293-300.

Berg, W.W.Jr. and Winchester, J.W. (1978) Aerosol chemistry of the marine atmosphere, in J.P.

Riley and R. Chester. (eds.), Chemical Oceanography 7, Academic Press, London, pp. 173-231.

Bloechl, E., Keller, M., Waecterhaeuser, G. and Stetter, K.O. (1992) Reactions depending on iron sulfide and linking geochemistry with biochemistry. Proc. Nat. Acad. Sci. U.S.A. 89, 8117-8120.

Boettcher, A.L., Mysen, B.O. and Modreski, P.J. (1975) Melting in the mantle: phase relationships in natural and synthetic peridotite-H_2O and peridotite-H_2O-CO_2 systems at high pressures, in L.H. Ahrens, J.B. Dawson, A.R. Duncan, and A.J. Erlank (eds.), Physics and Chemistry of the Earth, Pergamon Press, Oxford, pp. 855-867.

Borowska, Z.K. and Mauzerall, D.C. (1987) Efficient near ultraviolet light induced formation of hydrogen by ferrous hydroxide. Origins of Life 17, 251-259.

Borowska, Z.K. and Mauzerall, D.C. (1988) Photoreduction of carbon dioxide by aqueous ferrous ion: An alternative to the strongly reducing atmosphere for the chemical origin of life. Proc. Nat. Acad. Sci. USA 85, 6577-6580.

Borowska, Z.K. and Mauzerall, D.C (1991) Corrections and retractions. Proc. Nat. Acad. Sci U.S.A. 88 4564.

Braterman, P.S., Cairns-Smith, A.G. and Sloper, R.W. (1983) Photo-oxidation of hydrated Fe^{2+} - significance for banded iron formations. Nature 303,163-164.

Buick, R. and Dunlop, J.S.R. (1990) Evaporitic sediments of early Archaean age from the Warrawoona Group, North Pole, Western Australia. Sedimentology 37, 247-277.

Canuto, V.M., Levine, J.S., Augustsson, T.R. and Imhoff, C.L. (1982) UV radiation from the young Sun and oxygen and ozone levels in the prebiological palaeoatmosphere. Nature 296, 816-820.

Chameides, W.L. and Walker, J.C.G. (1981) Rates of fixation by lightning of carbon nitrogen in possible primitive atmospheres. Origin of Life 11, 291-302.

Chameides, W.L. and Davis, D.D. (1983) Aqueous-phase source of formic acid in clouds. Nature 304, 427-42.

Chang, S. (1979) Comets: cosmic connections with carbonaceous meteorites, interstellar molecules and the origin of life, in M. Neugebauer, D.K. Yeomans, and J.C. Brandt (eds.), in Space Missions to Comets, NASA Conference Publication 2089, NASA Scientific and Technical Information Division, Washington, DC, pp. 59-111.

Chang, S. (1988) Planetary environments and the conditions of life. Phil. Trans. Roy. Soc. (London A) 325, 601-610.

Chang, S., DesMarais, D., Mack, R., Miller, S.L. and Strathearn, G.E. (1983) Prebiotic organic synthesis and the origin of life, in J.W. Schopf (ed.), Earth's Earliest Biosphere, Princeton University Press, Princeton, New Jersey, pp. 53-92.

springs on the East Pacific Rise and their effluent dispersal. Nature 297, 187-191.

Egami, F. (1974) Minor elements and evolution. J. Mol. Evol. 4, 113-120.

Feely, R.A., Massoth, G.J., Baker, E.T., Cowen, J.P., Lamb, M.F. and Krogslund, K.A. (1990 The effect of hydrothermal processes on midwater phosphorus distributions in the northeast Pacific. Earth Planet. Sci. Lett. 96, 305-318.

Fegley, B., Jr., Prinn, R.G., Hartman, H. and Watkins, G. (1986) Chemical effects of large impacts on the Earth's primitive atmosphere. Nature, 319, 305-308.

Ferris, J.P. and Nicodem, D.E. (1972) Ammonia photolysis and the role of ammonia in chemical evolution. Nature 238, 268-269.

Flegmann, A.W. and Tattersall, R. (1979) Energetics of peptide bond formation at elevated temperatures. J. Mole. Evol. 12, 349-355.

Flint, E.B. and Suslick, K.S. (1991) The temperature of cavitation. Science 253, 1397-1399.

Folsome, C.E., Brittain, A., Smith, A., and Chang, S. (1981) Hydrazines and carbohydrazides produced from oxidized carbon in Earth's primitive environment. Nature 294, 64-65.

Garrett, W.D. (1967) The organic chemical composition of the ocean surface. Deep Sea Res. 14, 221-227.

Getoff, N.G., Scholes, G. and Weiss, J. (1960) Reduction of carbon dioxide in aqueous solution under the influence of radiation. Tetrahedron Lett. 1960, 17-23.

Groves, D.I., Dunlop, J.S.R., and Buick, R. (1981) An early habitat of life. Scientific Amer. 245, 64-73.

Halmann, M. and Aurian-Blajeni, B. (1981) Photoassisted carbon dioxide reduction and formation of two- and three-carbon compounds, in Y. Wolman (ed.) Origins of Life, D. Reidel Publishing Company, Dordrecht, pp. 143-150.

Hartman, H. (1975) Speculations on the origin and evolution of metabolism. J. Mol. Evol. 4, 359-370.

Hartmann, W.K. and Davis, D.R. (1975) Satellite sized planetesimals and lunar origin. Icarus 24, 504-515.

Hedges, J.I. and Hare, P.E. (1987) Amino acid adsorption by clay minerals in distilled water. Geochim. Cosmochim. Acta 51, 255-259.

Henderson-Sellers, A. and Henderson-Sellers, B. (1988) Equable climate in the early Archaean. Nature 336, 117-118.

springs on the East Pacific Rise and their effluent dispersal. Nature 297, 187-191.

Egami, F. (1974) Minor elements and evolution. J. Mol. Evol. 4, 113-120.

Feely, R.A., Massoth, G.J., Baker, E.T., Cowen, J.P., Lamb, M.F. and Krogslund, K.A. (1990 The effect of hydrothermal processes on midwater phosphorus distributions in the northeast Pacific. Earth Planet. Sci. Lett. 96, 305-318.

Fegley, B., Jr., Prinn, R.G., Hartman, H. and Watkins, G. (1986) Chemical effects of large impacts on the Earth's primitive atmosphere. Nature, 319, 305-308.

Ferris, J.P. and Nicodem, D.E. (1972) Ammonia photolysis and the role of ammonia in chemical evolution. Nature 238, 268-269.

Flegmann, A.W. and Tattersall, R. (1979) Energetics of peptide bond formation at elevated temperatures. J. Mole. Evol. 12, 349-355.

Flint, E.B. and Suslick, K.S. (1991) The temperature of cavitation. Science 253, 1397-1399.

Folsome, C.E., Brittain, A., Smith, A., and Chang, S. (1981) Hydrazines and carbohydrazides produced from oxidized carbon in Earth's primitive environment. Nature 294, 64-65.

Garrett, W.D. (1967) The organic chemical composition of the ocean surface. Deep Sea Res. 14, 221-227.

Getoff, N.G., Scholes, G. and Weiss, J. (1960) Reduction of carbon dioxide in aqueous solution under the influence of radiation. Tetrahedron Lett. 1960, 17-23.

Groves, D.I., Dunlop, J.S.R., and Buick, R. (1981) An early habitat of life. Scientific Amer. 245, 64-73.

Halmann, M. and Aurian-Blajeni, B. (1981) Photoassisted carbon dioxide reduction and formation of two- and three-carbon compounds, in Y. Wolman (ed.) Origins of Life, D. Reidel Publishing Company, Dordrecht, pp. 143-150.

Hartman, H. (1975) Speculations on the origin and evolution of metabolism. J. Mol. Evol. 4, 359-370.

Hartmann, W.K. and Davis, D.R. (1975) Satellite sized planetesimals and lunar origin. Icarus 24, 504-515.

Hedges, J.I. and Hare, P.E. (1987) Amino acid adsorption by clay minerals in distilled water. Geochim. Cosmochim. Acta 51, 255-259.

Henderson-Sellers, A. and Henderson-Sellers, B. (1988) Equable climate in the early Archaean. Nature 336, 117-118.

Hennet, J-C., Holm, N.G. and Engel, M.H. (1992) Abiotic synthesis of amino acids under hydrothermal conditions and the origin of a life: a perpetual phenomenon? Naturwiss. 79, 361-365.

Holland, H. (1973) The oceans: apossible source of iron in iron-formations. Econ. Geol. 68, 545-557.

Holland, H.D. (1984) The Chemical Evolution of the Atmosphere and Oceans. Princeton Unversity Press, Princeton.

Holm, N. (1985) New evidence for a tubular structure of b-iron(III) oxide hydroxide - akaganeite. Origins of Life 15, 131-139.

Ingmanson, D.E. and Dowler, M.J. (1977) Chemical evolution and the evolution of the Earth's crust. Origins of Life 8, 221-224.

Javoy, M., Pineau, F. and Allegre, C.J. (1982) Carbon geodynamic cycle. Nature 300, 171-173,

Kasting, J.F., Pollack, J.B. and Crisp, D. (1984) Effects of high CO_2 levels on surface temperature and atmospheric oxidation state of the early Earth. J Atmos. Chem. 1, 403-428.

Kasting, J.F., Zahnle, K.F., Pinto, J.P. and Young, A.Y. (1989) Sulfur, ultraviolet radiation, and the early evolution of life. Origins of Life Evol. Bios. 19, 95-108.

Kasting, J.F. (1990) Bolide impacts and the oxidation state of carbon in the Earth's early atmosphere. Origins of Life Evol. Bios. 20, 199-231.

Kastings, J.F., Pollack, J.B., and Crisp, D. (1984) Effects of high CO_2 levels on surface temperatures and atmospheric oxidation state on the early Earth. J. Atmos. Chem. 1, 403-428.

Katsura, T. and Ito, E. (1990) Melting and subsolidus phase relations in the $MgSiO_3$-$MgCO_3$ system at high pressures: implications to evolution of the Earth's atmosphere. Earth Planet. Sci. Lett. 99, 110-117.

Kempe, S., and Degens, E.T. (1985) An early soda ocean? Chem. Geol. 53, 95-108.

Koch, A.L. (1985) Primeval cells: Possible energy-generating and cell-division mechanisms. J. Mole. Evol. 21, 270-277.

Lahav, N. (1991) Prebiotic co-evolution of self-replication and translation or RNA world? J. Theor. Biol. 151, 531-539.

Lahav, N. and Chang, S (1976) The possible role of solid surface area in condensation reactions during chemical evolution: reevaluation. J. Mol. Evol. 8, 357-380.

Lange, M.A. and Ahrens, T.J. (1982) The evolution of an impact-generated atmosphere. Icarus 51, 96-120.

Lerman, L. (1986) Potential role of bubbles and droplets in primordial and planetary chemistry: exploration of the liquid-gas interface as a reaction zone for condensation reactions. Origins of Life 16, 201-202.

Liss, P.S. (1975) Chemistry of the sea surface microlayer, in J.P. Riley and G. Skirrow (eds.), Chemical Oceanography, Academic Press, London, pp. 193-243.

MacIntyre, F. (1974) The top millimeter of the ocean. Scientific Amer. 230, 62-77.

MacIntyre, F. (1974) Chemical fractionation and sea-surface microlayer processes, in E.D. Goldberg (ed.), The Sea 5, John Wiley and Sons, New York, pp. 245-299.

MacIntyre, F. and Winchester, J.W. (1969) Phosphate ion enrichment in drops from breaking bubbles. J. Phys. Chem. 73, 2163-2169.

Mackinnon, I.D.R. and Rietmeijer, F.J.M. (1987) Mineralogy of chondritic interplanetary dust particles. Rev. Geophys 25, 1527-1553.

Maher, K.A. and Stevenson, J.D. (1988) Impact frustration of the origin of life. Nature 331, 612-614.

Mathez, E.A. (1984) Influence of degassing on oxidation states of basaltic magmas. Nature 310, 371-375.

Matsui, T. and Abe, Y. (1986) Evolution of an impact-induced atmosphere and magma ocean on the accreting Earth. Nature 319, 303-308.

Miller, S.L. (1953) A production of amino acids under possible primitive Earth conditions. Science 117, 527-528.

Miller, S.L. (1957) The mechanism of synthesis of amino acids by electric discharges. Biochim. Biophys. Acta 23, 480-489.

Miller, S.L. and Van Trump, J.E. (1981) The Strecker synthesis in the primitive ocean, in Y. Wolman (ed.) Origins of Life, D. Reidel Publishing Company, Dordrecht, pp. 135-141.

Miller, S.L. and Bada, J.L. (1988) Submarine hot springs and the origin of life. Nature 334, 609-611.

Morowitz, H.J. (1992) Beginnings of Cellular Life, Yale University Press, New Haven.

Morowitz, H.J., Heinz, B., and Deamer, D.W. (1988) The chemical logic of a minimum protocell. Origins of Life Evol. Bios. 18, 281-287.

Mukhin, L.M., Gerasimov, M.V. and Safonova, E.N. (1989) Origin of precursors of organic molecules during evaporation of meteorites and mafic terrestrial rocks. Nature 340, 46-48.

Nissenbaum, A., Kenyon, D.H. and Oro, J. (1975) On the possible role of organic melanoidin polymers as matrices for prebiotic activity. J. Mole. Evol. 6, 253-270.

Oberbeck, V.R. and Fogelman, G. (1989) Impacts and the origin of life. Nature 339, 434,

Oberbeck, V.R., McKay, C.P., Scattergood, T.W., Carle, G.C., and Valentin, J.R. (1989) The role of cometary particle coalescence in chemical evolution. Origins of Life and Evol. Biosph. 19, 39-56.

Oberbeck, V.R. and Fogelman, G. (1990) Impact constraints on the environment for chemical evolution and the continuity of life. Origins of Life Evol. Bios. 20, 181-195.

Oberbeck, V.R., Marshall, J. and Shen, T. (1991) Prebiotic chemistry in clouds. J. Mole. Evol. 32, 296-303.

Oberbeck, V.R. and Aggarwal, H. (1993) Comet impacts and chemical evolution on the bombarded Earth. Origins of Life Evol. Bios. 21, 317-338.

Oro, J. (1961) Comets and the formation of biochemical compounds on the primitive Earth. Nature 190, 389-390.

Owen, T., Cess, R.D., and Ramanathan, V. (1979) Enhanced CO_2 greenhouse to compensate for reduced solar luminosity on early Earth. Nature 277, 640-642.

Pinto, J.P., Gladstone, G.R., and Yung, Y.L. (1980) Photochemical reduction of carbon dioxide to formaldehyde in the Earth's primitive atmosphere. Science 210, 183-185.

Pohorille, A. and Benjamin, I. (1991) Molecular dynamics of phenol at the liquid-vapor interface of water. J. Chem. Phys. 94, 5599-5605.

Prinn, R.G. and Fegley, B., Jr. (1987) Bolide impacts, acid rain, and biospheric traumas at the Cretaceous-Tertiary boundry. Earth Planet. Sci. Lett. 83, 1-15.

Rubey, W.W. (1951) Geologic history of sea water: an attempt to state the problem, Geol. Soc. Am. Bull. 62, 1111-1147.

Sackett, W. (1978) Suspended matter in sea-water, in J.P. Riley and R. Chester (ed.), Chemical Oceanography 7, Academic Press, London, pp. 127-171.

Sagan, C. and Khare, B.N. (1971) Long-wavelength ultraviolet photoproduction of amino acids on the primitive Earth. Science 173, 417-420.

Sagan, C. and Mullen, G. (1972) Earth and Mars: Evolution of atmospheres and surface temperatures. Science 177, 52-56.

Sanchez, R.A., Ferris, J.P. and Orgel, L.E. (1966) Conditions for purine synthesis: did prebiotic synthesis occur at low temperatures? Science 153, 72-73.

Scattergood, T.W., McKay, C.P., Borucki, W.J., Giver, L.P., and Van Ghyseghem, H. (1989) Production of organic compounds in plasmas: a comparison among electric sparks, laser-induced plasmas, and UV light. Icarus 81, 413-428.

Schrauzer, G.N., Guth, T.D., and Palmer, M.R. (1979) Nitrogen reducing solar cells, in R.R. Hautala, R.B. King, and C. Kutal (eds.) Solar Energy: Chemical Conversion and Storage, Humana Press, Clifton, New Jersey, pp. 261-269.

Schrauzer, G.N., Strampach, N., Hui, L.N. Palmer, M.R. and Salehi, J.(1980) Nitrogen photoreduction on desert sands under sterile conditions. Proc. Nat. Acad. Sci. U.S.A. 80, 3873-3876.

Shock, E.L. (1990) Geochemical constraints on the origin of organic compounds in hydrothermal systems. Origins of Life 20, 331-367.

Shock, E.L. (1990a) Do amino acids equilibrate in hydrothermal fluids?. Geochim. Cosmochim. Acta 54, 1185-1189.

Shock, E.L. (1992) Chemical environments of submarine hydrothermal systems. Origins of Life Evol. Bios. 22, 67-108.

Siskin, M. and Katritzky, A.R. (1991) Reactivity of organic compounds in hot water: Geochemical and technological implications. Science 245, 231-237.

Sleep, N.H., Zahnle, K.J., Kasting, J.F. and Morowitz, H.J. (1989) Annihilation of ecosystems by large asteroid impacts on the early Earth. Nature 342, 139-142.

Snetsinger, K.G., Ferry, G.V., Russell, P.B., Pueschel, R.F. , Oberbeck, V.R., Hayes, D.M. and Fong, W. (1987) Effects of El Chichon on stratospheric aerosols late 1982 to early 1984. J. Geophys. Res. 92, 14761-14771.

Stevenson, D.J. (1983) The nature of the Earth prior to the oldest known rock record: the Hadean Earth, in J.W. Schopf (ed.) Earth's Earliest Biosphere, Princeton University Press, Princeton, New Jersey, pp. 14-29.

Stillwell, W. (1980) Facilitated diffusion as a method for selective accumulation of materials from the primordial oceans by a lipid-vesicle protocell. Origins of Life 10, 277-292.

Stribling, R. and Miller, S.L. (1987) Energy yields for hydrogen cyanide and formaldehyde syntheses: the HCN and amino acid concentrations in the primitive ocean. Origins of Life17, 261-273.

Suslick, K.S. (1989) The chemical effects of ultrasound. Scientific Ameri. 260, 80-86.

Thomsen, L. (1980) ^{129}Xe on the outgassing of the atmosphere. J. Geophys. Res. 85, 4374-4378.

Tingle, T.N., Green, H.W. and Finnerty, A.A. (1988) Experiments and observations bearing on the solubility and diffusivity of carbon in olivine. J. Geophys. Res. 93, 15289-15304.

Tseng, R.-S., Viechnicki, J.T., Slop, R.A. and Brown, J.W. (1992) Sea-to-air transfer of surface-active organic compounds by bursting bubbles. J. Geophys. Res. 97, 5201-5206.

Veizer, J. (1983) Geologic evolution of the Archean-Early Proterozoic Earth, in J.W. Schopf (ed.) Earth's Earliest Biosphere, Princeton University Press, Princeton, New Jersey, pp. 240-259.

Virgo, D., Luth, R.W., Moats, M.A. and Ulmer, G.C. (1988) Constraints on the oxidation state of the mantle: an electrochemical and ^{57}Fe Mossbauer study of mantle-derived ilmenites. Geochim. Cosmochim. Acta 52, 1781-1794.

Von Damm, K.L., Edmond, J.M., Grant, B. and Measures, C.I. (1985) Chemistry of submarine hydrothermal solutions at 21° N, East Pacific Rise. Geochim. Cosmochim. Acta 49, 2197-2220.

Wachterhauser, G. (1988) Before enzymes and templates: theory of surface metabolism. Microbio. Rev. 52, 452-484.

Wachterhauser, G. (1988a) Pyrite formation, the first energy source for life: a hypothesis. System. Applied Microbio. 10, 207-210.

Wachterhauser, G. (1990) Evolution of the first metabolic cycles. Proc. Nat. Acad. Sci. USA 87, 200-204.

Walker, J.C.G. (1977) Evolution of the Atmosphere, Macmillan, London, p. 205.

Walker, J.C.G. (1983) Carbon geodynamic cycle. Nature 303, 730-731.

Walker, J.C.G., Klein, C., Schidlowske, M., Schopf, J.W., Stevenson, D.J. and Walter, M.R. (1983) Environmental evolution of the Archean-Early Proterozoic Earth, in J.W. Schopf (ed.) Earth's Earliest Biosphere, Princeton University Press, Princeton, pp. 260-289.

Walker, J.C.G. (1985) Carbon dioxide on the early Earth. Origins of Life 16, 117-127.

Walker, J.C.G. and Brimblecombe, P. (1985) Iron and sulfur in the pre-biologic ocean. Precambrian Res. 28, 205-222.

Wallace, G.T.Jr. and Duce, R.A. (1978) Open-ocean transport of particulate trace metals by bubbles. Deep Sea Res. 25, 827-835.

Weber, A.L. (1987) The triose model: glyceraldehyde as a source of energy and monomers for prebiotic condensation reactions. Origins of Life 17, 107-119.

Welhan, J.(1988) Origins of methane in hydrothermal systems. Chem. Geol. 71, 183-198.

Wen, J.-S., Pinto, J.P. and Yung, Y.L. (1989) Photochemistry of CO and H2O: analysis of laboratory experiments and applications to the prebiotic Earth's atmosphere. J. Geophys. R. 94, 14957-14970.

Wetherill, G.W. (1980) Formation of the terrestrial planets. Ann. Rev. Astrophy. 18, 77-113.

Whitesides, G.M., Mathias, J.P. and Seto, C.T. (1991) Molecular self-assembly and nanochemistry: a chemical strategy for the synthesis of nanostructures. Science 254, 1312-1319.

Wilson, M.A. and Pohorille, A.(1991) Interaction of monovalent ions with the water liquid-vapor interface: a molecular dynamics study. J. Chem. Phys. 95, 6005-6013.

Woese, C.R. (1979) A proposal concerning the origin of life on the planet Earth. J. Mol. Evol. 13, 95-101.

Wolery, T.J. and Sleep, N.H. (1976) Hydrothermal circulation and geochemical flux at mid-ocean ridges. J. Geol. 84, 249-275.

Wood, BJ. and Virgo, D. (1989) Upper mantle oxidation state: ferric iron contents of lherzolite spinels by ^{57}Fe Mossbauer spectroscopy and resultant oxygen fugacities. Geochim. Cosmochim. Acta 53, 1277-1291.

Wu, J. (1981) Evidence of sea spray produced by bursting bubbles. Science 212, 324-326.

Yamagata, H. and Mohri, T. (1982) Formation of cyanate and carbamyl phosphate by electric discharges of model primitive gas. Origins of Life 12, 41-44.

Yamagata, H., Watanabe, M.S. and Namba, T. (1991) Volcanic production of polyphosphates and its relevance to prebiotic evolution. Nature 352, 516-519,

Yanagawa, H., Ogawa, Y. Kojima, K. and Ito, M. (1988) Construction of protocellular structures under simulated primitive Earth conditions. Origins of Life Evol. Bios. 18, 179-207.

Yanagawa, H. and Kobayashi, K. (1992) An experimental approach to chemical evolution in submarine hydrothermal systems. Origins of Life Evol. Bios. 22, 147-160.

Zahnle, K.J. (1986) Photochemistry of methane and the formation of hydrocyanic acid (HCN) in the Earth's early atmosphere. J. Geophys. Res. 91, 2819-2834.

Zahnle, K.J.(1990) Atmospheric chemistry by large impacts, in V. Sharpton and P. Ward (eds.), Global Catastrophes in Earth History, Geological Society of America, Special Paper 247, pp. 271-288.

Zahnle, K.J., Kasting, J.F. and Pollack, J.B. (1988) Evolution of a steam atmosphere during Earth's accretion. Icarus 74, 62-97.

Zhao, M. and Bada, J.L.(1989) Extraterrestrial amino acids in Cretaceous/Tertiary boundary sediments at Stevns Klint, Denmark. Nature 339, 463-465.

Prebiotic Synthesis on Minerals: RNA Oligomer Formation

James P. Ferris
Department of Chemistry
Rensselaer Polytechnic Institute
Troy, NY 12180-3590

ABSTRACT: It is proposed that the mineral or metal ion catalyzed formation of RNA oligomers may have proceeded on the primitive earth. A variety of minerals including iron oxide hydroxides, apatite and montmorillonite bind nucleotides but so far only montmorillonite has been observed to catalyze the formation of RNA oligomers. The polymerization of aminoacyl adenylates is also observed to take place on montmorillonite with the formation of polypeptides. These findings are consistent with the hypothesis that the prebiotic formation of biological polymers took place on mineral surfaces. These polymeric species may have been the precursors to the RNA World.

1. INTRODUCTION

It has been proposed that the RNA World was one of the stages in the evolution of life on Earth. It is generally believed that the RNA oligomers, which play a central role in the proposed RNA World, are too complex to have arisen directly from the simple molecules formed by prebiotic processes but rather they evolved after life originated (see Schwartz, this volume). The possible direct formation of the these oligomers directly from simple monomers in mineral catalyzed reactions will be discussed in this chapter. The interaction of biological molecules with minerals will be described

J. M. Greenberg et al. (eds.), The Chemistry of Life's Origins, 301–322.

and examples of mineral catalyzed polymerization reactions will be described.

2. Why Minerals?

Clays and other minerals were first proposed by Bernal (1949) as the surfaces on which prebiotic reactions took place. His proposals, which were first presented in a lecture in 1947, six years before the Miller-Urey experiment, are still valid today. He suggested that minerals were necessary to:

1. Concentrate the organics present in a dilute ocean by "...adsorption on fine clay deposits."
2. Protect these organics from destruction by ultraviolet light.
3. Catalyze the polymerization of adsorbed organics.
4. Induce the chirality observed in contemporary biomolecules. He suggested that preferential adsorption and reaction of one enantiomer on an optically active mineral such as quartz may have led to the polymerization of monomers of the same chirality.

Experiments performed over the past forty years directed to the prebiotic synthesis of biopolymers in aqueous solution in the absence of catalysts have, in general, been unsuccessful. Those that have achieved a modest degree of success have been where solutions of monomers were evaporated to near dryness to achieve a sufficiently high concentration so reaction occurs. These experiments did show that the requisite polymers were formed on the primitive earth only when reactants were concentrated. The low yields in many of these reactions suggest that appropriate catalysts were also required to facilitate the desired condensation reactions.

Insight into the requirements for polymerization on the primitive earth can be obtained by considering similar processes in contemporary biological systems. Today biological polymerizations proceed in an aqueous environment on the surfaces of structurally defined enzymes, ribozymes and polynucleotides. Polymeric enzymes and ribozymes orient and catalyze the reaction of the adsorbed monomers and then release the polymers from their surfaces. The same scenario probably took place on the primitive earth with the exception that minerals served to adsorb and orient the monomers and catalyze the polymerization reactions. In general, the specificity and catalysis due to minerals is probably much lower than that of the highly evolved contemporary biochemical systems but was probably adequate to initiate the processes leading to the origin of life on earth.

The need for catalysis in prebiotic reactions can be illustrated by consideration of the simple condensation of a nucleoside and a nucleotide to a 3',5'-linked dinucleotide. Uncatalyzed chemical synthesis requires the extensive use of blocking groups to form the dinucleotide ApA containing a 3',5'-phosphodiester bond (Fig. 1). Six precisely located blocking groups are required. If the condensation reaction is attempted without the use of blocking groups, a mixture of products will be obtained and the 3',5'-linked dimer will be only one of many reaction products. Since it is difficult to envision a prebiotic scenario where specific blocking groups were put in place and then removed, it is more likely that unblocked monomers were polymerized while complexed to metal ions (Sawai and Orgel, 1975; Sleeper and Orgel, 1979; Sawai, Kuroda and Hojo, 1989) or on mineral surfaces. The regiochemistry may have been controlled by specific orientation of the organic compound on the mineral and was followed by catalysis of the condensation reaction by the mineral.

2.1. Properties of Some Selected Minerals

The structural and surface properties of three minerals used in prebiotic experiments will be outlined to illustrate the possible role of minerals in adsorption and catalysis of chemical reactions.

2.2. Iron Oxide Hydroxides (FeOOH)

Goethite is an example of an iron oxide hydroxide which may have been prevalent on the primitive earth. When Fe^{2+} is oxidized to Fe^{3+} it forms an amorphous iron hydroxide precipitate which changes to goethite, one of the most common forms of FeOOH (Murray, 1979). The bulk structure of goethite consists of a central Fe^{3+} with octahedrallyoordinated hydroxyl and oxygen groups. The oxygens are coordinated to two Fe^{3+} groups thereby linking the octahedrons

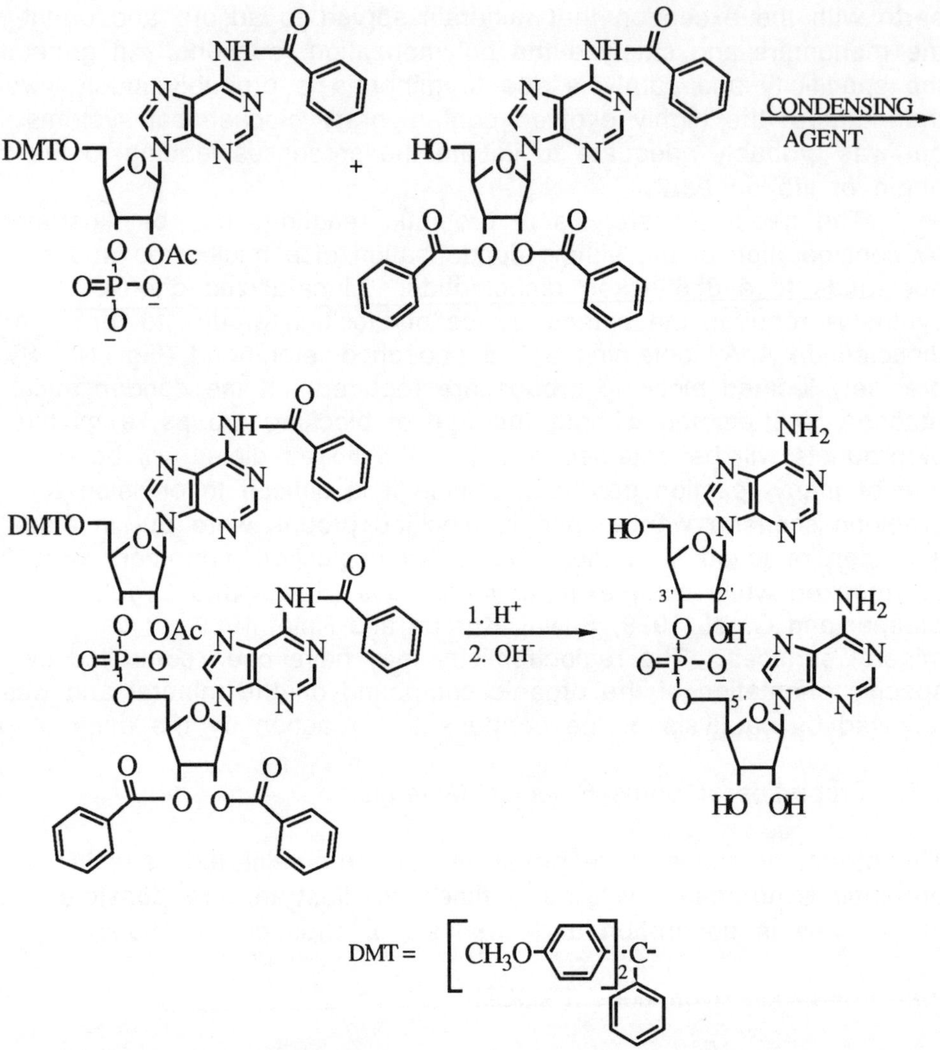

Fig. 1 CHEMICAL SYNTHESIS OF 3',5'-ApA. Six blocking groups are needed to achieve the synthesis of exclusively 3',5'-linked ApA. Two hydrolysis steps are required to remove these groups. Additional steps are needed to place the phosphate group on the 5'-terminus to give 3',5'-pApA, the dimer of pA. Our goal is the one step synthesis of 3',5'-pApA and higher oligomers without using blocking groups since complex blocking groups are not likely to have been part of prebiotic processes.

into a polymeric matrix. Hydrogen bonds between the hydroxyl hydrogens and the oxide bridges adds to the stability of the crystalline matrix. The Fe^{3+} on the surface of this matrix is also octahedrally coordinated but some of the oxygens and hydroxyls are not linked to the bulk structure. These groups are subject to acid-base reactions depending on the pH of the medium (Herbillon, 1988). They can be protonated in acid or the protons can be removed by base (Fig. 2). The protons are removed in a stepwise manner so that it is possible to have both positive and negative charges on the surface of the mineral.

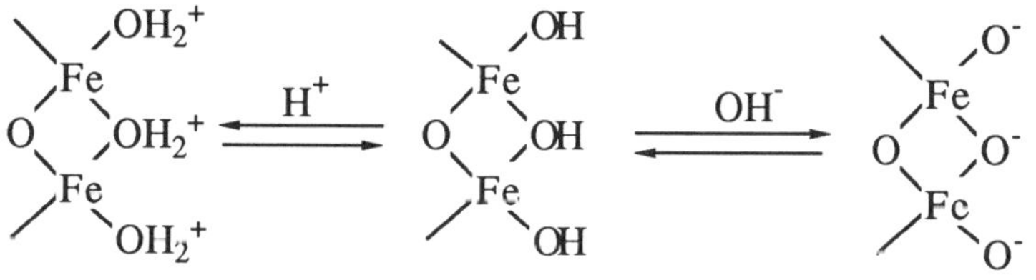

Fig. 2 THE SURFACE CHARGE ON GOETHITE IS pH DEPENDENT.
(Adapted from Newman,1987)

The positively charged hydroxyl groups on the surface of goethite are displaced by oxyanions like orthophosphate (HPO_4^{2-}) so the mineral will bind inorganic and organic phosphates (Herbillion, 1988) (Fig. 3). The capacity for oxyanion binding is dependent on the pH. The positive charge on the surface decreases in more basic solution and the adsorption of phosphate reaches a minimum when

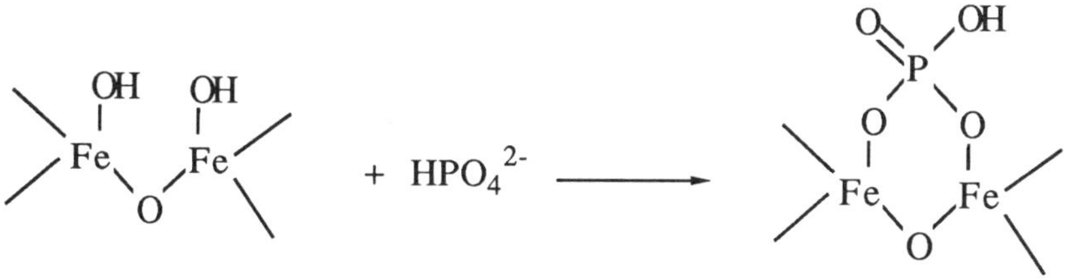

Fig. 3 THE REACTION OF GOETHITE WITH PHOSPHATE
(Adapted from Newman, 1987)

there is no net surface charge (the isoelectric point).

2.3. Hydroxyapatite

Hydroxyapatite, $Ca_5(OH)(PO_4)_3$, is one representative of a class of insoluble calcium phosphates. Those Ca^{2+} ions on the surface of the mineral can bind to inorganic or organic phosphates (Wells, 1962). This binding is the basis for the use of hydroxyapatite as a chromatographic support for the separation of nucleotides (Bernardi, 1971). This phosphate binding suggests that hydroxyapatite may have served to concentrate the organophosphate derivatives from aqueous solution on the primitive earth.

2.4. Montmorillonite

Montmorillonite is a dioctahedral smectite mineral with the ideal empirical formula $Si_4Al_2O_{10}(OH)_2$ (Fig. 4) (Van Olphen, 1977; Newman, 1987). However, all naturally occurring montmorillonites have Mg^{2+} substituted in some of the aluminum sites and occasionally some Al^{3+} substituted in the silicon sites. In addition, varying amounts of Fe^{3+} and Fe^{2+} are also incorporated into the structure in place of aluminum. These substitutions lead to the following formula for one of the montmorillonites used in our research; $(Si_{3.89}, Al_{0.12})$ $(Al_{1.40}, Fe_{0.32}, Mg_{0.31})$ $O_{10}(OH)_2$. Substitution of Mg^{2+} for Al^{3+} and Al^{3+} for Si^{4+} results in an excess negative charge on the montmorillonite lattice which is balanced by associated cations such as Na^+, K^+, Ca^{2+} and Mg^{2+}

Montmorillonite has a layered structure composed of two tetrahedral silicate sheets bound to a central octahedral alumina sheet. It is common for the layers to stack with exchangeable cations and water molecules occupying the space between the layers (Fig. 4). Positively charged organic compounds can replace the exchangeable inorganic cations. Alternatively, negatively charged organics, which coordinate to the exchangeable metal cations, can also bind in the interstices of the layers.

The layer edges carry a positive charge due to the binding of water molecules to the edge Al^{3+} groups not coordinated to the bulk montmorillonite structure (Fig. 5). Oxyanions such as phosphate and polyphosphate bind to the edges by displacement of the water bound

COMPOSITION

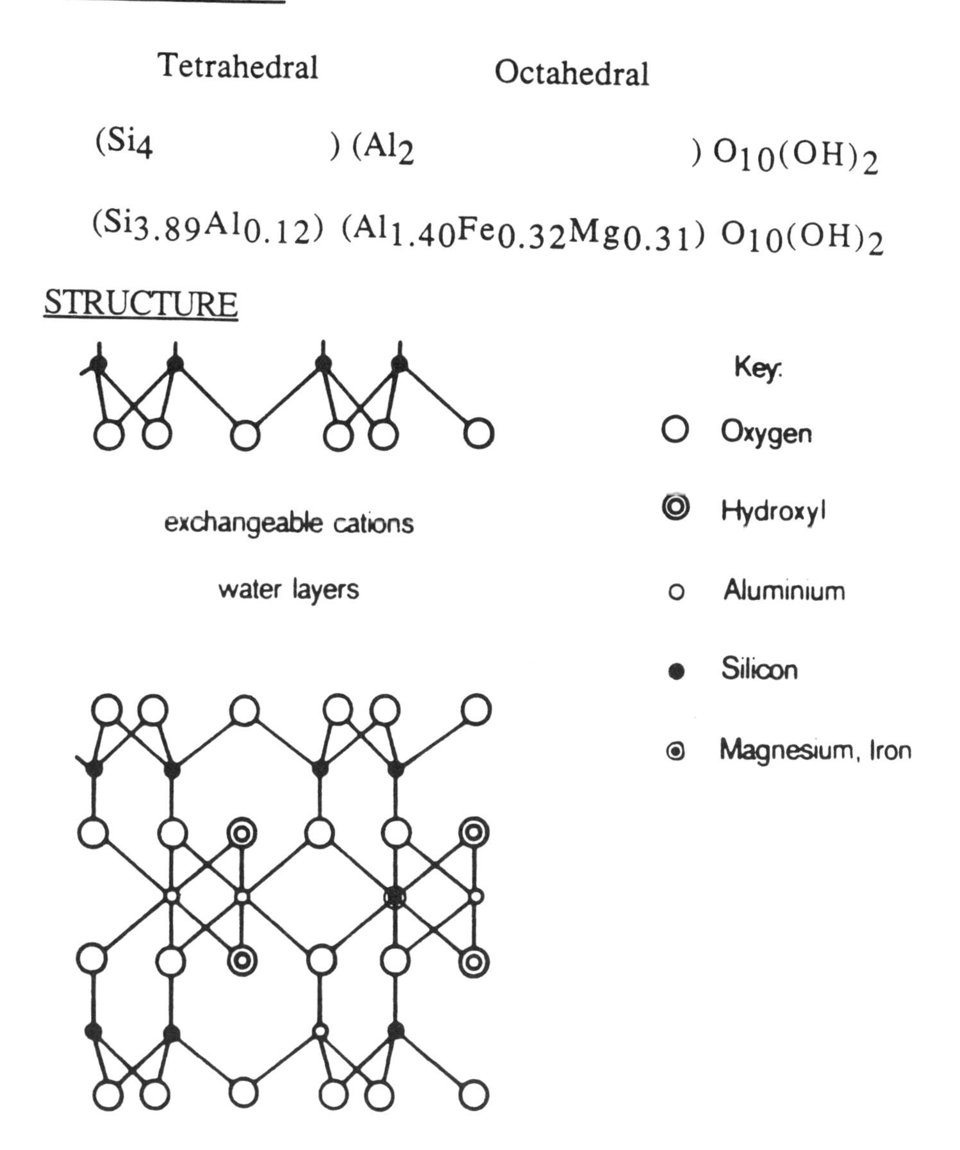

Fig. 4. THE COMPOSITION AND STRUCTURE OF MONTMORILLONITE
(adapted from Theng, 1974

to the Al^{3+} (Van Olphen, 1977).

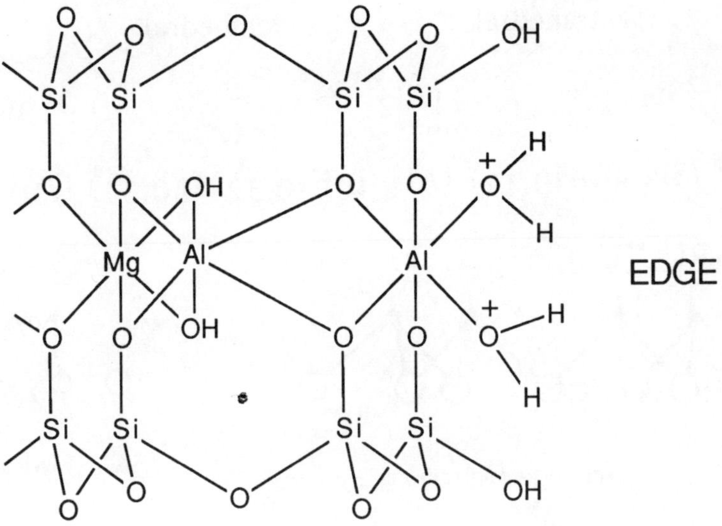

FIG. 5 A HYDRATED Al^{3+} ON THE EDGE OF A MONTMORILLONITE LAYER

Water coordinated at the edges contributes to the acidity of the montmorillonite. Solvated cations associated with the negatively charged layer surfaces are also a source of acidity. In both instances

$$M^{m+}(H_2O) = M^{m+}(H_2O)_{n-1}(OH) + H^+$$

the acidity is a result of the dissociation of the water molecules strongly polarized by the positively charged metal ions or lattice bound Al^{3+}.

3. Polypeptide Formation From Aminoacyl Adenylates on Montmorillonite

One of the earliest, and perhaps most controversial, experiments performed on a mineral surface was the reported formation of polypeptides by the reaction of aminoacyl adenylates on montmorillonite (Paecht-Horowitz et al., 1970) (Fig. 6). These workers reported that peptides were obtained in the absence of montmorillonite (Lewinsohn et al., 1967) but that substantially

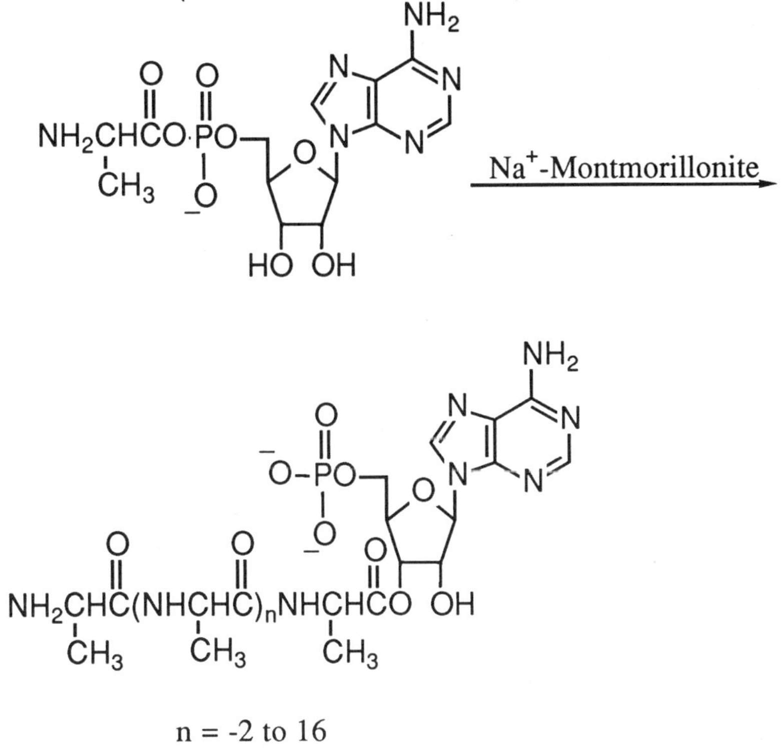

Na^+-Montmorillonite

n = -2 to 16

Figure 6. POLYPEPTIDE FORMATION ON MONTMORILLONITE
The formation of polypeptides by the condensation of aminoacyl adenylates on montmorillonite in aqueous solution (Paecht-Horowitz and Eirich, 1988).

Higher yields and longer chain lengths were observed in its presence. Later it was reported that the zeolite, Decalso F, catalyzed the formation of formation of aminoacyl adenylates from ATP and amino acids (Paecht-Horowitz and Katchalsky, 1973). It was not possible to duplicate this aminoacyl adenylate synthesis (Warden et al., 1974) and apparently, Paecht-Horowitz was also unable to repeat the aminoacyl adenylate synthesis using a new batch of Decalso F. None of the original batch of this zeolite was saved (Warden et al., 1974). There are conflicting claims concerning the polymerization of aminoacyl adenylates on montmorillonite. Warden et al. (1974) were able to repeat this reaction while Brack (1970) was not. A more recent report by Paecht-Horowitz and Eirich (1988) states that Dr. Y. Honda, a postdoctoral working in the laboratory of Professor Cyril Ponnamperuma, confirmed the catalytic action of sodium montmorillonite by forming "...oligo- and polyalanine of DP's (degree of polymerization) of 16 and over." In the same report it was noted that it was important to prepare the Na^+-montmorillonite precisely as they describe in their papers.

4. Template Directed Synthesis

The first demonstration of template-directed synthesis in which the template was bound to a surface was performed on apatite (Gibbs et al., 1980). Poly(U) was bound to the mineral and a solution of ImpA (structure given in Fig. 14) was added to the adsorbed template. The yield of oligomers was found to be the same in the presence or absence of hydroxyapatite. However, the proportion of higher molecular weight oligomers was greater in the presence of hydroxyapatite. This is believed to be due to the stronger binding of the longer oligomers to the apatite than the shorter ones. Consequently, the longer oligomers are selectively removed from solution so a greater proportion of shorter oligomers are available to bind to the template and are elongated by reaction with ImpA.

Template-directed synthesis of oligo(G's) (see Schwartz, this volume) was observed in the condensation reaction of 2-MeImpG (structure given in Fig. 7) on poly(C) bound to hydroxyapatite (Schwartz and Orgel, 1985). No preponderance of higher molecular weight oligomers was observed in this study.

Similar template-directed syntheses were performed on goethite (Holm et al., 1992) and ferrihydrate [$Fe(OH)_3$] (Schwartz and Orgel, 1985). Poly(U) bound to goethite undergoes reversible hydrogen bonding to adenosine. Binding is greater at 6°C than at 20°C as expected for Watson-Crick hydrogen bonding. The occurrence of Watson-Crick hydrogen bonding was further demonstrated by the template directed synthesis of oligomers of G by the self-condensation of 2-MeImpG in the presence of poly(C) bound to goethite (Fig. 8). A greater proportion of the higher molecular weight

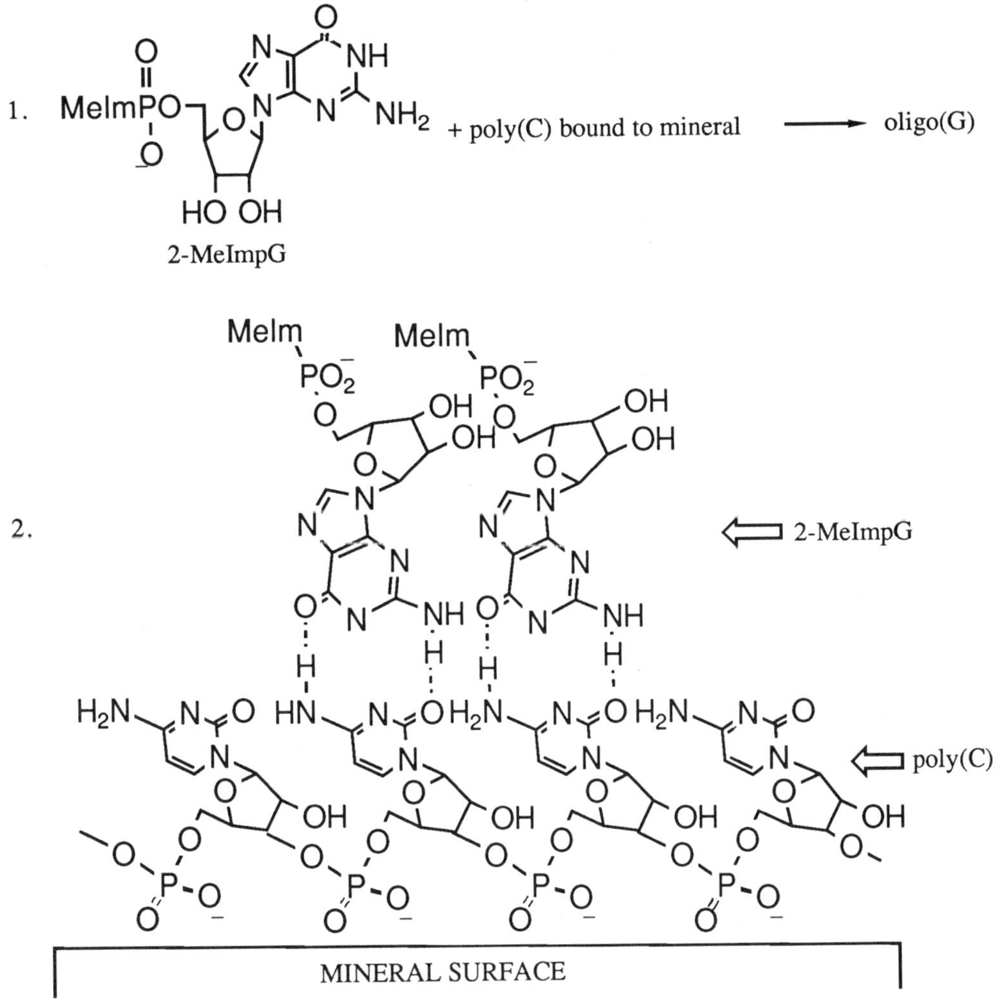

1. MeImpG + poly(C) bound to mineral ⟶ oligo(G)

2-MeImpG

2. 2-MeImpG

poly(C)

MINERAL SURFACE

Fig. 7 TEMPLATE DIRECTED SYNTHESIS ON MINERALS. 1. The reaction of 2-MeImpG with poly(C) to give oligo(G). 2. A schematic representation of 2-MeImpG hydrogen bonding to poly(C) which in turn is bound to the positives sites on a mineral surface .The negatively charged phosphate groups bind poly(C) to the Ca^{+2} on hydroxyapatite surface and to the positively charged $Fe(H_2O)^+$ on the Goethite surface. The activated 2-MeImpG hydrogen bonds to the poly(C) and condenses to form oligo(C) (Gibbs, Lohrmann and Orgel, 1980; Schwartz and Orgel, 1985).

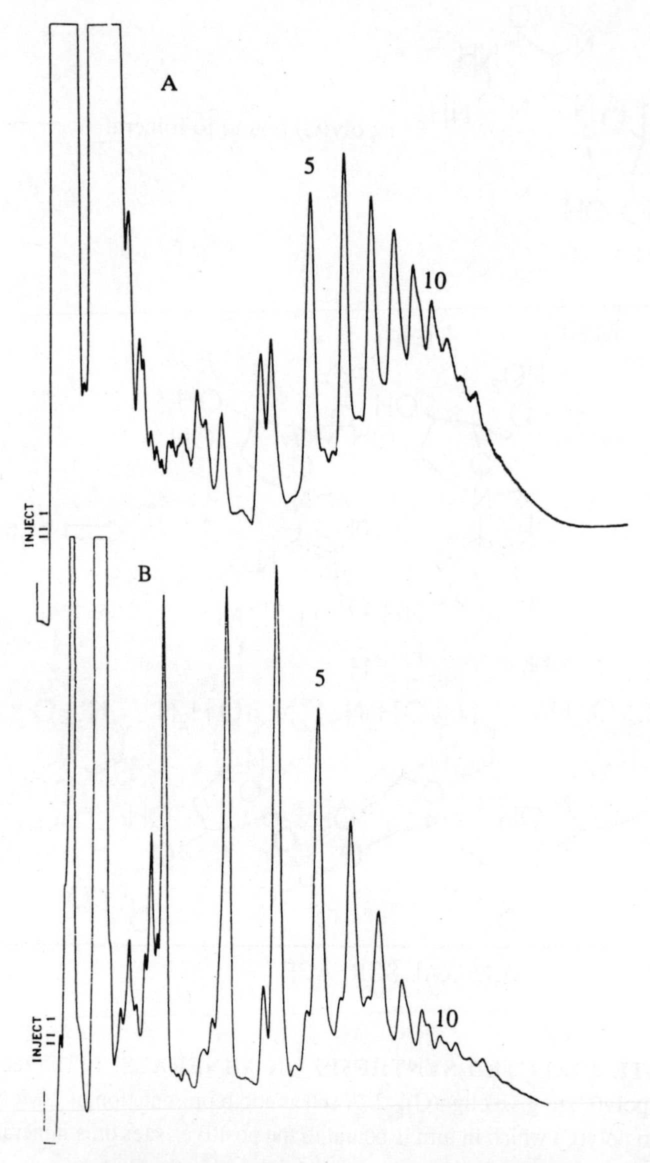

Fig. 8 HPLC ANALYSIS OF G OLIGOMERS FORMED ON POLY(C).
(A) Oligomers formed on a poly(C) template bound to goethite. (B) Oligomers formed in the absence of goethite. A greater proportion of higher molecular weight oligomers is formed in the presence of goethite (chromatogram A)

oligomers was observed when the template was bound to goethite (Fig. 8A) than was formed in the absence of the mineral (Fig. 8B).

Template-directed synthesis of oligo(G)'s was observed in the reaction of 2-MeImpG on poly(C) bound to ferric hydroxide, attapulgite clay, manganate and akaganeite (B-FeOOH) (Schwartz and Orgel, 1985). No appreciable enhancement of the formation of longer oligomers was reported.

5. Oligonucleotide Formation on Montmorillonite

The initial discovery of the catalytic activity of montmorillonite in our laboratory resulted from an unsuccessful attempt to enhance the rate of the oligomerization of HCN (Ferris et al., 1979). Instead, the condensation reaction was strongly inhibited by montmorillonite. Further investigation revealed that the inhibition was the result of the oxidation of diaminomaleonitrile (DAMN) to diiminosuccinonitrile (DISN) (Fig. 9) by the Fe^{3+} in the montmorillonite (Ferris et al, 1982). DAMN is an intermediate in oligomer formation. The facile formation of DISN from DAMN prompted investigation of its role as a potential prebiotic condensation agent. It was observed that DISN effected the cyclization of 3'-adenylic acid (3'-AMP) to 2',3'-cAMP in solution (Ferris et al., 1984) and that this reaction proceeded with higher efficiency in the presence of montmorillonite (Ferris et al., 1988). Finally, we were able to demonstrate the conversion of 3'-AMP to 2',3'-cAMP using DAMN and montmorillonite. DAMN was first

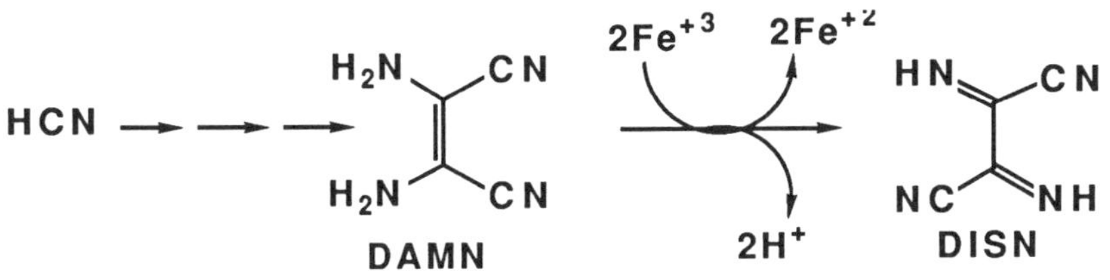

Fig. 9 FORMATION OF DISN, A CONDENSING AGENT, IN AQUEOUS SOLUTION. HCN spontaneously forms a tetramer (DAMN) in aqueous solution at pH 7-10 which is oxidized by Fe^{+3} to form a nucleotide condensing agent (DISN (Ferris et al., (1982).

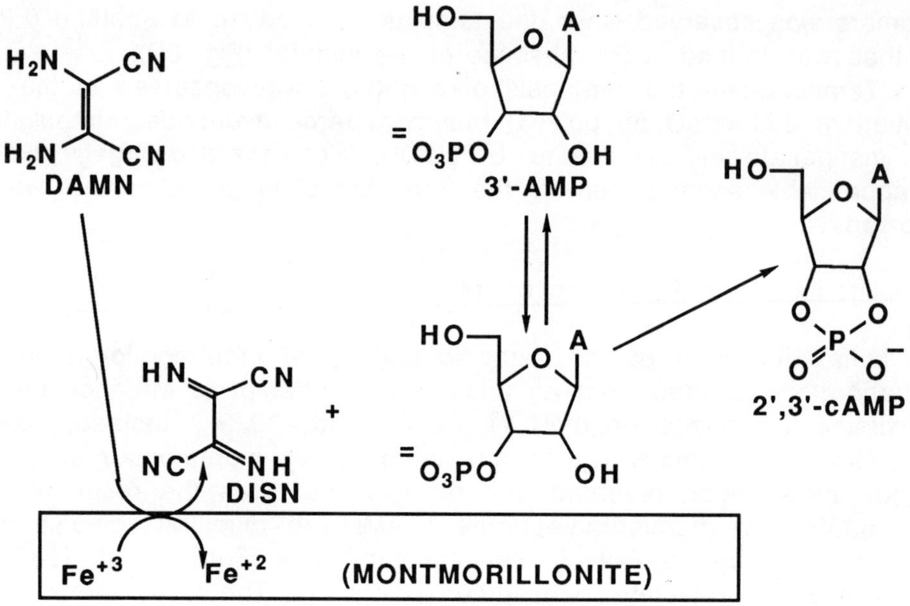

Fig. 10 THE FORMATION OF 2',3'-CYCLIC AMP ON MONTMORILLONITE. The in situ generation of a condensing agent (DISN) by Fe^{+3} in montmorillonite and the montmorillonite-catalyzed conversion of 3'-AMP to 2',3'-cyclic AMP on montmorillonite (Ferris et al., 1988).

converted to DISN which in turn effected the cyclization reaction (Fig. 10) (Ferris et al., 1988).

Attempts to extend this system to the reaction of 5'-adenylic acid (5'-AMP) were unsuccessful. The predominant reaction of DISN is with the 2'- and 3'-hydroxyl groups of 5'-AMP (Ferris and Yanagawa, 1984).

Treatment of 5'-AMP with a water soluble carbodiimide [1-ethyl-3-(3-dimethylaminopropyl)carbodiimide] (EDAC; structure given in Fig.11) did result in the formation of dimers from 5'-AMP (Fig. 11) and trimers and possibly tetramers from 5'-dAMP when Na^{+}-montmorillonite was present (Fig. 12) (Ferris et al., 1989; Ferris and Kamaluddin, 1989; Ferris et al., 1990).

Surprisingly, the use of dimers as starting materials did not result in the formation of tetramers and hexamers (Ferris and Peyser, 1992). Instead, the reaction of dpApA with EDAC on montmorillonite

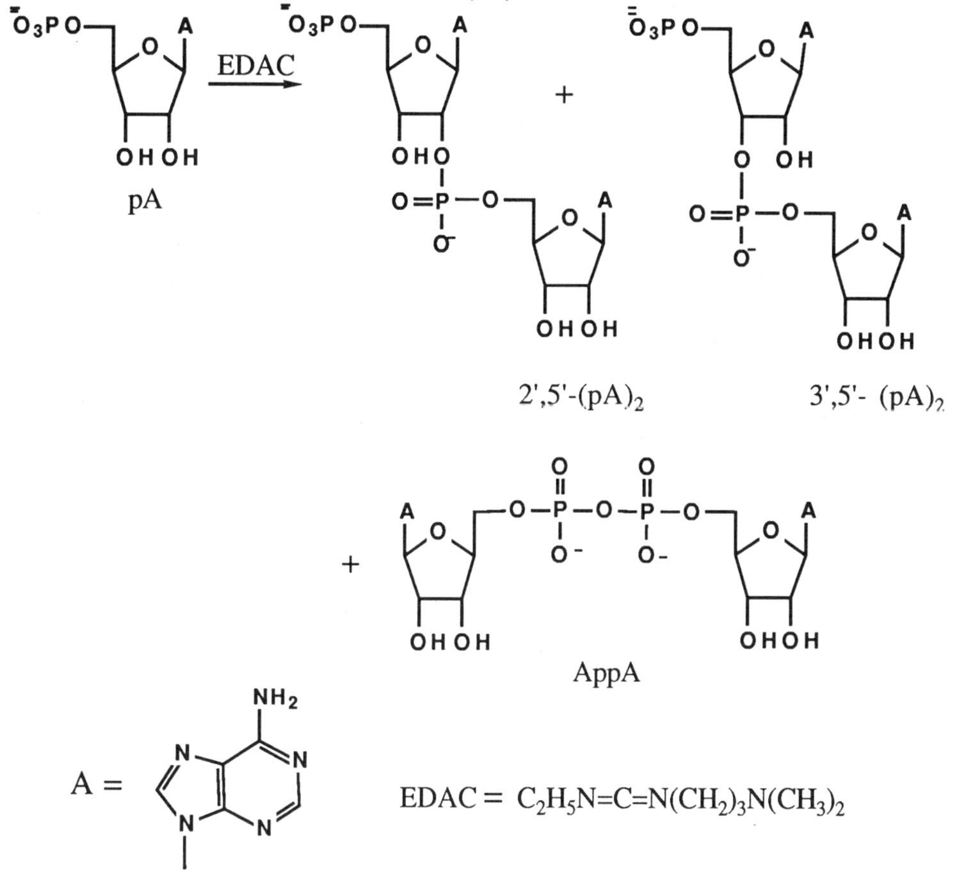

Fig. 11 5'-AMP + EDAC + MONTMORILLONITE. Formation of RNA dimers in the reaction of 5'-AMP with a water soluble condensing agent on montmorillonite (Ferris et al., 1989).

yielded the cyclic dimer in high yield (Fig. 13) while mainly starting material was recovered from the reaction of dApAp.

The promising results obtained using EDAC as a condensing agent prompted experiments on the 5'-phosphorimidazolide of adenosine (ImpA), a compound which can be synthesized under potentially prebiotic conditions from 5'-ATP and imidazole (Lohrmann, 1987). Reaction of ImpA in aqueous solution on montmorillonite yields oligomers (Fig. 14) which contain up to ten monomer units as shown

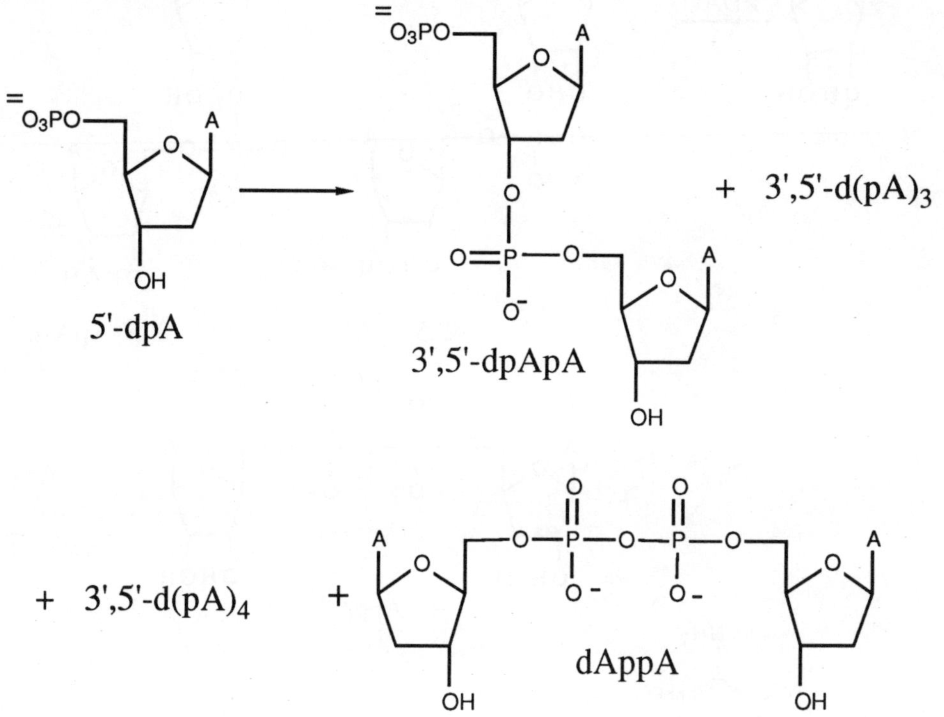

$5'$-dpA → $3',5'$-dpApA + $3',5'$-$d(pA)_3$ + $3',5'$-$d(pA)_4$ + dAppA

Fig.12 THE CONDENSATION OF 5'-dAMP TO OLIGOMERS ON MONTMORILLONITE (Ferris and Kamaluddin, 1990).

by HPLC anion exchange chromatography (Ferris and Ertem, 1992). Reaction of 9:1 ImpA:AppA also resulted in oligomer formation but with greater than 80% formation of 3',5'-phosphodiester bonds (Ferris and Ertem, 1992a).

Structure analysis was performed on the trimer, tetramer and pentamer fractions isolated from the HPLC column. Selective enzymatic hydrolyses and HPLC separation of the reaction products

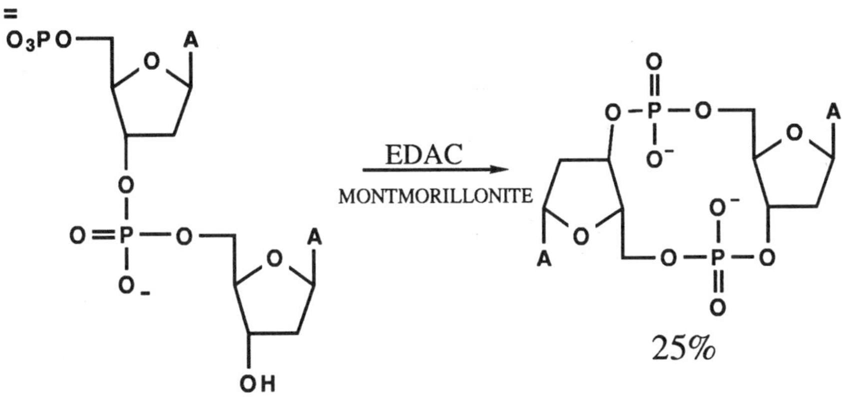

EDAC
MONTMORILLONITE

25% (1)

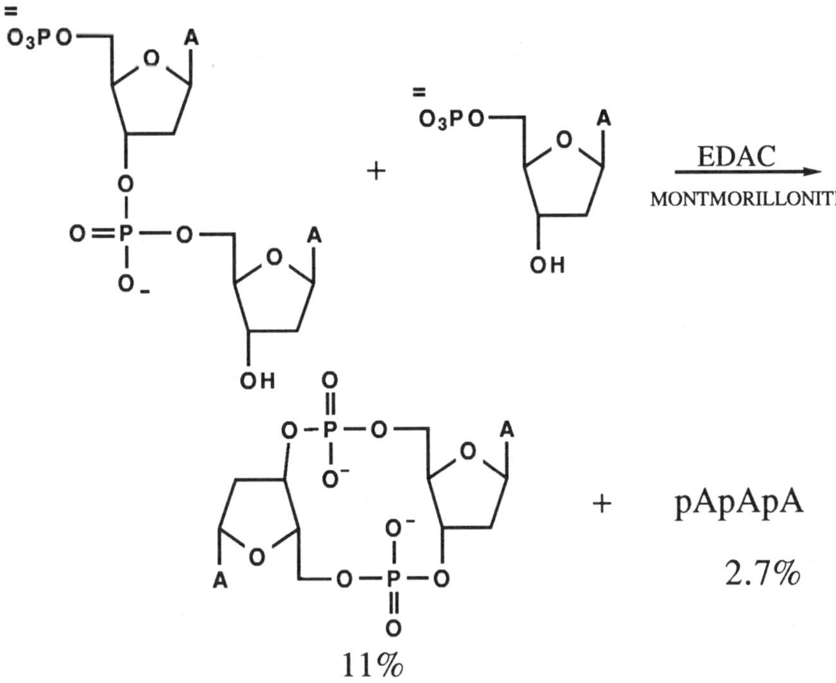

+ EDAC
MONTMORILLONITE

11% + pApApA (2)
2.7%

Fig. 13 THE MECHANISM OF OLIGOMER FORMATION ON MONTMORILLONITE. Conversion to the cyclic dimer is the only process observed in the reaction of dpApA with EDAC shown in equation 1. Trimer formation and diminished cyclic dimer formation is observed in thereaction of pApA with pA in equation 2. A 0.9% yield of pApApA was observed when only pA and EDAC react in the presence of montmorillonite. These experiments show that oligomer formation proceeds by the reaction of monomer with higher oligomers and not by the condensationof oligomers (Ferris and Peyser, submitted).

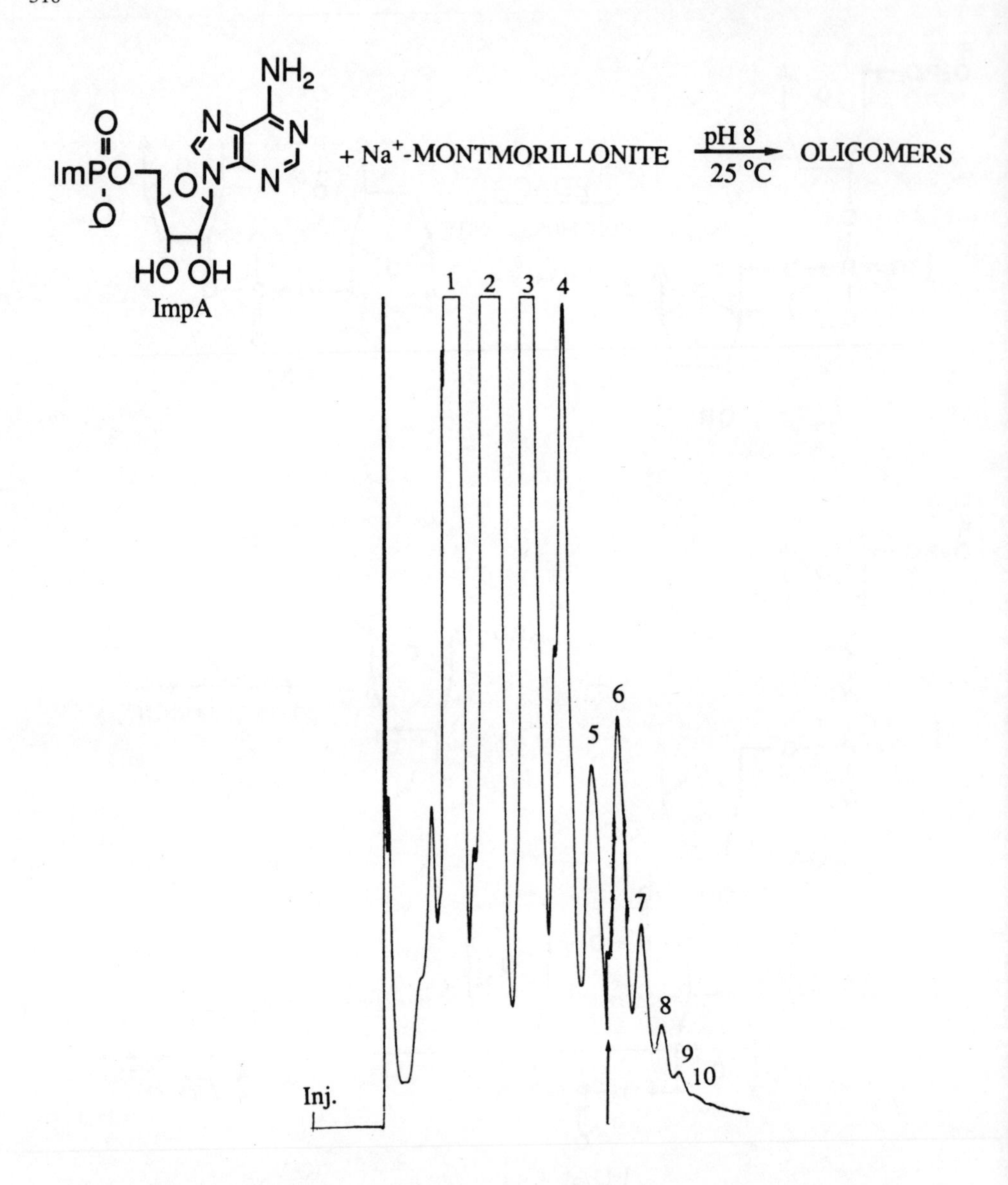

Fig. 14 HPLC ANALYSIS OF OLIGOMERS FORMED BY THE REACTION OF ImpA ON MONTMORILLONITE

Table. 1 Oligomers Present In The Trimer, Tetramer, and Pentamer Fractions In the Reaction of ImpA on Montmorillonite.

TRIMER	(%)
3',5'-pApApA	46
$pA^{3'}pA^{2'}pA$	27
$AppA^{3'}pA^{3'}pA$	17
$AppA^{3'}pA^{2'}pA$	5
TETRAMER	
3',5'-pApApApA	8
$pA^{3'}pA^{3'}pA^{2'}pA$	14
Other $(pA)_4$ isomers	21
$AppA(pA)_3$ isomers	3
$(Ap)_mAppA(pA)_n$ isomers; (m+n=3)	35
PENTAMER	
$(pA)_5$ isomers	48
$AppA(pA)_4$ isomers	10
$(Ap)_mAppA(pA)_n$	42

led to the compositions shown in Table 1 (Ferris and Ertem, 1992a).

The self condensation of ImpA on montmorillonite is consistent with the hypothesis that RNA oligomers formed on the primitive earth by the mineral-catalyzed oligomerization of activated mononucleotides. These oligomers form in high yield and high regiospecificity (80% 3',5'-bond formation) in aqueous medium of pH 8, at 2°C to 37°C. Template- directed synthesis on these oligomers would have resulted in the formation of the complementary RNA sequences. These template- directed condensations may have occurred with the template bound to mineral surfaces. Since 80% of the bonds are 3',5'-linked, these oligomers may be amenable to elongation using ribozymes (Young and Cech, 1989). These and other elongation processes may have resulted in the formation of RNA containing with sufficient information for the origin of life to occur. Alternatively, minerals other than montmorillonite may have been more effective catalysts for the formation of longer oligomers.

Acknowledgement: These ideas in this proposal were developed as a result of discussions with coworkers cited in the references. Their important contributions to this and other research in my laboratory is greatly appreciated. The research was supported by NSF Grant CHE-9000187 and NASA Grant NAGW-2781.

6. References

Bernal, J. D. (1949) *The Physical Basis of Life,* Proc. Roy. Soc. A. **62**, 537-558.

Bernardi, G. (1971) *Methods in Enzymology* , Grossman, L. and Moldave, K., (ed.), Vol. 21, Academic, N.Y., pp. 95-139.

Brack, A. (1970) *Polymerization en Phase Aqueuse D'acids Amines sur Argiles*, Clay Minerals **11**, 117-119.

Ferris, J. P. and Ertem, G. (1992a) *Oligomerization Reactions of Ribonucleotides on Montmorillonite: Reaction of the 5'-Phosphorimidazolide of Adenosine*, Science, **257**,1387-1389.

Ferris, J. P. and Ertem, G. (1992b) *Oligomerization Reactions of Ribonucleotides: The Reaction of the 5'-Phosphorimidazolides of Nucleosides on Montmorillonite and Other Minerals,* Origins Life Evol. Biosphere 22, 369-381.

Ferris, J. P. and Peyser, J. R. (1992) unpublished.

Ferris, J. P. and Yanagawa, H. (1984) *Reaction of Uridine and Uridine 5'-Phosphate with Diiminosuccinonitrile and Cyanogen Bromide in Aqueous Solution. Direct Synthesis of the 2,2'-Anhydronucleoside Linkage at 2°C*, J. Org. Chem. **49**, 2121-2125.

Ferris, J. P., Edelson, E. H., Mount, N. M. and Sullivan, A. E. (1979) *The Effect of Clays on the Oligomerization of HCN,* J. Mol. Evol. **13**, 317-330.

Ferris, J. P., Ertem, G. and Agarwal, V. K. (1989) *Mineral Catalysis of the Formation of Dimers of 5'-AMP in Aqueous Solution: The Possible Role of Montmorillonite Clays in the Prebiotic Synthesis of RNA*, Origins Life Evol. Biosphere **19**, 165-178.

Ferris, J. P., Hagan, W. J., Jr., Alwis, K. W. and McCrea, J. *Chemical Evolution 40. Clay-Mediated Oxidation of Diaminomaleonitrile,* (1982) J. Mol. Evol. **18**, 304-309.

Ferris, J. P., Huang, C.-H., Hagan, W. J., Jr. (1988) *Montmorillonite: A Multifunctional Mineral Catalyst for the Prebiological Formation of Phosphate Esters,* Origins Life Evol. Biosphere **18**, 121-133.

Ferris, J. P., Kamaluddin and Ertem, G. (1990) *Oligomerization Reactions of Deoxyribonculeotides on Montmorillonite Clay: The Effect of Mononucleotide Structure, Phosphate Activation and Montmorillonite Composition on Phosphodiester Bond Formation,* Origins Life Evol. Biosphere **20**, 279-297.

Ferris, J. P., Yanagawa, A., Dudgeon, P. A., Hagan, W. A., Jr., and Mallare, T. E. (1984) *The Investigation of the HCN Derivative Diiminosuccinonitrile as a Prebiotic Condensing Agent. The Formation of Phosphate Esters*, Origins of Life **15**, 29-43.

Gibbs, D., Lohrmann, R. and Orgel, L. E. (1980) *Template-Directed Synthesis and Selective Adsorption of Oligonucleotides on Hydroxyapatite,* J. Mol. Evol. **15**, 347-354.

Herbillion, A. J. (1988) *Introduction to the Surface Charge and Properties of Iron Oxides and Oxidic Soils* in J. W. Stucki, B. A. Goodman and U. Schwertmann (eds.) *Iron in Soils and Clay Minerals* , D. Reidel, Dordrecht, pp.251-266

Holm, N., Ertem,. G. and Ferris, J. P. (1992) *The Binding and Reactions of Nucleotides and Polynucleotides on Iron Oxide Polymorphs,* Origins Life Evol. Biosphere **23**, in press.

Lewinsohn, R., Paecht-Horowitz, M. and Katchalsky, A. (1967) *Polycondensation of Amino Acid Phosphoanhydrides III. Polycondensation of Alanyl Adenylate,* Biochem. Biophys. Acta **140**, 24-36.

Lohrmann, R.: 1977 *Formation of Nucleoside 5'-Phosphoramidates Under Potentially Prebiotic Conditions*, J. Mol. Evol. **10**, 137-154.

Murray, J. W. (1979) in Burns, R. G. (ed.) *Marine Minerals* Mineralogical Soc. Amer., pp. 47-98.

Newman, A. C. D. (ed.) (1987) *Chemistry of Clays and Clay Minerals*, Wiley, New York.

Paecht-Horowitz, M. and Katchalsky, A. (1973) *Synthesis of Amino Acyl-Adenylates Under Prebiotic Conditions,* J. Mol. Evol. **2**, 91-97.

Paecht-Horowitz, M., Berger, J. and Katchalsky, A. (1970) *Prebiotic Synthesis of Polypeptides by Heterogeneous Polycondensation of Amino-Acid Adenylates*, Nature **228**, 636-639.

Sawai, H. and Orgel, L. E. (1975) *Oligonucleotide Synthesis Catalyzed by the Zn^{2+} Ion*, J. Am. Chem. Soc. **97**, 3532-3533.

Sawai, H., Kuroda, K. Hojo, T. (1989) *Uranyl Ion as a Highly Efficient Catalyst for Internucleotide Bond Formation,* Bull. Chem. Soc. Jpn. **62**, 2018-2023.

Sleeper, H. L., Orgel, L. E. (1979) *The Catalysis of Nucleotide Polymerization by Compounds of Divalent Lead,* J. Mol. Evol. **12**, 357-364.

Schwartz, A. and Orgel, L. E. (1985) *Template-Directed Polynucleotide Synthesis on Mineral Surfaces*, J. Mol. Evol. **21**, 299-300.

Theng, B. K. G. (1974) *The Chemistry of Clay-Organic Reactions* John Wiley, New York, p. 10.

Van Olphen, H. (1977) *An Introduction to Clay Colloid Chemistry* 2nd Ed. Wiley-Interscience, New York.

Warden, J. T., McCullough, J. J., Lemmon, R. M. and Calvin, M. (1974) *A Re-examination of the Zeolite-Promoted, Clay-Mediated Peptide Synthesis,* J. Mol. Evol. **4**, 189-300.

Wells, A. F. (1962) *Structural Inorganic Chemistry* 3rd Ed Oxford University Press, London, pp. 562.

Young, B. and Cech, T. R. (1989) *Specificity for 3',5'-Linked Substances in RNA-Catalyzed RNA Polymerization*, J. Mol. Evol. **29**, 480-485.

Biology and Theory: RNA and the Origin of Life

Alan W. Schwartz
Laboratory for Evolutionary Biology
Faculty of Science
University of Nijmegen
Toernooiveld, 65256 ED Nijmegen
The Netherlands

ABSTRACT. Most current thinking concerning the origin of life has become rooted in a theory which is referred to as the "RNA World". The origins and consequences of the theory will be explained and recent research concerning the possible origins of the first RNA molecules will be reviewed.

1 Introduction: RNA and Biology

A modern phylogenetic tree based on nucleic acid sequences, such as that shown in Fig. 1, illustrates the point that all life on earth is descended from a single common ancestor.

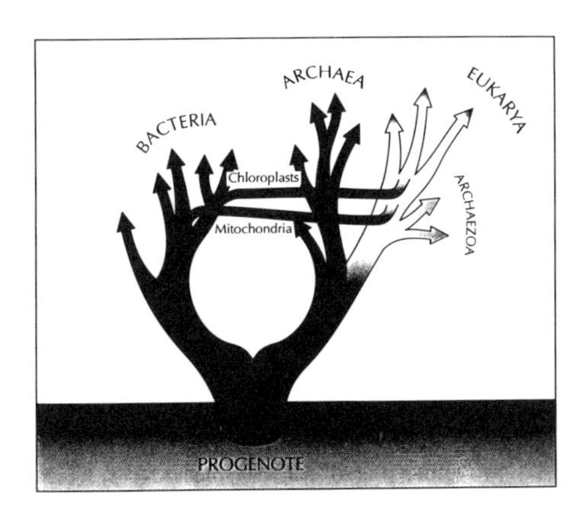

Figure 1. The evolution of the major kingdoms of living organisms as deduced from RNA sequences. Reproduced with permission from Doolittle (1991).

J. M. Greenberg et al. (eds.), The Chemistry of Life's Origins, 323–344.

It is important to emphasize that that ancestor, the "progenote" in the terminology of Carl Woese (1981), was certainly not the first form of life. It is probable that another, unseen phylogenetic tree preceded the one we can now reconstruct, and led from the first form of life to the progenote. The progenote was necessarily an organism whose genetic system was based on DNA. According to many biologists, however, a still earlier tree and the first form of life may have been based on evolving RNA molecules. To understand the reasons for this hypothesis, which has become known as the "RNA world," we will first have to review some basic biological principles. The flow of information in the cell can be represented as follows:

DNA $\leftrightarrows$ RNA → Protein

Genetic information is transcribed from its storage form as DNA into an RNA strand, or messenger. The three-letter genetic code, which specifies the assembly of amino acids into protein, is translated into protein in RNA-protein structures called ribosomes (Fig. 2).

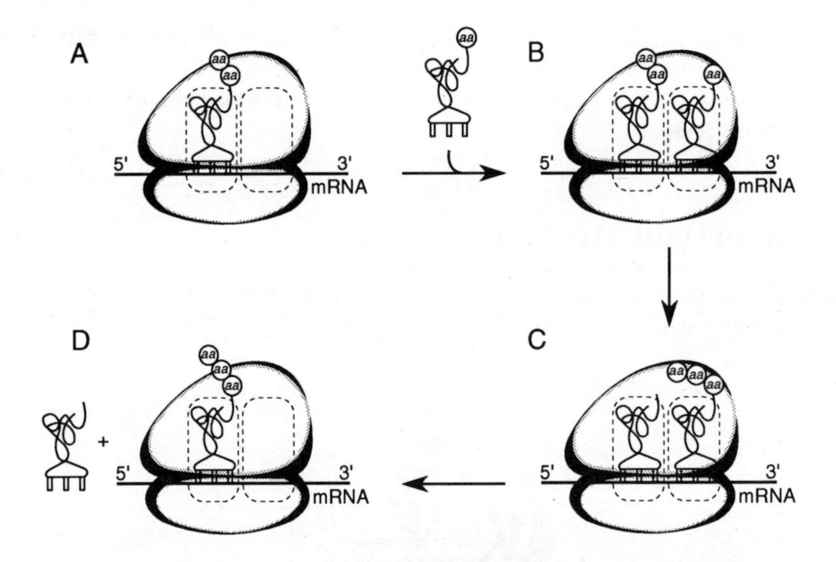

Figure 2. The aminoacyl-tRNA shuttle operating in a ribosome. A) A polypeptide chain attached to a tRNA molecule, which in turn is bound by three sets of Watson-Crick base pairs to an amino acid codon of an mRNA strand. The adjacent site within the ribosome is vacant. B) A new aminoacyl-tRNA enters the ribosome and binds to the adjacent position. C) A new peptide bond is formed between the amino acid attached to the "A site" and the polypeptide, which is transferred from the P to the A site. D) The polypeptide-tRNA is now transferred to the P site, expelling the free tRNA molecule and making the A site available for binding the next aminoacyl-tRNA. A new cycle of peptide bond synthesis can now take place.

Another type of RNA molecule plays a key role in this process; transfer-RNA (tRNA). Amino acids are converted into an activated form and attached to specific tRNAs by enzymes which are called aminoacyl synthetases. Assembly of protein takes place in the ribosomes, where tRNAs operate in pairs so that the growing polypeptide chain becomes attached to each new amino acid when its tRNA binds to the appropriate place on the messenger. An important point is that although proteins are present in the ribosomes, they do not appear to have a necessary catalytic function in the formation of the growing polypeptide chain. Once the aminoacyl-tRNA combination has been synthesized externally to the ribosome, it seems that bringing each amino acid into the proper position next to the peptide is all that is required for a new peptide bond to form. RNA therefore plays a central role in the protein synthesis apparatus. Even the synthesis of DNA requires a small RNA primer to start the copying of a DNA strand. Other reasons for thinking that RNA is more ancient than DNA involve the roles of ribonucleotides in biochemical pathways. Many of the coenzymes, which are small molecules that operate together with enzymes to carry out biochemical transformations, are in fact ribonucleotides or are related to ribonucleotides.

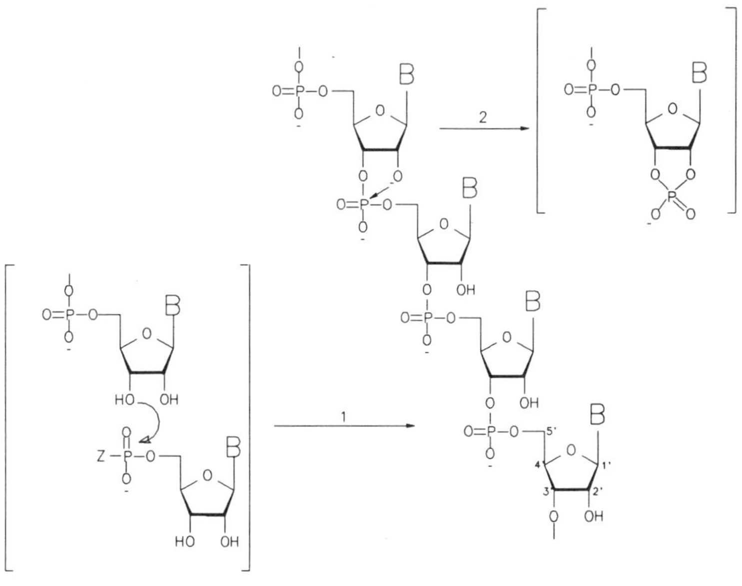

Figure 3. The backbone structure of RNA (see Figure 8 for the structures of the bases, B). In DNA the 2'-hydroxyl groups are absent and replaced by H). Reaction 1 shows the formation of a 3'-5' phosphodiester bond between two nucleotide units, in which 3'-hydroxyl group displaces an activating group Z. In reaction 2 the alkaline hydrolysis of a phosphodiester linkage to liberate a cyclic phosphate is shown. The neighboring 2'-hydroxyl group participates in this reaction.

Even the biosynthesis of the deoxynucleotides required for DNA synthesis involves the secondary conversion of a ribonucleotide to a deoxyribonucleotide. Therefore it appears that although DNA has been selected as the long-term storage form for genetic information, the original genetic material was RNA itself. Additional arguments for the primacy of RNA can be found in the chemical properties of the two classes of nucleic acids. The presence of an additional hydroxyl group in RNA has important consequences for both the formation and breaking of internucleotide phosphodiester linkages. The neighboring hydroxyl groups of ribose are far easier to phosphorylate than the single hydroxyl group of deoxyribose. Once formed, the internucleotide linkage in DNA is much harder to break by simple hydrolysis than the equivalent linkage in RNA. DNA is made to last, while RNA strands can be taken apart and reassembled much more easily in aqueous solution. As we shall see, this may have had advantages in early evolution.
The most compelling reasons for suspecting that RNA may have played a central role in the origin of life are related to a fascinating set of experiments utilizing an enzyme isolated from virus-infected bacteria as a tool for studying the replication of RNA molecules. In contrast to the multiple enzyme systems needed to carry out DNA replication, RNA replication can be performed by a single enzyme. One such enzyme has been the subject of many *in vitro* studies. When E. coli bacteria are infected with the virus $Q\beta$, a single enzyme carries out the task of replication of the viral RNA strand. This enzyme is known as the $Q\beta$ replicase.

Eigen and his associates have made use of the properties of the enzyme to study the replication of RNA molecules in cell-free systems. They have shown that conditions can be established which permit random fluctuations in base sequence (mutations) to be amplified and selected (Eigen et al., 1981). The $Q\beta$ RNA, as well as many biologically active RNA molecules, contains a fairly high proportion of self-complimentary regions, that is, stretches of bases which can form Watson-Crick pairs with other regions of the chain. The formation of these reversible sets of base-pairs accounts for the multiloop-like structure of many RNA molecules. In a sense, the three-dimensional structure of an RNA molecule corresponds to the phenotype of the molecule, just as a living organism is the phenotype of its genetic information (the genotype). In one set of experiments, an agent was added to the reaction system which can break linkages between nucleotides more easily in single-stranded than in double-stranded regions of RNA.

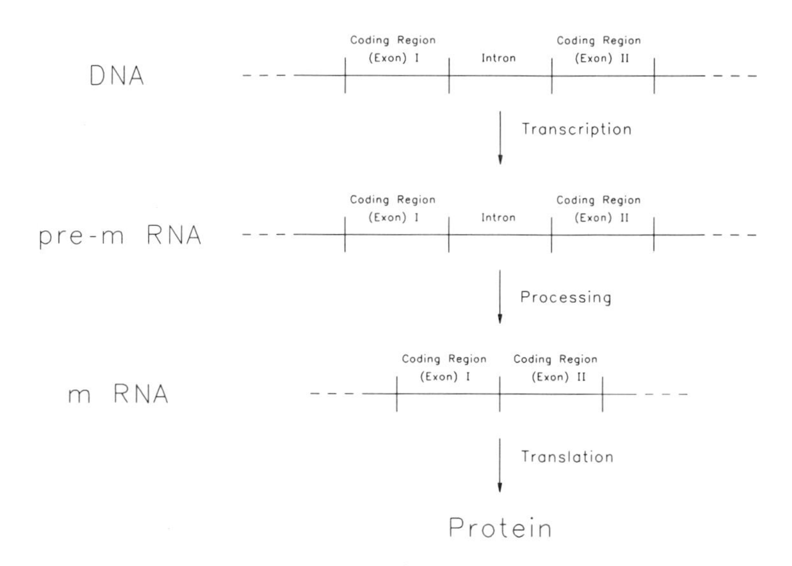

Figure 4. Processing of many kinds of RNA occurs *via* a self-splicing process. The diagram illustrates the manner in which two amino acid coding regions (exons) are spliced together by the intervening, non-coding sequence (intron).

After many replications under these conditions, it was found that a rather uniform set of RNA sequences emerged which had a high degree of self-complimentarity, and therefore assumed a folded conformation in solution which was optimally resistant to hydrolysis. In fact, the new form of RNA which had emerged was not merely a result of the presence of a hydrolytic reagent, its persistence was *dependent* on its presence. These experiments can truly be said to demonstrate evolution in a test tube.

2 The RNA World

Protein synthesis in cells and evolution experiments with RNA in the test tube both involve enzymes. While RNA and DNA are perfectly designed to carry genetic information and transfer that information from one molecule to the next by means of base-pairing, nucleic acids are not catalytically active and therefore require enzymes for their own replication as well as for the synthesis of everything else. The enzymes, on the other hand, cannot reproduce themselves without the blueprint provided by RNA and DNA. This dichotomy in biological function has long been a serious serious problem in theories of the origin of life. The paradox has been resolved, at least in principle, by the discovery of catalytic RNA, which led to a 1989 Nobel prize in Chemistry for Cech and Altman.

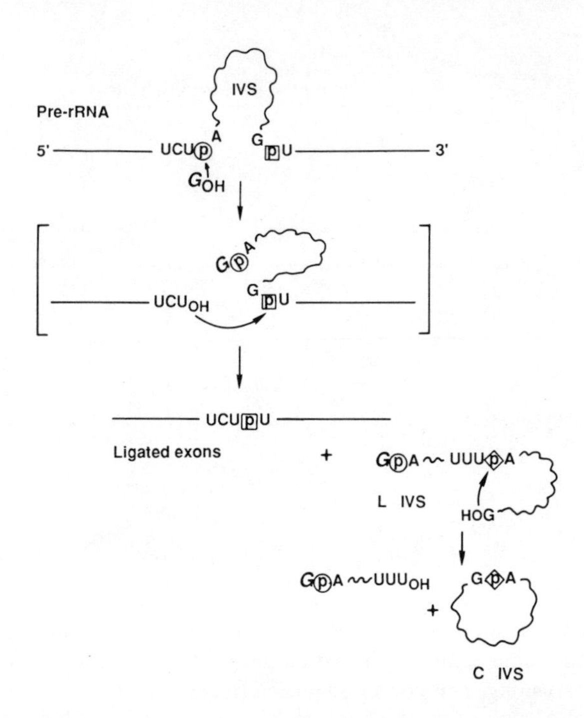

Figure 5. The self-splicing intron (IVS) of the *Tetrahymena* pre-rRNA cuts itself loose from the rRNA and splices the exons together to produce the complete rRNA. The liberated intron then proceeds to cyclize itself. Reproduced with permission from Cech (1987; Copyright by the AAAS).

This discovery has even wider implications for the question of the origin of life than may be apparent from the above. The catalytic RNA discovered by Cech and Altman is an example of a self-splicing intron (Kruger et al., 1982). During translation of the genetic information stored in DNA, long regions of the DNA are first transcribed into RNA. Some of this initial RNA is destined to remain in the form of RNA, and to supply the tRNAs and rRNAs needed for protein synthesis. Other segments of the initial RNA message contain coding regions for the synthesis of proteins. In all cases, however, some additional processing of the RNA usually takes place, since the coding regions are frequently interrupted by stretches of non-coding DNA. This processing involves a kind of surgery whereby segments of the RNA (intervening sequences, or introns) are cut out and the ends of the remaining message (the exons) are rejoined to form the final sequence. It was originally thought that enzymes were necessary to carry out the cutting and splicing. Attempts to isolate such enzymes revealed that RNA was also present in the enzymatically active complex responsible for the splicing. Attempts to purify the enzymatic activity led to the elimination of increasing amounts of the protein. It is now clear that in many cases the RNA intron is responsible for its own processing. Once the first self-splicing RNA enzyme (now referred to as a ribozyme) was discovered, the molecular biologists were able to apply their tools to a study of the possibilities of such molecules. Several variants have now been isolated, one of which is able to join together short oligonucleotides to produce longer chains.

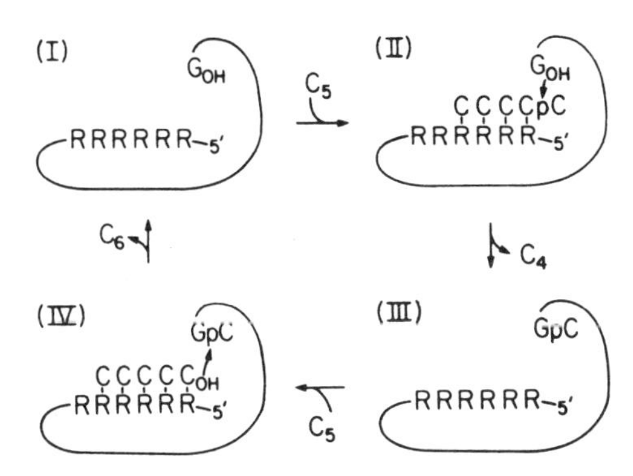

Figure 6. The *Tetrahymena* ribozyme is able to act as a polymerase by extending short oligonucleotides (pentacytidylic acids) to produce longer chains (C_5 to C_6). This reaction is performed at the expense of other oligonucleotides, which are broken down (C_5 to C_4). There is overall conservation of the number of phosphodiester bonds. Reproduced with permission from Cech (1986).

The ability to carry out splicing of this kind depends on the presence of a section of the ribozyme called the "internal guide sequence." This is a template which binds, and therefore specifies the oligonucleotides which are to be joined together, or oligomerized. By removing the original template the researchers were able to create conditions under which any of a large number of different templates could be employed. In effect, it became possible to link together a wide variety of oligomers. In principle, therefore, a random assortment of short oligomers could be sorted by the ribozyme so as to select those which are complimentary to the template. These then would be linked together. We can now imagine replacing the viral enzyme used by Eigen in his evolution experiments by the RNA molecule itself. Life may have started, therefore, with RNA molecules alone. The coded synthesis of proteins according to this theory of the "RNA world," may have been a later development in evolution. We will return to this question later. Introns which interrupt protein coding sequences may have a special role in evolution. It has been observed that in many cases introns divide proteins into structural "domains," or regions which code for particular three-dimensional structures such as α-helixes, β-sheets, etc. This suggests the possibility that introns can not only cut and splice intramolecularly, but also intermolecularly, so as to create new combinations of domains. This mechanism would provide an efficient way of rejoining functional regions of proteins during evolution so as to create new combinations and therefore, new enzymes. This process has been referred to as "exon shuffling." It has even been suggested that the introns are remnants of the original, self-replicating RNA molecules which populated the RNA world.

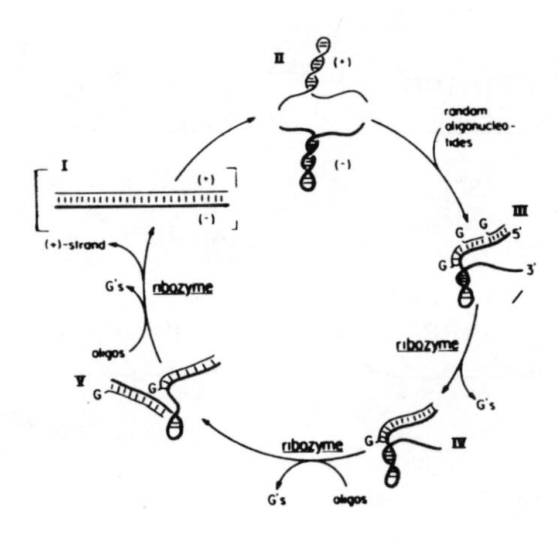

Figure 7. A scheme for the self-replication of a ribozyme. Reproduced with permission from Cech (1989; Copyright Macmillan Magazines Ltd.).

3 The Origin of Life

What in fact would be necessary to constitute a minimal living system? The first sensible answer to this question was formulated by Oparin in his influential book on the origin of life (1938). Oparin viewed cellular reproduction and metabolism as being central to the problem and developed a theory based on these phenomena. Since the discovery of the central role of nucleic acids in biology it has been realized that a primitive genetic system of some kind may have been a characteristic of the earliest living systems. During reproduction and cell growth, two functionally distinct processes can be recognized. In transcription, one chain of the DNA double-helix is copied to produce a messenger RNA molecule. The messenger is then transported to the ribosome where translation of the code into protein takes place. Both processes are highly complex, and involve the participation of many enzymes. However it is the process of transcription that provides a hint as to a possible evolutionary precursor. In experiments which make use of an isolated polymerase enzyme, a given polynucleotide will, in the presence of a supply of nucleoside triphosphates, be endlessly transcribed into its complement and back again, thereby replicating itself many times. This is essentially what happens during the PCR (polymerase chain reaction) process of DNA multiplication which is the basis of the DNA "fingerprinting" technique. Because the original polynucleotide and its complement bind very tightly to form a Watson-Crick double helix, it is necessary to raise the temperature periodically so as to "melt" (separate) each molecule from its complement to permit further replication. The polymerase, in catalyzing the formation of phosphodiester linkages between monomers, is fundamentally performing a very simple job, albeit at a high rate of speed. According to the theory of the "RNA world," the first self-replicating RNA molecule would have acted as its own polymerase. Theoretically a nonenzymatic reaction might be found which would accomplish the same task performed by the polymerase.

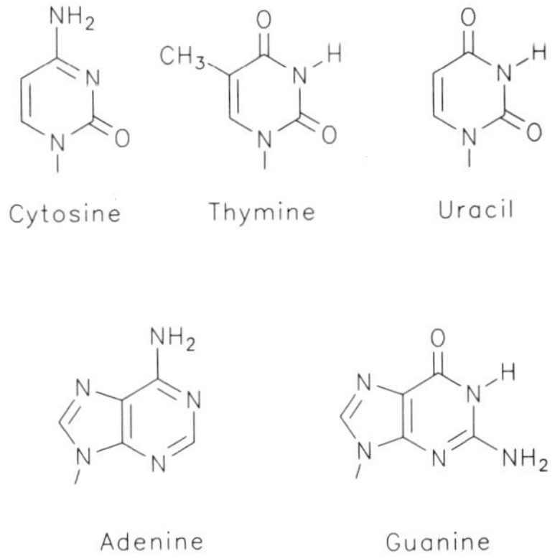

Figure 8. The nucleic acid bases. Adenine, guanine and cytosine occur in both RNA and DNA. Uracil is found in RNA but is replaced by thymine in DNA. In nucleic acid replication and translation, specific hydrogen-bonded pairs are formed between adenine and uracil (or thymine), and between guanine and cytosine.

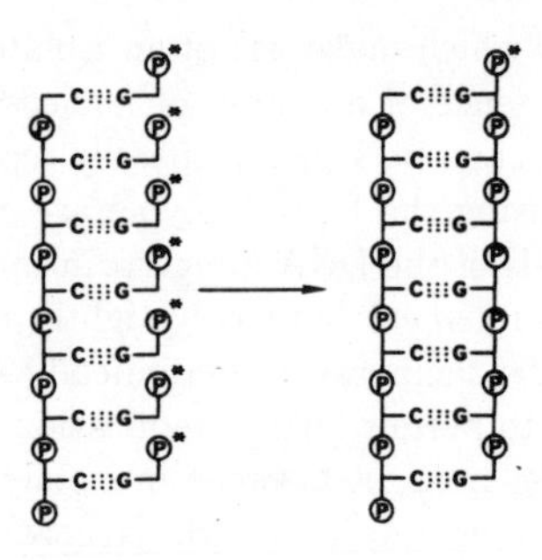

Figure 9. Schematic diagram illustrating the principle of template- directed oligomerization. A poly(C) template forms a double-stranded Watson-Crick helix with the activated monomer guanosine monophosphate, whereupon phosphodiester linkages form to produce a chain of poly(G) (From Orgel, 1986).

This has been the goal of the "template-directed" oligomerization of mononucleotides developed in the Salk Institute laboratory of Leslie Orgel. The starting point for this method is the self-organizing behavior of polynucleotides.

In DNA and RNA, a combination of hydrogen-bonds between complimentary base pairs (Fig. 8) and the tendency of the purines and (to a lesser extent) pyrimidines to form stacks in aqueous solution, accounts for the stability of the helix formed between two chains with complimentary nucleotide sequences. It has long been known, however, that under certain conditions double-helices can be formed with one chain consisting of a sequence of pyrimidines, such as poly(C), and a second "chain" consisting of only monomers such as the mononucleotide pG (Fig. 9). Using a simple organic chemical group to "activate" the phosphate of a pG, Orgel has demonstrated that a poly(C) will catalyze the synthesis of long chains of oligoguanylic acid, which form a very stable double helix with the poly (C) template.

The details of this and related oligomerizations have been worked out in exquisite detail in Orgel's laboratory (Inoue and Orgel, 1982; 1983). The reaction is highly dependent on the nature of the activating group which is used, as well as the chemical nature of the individual nucleotides. The most efficient reactions and the longest products are obtained in the oligomerization of 2-MeImpG on poly(C). Unfortunately, the next step which is theoretically required to complete a self-replication cycle - the synthesis of poly(C) using poly(G) as template - cannot be achieved. The reasons for this failure are well understood, and are related to the tendency of poly(G) to form multichain complexes with itself, as well as with the much smaller tendency for cytidylic acid to self- associate than is the case with guanylic acid.

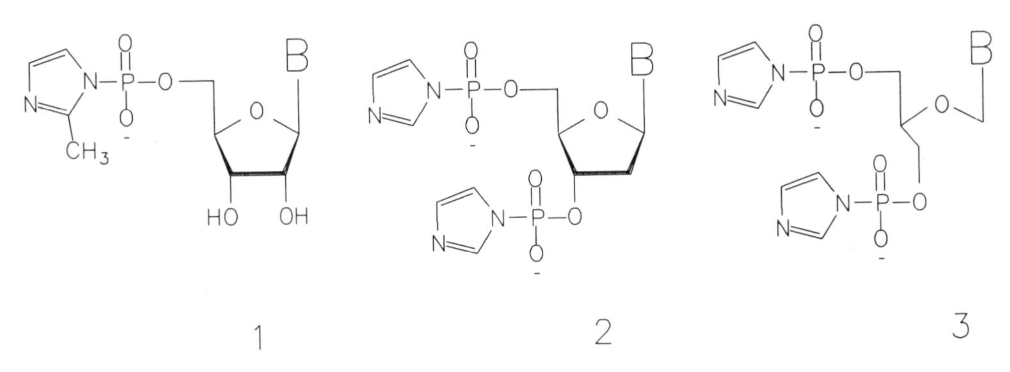

Figure 10. Activated monomers used in template-directed syntheses of oligonucleotides and their analogs. 1) 2-MeImpB, a ribonucleotide (pB) in which the 5'-phosphate group is activated by attachment of a 2-methyl imidazole group. 2) ImpdBpIm, a deoxynucleoside diphosphate in which the 3'- and 5'-phosphate groups have been activated by attachment of imidazole. 3). An acyclic analog of 2, in which the ribose ring has been replaced by an open chain structure based on glycerol and formaldehyde. B represents one of the nucleic acid bases shown in Fig. 8.

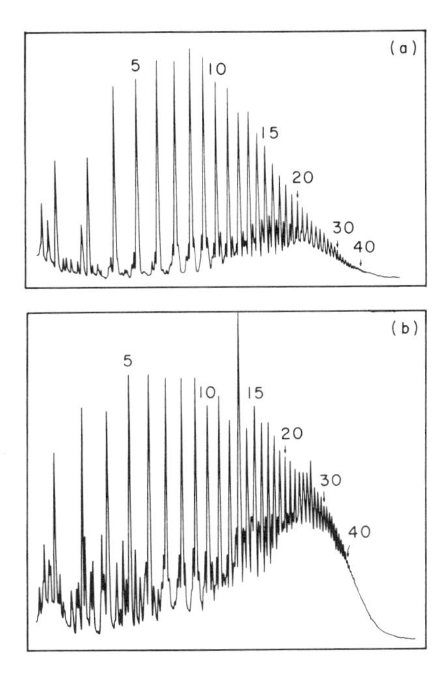

Figure 11. HPLC analysis showing the series of oligomers of guanylic acid produced by the oligomerization the 2-MeImpG on a poly(C) template. Two sets of products are shown; the lower reaction was carried out using higher concentrations of reactants. Reproduced with permission from Inoue and Orgel (1982).

A second problem is encountered in attempts to polymerize adenylic acid using a poly(U) template. Under the conditions of the synthesis two chains of poly(U) form a triple complex with the monomeric adenylic acid, resulting in a poor reaction in which short chains containing both 2'-5' and 3'-5' linkages are formed. The situation can be improved to some extent by incorporating U, A and G into a chain composed largely of C, which is then capable of catalyzing the synthesis of complementary oligomers.

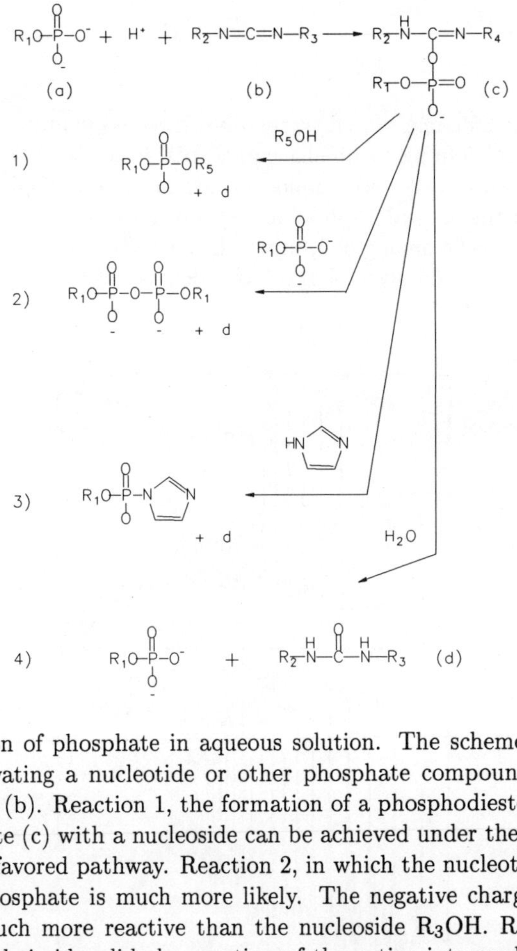

Figure 12. Activation of phosphate in aqueous solution. The scheme shows some of the possibilities for activating a nucleotide or other phosphate compound (a) with a water-soluble carbodiimide (b). Reaction 1, the formation of a phosphodiester by reaction of the activated intermediate (c) with a nucleoside can be achieved under the most favorable conditions, but is not a favored pathway. Reaction 2, in which the nucleotide reacts with itself to produce a pyrophosphate is much more likely. The negative charge on the phosphate makes this group much more reactive than the nucleoside R_3OH. Reaction 3 shows the formation of a phosphoimidazolide by reaction of the active intermediate with imidazole. This compound, although an active phosphorylating agent, is more stable than (c), which reacts quite rapidly with water to reform the starting material. Phosphoimidazolides are useful compounds for studying template-directed oligomerizations of nucleotides.

However, since C must make up more than 50% of the residues in the chain in order to have a satisfactory reaction, a second stage of replicating the synthetic oligomers is once again impossible, since they now contain more than 50% G. It may already be obvious that the requirements for obtaining efficient nonenzymatic replication of a polynucleotide are very restrictive. The problem appears to become even more difficult when it is realized that the initial stages in chemical evolution must have occurred in a mixture of closely related compounds. It is certainly not likely that nucleotides could have been synthesized which consisted exclusively of the four bases of RNA attached in just the right way to just the right sugar in just the right configuration, etc. Indeed, synthetic experiments have demonstrated that the biological nucleotides are among the least likely compounds to form spontaneously. It is worth looking into the reasons for this statement.

4 Prebiotic Syntheses

The starting materials for any "prebiotic" synthesis must be compatible with some geologically reasonable model of the conditions on the prebiotic earth. We have only indirect evidence concerning these conditions. The earth was formed 4550 My (10^6 years) ago and fossil remains as well as other evidence show that life was already present at 3.5 My. The oldest rocks in existence, which are 3.75 My of age show clearly that liquid water was present at this time, and the carbon isotope data can be interpreted as supporting the presence of life (see Schidlowski). At the very least this suggests that life emerged on earth perhaps 4 My ago under conditions not very different from today, except for the absence of significant concentrations of free oxygen. The early atmosphere would have contained N_2, CO_2 and an uncertain but low concentration of CH_4. The presence or absence of methane is the primitive atmosphere is a very important problem, as the formation of HCN (an important precursor of biologically significant products) is highly dependent on its presence (or of an equivalent amount of H_2).

Ever since the success of the "Miller Experiment" and its followers (Miller and Orgel, 1974), chemical evolutionists have been tempted to invoke the "dilute soup" of Haldane in which all of the presumed ingredients for the first organisms, or at least their macromolecules, would have been present. The model chosen by Miller, based on a highly reducing atmosphere which was postulated by both Haldane and Oparin, contained a large concentration of CH_4. The "myth" of the prebiotic soup has been attacked a number of times, notably by the geochemist Sillen, and it has become fashionable in recent years to ridicule the Oparin-Haldane Hypothesis. An important reason for the disrepute into which the soup has fallen is the realization that the early atmosphere of earth is more likely that have been "redox-neutral" (i.e., CO_2 and N_2) rather than reducing. Recently the appreciation of the probable contributions of cometary impacts on the early Earth (Greenberg, 1984; Greenberg and Mendoza-Gomez, 1992; Chyba and Sagan, 1992) have led to a kind of critical relaxation in this regard, the temptation being to suggest that what we can't make in the early atmosphere we can get from comets. However, for reasons which I hope to make clear, there is really no sound basis for assuming that comets, whatever their organic content, can solve some of the more vexing problems involved in creating a suitable environment for the origin of life. It will be beneficial to consider the possible sources of the building blocks required for the synthesis of mononucleotides.

Some degree of confusion appears to exist regarding the synthesis of purines and pyrimidines. The purine adenine is unquestionably formed in low yield (up to 0.1%) under mild conditions from HCN. Guanine, in spite of frequent statements to the contrary, has never been reliably identified as a product of HCN oligomerization, nor has any prebiotically acceptable synthesis of this key purine been reported. It can be synthesized from intermediates formed from reactions of HCN, but requires high concentrations of NH_3. The pyrimidines cytosine and uracil have been synthesized, although the formation of these pyrimidines is generally viewed as being somewhat less probable than that of adenine, due to the number of steps involved. However, the synthesis of uracil from cyanoacetylene (after HCN the most important product from a spark discharge in N_2-CH_4 mixtures) and urea (a product of the HCN oligomerization) seem quite reasonable. In contrast to the difficulties which have been experienced in laboratory simulations, investigations of carbonaceous

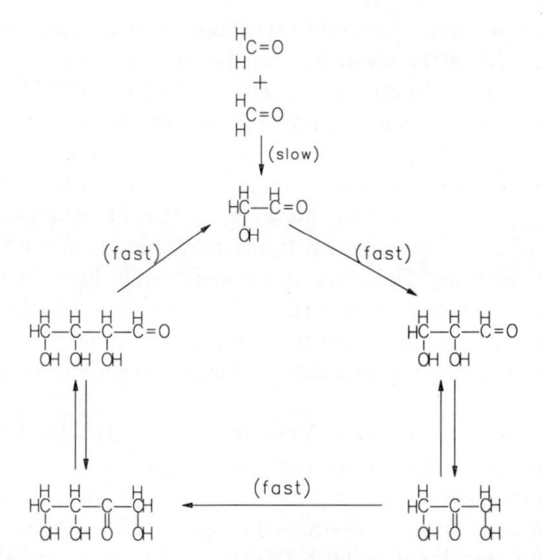

Figure 13. The "formose" reaction is the autocatalytic condensation of formaldehyde (CH_2O) to produce sugars (of the general formula $C_nH_{2n}O_n$). Only the autocatalytic cycle is shown. Although many more reactions take place (indicated in part by the arrows proceeding outward from the cycle), the few reactions shown account for the autocatalytic kinetics. The rate-limiting step appears to be the condensation of two molecules of formaldehyde to produce one molecule of glycolaldehyde (a). The subsequent additions of formaldehyde and rearrangements shown, which take place much more rapidly than the formation of glycolaldehyde, culminate in the disassociation of a tetrose (e) into two molecules of glycolaldehyde. This doubling of the initial concentration of key reactants results in a reaction which continues to accelerate as long as the concentration of formaldehyde remains high.

chondrites have revealed the presence of the biological purines adenine and guanine, as well as the related purines hypoxanthine and xanthine. In addition, uracil was also detected, but in much lower concentrations. Neither cytosine nor thymine have been identified. Summarizing, a cometary/meteoritic source of purines seems reasonable, with possible terrestrial synthesis of pyrimidines. As in all syntheses of such compounds, however, they are inevitably accompanied by other structures containing chemically similar groups.

The only source of ribose that we know of is the "formose" reaction, which is the base-catalyzed condensation of formaldehyde. This is an autocatalytic reaction which rapidly converts this eminently plausible starting material into a mixture containing just about every possible carbohydrate. The success of this reaction is its undoing, since all simple sugars (such as ribose) tend to be converted to more complex molecules with increasing time. A more serious problem is that all of the products have very nearly the same chemical properties (all sugars, being oligomers of formaldehyde, have the empirical formula $C_nH_{2n}O$). Therefore, all of the isomers tend to react in similar ways with other compounds.

There are some ways of influencing the outcome of the formose reaction, although the conditions involved are fairly arbitrary. For example, careful manipulation of the reaction conditions can result in the selective formation of highly branched sugars. Certain branched sugars such as pentaerythritol can become major products under photochemical conditions in highly concentrated formaldehyde solutions (Shigemasa et al., 1977). We have recently confirmed this result and demonstrated that pentaerythritol is formed with great selectivity even in dilute solution. These syntheses suggest some interesting alternative possibilities to ribose, but do not yet seem to explain how the first nucleotides might have arisen (see, however Schwartz and Graaf, 1993a and b). Work from Eschenmoser's laboratory has demonstrated that a selective condensation of glycolaldehyde phosphate with itself and with formaldehyde can be achieved, producing a fairly selective synthesis of ribose-2,4-diphosphate. However, there is no known prebiotic synthesis of glycolaldehyde phosphate.

A possible way out of the dilemma caused by the complexity of the mixtures of products which result from reasonable prebiotic synthesis could conceivably exist in an evolutionary process, by means of which a polynucleotide template could select for the "correct" structures. Let us for the sake of argument assume that in spite of the problems which have already been described, somehow a nonenzymatically self-replicating polynucleotide somehow or other was formed. Might this structure not be capable of selecting only one set of monomers out of a mixture of various structures? Experiments have been conducted to test this hypothesis. Specifically, the ability of poly(C) to discriminate between the D- and L- optical isomers of guanylic acid was tested. This is a critical test, since while one might hypothesize that selective chemical syntheses might have occurred spontaneously to form more of one sugar than another, this would not solve the problem of optical isomerism. Since the chemical properties of optical isomers are otherwise identical, only an optically active structure can distinguish between them. In the test mentioned, it was found that poly(C) (which was the enzymatically synthesized polymer composed exclusively of D-cytidylic acid) catalyzed the oligomerization of the D-monomer and, as expected, not the L-monomer. However, when a racemic (50% D + 50% L) mixture of the monomers was used, the results were quite discouraging. A heterogeneous mixture of short oligomers was obtained, consisting of several kinds of linkages. Apparently, the template could not adequately distinguish between the isomers, so that incorporation of the L-form did occur, distorting the interaction between template and monomer, and resulting in both linkage errors and premature termination of the oligomerization.

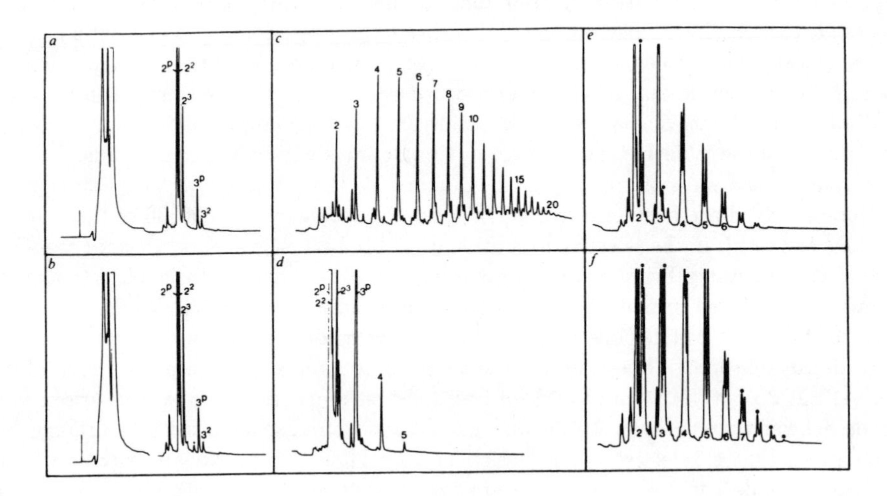

Figure 14. Enantiomeric cross-inhibition. a) and b) show the products (mostly dimers) of the oligomerization of the D- and L- forms of 2-MeImpG, respectively. c) Demonstrates the oligomerization of D-monomer on a (D) poly(C) template, and d) shows the virtual lack of oligomerization of the L-monomer on the same template. When a racemic mixture of D- and L- isomers was oligomerized on poly (C), the products shown in e) were formed (f is the same reaction as e, but at 2X higher concentration). Reproduced with permission from Joyce et al. (1984).

5 Primitive Precursors of RNA?

In the light of these results, it might be asked whether the concept of an RNA world makes sense. If RNA molecules can only reproduce themselves in a perfectly attuned environment, how could they have arisen in the first place? It is for this reason that merely having a supply of randomly synthesized organic molecules (for example, transported to Earth by comets), still leaves an enormous problem unsolved; namely, what was the *selection process* which permitted the synthesis of polynucleotides and polypeptides? One possibility which has been suggested is that self-replication and information storage arose in another, simpler molecule, which later evolved into RNA. A possible example of such a precursor is shown in Figure 10. Here the five carbon sugar ribose has been replaced by a three carbon sugar-alcohol (glycerol), combined with a molecule of formaldehyde. This structure is interesting because the monomers are completely symmetrical and therefore do not display optical isomerism. Another interesting feature of such glycerol-based nucleotide analogs is that the linkage between the units can be formed by the reaction of phosphate groups with each other to form pyrophosphate linkages. This reaction is known to proceed more readily than the formation of a phosphodiester linkage. To establish this point, the monomer shown as 2 in Fig. 10 was oligomerized on a poly(C) template. Although the deoxynucleoside diphosphate 2 is larger and seems more complex than the activated ribonucleotide 1, it is in a sense chemically simpler, since the number of ways this compound can react are restricted (The only bonds which can be formed are between the two phosphate groups).

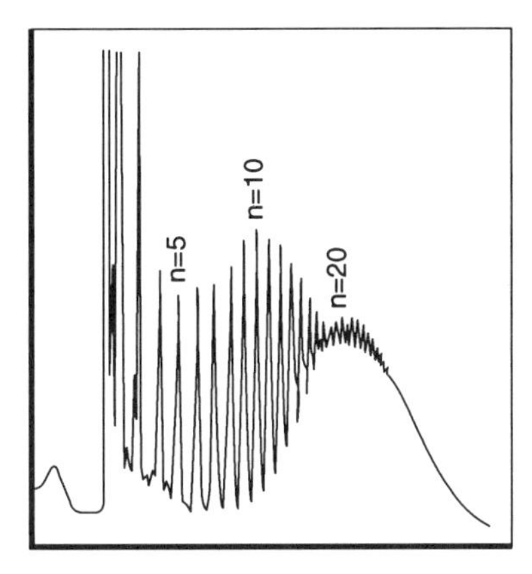

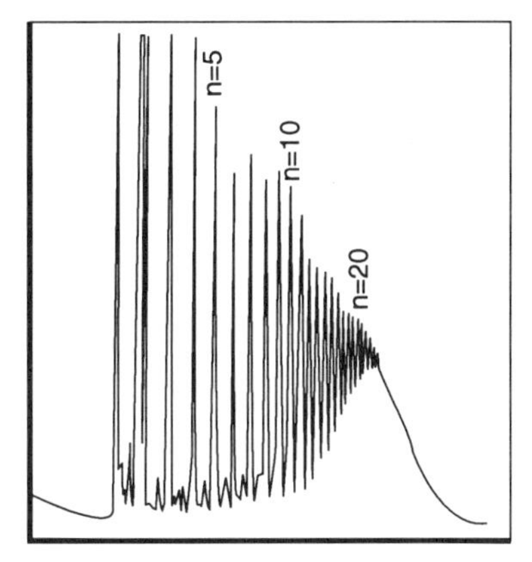

Figure 15. Oligomerization of deoxynucleotide diphosphates on polynucleotide templates. A) Oligomerization of ImpdApIm in the presence of poly(U). B) As A), but without template. C) Oligomerization of ImpdGpIm in the presence of poly(C). D) as C), but without template (from Schwartz et al., 1987).

The products, which are based on long chains of pyrophosphates, must form double-helical structures with the poly(C) in order to account for the catalytic effect. This experiment, although conducted with the aid of analogs based on deoxynucleotides (which are thought to have evolved later than ribonucleotides), demonstrated the possibility that other, even simpler structures might be employed in template-directed reactions. The next step was to substitute a glycerol-formaldehyde unit for ribose (3 in Fig. 10). These monomers also polymerize readily on polynucleotide templates (Fig. 16), although the extent of the reaction is somewhat less than that which was achieved using deoxynucleoside diphosphates as monomers. This difference is at least partially due to the fact that the more flexible glycerol structure allows the two phosphate groups to react with each other to form an internal pyrophosphate, as well as to form internucleotide linkages. A good part of the starting material is consumed in internal cyclization.

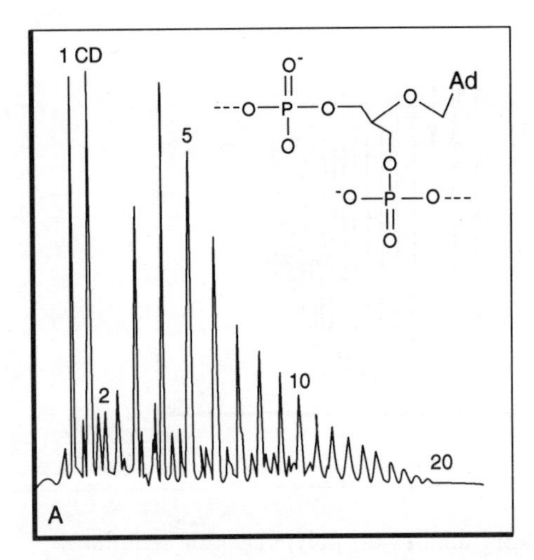

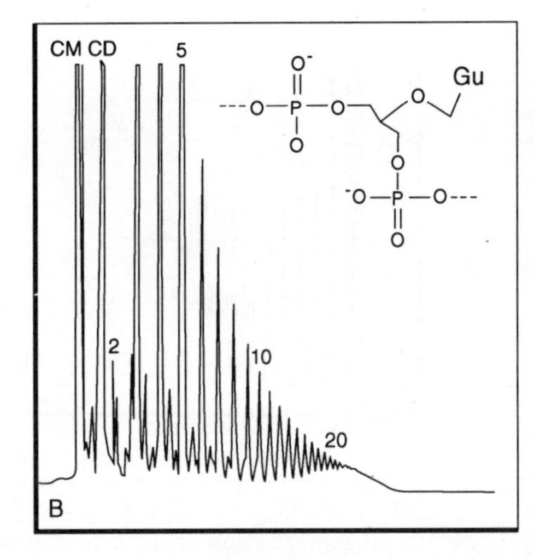

Figure 16. The oligomerization of acyclic (glycerol-based) nucleotide analogs on polynucleotide templates. A) Products of the oligomerization of ImpÃpIm on poly(U). B) Products of the oligomerization of ImpG̃pIm on poly(C) (Visscher and Schwartz, 1988).

Up to this point, enzymatically synthesized polynucleotides had been used as templates. When the corresponding model templates based on cytosine-containing analogs were synthesized and tested for their effect on the oligomerization of guanine-containing monomers, the results shown in Figures 17 and 18 were obtained.

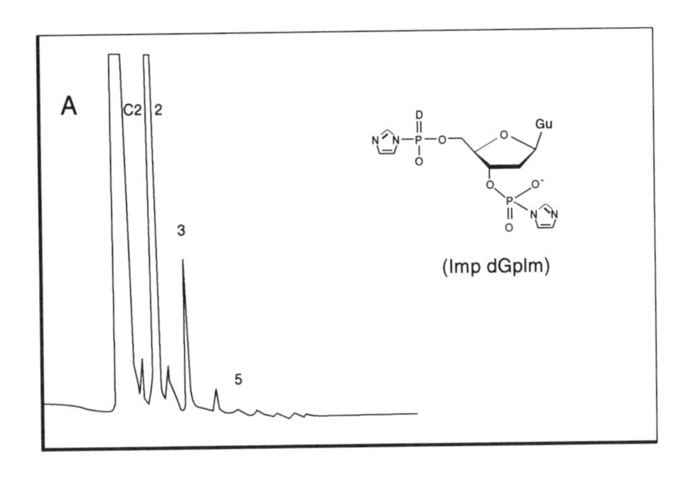

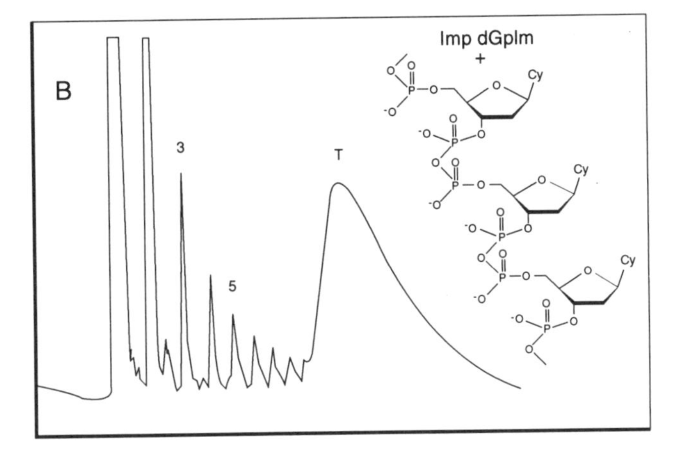

Figure 17. Oligomerization of the analog ImpdGpIm on a pyrophosphate-linked analog template. A) Products obtained from the monomer in the absence of a template. B) Products obtained in the presence of the template. Oligomers up to the 9mer are visible; higher products are obscured by the template (marked T) (from Visscher et al., 1989).

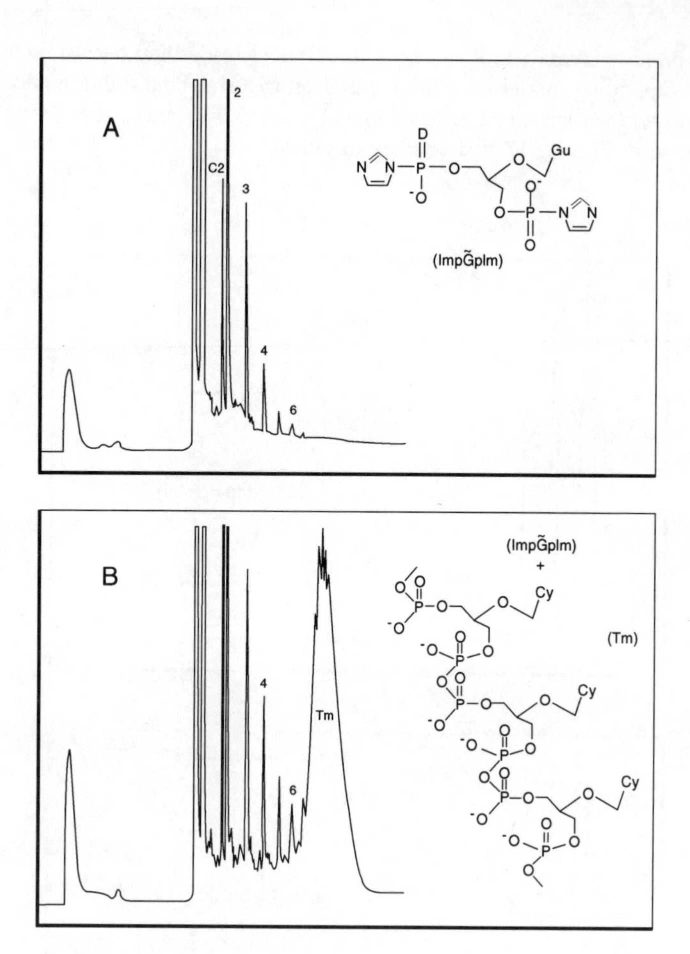

Figure 18. Oligomerization of the acyclic analog ImpG̃pIm on a pyrophosphate-linked, glycerol based template. A) Products obtained in the absence of a template. B) Products obtained in the presence of the template (marked Tm. From Visscher and Schwartz, 1990).

There is a definite catalytic effect observable in both reactions, although the effect in the case of the acyclic analog can be described as somewhat disappointing. There are several plausible explanations for these results. In the first place, the charge- repulsion between two backbones consisting of pyrophosphates may destablize the structure. Attempts to reduce the charge effect by making use of multivalent metal ions and of various polyamines related to spermine and spermidine have not solved the problem. It is also possible that the highly flexible glycerol-pyrophosphate backbone may provide too much freedom of vibration for it to form a truly stable double helix. If this explanation is correct, then we may need to think again about the kinds of structures which might have been ancestral to the RNA world.

References

Cech, T.R. (1986) A Model for the RNA-Catalyzed Replication of RNA. Proc. Natl. Acad. Sci USA 83, 4360-4363

Cech, T.R. (1987) The Chemistry of Self-Splicing RNA and RNA Enzymes. Science 236, 1532-1539.

Cech, T.R. (1989) Ribozyme Self-Replication ? Nature 339, 507-508.

Chyba, C.F. and Sagan, C. (1992) Endogenous Production, Exogenous Delivery and Impact Shock Synthesis of Organic Molecules: An Inventory for the Origins of Life. Nature 355, 125-132.

Doolittle, W.F. (1991) The Origins of Introns. Current Biology 1, 145.

Eigen, M., Gardiner, W., Schuster, P. and Winkler-Oswatitsch, R. (1981) The Origin of Genetic Information. Scientific American 244, 88-108.

Greenberg, J.M. (1984) Structure and Evolution of Inter- stellar Grains". Scientific American 250, 124-135.

Greenberg, J.M. and Mendoza-Gomez, C.X. (1992) The Seeding of Life by Comets. Adv. Space Res. 12, 169-180.

Inoue, T., Orgel, L.E. (1982) Oligomerization of (Guanosine 5'-Phosphor)-2-methylimidazolide on Poly(C). An RNA Polymerase Model. J. Mol. Biol. 162, 201-217

Inoue, T., Orgel, L.E. (1983) A Nonenzymatic RNA Polymerase Model. Science 219, 859-862.

Joyce, G.F., Visser, G.M., van Boeckel, A.A., van Boom, J.H., Orgel, L.E., Westrenen van, J. (1984) Chiral Selection in Poly(C)-Directed Synthesis of Oligo(G). Nature 310, 602-604.

Kruger, K., Grabowski, P.J., Zaug, A.J., Sands, J., Gottschling, D.E., Cech, T.R. (1982) Self-Splicing RNA: Autoexcision and Autocyclization of the Ribosomal RNA Intervening Sequence of Tetrahymena. Cell 37, 147-157.

Miller, S.L., Orgel, L.E. (1974) The Origins of Life on the Earth. Prentice-Hall, Englewood Cliffs, N.Y.

Oparin, A.I. (Translated by S. Margulis) (1938) The Origin of Life. Macmillan, N.Y.

Orgel, L.E. (1986) RNA Catalysis and the Origins of Life. J. Theor. Biol. 123, 127-149.

Schwartz, A.W., Graaf, R.M. de. (1993a) The Prebiotic Synthesis of Carbohydrates - A Reassessment. J. Mol. Evol. 36, 101-106.

Schwartz, A.W., Graaf, R.M. de. (1993b) Nucleotide Analogs Based on Pentaerythritol - An Hypothesis. Origins of Life 23, 185-194.

Schwartz, A.W., Visscher, J., Bakker, C.G., Niessen, J. (1987) Nucleic Acid-Like Structures II. Polynucleotide Analogues as Possible Primitive Precursors of Nucleic Acids. Origins of Life 17, 351-357.

Shigemasa, Y., Matsuda, Y., Sakazawa, C., Matsuura, T. (1977) Formose Reactions II. The Photochemical Formose Reaction. Bull. Chem. Soc. Japan 50, 222-226.

Visscher, J., Bakker, C.G., van der Woerd, R., Schwartz, A.W. (1989) Template-Directed Oligomerization Catalyzed by a Polynucleotide Analog. Science 244, 329-331.

Visscher, J., Schwartz, A.W. (1988) Template-Directed Synthesis of Acyclic Oligonucleotide Analogues. J. Mol. Evol. 28, 3-6.

Visscher, J., Schwartz, A.W. (1990) Template-Catalyzed Oligomerization with an Atactic, Glycerol-Based, Polynucleotide Analog. J. Mol. Evol. 31, 163-166.

Woese, C.R. (1981) Archaebacteria Sci. Am. 244, 98-122.

CHIRALITY AND THE ORIGINS OF LIFE

A. BRACK
Centre de Biophysique Moléculaire
1A, Avenue de la Recherche Scientifique
45071 - Orléans Cedex 2
France

ABSTRACT. Molecular asymmetry is examined in present life, and in prebiotic systems. The origin of chirality has been ascribed to a chance choice of the initial enantiomer or to a universal disymmetric force. The weak neutral current of the electro weak interaction provides a constant and uniform chiral agency which favors both L-amino acids and D-sugars. Amplification of a small excess in one enantiomer can be achieved in open systems far from equilibrium, by crystallization or via ordered structures of biopolymers. A tentative coherent approach to the problem of chirality in the course of evolution is presented in the form of an integrated model which includes all of the main homochiral biomolecular families (amino acids, sugars, lipids).

1. Introduction

In contrast to the atomic theory which can be traced back to the greek and latin philosophers, the notion of an asymmetry at the molecular level has appeared only during the last century. Although nature has provided many macroscopic asymmetrical examples such as the helical twist of a snail shell or of the pine cone scales, most living beings share a morphological symmetry either radial of bilateral. It was a great event when Pasteur discovered that living beings can use only one enantiomer. The mold grew well when fed with L-tartrate crystals, whereas it starved if the nutrient solution was made of the D-crystals.

J. M. Greenberg et al. (eds.), The Chemistry of Life's Origins, 345–355.

Pasteur established for the first time the rule that optical activity is the demarcation line between animate and inanimate kingdoms and expressed the feeling that life is dominated by disymmetric actions whose existence is cosmic.

Present terrestrial life is dominated by proteins which catalyze biochemical reactions, nucleic acids which carry genetic information and a lipidic micellar system which forms the cellular protecting membranes. Most of the constituents, i.e. amino acids, sugars and lipid polar heads contain at least one asymmetric carbon atom (Figure 1).

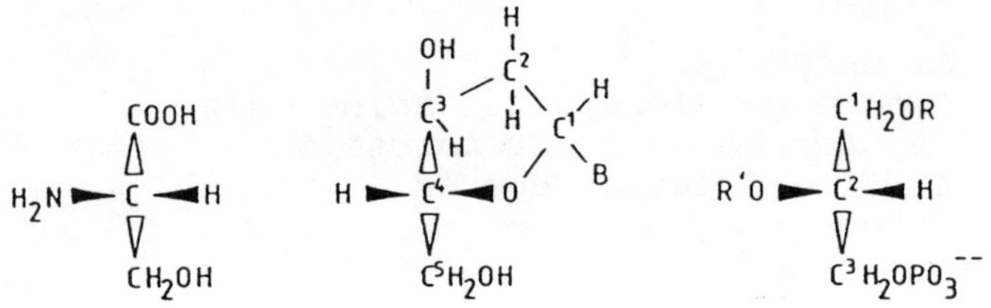

Figure 1 From left to right : L-serine, deoxy-β-D-ribofuranose, L-phosphatidic acid.

The nineteen chiral proteinaceous amino acids belong without exception to the L-configuration class, whereas the two sugars found in nucleic acids are related to the D-series. This assignment occured by a mere convention that consisted of connecting the configuration of the sugar asymmetric carbon atom which is most distant from the carbonyl group to the configuration of the amino acid asymmetric carbon atom through lactic acid by a series of chemical reactions. Had another sequence of reactions been selected, a uniform configuration could have been found for amino acids and sugars.

The biopolymers themselves form asymmetric helical structures and superstructures, the α and β conformations of polypeptides, the A,B and Z-forms of nucleic acids and the helical conformations of polysaccharides.

2. Molecular asymmetry before life

Any chemical reaction producing chiral molecules in statistically large numbers which is run in a symmetrical environment yields a racemic mixture, i.e. a mixture of equal quantities of right and left handed enantiomers. This was already the case when the first fundamental constituents of biomolecules were formed.

However, in view of the importance of optical purity in present life, it is difficult to believe that, at the beginning, a completely racemic life using simultaneously biomolecules of both configurations in the same protocell arose. For racemic life, the diversity of the biomolecules would be much higher ; the number of molecules with n asymmetric centers would be increased by a factor 2^n. Thus, the problem set has two parts, the origin of a prevalent enantiomer and its further amplification until the appearance of life.

Theoretical models of the origin of chirality on Earth can be divided into two classes, those which call for a chance mechanism and those which call for a determinate mechanism resulting from an asymmetrical environment originating from the Universe or from the Earth.

3. Chance mechanism

The proponents of the chance mechanism argue that the notion of equimolarity of a racemic mixture is only relative. For a relatively small number of molecules, random fluctuations may favor one enantiomer over the other. For instance, for a population of ten million molecules, which is about the amount of chiral constituents of the smallest living cell, there is a 50 % chance to find a 0.02 % or more excess in one enantiomer (Spach and Brack, 1983 ; Spach and Brack, 1988). But it is not yet known whether such a small amount is sufficient for further amplification by mechanisms to be described later.

In a rather simple kinetic model proposed by Frank (1953), an open flow reactor, run in far from equilibrium conditions, is fed by achiral compounds and forms two enantiomers reversibly and autocatalytically. If the two enantiomers can react to form an irreversible combination flowing out the ractor (by precipitation, for instance), and if certain conditions of fluxes and concentrations are reached, the racemic production may become metastable and the system switches permanently toward the production of either one or the other enantiomer, depending on a small excess in one enantiomer. The model was generalized and detailed by Morozov (1979) and Avetisov et al (1991).

Spontaneous resolution on crystallization represents the most effective means of chiral symmetry breaking since optically pure enantiomers can be isolated on scales ranging from grams to tons with the help of homochiral seeds. Havinga first described in 1941 and 1954 a system in which two enantiomeric forms of a compound are in rapid equilibrium in solution and only in solution. When the solution is supercooled, a spontaneous or induced precipitation of one of the two forms may occur until the supercooling of the solution stops. Another nice example was reported by Pincock (1971). Enantiomers of 1,1'-binaphtyl interconvert in solution by rotation around a C-C bond at a very low rate at room temperature but very rapidly in the melt above 158 C. By heating in sealed tubes at 180 C in the absence of any chiral contaminant, then cooling at 130 C, the compound remains supercooled for many hours and the crystallization can be initiated by touching the tube with a tip of dry ice. The crystals were always found to be enriched in one enantiomer but the average after many experiments was, of course, strictly 50 % left-handed and 50% right-handed enantiomers. The nucleation is merely determined by chance. More recently, Kondepudi et al (1990) demonstrated another total spontaneous resolution by crystallization. Sodium chlorate crystals are optically active although the molecules of the compound are not chiral. When crystallized from an aqueous solution while the solution is not stirred, statistically equal numbers of L and D crystals were found. When the solution was stirred, almost all the $NaClO_3$ crystals (99.7 percent) in a given sample had the same chirality, either L or D. Did terrestrial quartz, known as an asymmetric adsorbent and catalyst, undergo such a spontaneous resolution during crystallization ? Close examination of over 2700 natural quartz crystals gave 49.83% L and 50.17 % D. The even distribution is supported by the fact that the chirality of a quartz crystal has no influence on its sale price !

As far as we know today, clays do not help the separation of enantiomers.

4. Determinate mechanisms

Parity non conservation has raised many hopes and caused many disappointments. This fundamental asymmetry of matter has been examined from various aspects such as circularly polarized photons emitted by the slowing down of longitudinally polarized electrons (Bremsstrahlung), inducing degradation reactions or stereoselective crystallization of racemic mixtures. No experiment has convincingly supported these theoretical considerations for the origin of a dominant enantiomer on Earth. Either the results were shown to be artefacts or to be so weak that they are doubtful (Bonner, 1991).

Parity is also violated in the weak interactions mediated by neutral bosons Z°. All electrons are intrinsically left-handed, their momentum and spin being more likely antiparallel. The antimatter counterpart, the positrons, are intrinsically right-handed. Therefore, L-serine, made of left-handed electrons, and D-serine, also made of left-handed electrons, are not true enantiomers but diastereoisomers. There is a very tiny parity violating energy difference in favor of L-amino acids in their preferred conformations in water and in favour of D-sugars (Mason and Tranter, 1985 ; Mac Dermott and Tranter, 1989). The energy difference is about 3×10^{-19} eV corresponding to one part in 10^{17} for the excess of L-molecules in a racemic mixture at thermodynamic equilibrium at ambient temperature.

Other chiral force fields that could have been acting on the Earth surface have been looked for. Asymmetric synthesis and degradation have been realized with success with circularly polarized light. An original approach using Earth gravity and macroscopic vortex has been tested (Dougherty, 1981).

Unfortunately, the classical electromagnetic interactions such as circularly polarized light or other fields that can be imagined acting on Earth would probably never result in a very high yield of optically pure compounds. They would also probably cancel on a time and space average.

Recently, Engel et al (1990) reported an excess of L-type alanine in the Murchison meteorite (D/L of 0.85). If the data of Engel are valid, a meteoritic preference for L-amino acids would push the problem of the origin of biological chirality out into the cosmos. Among the possible extraterrestrial sources of circularly polarized light, sunlight is generally discounted as being probably too weak and not of a consistent handedness for sufficiently long periods. According to Bonner and Rubenstein (1987) synchrotron radiation from the neutron star remnants of supernova events is a better candidate. Interaction of neutron star circularly polarized light with interstellar grains in dense clouds could produce chiral molecules in the organic mantles by partial asymmetric photolysis of mirror image molecules (Bonner, 1991 ; Briggs et al, 1992 ; Greenberg et al, 1993). The delivery of the chiral molecules to the Earth may have been achieved via comets or/and asteroids.

5. Enantiomer enrichment processes

5.1. AMPLIFICATION OF ELECTROWEAK ADVANTAGE

The problem of amplification of the electroweak advantage of L-amino acids (3.10^{-19} eV) over D-amino acids has been considered by Kondepudi and Nelson (1985) following the seminal ideas of Frank (1953). Using an autocatalytic mechanism and the theory of delayed bifurcations, they calculated that a lake of 1 km^2 and 4 m deep would be large enough to enrich the favored enantiomer by up to 98 % in 10^4 years. This possibility was disputed by Goldanskii (Avetisov et al, 1987) who argued that the time required to pass through the critical bifurcation point is too long : fluctuations could throw the system into the opposite handedness.

Starting from Z° interactions, Salam (1991) speculated on enantiomeric amplification in terms of quantum mechanical cooperative and condensation phenomena which would give rise to D to L transformations below a critical temperature T_c. The number of D and L enantiomers in any given energy level is a function of temperature. Below T_c, there are not enough quantum states available to accomodate all the enantiomers. Below T_c the system adjusts by taking all the particles which cannot be accomodated by the distribution formula (surplus particles) and putting them in the single quantum state which has the lowest energy (Leggett, 1990). Thus, at low temperature, a macroscopic number of molecules occupy the single quantum state of lowest energy, i.e. L-amino acids.This phenomenon is known as Bose condensation. Ideally, one should be able to compute T_c when electroweak theory is fully worked out. At present, the best way to determine T_c for a given amino acid is by experimentation. Numerical values are eagerly anticipated.

5.2 AMPLIFICATION VIA ORGANIC CRYSTALS

Glycine, the only symmetrical natural amino acid, gives centrosymmetric crystals with a bipyramidal symmetry. When adding a racemic D,L amino acid, D-enantiomers occlude on one face whereas L-amino acids crystallize on the opposite face. By addition of a few hydrophobic D-leucine to glycine crystals, the crystals orient at the air/water interface pointing the D-leucine site toward the air and the L-occluding site toward the aqueous phase. In the presence of racemic D,L glutamic acid, the asymmetric floating crystals occlude preferentially L-glutamic acid up to 100 % (Weissbuch et al, 1984, 1988).

5.3 AMPLIFICATION VIA BIOPOLYMERS

The stereoselective formation of covalent compounds has been approached through polymerization of chiral monomers such as activated α-amino acids, as first suggested by Wald in 1957 (Lundbeerg and Doty, 1957). For instance, a right handed α helical seed made of L-amino acids incorporates L-amino acid monomers 18 times faster than D-amino acids (Brack and Spach, 1971). A slight dissymmetry existing in a racemic mixture of leucine N-carboxy anhydride can be amplified in the α helical polypeptide on partial polymerization. Leucine polymers were also found to hydrolyze stereoselectively, providing for additional enantiomeric enhancement (Bonner et al, 1981).

We have shown that homochiral synthetic polypeptides having alternating hydrophobic (leucyl) and hydrophilic (lysyl) residues adopt a β-sheet conformation (Brack and Orgel, 1975). The progressive introduction, at random, of amino acids of opposite configuration in the alternating sequences diminishes progressively the proportion of β-structures. The result of this is to generate homochiral β-sheet cores attached to random coil portions of the polymer containing both L- and D-residues (Brack and Spach, 1979). On partial hydrolysis of such unbalanced polymers, it was found that the random coil portions hydrolyzed more readily allowing the isolation of unhydrolyzed β-sheet fractions enriched in the dominant enantiomer (Brack and Spach, 1980).

The prebiotic formation of nucleotides and polynucleotides is still a challenging problem, from both the chemical and configurational points of view, complicated by the presence of several asymmetric centers in the sugar moiety. Chiral selection has been observed by Joyce et al(1984) during poly(C)-directed oligomerization of activated guanosine mononucleotides. However, in template directed reactions with a racemic mixture, monomers of the opposite handedness to the template are incorporated as chain terminators, thus inhibiting further growth of the chain (see Schwartz, this volume). Glycerol has been considered as a chirally simplified analogue of ribose (Spach, 1984 ; Joyce et al, 1987) although it will introduce more flexibility into the polymer chain. Glycerol is found today in the phosphoric acid polyester backbone of natural teichoic acids, grafted with aminoacids and sugars. Acyclic and cyclic prochiral glycerol phosphate derivatives substituted by nucleic bases are presently studied as monomers in different laboratories (see Schwartz, this volume).

Limited attention has been given to the optical purity of lipids and to the possible stereoselective role of membranes. The idea of spontaneous two dimensional resolution of chiral lipids as a mean to provide an asymmetric and stereoselective barrier was first expressed by Arnett and Thomson (1981). The experiments revealed, however, that dipalmitoylphosphatidyl choline molecules behave as if they were non chiral. The asymmetric carbon atom of the polar head appears to be buried in the rest of the molecule. But evidence for a partial resolution in a chiral monolayer has been provided with a stearamide bearing an optically active amide moiety (Arnett and Gold, 1982). The interaction of D- and L-alanine with an optically active model membrane system has been examined by Tracey and Diehl (1975). The NMR study showed clearly that the D- and L-amino acids were ordered differently in a decyl-2-sulfate phase.

6. Conclusion

The enrichment experiments reported in the literature, although very significant, are generally restricted to one family of compounds, mainly amino acid derivatives. One has now to elaborate a model (Spach and Brack, 1988) that tends to integrate all biomolecules and which extends the previous ideas of Wald (1957). The formation of small asymmetric volumes bounded by a semi-permeable membrane composed of chiral molecules is suggested as a stereoselecting and concentrating device. It is likely that living organisms inherited optical activity as a result of molecular selection processes which formed structures of higher order and complexity with gain of stability. The increased stability helped to resist spontaneous racemization, the half-time of which lies about 10^5-10^6 years at ambiant temperature for amino acids (Bada and Miller, 1987). Other calamities were also occuring on the primitive Earth such as meteoritic and cometary impacts. According to Oberbeck and Fogleman (1989), intervals between life-annihilating impacts were about 2.5 to 11 million years, 3.8 Gyr ago.

Perhaps two mirror-image populations of primitive living systems developed and competed at the beginning, until one of them won the fight. Most probably, there was never any racemic life using both L and D monomeric units in the chains.

In the context of life's origin, the most relevant aspect of chirality is homochirality. Thus, efforts must be focused on the search for stereoselective processes leading to homochirality, leaving the choice of sign (L or D) to a chance event.

7. References

Arnett, E.M. and Thompson, O. (1981) 'Chiral aggregation phenomena-2. Evidence for partial "two-dimensional resolutions" in a chiral monolayer', J. Amer. Chem. Soc. 103, 968-970.

Arnett, E.M. and Gold, J.M. (1982) 'Chiral aggregation phenomena-4. Search for stereospecific interactions between highly purified enantiomeric and racemic dipalmitoyl phosphatidylcholines and other chiral surfactants in monolayers, vesicles and gels', J. Amer. Chem. Soc. 104, 636-639.

Avetisov, V.A., Kuz'min, V.V. and Anikin, S.A. (1987) 'Sensitivity of chemical chiral systems to weak asymmetric factors', Chem. Phys. 112, 179-187.

Bada, J.L. and Miller, S.L. (1987) 'Racemization and the origin of optically active organic compounds in living organisms', Biosystems 20, 21-26.

Bonner, W.A., Blair, N.E. and Dirbas, F.M. (1981) 'Experiments on the abiotic amplification of optical activity', Origins of Life 11, 119-134.

Bonner, W.A. and Rubenstein, E. (1987) 'Supernovae, neutron stars and biomolecular chirality', BioSystems 20, 99-111.

Bonner, W.A. (1991) 'The origin and amplification of biomolecular chirality' Origins of Life 21, p 59-111.
Brack, A. and Spach, G. (1971) 'New example of helical polymerization', Nature 229, 124-125.
Brack, A. and Orgel, L.E. (1975) 'β-structures of alternating polypeptides and their possible prebiotic significance', Nature 256, 383-387.
Brack, A. and Spach, G. (1979) 'β-sptructures of polypeptides with L- and D-residues Part I. Synthesis and conformational studies', J. Mol. Evol. 13, 35-46.
Brack, A. and Spach, G. (1980) 'β-structures of polypeptides with L- and D-residues Part III. Experimental evidences for enrichment in enantiomer', J. Mol. Evol. 15, 231-238.
Briggs, R., Ertem, G., Ferris, J.P., Greenberg, J.M., McCain, P.J., Mendoza-Gomez, C.X. and Schutte, W., 1992 'Comet Halley as an aggregate of interstellar dust and further evidence for the photochemical formation of organics in the interstellar medium' Origins of Life 22, p 287-307.
Dougherty, R.C. (1981) 'Chemical geometrodynamics : physical fields can cause asymmetric synthesis', Origins of Life 11, 71-84.
Engel, M.H., Macko, S.A. and Silfer, J.A. (1990) 'Carbon isotope composition of individual amino acids in the Murchison meteorite', Nature 348, 47-49.
Frank, F.C. (1953) 'On spontaneous asymmetric synthesis', Biochim. Biophys. Acta 11, 459-463.
Greenberg, J.M., Kouchi, A., Niessen, W., Van Paradys, J. and de Groot, M.S. (1993) 'Interstellar dust, chirality, comets and the origins of life : life from dead stars', in preparation.
Havinga, E. (1954) 'Spontaneous formation of optically active substances', Biochem. Biophys. Acta 13, 171-174.
Joyce G.F., Visser, G.M., Van Boeckel, C.A.A., Van Boom, J.H., Orgel, L.E. and Van Westrenen, J. (1984) 'Chiral selection in poly(C)-directed synthesis of oligo(G)', Nature 310, 602-604.
Joyce, G.F., Schwartz, A.W., Miller, S.L. and Orgel, L.E. (1987) 'The case for an ancestral genetic system involving simple analogues of the nucleotides', Proc. Natl. Acad. Sci. USA 84, 4398-4402.
Kondepudi, D.K. and Nelson, G.W. (1985) 'Weak nentral currents and the origin of biomolecular chirality', Nature 314, 438-441.
Kondepudi, D.K., Kaufman, R.J. and Singh, N. (1990) 'Chiral symmetry breaking in sodium chlorate crystallization', Science 250, 975-976.

Leggett, A. (1990) in "The new physics", Davies P. (ed.), Cambridge Univ. Press, New-York, p. 276.
Lundberg, R.D. and Doty, P. (1957) 'A study of the kinetics of the primary amine-initiated polymerization of N-carboxyanhydrides with special reference to configurational and stereochemical effects', J. Amer. Chem. Soc. 79,3961-3972.
Mac Dermott, A.J. and Tranter, G.E. (1989) 'Electroweak bioenantioselection' Croatica Chem. Acta 62 (2A), 165-187.
Mason, S.F. and Tranter, G.F. (1985) 'The electroweak origin of biomolecular handedness' Proc. R. Soc. Lond. A 397, 45-65.
Morozov, L. (1979) 'Mirror symmetry breaking in biochemical evolution', Origins of Life 9, 187-217.
Oberbeck, V.R. and Fogleman, G. (1989) 'Estimates of the maximum time required to originate Life', Origins of Life 19, 549-560.
Pincock, R.E., Perkins, R.R., Ma, A.S. and Wilson, K.R. (1971) 'Probability distribution of enantiomorphous forms in spontaneous generation of optically active substances', Science 174, 1018-1020.
Salam, A. (1991) 'The role of chirality in the origin of life', J. Mol. Evol. 33, 105-113.
Spach, G. and Brack, A. (1983) 'Reflections on molecular asymmetry and appearance of life', in C. Hélène (ed.), "Structure, dynamics, interactions and evolution of biological macromolecules", D. Reidel Pub. Cy, pp. 383-394.
Spach, G. (1984) 'Chiral versus chemical evolutions and the appearance of life', Origins of Life 14, 433-437.
Spach, G. and Brack, A. (1988) 'Chemical production of optically pure systems', in G. Marx (ed.), Bioastronomy. The next steps, Kluwer Acad. Pub., pp. 223-231.
Tracey, A.S. and Diehl, P. (1975) 'The interaction of D- and L-alanine with an optically active model membrane system' FEBS Letters 59, 131-132.
Wald, G. (1957) 'The origin of optical activity', Ann. N.Y. Acad. Sci. 69, 352-368.
Weissbuch, I., Addadi, L., Berkovitch-Yellin, Z., Gati, E., Lahav, M. and Leiserowitz, L. (1984) 'Spontaneous generation and amplifiction of optical activity in α-amino acids by enantioselective occlusion into centrosymmetric crystals of glycine', Nature 310, 161-164.

Weissbuch, I., Addadi, L., Leiserowitz, L. and Lahav, M. (1988) 'Total asymmetric transformations at interfaces with centrosymmetric crystals : role of hydrophobic and kinetic effects in crystallization of the system glycine α-amino acids ', J. Amer. Chem. Soc. 110, 561-567.

8. Review articles

Avetisov, V.A., Goldanskii, V.I. and Kuz'min, V.V. (1991) 'Handedness, origin of life and evolution', Physics today, July, 33-41.

Bonner, W.A. (1991) 'The origin and amplification of biomolecular chirality', Origins of Life 21, 59-111.

Hegstrom, R.A. and Kondepudi, D.K. (1990) 'The handedness of the Universe' Scientific American, January, 108-115.

Mason, S.F. (1984) 'Origins of biomolecular handedness', Nature 311, 19-23.

EARLY PROTEINS

A. BRACK
Centre de Biophysique Moléculaire
1A, Avenue de la Recherche Scientifique
45071 - Orléans Cedex 2
France

SUMMARY : The nature and function of contemporary proteins are described. Different attempts to synthesize peptides from their component amino acids under conditions that ressemble those of the primitive Earth have been made. Experimental evidence is given for selective condensation of amino acids in water via the intermediary formation of N-carboxyanhydride. Selective resistance toward degradation of β-pleated sheet conformation exemplifies a possible accumulation of homochiral sequences made of hydrophilic and strong hydrophobic residues. Peptides containing basic amino acids strongly accelerate the hydrolysis of oligoribonucleotides. The peptides bind to the oligonucleotides and undergo a coil to β-sheet transition which plays a determinant role in the observed chemical activity.

1. Introduction

For more than sixty years, the dominant hypothesis referred to the primordial soup proposed in 1924 by Oparin. By analogy with contemporary living systems, it was generally believed that primitive life emerged as a cell, thus requiring at least three families of organic compounds : boundary molecules able to isolate the system from the aqueous environment, informative molecules allowing the storage and the transfer of informations and molecules with chemical activity providing the chemical work of the system. It was also believed that chemists would be skilful enough to make small-scale versions of these molecules in order to reconstruct an artificial primitive cell in the laboratory. According to the primordial soup hypothesis, small organic molecules were formed in a reducing atmosphere dominated by methane. When reaching terrestrial liquid water, the organic molecules were processed and generated the constituants of the first living systems and their nutrients. The present paper overviews the small-scale versions of proteins which have been synthesized in the laboratory.

J. M. Greenberg et al. (eds.), The Chemistry of Life's Origins, 357–388.

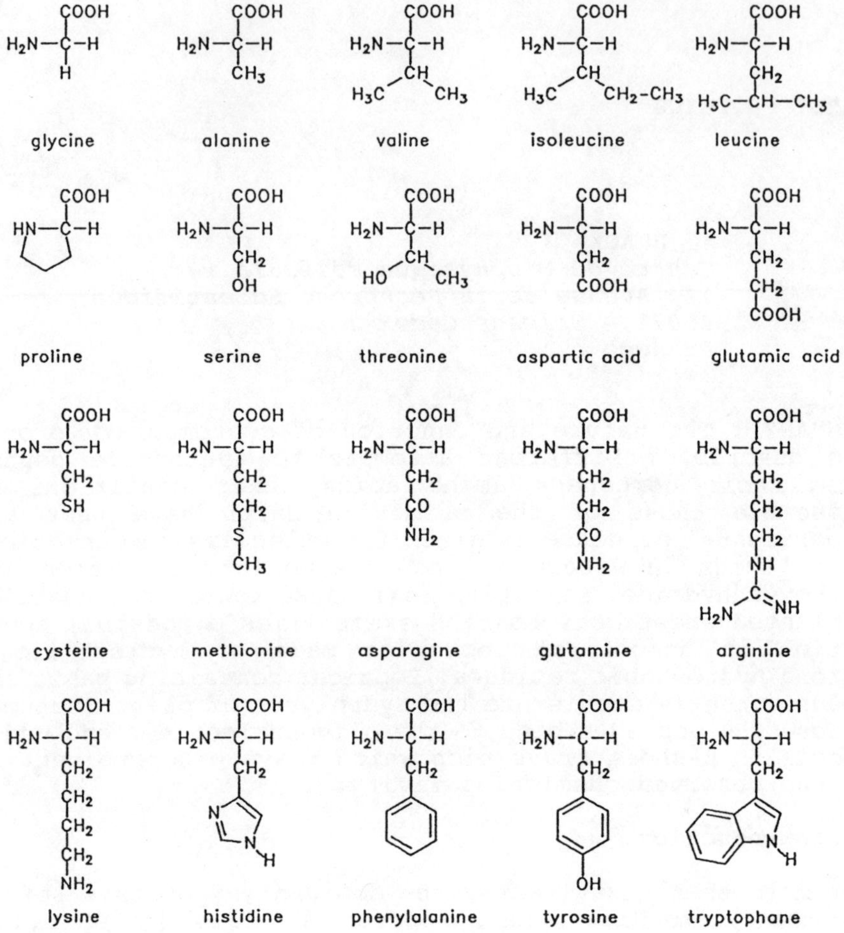

glycine	alanine	valine	isoleucine	leucine
proline	serine	threonine	aspartic acid	glutamic acid
cysteine	methionine	asparagine	glutamine	arginine
lysine	histidine	phenylalanine	tyrosine	tryptophane

Table 1 : The twenty amino acids found in proteins

2. Contemporary proteins

In a contemporary cell, proteins represent about 40 % of the dry weight. They fulfill a structural role and they are particularly helpful as chemical catalysts (enzymes). They can be represented as long words written with an alphabet of 20 different letters, the amino acids. The ordering of amino acids along the chain confers specific properties to the protein as the sequence of letters gives a sense to the words. Each amino acid is built around a central tetravalent carbon atom (Table 1). This carbon atom bears a hydrogen atom, an amino group $-NH_2$, a carboxylic acid function -COOH and a side-chain R. The twenty amino acids differ by the side-chain R (Table I). Some side-chains contain only hydrogen and carbon atoms. These side-chains are always branched. The hydrocarbon chains don't interact with water and escape water molecules (hydrophobes). They play an important role in the global geometry of the chain. Some side-chains bear a chemical function which is responsible for the catalytic activity of the enzyme.

The central carbon atom, except for glycine, is asymmetric. Therefore, each of the 19 natural amino acids exists under two enantiomers L and D, mirror images which are not superimposable ; proteins use only L-enantiomers (see the chapter devoted to the origin of chirality).

The protein chain results from the condensation of amino acids. Water molecules are removed between amino acids in the aqueous environment of the cell (about 75 % of water). The main chain construction involves the amino and carboxylic groups bound to the central carbon atom. The side-chain chemical functions never compete with these groups.

The protein chain adopts rigid asymmetric conformations. In fact, the peptide chain can be described as the succession of three atoms

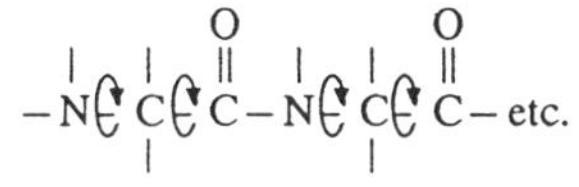

$$-N \circlearrowleft C \circlearrowleft \overset{O}{\overset{\|}{C}} - N \circlearrowleft C \circlearrowleft \overset{O}{\overset{\|}{C}} - \text{etc.}$$

The N-C and C-C bonds can be considered as rotation axis. However, the rotations around these axis are not totally free because of steric hindrance and only few couples of angles are allowed. When the same couple of rotational angles is repeated along the chain, the skeletum adopts a rigid helical conformation. Two types are commonly found in proteins. The right handed α helix contains 3.6 amino acids per turn. It is stabilized by H-bonds between the carbonyle CO and the NH groups of the same chain.

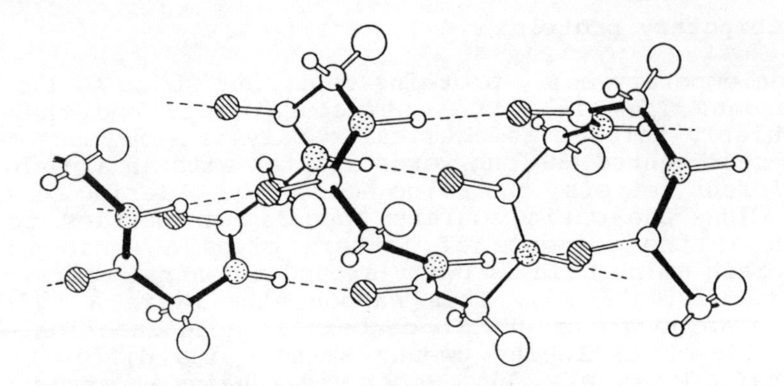

Figure 1. Right-handed α-helix build-up with L-amino acids. There are 3.6 residues per turn. The helix is stabilized by hydrogen bonds (----) between NH groups ( ) and CO groups ( ).

Helices of the second family contain only two amino acids per turn. The skeletum is extended and several strands aggregate to form a slightly pleated sheet, called β-pleated sheet conformation or β-structure.

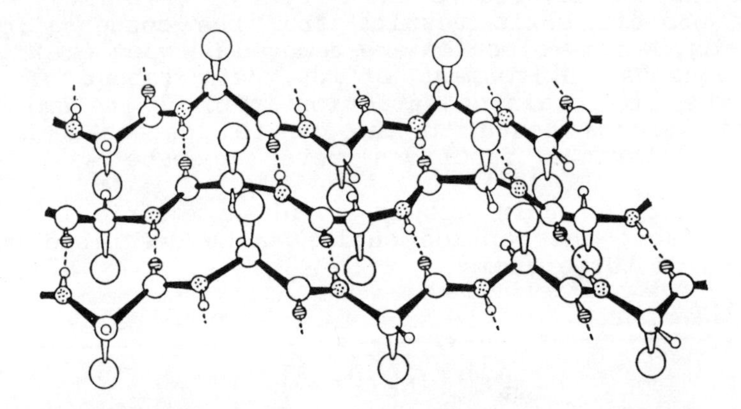

Figure 2. β-pleated sheet structure made of several strands relatd per hydrogen bonds (-----) between NH groups ( ) and CO groups ( )

Collagen represents a third family of helices with 3 amino acids per turn. Three of these helices aggregate to form a supercoiled bundle.

Proteins, even the smallest ones, are too sophisticated entities to be considered as the products of an organic chemistry working at random, without any chemical selection. The chemist has therefore to understand, with simple models, how primitive proteins were selected and how they began to exhibit chemical activity.

3. From amino acids to primitive peptides

Amino acids were most likely available on the primitive Earth. They have been synthesized in the laboratory under very simple conditions using different sources of energy like UV, X-rays, heat or electric discharges. They have also been found in meteorites. However, the mixture of organics is complex : it contains C_α-monosubstituted α-amino acids including proteinaceous ones (type A), ω amino acids (type B) and C_α-disubstituted α-amino acids (type C) (scheme 1). Moreover, in the Murchison meteorite for instance, the amino acids present in the range of 10-20 ppm are diluted with polymerized aromatic hydrocarbons, mono-carboxylic acids and aliphatic and aromatic hydrocarbons. How could selected peptides, i.e. short condensates of homochiral C_α-monosubstituted α-amino acids, emerge from this mixture of organic compounds in aqueous solution ?

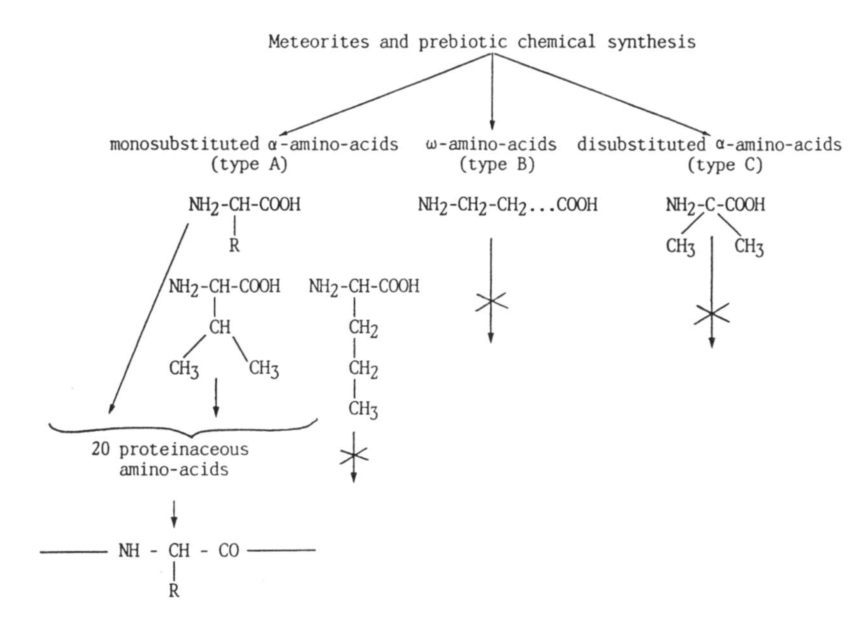

Scheme 1 : Present-day proteins use only one family of amino acids (type A). Within the selected family, the side-chain R must also fulfil certain conditions

Peptides are formed when water molecules are removed from amino and carboxylic acid functions in a milieu which is predominantly aqueous.

Peptide chemistry offers a whole range of activating agents to condense amino acids in organic solvents. In water, the number of condensing agents is restricted, especially when looking for prebiotically plausible compounds.

3.1. HCN DERIVATIVES (REVIEWED BY FERRIS AND HAGAN, 1984)

Carbodiimides, R-N=C=N-R, are commonly used in organic medium. They lead to the formation of a peptide bond by fixation of a water molecule. They can be used in water providing a careful choice of the substituent R and of experimental conditions. Under these conditions, Cavadore and Previero (1969) obtained long peptides in water up to the 30-mer. The simplest carbodiimide, H-N=C=N-H, can be considered as a tautomeric form of cyanamide NH_2-CN which can be obtained by UV irradiation of an aqueous solution of ammonium cyanide, NH_4CN. The formation of cyanamide requires the presence of iron which helps the absorption of UV energy. In fact, cyanamide is not stable and forms a dimer, dicyandiamide or cyanoguanidine NH_2-C(=NH)-NH-CN which is as reactive as carbodiimides. Peptides were obtained with cyanamide and cyanoguanidine. However, the reactions were very slow and did not proceed beyond the tetrapeptide (Halman 1968, Nooner *et al.* 1977, Hawker and Oro 1981, Steinman *et al.* 1965). With diamino maleonitrile, NC-C(NH_2)=C(NH_2)-CN) the formation of diglycine in 3.1 % yield has been observed (Chang *et al.* 1969).

3.2. CONDENSATIONS ON CLAYS (REVIEWED BY PONNAMPERUMA *et al.* 1992)

Mixed anhydrides can also be used to condense amino acids in water :

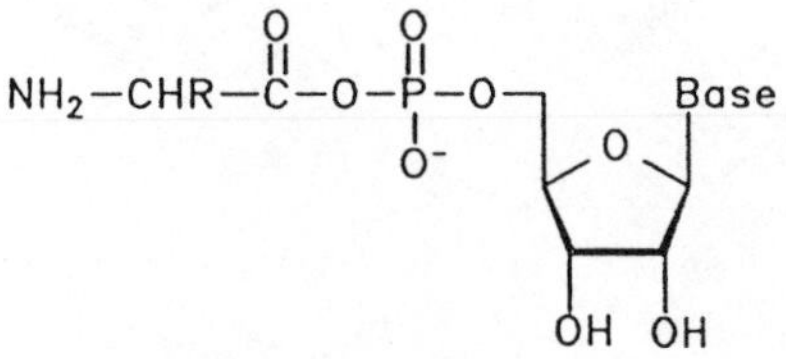

$$NH_2-CHR-\overset{O}{\overset{\|}{C}}-O-\underset{O^-}{\overset{O}{\overset{\|}{P}}}-O-\text{(ribose: OH, OH)}-\text{Base}$$

In an homogeneous aqueous solution, alanyladenylate condensed partially up to heptaalanine but desactivation via hydrolysis remained the main pathway. In the presence of clays, such as montmorillonite, hydrolysis was suppressed and discrete polyalanines were obtained (Paecht-Horowitz and Eirich, 1988 and references there in). However, montmorillonite-mediated polymerization of alanyladenylate did not lead to high oligopeptides in quantitative yield in other laboratories (Warden *et al.* 1974, Brack 1976).

Lahav *et al.* (1978) subjected mixtures of glycine and Na-kaolinite or Na-bentonite to wet-dry and temperature fluctuations (25-94 C) and observed the formation of oligopeptides up to five glycine residues in length. Only trace amounts of diglycine formed without clays. White and Erickson (1980) studied the effects of the dipeptide histidyl-histidine in the polymerization of glycine during fluctuating moisture and temperature cycles on kaolinite. A turnover of 52 was observed, i.e. each molecule of dipeptide helps the polymerization of 52 molecules of glycine.

3.3 POLYMERIZATION IN AGGREGATES

Surfactant aggregates have a spectacular effect on the polymerization of amino-acyladenylates as shown by Armstrong *et al.* (1978). Amino esters oriented in monolayers (Fukuda *et al.* 1967, Folda *et al.* 1982) or in micelles (Kunieda *et al.* 1981, Hanabusa *et al.* 1986) polymerize easily.

3.4 THERMAL CONDENSATION OF AMINO-ACIDS

Fox and co-workers (1977) have shown that dry mixtures of amino-acids polymerize when heated at 130 C to give proteinoids. In the presence of polyphosphates, the temperature can be decreased to 60 C. High molecular weights are obtained when an excess of acidic or basic amino acids are present. When heated in aqueous solutions at 130-180 C, the pro-teinoids aggregate spontaneously in microspheres of 1-2 μ, presenting an interface resembling the lipid bilayers of the living cells. In appropriate conditions, these microspheres increase slowly in size from dissolved proteinoids and are sometimes able to bud and to divide like bacteria. These microsphere catalyse the decomposition of glucose and work as esterases and peroxydases. The main advantage of proteinoids is their polymeric character and their organization into particles but they represent a dramatic increase in complexity. When heating a mixture of selected L-amino-acids, one gets a polycondensate which is only 50 % peptidic, the peptidic fraction is racemized, the peptide linkages are ambiguous since they include the α, β and γ functions of the dicarboxylic amino-acids and the sequences are multiple although not completely random.

3.5 SELECTIVE AQUEOUS CONDENSATION OF AMINO-ACIDS

N-carboxyanhydrides (II) are good candidates for the selective polymerization of proteinaceous amino-acids in water. Their formation requires an N-derivatization (I) - which is known to be difficult with disubstituted amino-acids due to steric hindrance - followed by a ring closure involving the α-carboxylic function. Ring-closure has proven to be possible with 5-membered rings (α-amino-acids) and with 6-membered rings to a lesser extent (β-alanine). It should not be effective with larger rings (γ-amino-acids, γ-carboxylic group of glutamic acid, dipeptides, etc).

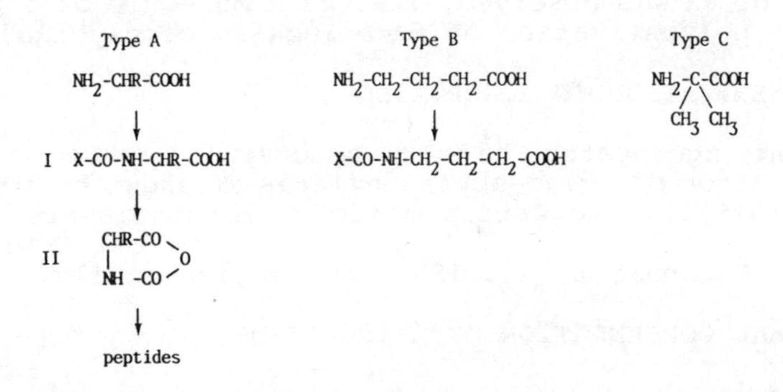

3.5.1. *Aqueous polymerization of esters.*

Aqueous polymerization of p-nitrophenylesters of proteinaceous amino-acids is much more efficient when run in the presence of sodium hydrogen carbonate. Infrared spectroscopy demonstrated the intermediary formation of N-carboxyanhydrides, probably via the carbamate (Brack, 1982) :

$$H_2N\text{-}CH[(CH_2)_2COO^-]\text{-}C(=O)\text{-}O\text{-}C_6H_4\text{-}NO_2 \xrightarrow[\text{or } CO_2]{HCO_3^-} {}^-OOC\text{-}HN\text{-}CH[(CH_2)_2COO^-]\text{-}C(=O)\text{-}O\text{-}C_6H_4\text{-}NO_2 \longrightarrow \text{N-carboxyanhydride} + {}^-O\text{-}C_6H_4\text{-}NO_2$$

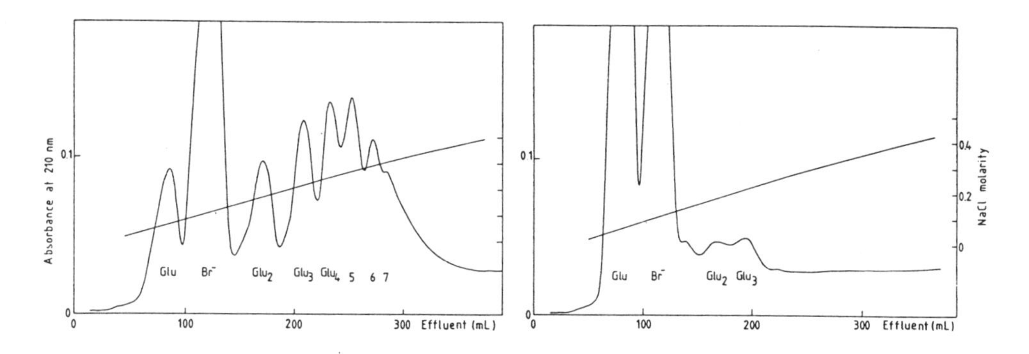

Figure 3. DEAE-cellulose gradient elution chromatography of Glu-oligomers obtained by polymerization of HBr, H-Glu-ONp in the presence of $NaHCO_3$ (left) and of NaOH (right).

With L-leucine p-nitrophenyl ester, optically pure oligoleucines with an average degree of polymerization of 10 precipitated in 77 % yield (Brack, 1984). Under the same conditions, the active ester of α-amino isobutyric acid (Aib) did not condense and the condensation of β-Ala-ONp was not affected by bicarbonate ions. When an equimolar mixture of p-nitrophenyl esters of Ala, Leu, Val, β-Ala, and Aib was subjected to polymerization, the precipitate was somewhat enriched in proteinaceous amino acids but not to a large extent (Table 2). Amino groups of Aib-ONp and β-Ala-ONp may react with N-carboxyanhydrides of Ala, Leu and Val and are incorporated at the C-termini of the growing chains.

	Ala	Leu	Val	βAla	Aib	Yield (%)
Absence of bicarbonate	12.4	33.5	20.6	27.1	6.4	26
Presence of bicarbonate	15.6	37.7	19.0	22.2	5.5	40

Table 1. Mole percents of different amino-acids found after complete hydrolysis of the precipitate obtained when a stoichiometric mixture of the p-nitrophenyl esters is left in water at pH 6.8.

Thioethyl α-ester of glutamic acid was also tested (Brack, 1987). Glu-SEt was dissolved in an imidazole buffer, at pH 7.5, in the presence and in the absence of bicarbonate ions. After 2 days at room temperature, in the absence of bicarbonate ions, ion exchange chromatography revealed the formation of cyclic dipeptide (substituted diketopiperazine, DKP) and of traces of the linear dimer (Figure 4a). In the presence of bicarbonate, the condensation was much more efficient and oligomers up to the pentamer could be detected in addition to DKP (Figure 4b). When starting with preformed Glu-Glu-SEt, no difference could be observed, the major product of the reaction being DKP in both cases. From this behaviour, one can argue that the condensation of Glu-SEt in the presence of bicarbonate does not proceed through Glu-Glu-SEt formation, otherwise the condensation reaction would end with DKP formation. The mechanism involving N-carboxyanhydride formation is favored, although it was not possible to detect the intermediate by infrared spectroscopy.

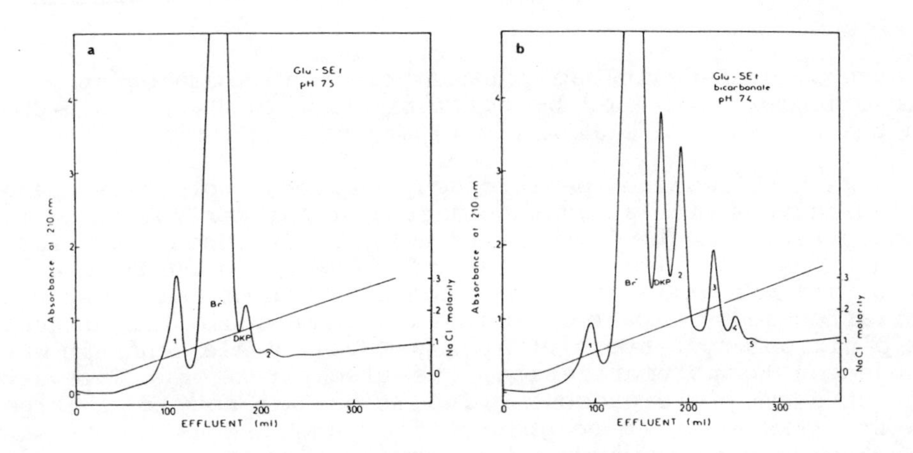

Figure 4. DEAE-cellulose gradient elution chromatography of Glu-oligomers obtained by polymerization of HBr, H-Glu-SEt; (a) polymerization in the absence of bicarbonate ions ; (b) polymerization in the presence of bicarbonate ions.

The key role of bicarbonate ions was investigated with simpler, more prebiotic esters. Alpha-methyl-L-glutamate was left in water at pH 6.8 for 5 days at room temperature in the absence and in the presence of bicarbonate ions. Ion exchange chromatography did not show any condensation at all in both cases.

3.5.2. *Aqueous polymerization by N,N'-carbonyldiimidazole.* Direct activation with N,N'-carbonyldiimidazole (CDI) in an imidazole buffer at pH 6.8 is also efficient, as already shown by Ehler and Orgel (1976).

Free glutamic acid was treated by CDI (Figure 5). Oligomers up to the 11-mer were identified. No γ-linked Glu-Glu, cyclo (Glu-Glu) or Pyroglu-Glu could be detected. The preformed dipeptide Glu-Glu did not condense under the same conditons (Brack, 1984; Brack, 1987).

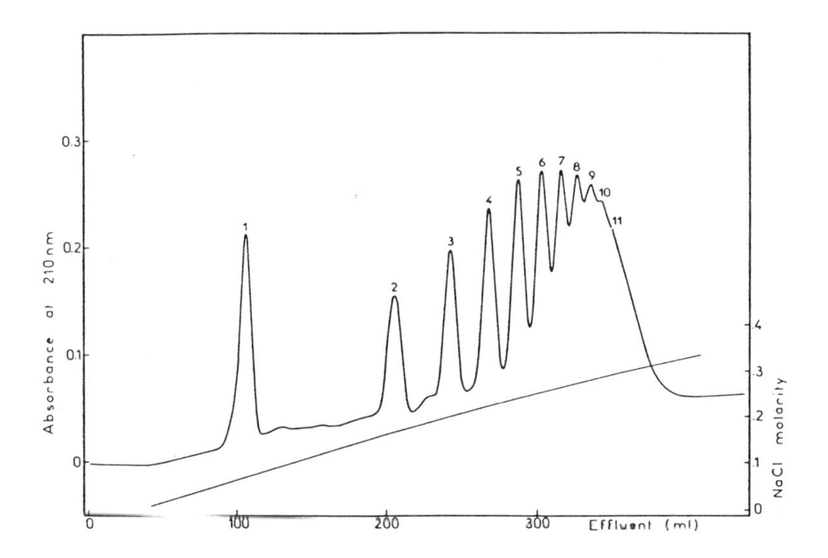

Figure 5. DEAE-cellulose gradient elution chromatography of Glu-oligomers obtained when a Glu solution was treated with N,N'-carbonyldiimidazole (2 equivalents) at pH 6.8.

L-leucine was treated by CDI. Optically pure oligo-L-leucines precipitated with an average molecular weight of about 8, in 70 % yield. Under the same conditions, γ-aminobutyric acid (γ-Abu) and α-aminoisobutyric acid (Aib) did not precipitate. The reaction mixtures were injected into an amino-acid analyzer. They contained 82 % γ-Abu, 12 % γ-Abu$_2$, 6 % γ-Abu$_3$ and 100 % Aib, respectively. The selective con-densation was exemplified with an aqueous mixture of proteinaceous amino-acids (Gly, Ala, Glu, Val) and non-proteinaceous amino-acids (Aib, isovaline, α-amino butyric acid, γ-Abu, β-Ala) in proportions close to that found in the Murchison meteorite. This was treated with N,N'-carbonyldiimidazole.

The polycondensate, isolated by column chromatography in 55 % yield was enriched in the monosubstituted α-amino-acids (proteinaceous amino-acids plus α-aminobutyric acid) as shown in Table 3 (Brack, 1987).

	Aib	Gly	Ala	Ival	Glu	γAbu	αAbu	βAla	Val
Starting mixture (in mole %)	28.3	27.1	12.2	7.8	6.6	6.5	4.6	4.4	2.9
After hydrolysis of the condensate (in mole %)	2.3	44.3	19.5	overlapping with Val	15.0	1.7	7.4	5.3	overlapping with Ival

Table 3. Aqueous polymerization via N,N'-carbonyldiimidazole of the nine most abundant amino-acids found in the Murchison meteorite.

To confirm the intermediary formation of N-carboxyanhydride and its role in the selection process, the polycondensations were followed by infrared spec-troscopy. After addition of CDI, no activation of the car-boxylic function could be detected at 1730 cm^{-1}. For illus-tration, acetic acid gave only 6 % of the expected acetyl-imidazole absorption band at 1730 cm^{-1}, calibrated with the commercial compound. On the other hand, an absorption band attributed to the N-imidazoyl-(1)-carbonyl group developped with leucine at 1690 cm^{-1} during the first minutes and then decreased slowly (Figure 6a). By continuous extraction of leucine mixtures with chloroform, it was possible to isolate a compound presenting two absorption bands characteristic of N-carboxyanhydrides (Figure 6b). The 1690 cm^{-1} absorption band appeared also with γ-Abu (Figure 6c) and α-Glu-Glu, but remained constant for hours. With Aib, this band did not form at all (Figure 6d). The polymerization reaction proceeds therefore according to the scheme :

$$NH_2\text{—CHR—COO}^- + \text{CDI} \rightarrow \text{Im—CO—NH—CHR—COO}^- \rightarrow \begin{array}{l}\text{CHR—CO}\diagdown \\ \;|\qquad\qquad\;\; \text{O} \\ \text{NH—CO}\diagup\end{array}$$

thus confirming our initial hypothesis.

N,N'-carbonyldiimidazole is not stable in water and could not accumulate in large quantities on the primitive Earth. To bypass this difficulty, glutamic acid dissolved in an imidazole buffer at pH 7.3 was left in contact with phosgene. The elution pattern showed the presence of oligo-

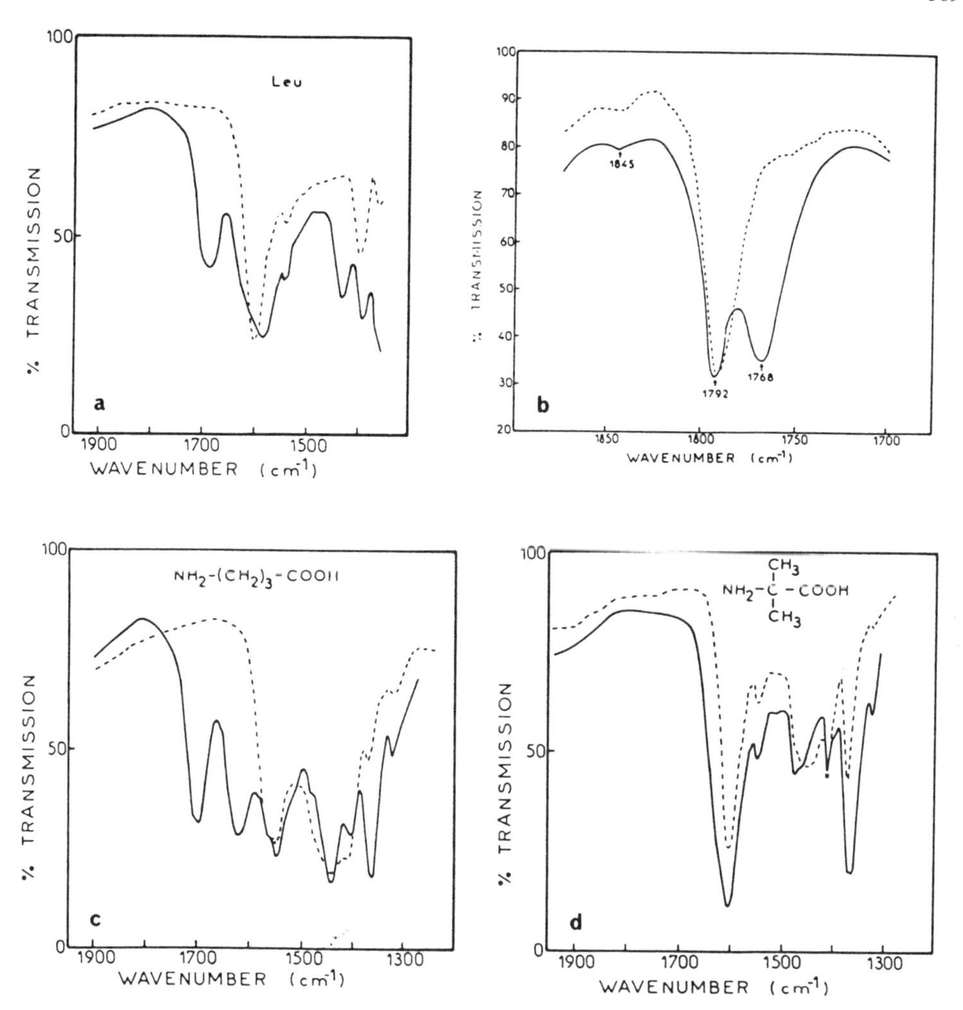

Figure 6. Infrared spectra of the solutions obtained when CDI was added to different amino acids in D_2O (a, c and d). For a, c and d, the dashed line represents the spectrum of the mixture prior to the addition of CDI. Spectrum b was obtained by continous extraction of leucine reaction mixture with chloroform. The absorption bands at 1792 and 1845 cm^{-1} are attributed to N-carboxyanhydrides. The dashed line represents the spectrum of pure leucine N-carboxyanhydride run under the same conditions.

mers up to 6, indicating that condensation occurs in aqueous phase with phosgene (Figure 7).

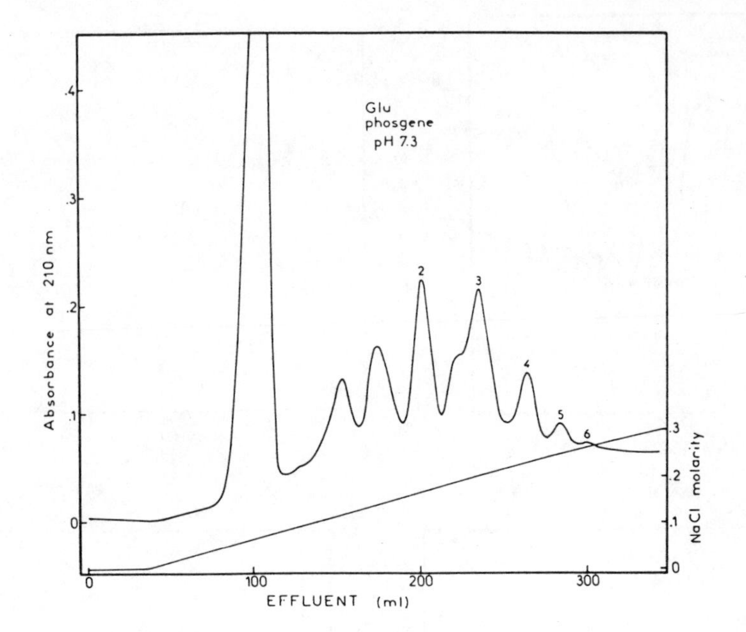

Figure 7. DEAE-cellulose gradient elution chromatography of Glu oligomers obtained when a Glu solution was left in the presence of phosgene at pH 7.3.

In conclusion, we have shown that C_α-monosubstituted α-amino acids including proteinaceous amino acids can be selectively polymerized in water from a background noise through N-carboxyanhydrides. In addition, condensation of trifunctional amino acids leads exclusively to α-linked peptides.

4. Selective resistance to degradation via homochiral β-sheet formation (Brack, 1987)

We have shown that copolypeptides with alternating hydrophilic and hydrophobic residues adopt a water-soluble β-pleated sheet structure because of hydrophobic side-chain clustering (Brack and Orgel, 1975 ; Brack and Caille, 1978).

Due to β-structure formation, alternating sequences resist chemical degradation, while α-helix forming random sequences are degraded much faster under the same conditions (Brack and Spach, 1981).

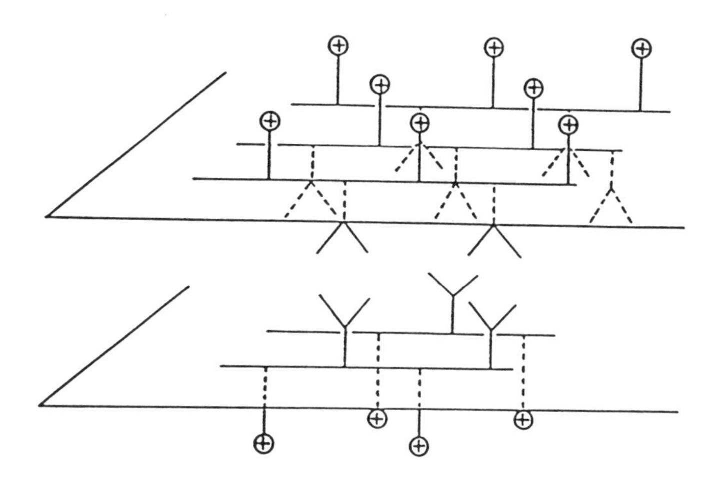

Figure 8. Double layer of alternating hydrophobic-hydrophilic polypeptides in the β-form.

We have used the high resistance of the β-sheet to degradation as a means by which certain sequences could be selected on the basis of the chainlength or of the hydrophobic character of the amino acid side-chains.

Mild acidic hydrolysis was carried out. Figure 7 clearly shows that poly(Val-Lys) and poly(Leu-Lys) kinetics exhibit two steps. The fast step has been attributed to the hydrolysis of the random-coiled chains and the slow step was attributed to the hydrolysis of the chains engaged in the β-pleated sheets.

Non-proteinaceous α-aminobutyric acid (α-Abu) and norvaline (Norval) were associated to lysine in alternating polypeptides. The polypeptides poly (αAbu-Lys) and poly (Norval-Lys) don't form β-pleated sheets and they were found to be 15 times more sensitive to mild hydrolysis (Table 4 and Figure 9) than poly(Leu-Lys), for instance. The unstability is related to the weak hydrophobic character of the linear aliphatic side-chains. This may explain why sequences containing these amino acids could not accumulate on the primitive Earth and were, therefore, not used by early proteins.

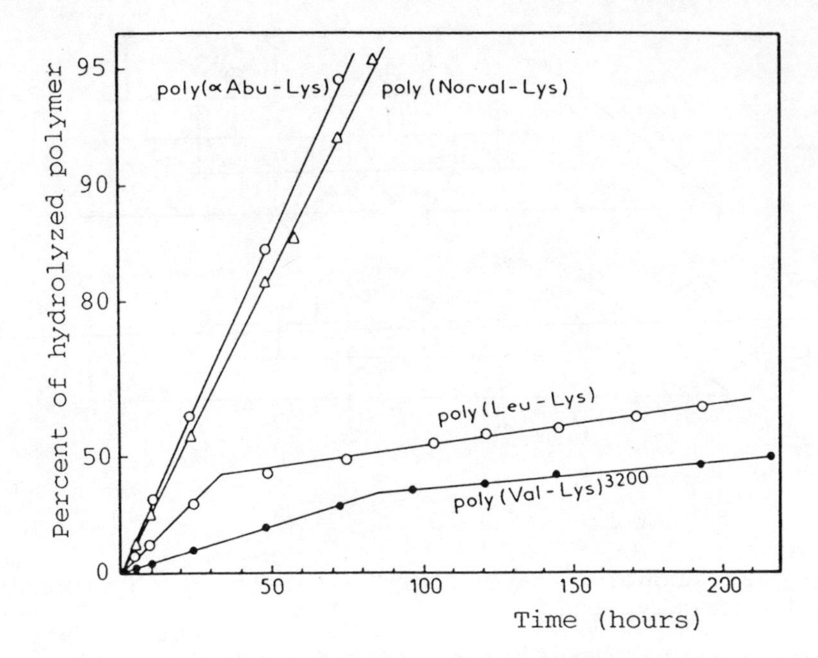

Figure 9. Hydrolysis kinetic of different alternating polypeptides. Percent of hydrolyzed polymer as a function of time.

	Mn	β-sheet formation	Rate constant in min^{-1}		Rate ratio (based on 4.3×10^{-5})
			Fast	Slow	
Poly (Val-Lys)10700	10700	+ + +	9.5×10^{-5}	2.8×10^{-5}	
Poly (Val-Lys)3200	3200	+ +	9.5×10^{-5}	2.7×10^{-5}	
Poly (Leu-Lys)	3200	+ +	2.8×10^{-4}	4.3×10^{-5}	7
Poly (αAbu-Lys)	7900	–	6.6×10^{-4}	–	15
Poly (Norval-Lys)	2800	–	5.9×10^{-4}	–	14

Table 4. Selective resistance to mild hydrolysis.

5. Enantiomer enrichment via β-sheets

Optical activity and life, as known on the Earth, are two properties of the matter which are intimately related. For instance, present proteins are built up with amino acids residues of identical chirality and their backbone itself is folded into regular chiral structures, α-helices and β-sheets, the ordering of which governs their biological activity. Yet, it is generally admitted that the first amino acids available on the primitive Earth were, because of their synthetic origin, racemic mixtures and that earlier polypeptides contained both enantiomers. Therefore, two questions arise : What is the origin of the dominant handedness of the fundamental constituents of the living matter ? Are there amplification mechanisms that can account for enrichment in one enantiomer, starting from a slightly unbalanced composition of the chiral compounds ?

From sterical reasons, sheet structures can only be built up with almost fully extended poly-D,L-peptide chains, but these are thought to be energetically not very stable and actually they have never been experimentally demonstrated. On the other hand, no β-sheet structures, with either parallel or antiparallel chains, are conceivable when the chains contain both L- and D- residues randomly distributed. This is clearly understood when examining atomic models : some side-chains would be forced into the plane of the sheet, which is sterically impossible. Alternative but peculiar possibilities are the rippled sheet structure valid for a mixture of poly-L- and poly-D-peptides, and the so-called polar β-structure valid for alternating poly-D-L-peptides.

The influence of D- and L-residues in polypeptide chain on the β-sheet formation was studied on model polymers giving stable β-structure in aqueous solution of high ionic strength. These are polypeptides with alternating hydrophilic-hydrophobic residues. L- and D-isomers were introduced at random in the chain and equally distributed among leucyl and lysyl residues (Brack and Spach, 1979).

Conformation analysis of infrared and circular dichroïsm spectra for different values of the ratio $x = L/(L+D)$ gave the total amount of β-structure (Figure 10) and the excess of L- over D-residues in that structure (Figure 11). These data were compared with theoretical curves calculated in the frame of two hypotheses : first, only segments containing residues of identical chirality can associate into antiparallel β-sheets, and second, these segments must have a minimum number of residues n_{β}. The best agreement was obtained when n_{β} was made equal to seven (figure 10) a very reasonable value in view of the minimum length of peptides or protein segments able to form β-sheets (Spach and Brack, 1979).

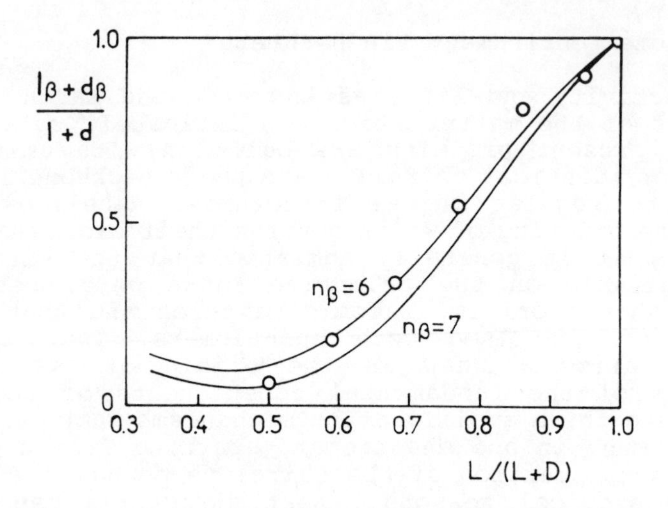

Figure 10. Total β-fraction, from infrared data, versus monomer composition. Comparison with theoretical curves.

Interestingly, the β-sheets are rapidly enriched in L-residues when the ratio x increases from 0.5 to 1.0. Statistical analysis indicates that for x equal to 0.75, the pool of β-sheets is almost optically pure.

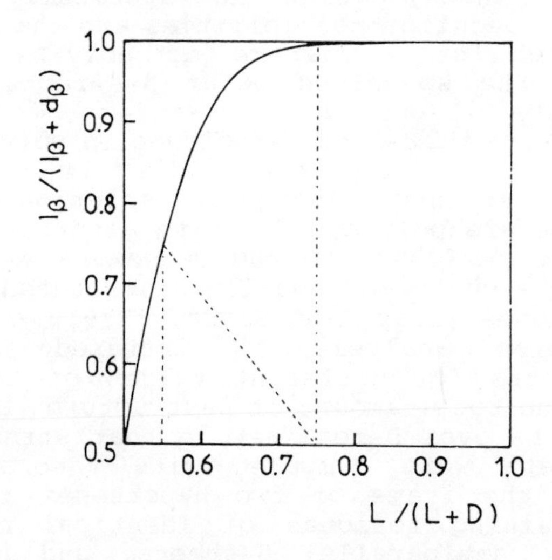

Figure 11. Fraction of L-residues engaged in the pool of β-structures versus stereomonomer composition. Pathway to enrichment in enantiomer

Thus, if the stereocomposition of the polymers departs from the racemic mixture, the molecules can be described as formed of β-sheet nuclei enriched in one enantiomer whereas the rest of the chains is in a disordered state (Figure 12). These samples were subjected to mild acidic hydrolysis. Kinetic measurements showed two pseudo first order rate constants, in agreement with the existence of two conformational species. Indeed, the unordered parts of the chains are hydrolyzed more rapidly, allowing the isolation of a β-fraction enriched in one enantiomer (Brack and Spach, 1980). The enrichment is not as complete as expected from the theory. However, were the process to be repeated, i.e. the isolated β-nuclei degradated, the amino acid mixture recovered and polymerized again, a second enrichment could be achieved (Figure 9). Such a process, though its yield is scanty, is a plausible one in prebiotic, aqueous conditions, but, of course, it does not explain the predominance of a given enantiomer.

Figure 12. Scheme of a poly-D,L-peptide chain with β-sheet nuclei (circled).

6. Catalytic activity of synthetic peptides and polypeptides

Many synthetic peptides and polypeptides have been found to exhibit catalytic activity. As already mentioned, small amounts of the dipeptide His-His increases by a factor three the condensation of glycine during wet-dry fluctuations on clay surfaces (White and Erickson, 1980). Polyglycine enhances the dimerization of alanine mediated by dicyanamide (Steinman and Cole, 1967). The tripeptide Lys-Trp-Lys is able to photosensitize the splitting of thymidine dimer in

DNA (Charlier and Hélène, 1975) and to recognize and cleave DNA at apurinic/apyrimidic sites (Behmoaras *et al.*, 1981 ; Ducker and Hart, 1982 ; Pierre and Laval, 1981). Catalysis of glycosidation by a decapeptide (Chakravarty *et al.*, 1973) and of DOPA oxydation by a poly L-lysine-Cu(II) complex (Nozawa and Hatano, 1971) have also been reported.

6.1 ESTER HYDROLYSIS

Much work has been devoted to the design and synthesis of peptides and polypeptides acting as artificial esterases. Some esterolytic activity has been reported for linear peptides such as Thr-Ala-Ser-His-Asp (Sheehan *et al.*, 1966), His-Gly-Asp-Ser-Phe (Kapoor *et al.*, 1970), Ser-Pro-Cys-Ser-Glu-Thr-Tyr (Fridkin and Goren, 1974), His-Ala-Asp-Gly-Cys (Petz and Schneider, 1976). Cyclic peptides have also been studied, like cyclo (Leu-His) by Kawaguchi *et al.* (1983), cyclo (His-Phe-Gly-Cys-DPhe-Ser-Gly-Glu-Cys) by Schultz *et al.* (1982) and cyclo (Asp-βAla-Gly-Ser-βAla-Gly-His-βAla-Gly) by Nishi *et al.* (1966). Polypeptides have also been tested. The most interesting were poly L-lysine complexed to Cu(II) which catalyzes the stereoselective hydrolysis of phenylalanine esters (Nozawa *et al.* 1972), poly (His-Ala-Glu) (Goren *et al.*, 1978), poly (Gly-Ser-Asp-His-Ala-Pro) (Trudelle, 1982) and copoly (Tyr,Glu,Ala) (Noguchi *et al.*, 1977). However, in most cases, catalytic activities of these compounds towards esters are not markedly improved as compared to the free amino acids in solution.

The approach published recently by Stewart (Hahn *et al.*, 1990) appears much more promising. They synthesized a bundle of four short parallel amphipathic helical peptides linked covalently at their carboxyl ends. The four chains bear the serine protease catalytic site residues serine, histidine, aspartic acid and glutamic acid at the amino end in the same spatial arrangement as in chymotrypsin. The bundle has affinity for chymotrypsin ester substrates similar to that of chymotrypsin and hydrolyzes them at rates 0.01 that of the natural enzyme ; total turnovers over 100 have been observed.

6.2 ARTIFICIAL RIBONUCLEASES

Polycationic polypeptides containing arginine or lysine and hydrophobic amino acids are able to accelerate oligoribonucleotide hydrolysis. The greatest effect was observed when the polypeptides are structured in β-sheets.

When mixing an aqueous solution of alternating poly-(Leu-Lys) to an aqueous solution of ApAp at pH 8, the solution becomes turbid reflecting the formation of a complex. The mixture was analyzed as a function of time by reversed-phase HPLC, after complete dissociation of the complex.

Poly(Leu-Lys) stimulates strongly the rate of hydrolysis as compared to the control run in the absence of polypeptide. From the pseudo-first-order kinetics it can be calculated that the hydrolysis rate is increased by a factor of about 150 (Barbier and Brack, 1988). HPLC chromatograms show that the hydrolysis produces 2':3' cyclic AMP (A > p) together with A2'p and A3'p. This suggests that the polypeptide accelerates the classical alkaline hydrolysis of RNA which is known to proceed in two steps : cleavage of the phosphodiester bond and subsequent formation of a 2':3' cyclic phosphate (A > p) followed by the opening of the cycle in a second step. The acceleration of the hydrolysis affects essentially the first step of the mechanism since it has been found that poly(Leu-Lys) increases the rate of A > p hydrolysis only by a factor 5 to produce A2'p and A3'p monomers in a 1.14 ratio in favor of A2'p.

The activity of poly(Leu-Lys) was extended to a mixture of oligo (A)s up to the 25-mer which is well resolved by HPLC on RPC5 and which can be cheaply obtained on a large scale from commercially available polyA. Again poly(Leu-Lys) accelerates strongly the hydrolysis (Figure 13), L-lysyl residues being more active than L-arginyl ones.

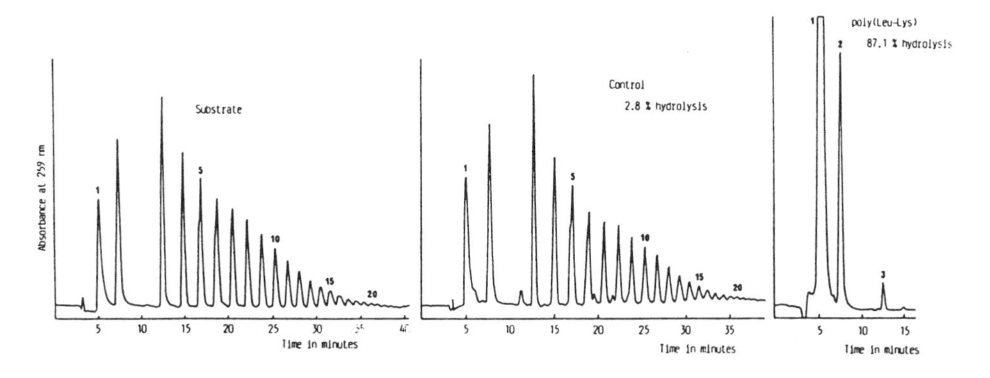

Figure 13. RCP5 HPLC elution profiles of the reaction products of the hydrolysis of 1-25 oligo(A)s : oligo(A)s used as substrate, control, in the presence of poly(Leu-Lys).

Poly(Leu-Lys) was compared to alternating poly(Ala-Lys). The activity increases with hydrophobicity. Poly(Leu-Lys) appears to be equally active on poly A, poly U, poly G, and poly C.

Basic copolypeptides with alternating hydrophilic and hydrophobic residues adopt a random coil conformation in pure water due to charge repulsions. In the presence of oligo(A)s, poly(Leu-Lys) transforms into a β-sheet structure which can be considered as the result of electrostatic interactions between phosphate and amino groups.

When increasing the ionic strength in the reaction mixture, the formation of the complexes becomes unlikely. For instance, a 2 M $NaClO_4$ salinity completely dissociates preformed complexes. With $NaClO_4$ molarity varying from 0 to 2 M, the percentage of hydrolyzed phosphodiester bonds dropped from 85 % to 7.3 %, suggesting that the cleavage reaction is directly related to the complex formation.

The geometry plays an important role in the hydrolysis since poly(Pro-Lys-Leu-Lys-Leu), which cannot adopt the β-sheet conformation because of the prolyl residues, is practically inactive as a hydrolytic catalyst. Racemic, alternating poly(D,L-Leu - D,L-Lys) was also found to have very little activity (Figure 14).

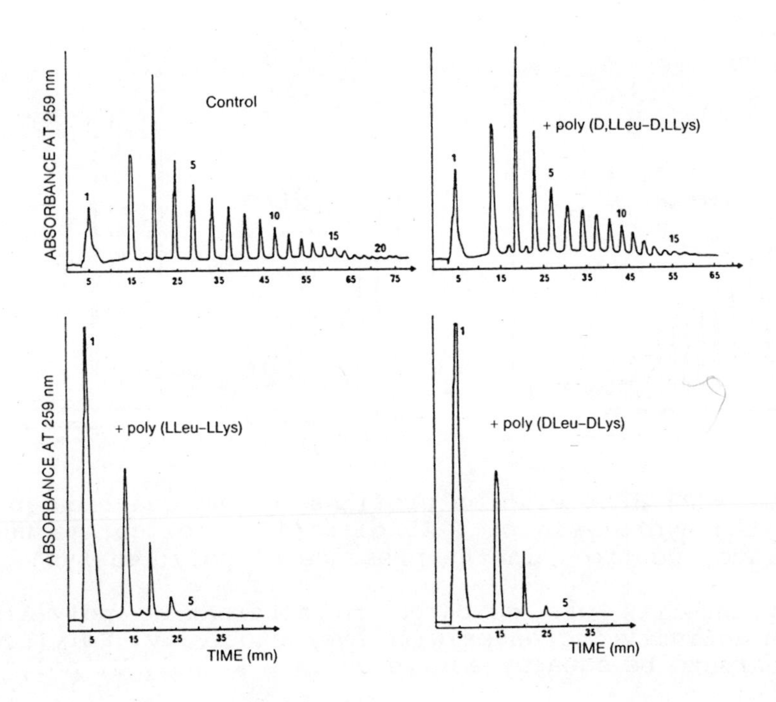

Figure 14. RPC5 elution profiles of the reaction products of the hydrolysis of 1-25 oligo (A)s under standard conditions.

A set of alternating poly(leucyl-lysyl) ranging from the racemic to homochiral poly(L-Leu - L-Lys) polymer (Brack and Spach, 1979) has been checked. When both L- and D-residues are present in the chains, the conformation can be described as a mixture of random coil and β-sheet conformations, the amount of β-sheet increasing with the proportion of L-residues in the polymers. The catalytic activities expressed as a percentage of hydrolyzed phospho-diester bonds, follow the proportion of β-sheets, indicating clearly that the β-sheets are involved in the catalysis. There is a linear relationship between the catalytic acti-vity and the proportion of β-sheets (Figure 15) (Brack and Barbier, 1990).

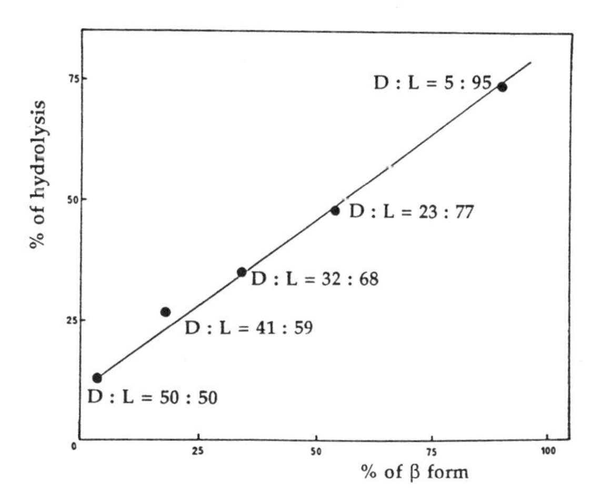

Figure 15. Variation of the catalytic activity (expressed as the percentage of hydrolyzed phosphodiester bonds) as a function of the proportion of β-sheets in the free polypeptides.

The alternating polypeptides tested in the hydrolysis experiments are polydisperse with regard to chain-length and average about 50 residues per chain. Shorter well-defined monodisperse peptides were prepared in order to evaluate the critical chain-length required for the hydrolytic activity. Oligopeptides, acetyl-$(Leu\text{-}Lys)_n$-ethylamide, were prepared with n = 1,3 and 5. The hydrolytic activity of the di-, hexa- and decapeptide represents respectively 4, 13 and 92 % of that of $(Leu\text{-}Lys)_{26}$ (Table 5 and Figure 16). None of the peptides adopts a β-sheet structure in saline aqueous solution. However, the decapeptide exhibits this conformation when complexed to the oligonucleotides (Brack and Barbier, 1990).

Peptide	Dipeptide	Hexapeptide	Decapeptide	Polypeptide	Control
Conformation in salt-containing aqueous solution	random coil	random coil	random coil	β-sheet	–
Conformation in the complex	random coil	random coil	β-sheet	β-sheet	–
% of hydrolyzed phosphodiester bonds	6.2	11.8	64.3	69.4	3.4

Table 5. Influence of the peptide chain length on the hydrolytic activity. Experimental conditions : 50 C, 7 days, Gly-Gly buffer 0.1 M pH 8.

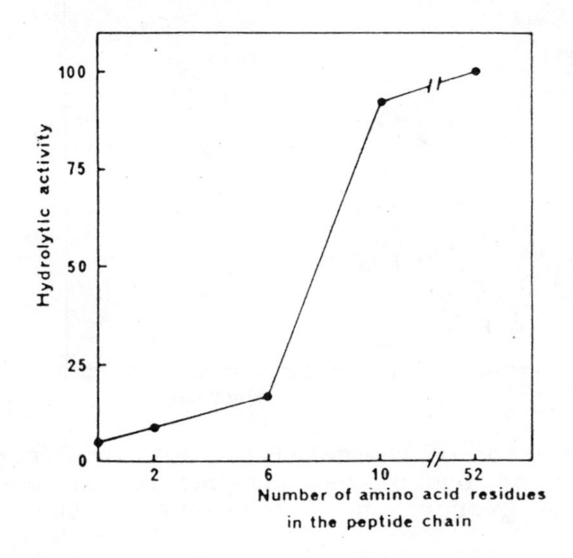

Figure 16. Variation of the relative hydrolytic activity (obtained by referring the percentages of hydrolyzed phosphodiester bonds to that obtained in the case of the poolypeptide) as a function of chain length.

In the β-sheet structure, the positive charges are regularly arranged along the strands. A similar situation can be achieved with an α-helix if lysyl doublets alternate with hydrophobic doublets as in poly(Leu-Lys-Lys-Leu).

Figure 18 shows the double row of charges twisted around the helix. $(\text{Leu-Lys-Lys-Leu})_{45}$ is indeed active although its activity represents only two thirds of that of $(\text{Leu-Lys})_{26}$. Poly(Pro-Lys-Lys-Leu) also has regular repeats of lysyl doublets but the presence of prolyl residues forbids α-helix formation. Its hydrolytic activity is very weak. Oligo acetyl-(Leu-Lys-Lys-Leu)$_n$-ethylamides were prepared with n = 1,2,3 and 4. The hydrolytic activity of the tetra-, octa-, dodeca- and hexadecapeptide represents respectively 2, 27, 54 and 77 % of that of the corresponding polymer and it is possible to correlate the activity to the propensity to form α-helices (Table 6 and Figure 19)(Perello *et al.*, 1991, Barbier and Brack, 1992).

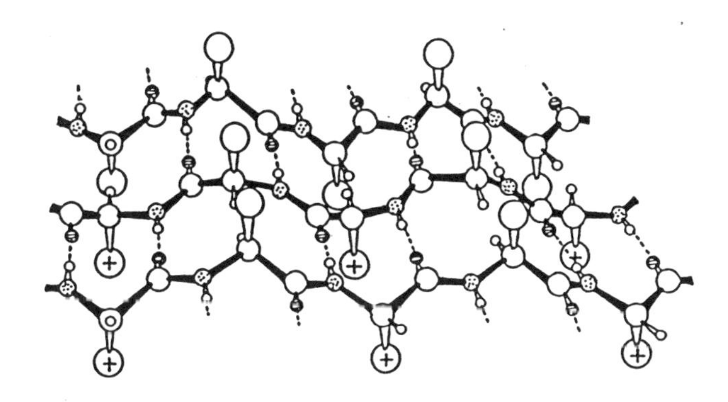

Figure 17. β-sheet obtained with alternating poly(Leu-Lys)

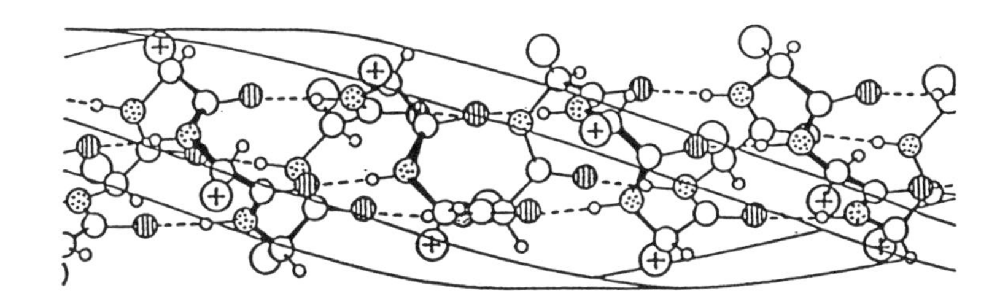

Figure 18. Twisted double row of positive charges in α-helical poly(Leu-Lys-Lys-Leu).

Samples Ac(LKKL)$_n$NHEt	% α-helix in 0.1 M $NaClO_4$	% of hydrolyzed phosphodiester bonds
control (no peptide added)	-	4
n = 1	3	5
n = 2	12	17.3
n = 3	60	31.1
n = 4	74	42.5
n = 45	100	54

Table 6. Comparison between the tendency to form an α-helix and the hydrolytic activity of Ac-(Leu-Lys-Lys-Leu)$_n$-NHEt and poly (Leu-Lys-Lys-Leu).

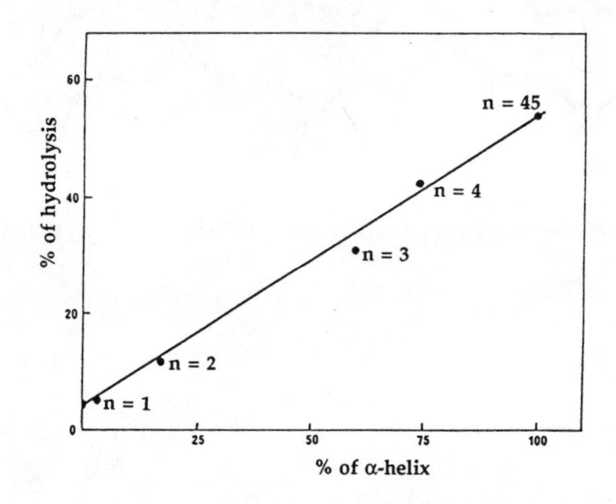

Figure 19. Variation of the catalytic activity (expressed as the percentage of hydrolyzed phosphodiester bonds) as a function of the percentage of α-helix exhibited by the free peptides.

7. Conclusion

Since the historical experiment of Stanley Miller in 1953, nearly all of the biogenic elementary building blocks have been successfully synthesized from methane (CH_4) or its derivatives, formaldehyde (HCHO) and - in the presence of nitrogen - hydrogen cyanide (HCN).

At the present time, however, geochemists favor an atmosphere produced by volcanic outgassing which is dominated by carbon dioxide, nitrogen and water vapor. In such an atmosphere, the production of biogenic building blocks appears to be inhibited. Thus, if carbon was available as carbon dioxide, there was a serious problem for the supply of biological molecules on Earth and other sources must be searched for.

The study of meteorites, particularly the carbonaceous chondrites that contain up to 5 % by weight or organic carbon, demonstrates that at least eight proteinaceous amino acids can be imported from space. The evidence for organic molecules in comet dust (Kissel and Krueger, 1987) and the chemical analysis of laboratory produced ultraviolet irradiated interstellar dust mantle ice analogs suggest that comets could contribute significantly to prebiotic chemistry (Greenberg, 1984, Briggs, et al., 1992, Greenberg and Mendoza-Gomez, 1992, Greenberg and Mendoza-Gomez and Greenberg, this volume).

To day, nucleotide chemists fail to demonstrate that accumulation of substantial quantities of relatively pure oligonucleotides on the primitive Earth was a plausible chemical event and suggest a new scenario.

According to the new scenario, the RNA world appears as an episode in biology but not at the very beginning of life.

The chemist has now to invent a chiral self-replicating system able to amplify a simple information, with a small error rate to allow selection, without the participation of RNA. Was the first amplifying machinery based on thioester chemistry, on organic molecules adsorbed on pyrite or on organic molecules associated with clays ?

Amino acids were most likely available on the primitive Earth, either home-made or imported from space. Starting from a mixture of amino acids in liquid water, peptides can be selected on the basis of chemical selection and chemical resistance toward degradation through simple pathways which have been demonstrated in the laboratory. Some simple well-defined peptides have also been shown to develop active sites with efficient chemical activity. Accumulation of chemically active peptides on the primitive Earth appears therefore plausible via thermostable and stereoselective β-sheets made of alternating polypeptides. Large avenues are thus offered to peptides and should lead to new results.

8. References

Armstrong, D.W., Seguin, R., Mc Neal, C.J., Mac Farlane, R.D. and Fendler, J.H. (1981) 'Spontaneous polypeptide formation from amino acyl adenytlates in surfactant aggregates', J. Amer. Chem. Soc. 100, 4605-4606..

Barbier, B. and Brack, A. (1988) 'Baric polypeptides accelerate the hydrolysis of ribonucleic acids', J. Amer. Chem. Soc., 110, 6880-6882.

Barbier, B. and Brack, A. (1988) 'Conformation-controlled hydrolysis of polyribonucleotides by sequential basic polypeptides', J. Amer. Chem. Soc., 110, 6880-6882.

Behmoaras, T., Toulmé, J.J. and Hélène, C. (1981) 'A tryptophanecontaining peptide recognized and cleaves DNA at apurinic sites', Nature, 292, 858-859.

Brack, A. and Orgel, L.F. (1975) 'β-structures of alternating polypeptides and their possible prebiotic significance', Nature 256, 383-387.

Brack, A. (1976) 'Polymerisation en phase aqueuse d'acides aminés sur des argiles', Clay Minerals, 11, 117-120.

Brack, A. and Caille, A. (1978) 'Synthesis and β-conformation of copolypeptides with alternating hydrophilic and hydrophobic residues', Int. J. Peptide Protein Res., 11, 128-139.

Brack, A. and Spach, G. (1979) 'β-structures of polypeptides with L- and D-residues Part I. Synthesis and conformational studies', J. Mol. Evol. 13, 35-46.

Brack, A. and Spach, G. (1980) 'β-structures of polypeptides with L- and D-residues Part III. Experimental evidences for enrichment in enantiomer', J. Mol. Evol. 15, 231-238.

Brack, A. and Spach, G. (1981) 'Multiconformational synthetic polypeptides', J. Amer. Chem. Soc., 103, 6319-6323.

Brack, A. (1982) 'Aqueous polymerization of L-amino acid active esters in bicarbonate solution via Leuchs anhydrides', Bio Systems 15, 201-207.

Brack, A. (1984) 'Prebiotic synthesis and organization of biopolymer-like macromolecules', Origins of Life, 14, 229-236.

Brack, A. (1987) 'Selective emergence and survival of early polypeptides in water', Origins of Life, 17, 367-379.

Brack, A. and Barbier, B. (1990) 'Chemical activity of simple basic peptides' Origins of Life, 20, 139-144.

Briggs, R., Ertem, G., Ferris, J.P., Greenberg, J.M., McCain, P.J., Mendoza-Gomez, C.X. and Schutte, W., (1992) 'Comet Halley as an aggregate of interstellar dust and further evidence for the photochemical formation of organics in the interstellar medium' Origins of Life 22, 287-307.

Cavadore, J.C. and Previero, A. (1969) 'Polycondensation d'acides aminés libres en solution aqueuse par l'intermédiaire d'une carbodiimide hydrosoluble' Bull. Soc. Chim. Biol. 51, 1245-1253.

Chakravarty, P.K., Mathur, K.B. and Dhar, M.M. (1973) 'The synthesis of a decapeptide with glycosidase activity', Experimentia, 29, 786-788.

Chang, S., Flores, J. and Ponnamperuma, C. (1969) 'Peptide formation mediated by hydrogen cyanide tetramer : a possible prebiotic process', Proc. Nat. Acad. Sci., 64, 1011-1015.

Charlier, M. and Hélène, C. (1975) 'Photosensitized splitting of thymine dimers in DNA by indole derivatives and tryptophane-containing peptides', Photochem. Photobiol., 21, 31-37.

Duker, N.J. and Hart, D.M. (1982) 'Cleavage of DNA at apyrimidic sites by Lys-Trp-Lys' Biochem. Biophys. Res. Com., 105, 1433-1439.

Ehler, K.W. and Orgel, L.E. (1976) 'N,N'-carbonyldiimidazole-induced peptide formation in aqueous solution', Biochim. Biophys. Acta, 434, 233-243.

Ferris, J.P. and Hagan Jr., W.J. (1984) 'HCN and chemical evolution : the possible role of cyano compounds in prebiotic synthesis', Tetrahedron 40, 1093-1120.

Folda, T., Gros, L. and Ringsdorf, H. (1982) 'Formation of oriented polypeptides and polyamides in monolayers and liposomes', Makromol. Chem., Rapid Commun., 3, 167-174.

Fox, S.W. and Dose, K.d (1977) in 'Molecular evolution and the origin of life' Marcel Dekker, New York.

Fridkin, M. and Goren, H.J. (1974) 'Synthesis and catalytic properties of the heptapeptide Ser-Pro-Cys-Ser-Glu-Thr-Tyr', Eur. J. Biochem., 41, 273-283.

Fukuda, K., Shibasaki, Y. and Nakahara, H. (1981) 'Polycondensation of long chain esters of α-amino acids in monolayers at air/water interface and in multilayers on solid surface', J. Macromol. Sci. Chem., A 15, 999-1014.

Goren, H.J., Fletcher, T., Fridkin, M. and Katchalski, E. (1978) 'Poly(His-Ala-GLu) II. Catalysis of p-nitrophenyl acetate hydrolysis', Biopolymers, 17, 1679-1692.

Greenberg, J.M., (1984) 'The structure and evolutioı interstellar grains' Scientific American 250, 124-135.

Greenberg, J.M. and Mendoza-Gomez, C.X., (1992) ' seedings of life by comets' Adv. Space Res. 12, 169-180.

Hahn, K.W., Klis, W.A. and Stewart, J.M. (1990) 'Des and synthesis of a peptide having chymotrypsin-l esterase activity', Science 248,1544-1547.

Halmann, M. (1968) 'Cyanamid-induced condensation re tions of glycine' Arch. Biochem. Biophys., 128, 8 810.

Hanabusa, K., Kato, K., Shirai, H. and Hojo, N. (19 'Synthesis of poly(α-amino acid) on the surface functional revrsed micell' J. Polymere Sci., Part 24, 311-317.

Hawker, J.R. and Oro, J. (1981) 'Cyanamide-media syntheses of peptides containing histidine and hyd phobic amino acids' 17, 285-294.

Kapoor, A., Kang, S.M. and Trimboli, M.A. (19 'Studies of enzyme active sites : synthesis and ca lytic properties of His-Gly-Asp-Ser-Phe, J. Pha Sci., 59, 1296130.

Kawaguchi, K., Tanikara, T. and Imanishi, Y. (19 'Catalytic hydrolysis of charged carboxylic a active esters with cyclic dipeptides carry hydrophobic and nucleophilic groups', Polym. J., 97-102.

Kissel, J. and Krueger, F.R., (1987) 'The organic c ponent in dust from comet Halley as measured by PUMA mass spectrometer on board Vega 1' Nature 3 755-760.

Kunieda, N., Watanabe, M., Okamoto, K. and Kinoshi M. (1981) 'Poly condensation of thioglycine S-dode ester hydrobromide in water', Makromol. Chem., 1 211-214.

Lahav, N., White, D.H. and Chang, S. (1978) 'Pept formation in the prebiotic era : thermal condensat of glycine in fluctuating clay environments' Scier 201, 67-69.

Nishi, N., Nakajima, B.I., Morishige, M. and Tokura (1986) "Hydrolysis of various types of ester s strates with linear, cyclic and polymeric pepti containing His, Ser and Asp residues' Int. J. Pept Protein Res., 27, 261-268.

Noguchi, J., Nishi, N., Tokura, S. and Murakami, (1977) 'Studies on the catalytic action of poly amino acids VII. Stereospecificity in the enzyme l hydrolysis of benzoyl-Arg-p-nitroanilides by cop (Cys, Glu)', J. Biochem., 81, 47-55.

Nooner, D.W., Sherwood, E., More, M.A. and Oro, J. (1977) 'Cyanamide-mediated syntheses under plausible primitive Earth conditions III. Synthesis of peptides', 10, 211-220.

Nozawa, T. and Hatano, M. (1971) 'The mecanism of the asymmetrical selective oxidation of 3,4-dihydroxyphenylalanine catalyzed by the poly-L-lysine copper (II) complex', Makromol. Chem., 141, 31-41.

Nozawa, T., Akimoto, Y. and Hatano, M. (1972) 'On the mecanism of the stereoselective hydrolysis of Phe esters catalyzed by poly(L-lysine)-copper (II) complexes', Makromol. Chem., 161, 289-291.

Paecht-Horowitz, M. and Eirich, F.R. (1988) 'The polymerization of amino acid adenylates on sodium-montmorillonite with preabsorbed polypeptides', Origins of Life, 18, 359-387.

Perello, M., Barbier, B. and Brack, A. (1991) 'Hydrolysis of oligoribonucleotides by α-helical basic peptides', Int. J. Peptide Protein Res., 38, 154-160.

Petz, D. and Schneider, F. (1976) 'Synthesis and catalytic properties of peptides with hydrolytic activity', Z. Naturforsch, 31C, 534-543.

Pierre, J. and Laval, J. (1981) 'Specific nicking of DNA at apurinic sites by peptides containing aromatic residues', J. Biol. Chem., 256, 10217-10220.

Ponnamperuma, C., Shimoyama, A. and Friebele, E. (1982) 'Clay and the origin of life', Origins of life, 12, 9-40.

Schultz, R.M., Huff, J.P., Anagnostaras, P., Olsher, U. and Blout, E.R. (1982) 'Synthesis and conformational properties of a synthetic cyclic peptide for the active site of α-chymotrypsin', Int. J. Peptide Protein Res., 19, 454-469.

Sheehan, J.C., Bennett, G.B. and Schneider, J.A. (1966) 'Synthetic peptide models of enzyme active sites III. Stereoselective estearase models', J. Amer. Chem. Soc. 88, 3456.

Spach, G. and Brack, A. (1979) 'β-structures of polypeptides with L- and D-residues. Part II. Statistical analysis and enrichment in enantiomer', J. Mol. Evol., 13, 47-56.

Steinman, G., Lemmon, R.M. and Calvin, M. (1965) 'Dicyandiamide : possible role in peptide synthesis during chemical evolution', Science 147, 1574.

Steinman, G. and Cole, M.N. (1967) 'Synthesis of biologically pertinent peptides under possible primordial conditions', Proc. Natl. Acad. Sci., 58, 735-742.

Trudelle, Y. (1982) 'Synthesis, conformation and re tivity towards p-nitrophenyul acetate of polypepti incorporating aspartic acid, serine and histidin Int. J. Peptide Protein Res., 19, 528-535.
Warden, J.T., Mc Cullough, J.J. Lemmon, R.M. Calvin, M. (1974) 'A re-examination of the zeoli promoted, clay-mediated peptide synthesis', J. M Evol., 4, 189-194.
White, D.H. and Rickson, J.C. (1980) 'Catalysis peptide bond formation by histidyl-histidine i fluctuating clay environment', J. Mol. Evol., 279-290.

THE BEGINNINGS OF LIFE ON EARTH: EVIDENCE FROM THE GEOLOGICAL RECORD

M. SCHIDLOWSKI
Max-Planck-Institut für Chemie
Saarstrasse 23
D-6500 Mainz (FRG)

ABSTRACT. With the currently available geological record at hand, there is no doubt that microbial (prokaryotic) ecosystems have been prolific on the Archaean Earth since 3.5, if not 3.8 Gyr ago. While the information encoded in the oldest record (>3.5 Gyr) is blurred by a metamorphic overprint, the paleontological and biogeochemical evidence pertinent to the existence of life at times < 3.5 Gyr is so firmly established as to be virtually unassailable. In spite of the marked impairment of the oldest sedimentary record, the residual evidence preserved is more than adequate to build a cogent case for the initiation of life processes as early as 3.8 Gyr ago, with the concomitant establishment of a biogeochemical carbon cycle based on the operation of (photo)autotrophic carbon fixation.

1. Introduction

There is by now general agreement that life has emerged at a certain stage of either cosmic or planetary evolution as an intrinsically new property of matter. In particular, life processes are characterized by (1) their preferential reliance on a limited suite of chemical elements (notably C,O,H,N,S,P) all of which being among the most abundant elements in the Universe, (2) their persistence as dynamic states removed from thermodynamic equilibrium to markedly lower entropy levels, (3) their morphological expression in the form of discrete structures or "compartments" culminating in the formation of the cell as the basic morphological and functional entity of life, and (4) their capability of proliferation by means of identical reproduction.

There is likewise little doubt that the beginnings of organic chemistry and the early stages of prebiological (chemical) evolution took place already in astrophysical environments such as the interstellar medium. From these environments, radio astronomical molecular spectroscopy has confirmed the existence of an impressive

J. M. Greenberg et al. (eds.), The Chemistry of Life's Origins, 389–414.

array of organic molecules that include species known to function as intermediates in the *prebiotic* synthesis of sugars, proteins and nucleic acids (Irvine and Knacke, 1989). Hence, we may reasonably assume that the basic molecular building blocks of living matter had been continuously synthesized in interstellar clouds since the very beginnings of the Universe some 20 Gyr ago. It has been estimated that specifically photoprocessing by cosmic UV of the interstellar medium has given rise to an impressive intergalactic reservoir of organic substances on the order of about one permil of the total mass of the Milky Way (Greenberg, 1984), with the bulk being tied up in the refractory organic mantles of common interstellar grains.

Such cosmic scenarios for the initiation of prebiological organic chemistry may have counterparts on the planetary level which preferentially involve (1) high-energy processing, mainly by solar UV or electric discharges, of reduced or weakly oxidized gases (H_2, CH_4, NH_3, CO) released during early degassing of the planetary mantle (cf. Miller et al., 1976; Oró et al., 1990), and (2) chemical processes in high-temperature environments associated with volcanic activity (cf. Corliss et al., 1981). Notably, the pioneering work by Miller (1955) had shown that reactions among the components of a reducing primordial atmosphere may give formidable yields of amino acids as well as of hydrogen cyanide and cyanoacetylene (the two latter figuring as reactants in the prebiotic synthesis of nucleoside bases and pyrimidines, respectively). In concert with complementary syntheses powered by either solar UV or local sources of geothermal energy (e.g., Fischer-Tropsch-type reactions), processes of the above type could have laid the physicochemical base for prebiotic organic chemistry at or near a planetary surface such as the Earth's. Parenthetically, mention should be made also of the potential "seeding" function of organic-rich cometary materials transferred to previously sterile planetary surfaces, resulting in rigorous reactions among the highly unsaturated organic compounds of cometary dust particles upon the encounter of aqueous environments (cf. Kissel and Krueger, 1987; Greenberg et al., 1989).

With the beginnings of a primitive organic chemistry secured by reactions of the above and related type, subsequent *chemical evolution* can be expected to have finally culminated in the emergence of the first self-replicating life-like systems ("protobionts"). These, in turn, provided the link between chemical and *biological ("Darwinian") evolution* that consequently gave rise to the abundance and morphological variety of life as manifest in the contemporary biosphere.

2. The Geological Record of Life

With the geological rock record constituting the only source of paleontological and biogeochemical information, empirical evidence as to the presence of life on our home planet cannot predate the oldest terrestrial sediments with an age of about 3.8 Gyr (Moorbath et al., 1973). Since this record has hitherto failed to disclose unequivocal vestiges of *prebiotic* organic chemistry, both early chemical and the beginnings of biological evolution must have proceeded in the largely undocumented "Hadean" era bracketing the time interval between the Earth's formation (~ 4.5 Gyr) and the appearance of the oldest sedimentary rocks 3.8 Gyr ago (Fig. 1). As this time segment is completely devoid of geological documents, current speculations about early chemical evolution on the juvenile planet are necessarily extrapolations from the cosmic and planetary scenarios outlined above. Accordingly, the initiation of life processes on the ancient Earth remains shrouded in mystery, leaving even room for controversial standpoints with regard to an autochthonous (planet-based) vs. an extraneous (cosmic) provenance of terrestrial life (cf. Schidlowski, 1990). Since a good case can be made for the operation of a biologically modulated carbon cycle as early as 3.8 Gyr ago, any Earth-based origin of life must have necessarily been accomplished within an uncomfortably short time interval of 0.7 Gyr.

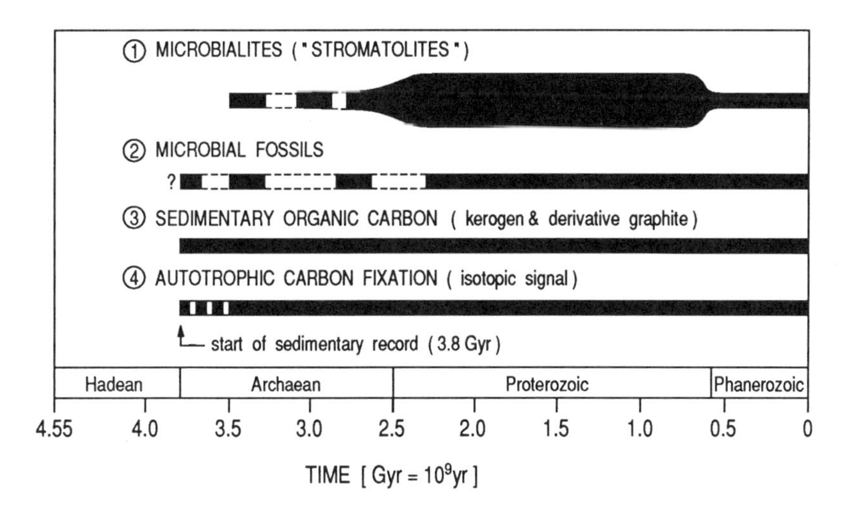

Figure 1. Selected categories of paleontological (1,2) and biogeochemical (3,4) evidence over 3.8 Gyr of Earth history. The stromatolite diagram (1) reflects the proliferation of microbial ecosystems after the establishment of stable marine shelves during the Proterozoic and their subsequent sharp decline after the advent of heterotrophic (animal) grazers 0.6-0.7 Gyr ago. The extension to 3.8 Gyr of the record of microbial fossils (2) is contingent on the biogenicity of the principal cell-like morphotype (*Isuasphaera isua*) from the Isua suite. The isotopic signature of biological (autotrophic) carbon fixation as preserved in sedimentary organic carbon bears a metamorphic overprint as from t > 3.5 Gyr (broken line), whose direction and magnitude is currently well understood.

In any case, students of early terrestrial life had been intrigued over the last three decades by the evidence continuously piling up in support of biological activity already in Archaean times (Fig.1). While the morphological record of microbial (prokaryotic) life had been shown to hold for 3.5, if not 3.8 Gyr of geological history, the isotopic composition of the carbon constituents of the oldest sediments had furnished fair proof of the presence of biologically mediated isotope fractionations as from the very onset of the currently known sedimentary record. In concert, these findings give eloquent testimony to the existence of prolific microbial ecosystems not long after the Earth's formation and the attendant operation of a biogeochemical carbon cycle over almost 4 Gyr of planetary history.

The following discourse offers a summary of the currently available empirical evidence that bears on the antiquity of life processes on Earth. Notably in the biogeochemical part, the argument draws in full on previous statements of the subject by the author (Schidlowski et al., 1979, 1983; Schidlowski, 1987, 1988).

2.1. THE PALEONTOLOGICAL RECORD

Given suitable conditions in their primary sedimentary burial grounds, dead organisms are susceptible to fossilisation and thus apt to leave a morphological record in sedimentary rocks. Though in part selective, this record may withstand the onslaught of an annealing geological history, persevering through billions of years before being erased, in the fullness of time, by the metamorphic and anatectic reconstitution of the host rock. This holds for both higher multicellular life forms (Metaphyta and Metazoa) and for microorganisms such as algae and cyanobacteria.

Being referred to as the "age of microorganisms", the Precambrian (Archaean and Proterozoic) segment of Earth history (Fig.1) was, in fact, principally characterized by prokaryotic and eukaryotic microbial ecosystems which held dominion over the Earth from the onset of the record until the advent of the first Ediacara-type metazoan faunas between 0.6 and 0.7 Gyr ago (Glaessner, 1983, 1984). While the oldest microbial life on Earth was likely to be of archaebacterial affinity, there is firm evidence that prokaryotic microorganisms (still lacking membrane-bound cell nuclei and assembling DNA in simple strands and loops instead of paired rod-like chromosomes) had dominated the scene as from at least 3.5 Gyr ago. Subsequent evolution of this early microbial world reached a climax during the Proterozoic with the emergence of the eukaryotic (nucleated) cell some 1.4 Gyr ago (Cloud, 1976; Schopf and Oehler, 1976; for dissent in favour of the rise of eukaryotes already some 2 Gyr ago see Kazmierczak, 1979, and Pflug and Reitz, 1985). As is generally agreed upon, the advent of the eukaryotic cell (characterized by a membrane-bound nucleus, organelles, sexual reproduction and mitotic cell division) had set the stage for all subsequent diversification of life. This holds particularly for the so-called "Cambrian explosion" resulting in the almost instantaneous appearance of all major animal phyla at the dawn of the Phanerozoic ($\sim$ 0.56 Gyr ago).

Specifically the aqueous (marine) habitats of the Proterozoic era (2.5–0.6 Gyr) were shown to have been replete with exuberant microbial communities of both the benthic (bottom-living) and planktonic (buoyant) type that have left impressive records of cellularly preserved microfossils notably in siliceous formations ("cherts"), with the remarkably differentiated microfloras of the Australian Bitter Springs Formation ($\sim$ 0.9 Gyr) and the $\sim$ 2.0 Gyr-old Gunflint Iron Formation of Canada as most outstanding examples (cf. Schopf, 1968; Barghoorn and Tyler, 1965; Cloud, 1965). Moreover, an abundance of biosedimentary structures in the form of petrified microbial carpets ("stromatolites") that preserve the matting behaviour of benthic microorganisms at the sediment-water interface has been reported specifically from Proterozoic terranes (Hofmann, 1973; Walter, 1983), all this testifying that the early microbial world had climaxed during this time period.

2.1.1. *The Oldest (Archaean) Microfossil Record.* As we go back from the Proterozoic to the Archaean (3.8–2.5 Gyr), the fossil record becomes increasingly scant (Awramik, 1982; Schopf and Walter, 1983). This is particularly true for cellular fossils of cyanobacterial affinity that had been abundantly preserved in Proterozoic cherts. Quite obviously, both the increasing diagenetic alteration and metamorphic reconstitution of the progressively older host rocks tend to blur the primary morphologies of delicate microstructures, resulting in a large-scale loss of contours and other critical morphological detail. At the end of such alteration series stand so-called "dubiofossils" of variable or sometimes questionable confidence levels.

In spite of the blatant impoverishment of the oldest record (cf. Fig.1) there are, however, single reports of well-preserved microfossils which may pass as cellular evidence of Archaean life. Prominent among these finds are respective assemblages from the Warrawoona Group of the Pilbara Block in Western Australia (3.3–3.5 Gyr) whose lowermost fossiliferous sequence closely approaches the 3.5 Gyr-mark. While first reports of this remarkably differentiated microflora (Awramik et al., 1983) had been disputed mainly on grounds of unprecisely constrained rock relationships involving cherty fissure fillings (Buick, 1984), the respective chert lithology has proved fossiliferous also in places where it undoubtedly does not constitute mere fissure fillings, but figures as part of a normally bedded sedimentary series (Schopf and Packer, 1987). Conspicuous within the Warrawoona microfossil assemblage are both the coccoidal and filamentous morphotypes (Figs. 2 and 3) that were abundantly present in the cyanobacterial communities of Proterozoic chert formations, this probably demonstrating an astounding degree of evolutionary conservatism in prokaryotic morphology and community composition through time. While the septate filamentous morphologies described by both Awramik et al. (1983) and Schopf and Packer (1987) stand for fossil trichomes that could be attributed to both filamentous cyanobacteria or more primitive prokaryotes (e.g. Chloroflexaceae), the coccoidal unicell aggregates depicted by Schopf and Packer (1987) have been claimed to strictly exclude other than cyanobacterial affinities.

As cyanobacteria entertain the water-splitting variant of the photosynthetic process (cf. Section 2.2.1, Eq.3), a well-established cyanobacterial connection of the Warrawoona microflora would have crucial consequences for the advent of oxygenic photosynthesis, supporting conjectures of a very early emergence of this process based on geochemical inferences (Schidlowski, 1978; Walker et al., 1983). Altogether, the Warrawoona microflora gives proof of the existence on Earth as from at least 3.5 Gyr ago of microbial communities characterized by a remarkable degree of morphological and physiological diversification that had operated on the organisational level of the prokaryotic cell.

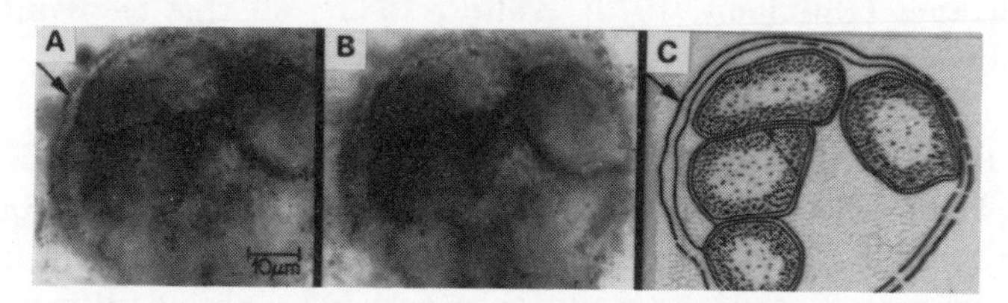

Figure 2. Photomicrographs of a sheath-enclosed cell colony of presumably cyanobacterial (chroococcalean) affinity from the Warrawoona Group, Australia (3.3–3.5 Gyr) taken at different focal depths (A,B), along with a corresponding reconstruction (C). Arrows indicate remnants of original sheath. – Graphs by courteousy of J.W. Schopf (cf. also Schopf and Packer, 1987).

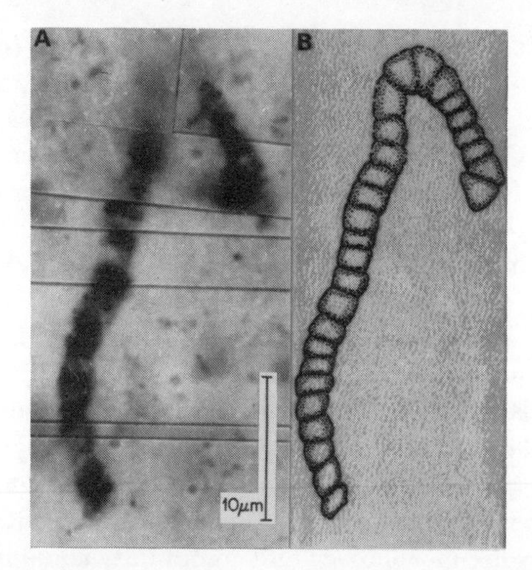

Figure 3. Septate filament from the Warrawoona Group (3.3–3.5 Gyr) resemblingh trichomes of prokaryotic microorganisms of either cyanobacterial or flexibacterial affinities. A = original (photomontage) from Schopf and Packer (1987), B = reconstruction.

Since coeval microfossil assemblages such as those described from the Swaziland Supergroup, South Africa (Muir and Grant, 1976; Knoll and Barghoorn, 1977; Walsh, 1992) are dominated by evidence on the dubiofossil level, the Warrawoona microflora holds a benchmark function in the currently known evolutionary sequence of early terrestrial life. Considering the remarkable degree of organisational complexity of the prokaryotic cell in general and the conspicuous diversification of the Warrawoona community in particular, we must necessarily infer that the lineages of the principal morphotypes of this microflora must have emerged well before Warrawoona times. Therefore, it seems almost certain that precursor floras had been extant prior to 3.5 Gyr where the preserved rock record becomes patchy and progressively metamorphosed. In this context, cell-like morphologies described from the $\sim$ 3.8 Gyr-old amphibolite-grade metasediments from Isua, West Greenland (Pflug 1978, 1987; Pflug and Jaeschke-Boyer, 1979; Bridgwater et al., 1981; Roedder, 1981) were bound to attract considerable attention.

Central to this debate is a globular to elongate, presumedly sheath-enclothed microstructure hosted by a metamorphosed chert facies. Described as *Isuasphaera isua* (Pflug, 1978), the biogenicity of this morphotype (Fig. 4) had been violently disputed, specifically on grounds of the improbability of survival of cellular fossils during the amphibolite-grade metamorphism of the host rock (Bridgwater et al., 1981). Meanwhile, however, there has accrued ample evidence that fossils in general and microfossils in particular may readily withstand obliteration in medium-grade metamorphic rocks to a degree permitting even micropaleontological approaches to the study of metamorphosed sediments (Pflug and Reitz, 1992).

Since the structurally more differentiated filamentous and coccoid morphotypes that characterize the younger Warrawoona assemblage are largely absent from the Isua community, the morphological evidence is gravely empoverished, with part of this probably due to the loss of microstructural detail during the amphibolite-grade reconstitution of the host rock. However, in spite of the apparent reduction in morphotype inventory and a severe impairment of the structural evidence, there is a striking resemblance of *Isuasphaera isua* as the dominant morphotype to a possible counterpart in the younger (Proterozoic) record described as *Huroniospora* sp. whose biogenic interpretation goes largely unchallenged. The near perfect morphological match between the two forms (Fig. 4) was shown to be paralleled also by selected microchemical characteristics as revealed by Laser Raman spectroscopy (Pflug, 1987). Hence, even with due application of the critical standards called for when dealing with a metasedimentary sequence, a reasonable case can be made for the occurrence in the Isua suite of microfossils qualifying at least for dubiofossil status since they stand in the continuity of the younger record. If confirmed by subsequent scrutiny, the additional report from the Isua banded iron-formation of presumably aggregate-forming iron bacteria characterized by intricate external cell-wall structures (Robbins, 1987) might further testify to the existence of microbial life already in Isua times.

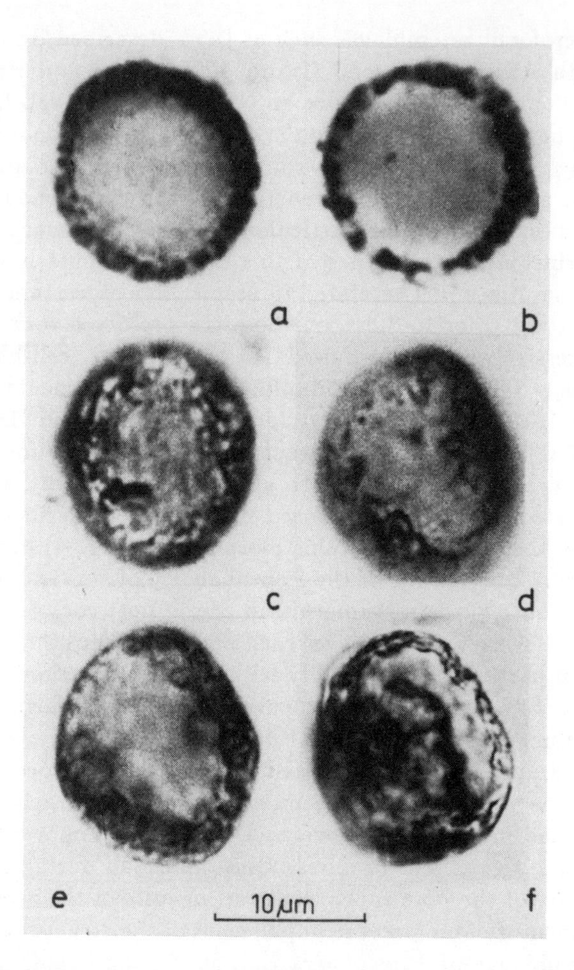

Figure 4. Comparison of *Huroniospora* from the Proterozoic (2.0 Gyr) Gunflint iron-formation, Ontario (a-c) with *Isuasphaera* from the 3.8 Gyr-old Isua metasedimentary suite, West Greenland (d-f). The optically distinctive marginal rim may be explained as relic of the original cell wall. From Pflug (1987).

In conclusion, the statement seems warranted that the presently available evidence would not *a priori* exclude microbial affinities for at least some of the cell-like microstructures reported from the 3.8 Gyr-old metasedimentary Isua suite. In spite of the uncertainty surrounding a fair number of the morphotypes described (Bridgwater et al., 1981), and of occasional convergences with mineralogical features such as limonite-stained dissolution cavities (Roedder, 1981), there is a reasonable chance that the microstructure inventory as a whole includes at least some elements of a structurally degenerated microfossil assemblage such as might result from intense metamorphic alteration of a Warrawoona-type microflora. As will be detailed further below, the existence in Isua times of microbial ecosystems would be consistent with, if not conditional for, the actually observed content and isotopic composition of reduced (organic) carbon in the Isua suite. Considering the inherent shortcomings of the morphological evidence notably in metamorphosed rocks, a microchemical approach to putatively biogenic microstructures may also hold considerable promise. While first endeavours in this respect had been initiated successfully during the last decade (cf. Pflug, 1987), a specifically targeted, large-scale application of the impressive array of microanalytical techniques currently available is likely to furnish further crucial constraints for assessing the biogenicity of individual disputed micromorphologies.

2.1.2. *The Early Record of Biosedimentary Structures ("Stromatolites")*. Another category of paleontological evidence that gives testimony to the existence of microbial life on the juvenile planet is represented by laminated organosedimentary structures of the "stromatolitic" type. Stromatolites had been viewed for long as the only undisputable manifestations of life in Archaean times, constituting visible (macroscopic) relics of ancient microbial mat communities that monitor the geological history of mat-forming cyanobacterial and algal microbenthos over a time interval of some 3.5 Gyr (Fig.1).

Stromatolites in the widest sense (Fig. 5) represent fossilized stacks of successively superimposed microbial (mostly prokaryotic) mats that had once thrived as organic films at the sediment-water interface of aquatic habitats. Deriving ultimately from the interaction of a benthic microbial layer with the underlying sediment, they may be aptly referred to as "microbialites" (Burne and Moore, 1987), their lithification resulting from either trapping, binding, or biologically mediated precipitation of

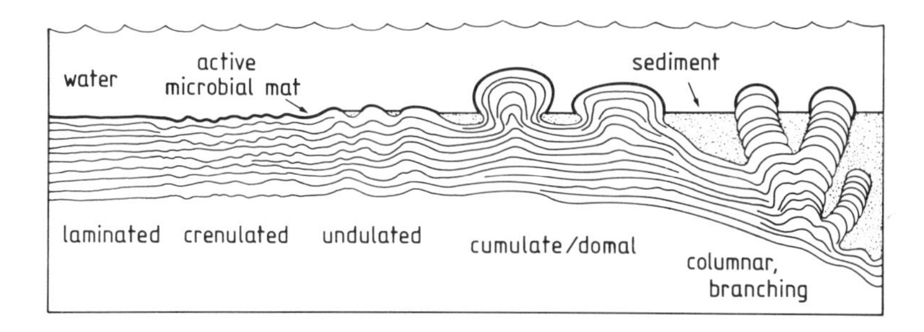

Figure 5. Main morphological inventory of microbial mats thriving at the sediment-water interface. The fossil manifestations of these laminated microbial (mostly cyanobacterial) colonies are called "stromatolites" (see also Fig. 6).

selected mineral constituents (cf. Hofmann, 1973; Walter, 1977; and others). The individual laminae of a stromatolitic structure preserve a record of successive stages in the growth of the parent microbial ecosystem, the complex interplay of the organic film with the sedimentary environment having usually given rise to a wealth of accretional morphologies (stratiform, domal, columnar, branched, etc.; see Fig. 5).

It is, meanwhile, firmly established that the stromatolite record goes back to 3.5 Gyr ago. Conspicuous among the oldest finds are occurrences within the Warrawoona Group (3.3–3.5 Gyr) of the Pilbara Block of Western Australia (Dunlop et al., 1978; Lowe, 1980; Walter et al., 1980) which leave no doubt that laminated communities of benthic prokaryotes were indeed extant (and probably abundant) in aquatic environments of this age. Occurring in siliceous (cherty) sediments that had been shown to also host a remarkably differentiated microflora (see Section 2.1.1), the morphological inventory of the Warrawoona stromatolites looks fairly modern, attesting to the fact that microbial mats and related biosedimentary structures figure among the most conservative features of the paleontological record.

Figure 6. Archaean microbialite ("stromatolite") from the Bulawayan Group (about 2.7 Gyr) of the Rhodesian schist belt series (Zimbabwe). The bun-shaped, interfering laminae are lithified microbial mats generated successively by prokaryotic microbenthos of preferentially cyanobacterial affinity.

Apart from the Warrawoona discoveries, roughly coeval finds have been described from several other Archaean formations, notably the ~ 3.5 Gyr-old Sebakwian Group of Zimbabwe (Orpen and Wilson, 1981), and the Fig Tree Group (3.3-3.5 Gyr) of the Barberton Mountain Land, South Africa (Byerly et al., 1986). Stromatolite reports from younger Archaean rocks are numerous including, *inter alia*, the stromatolite occurrences within the ~ 3.0 Gyr-old Pongola Group of South Africa (Mason and von Brunn, 1977) as well as the famous Huntsman Quarry "algal limestones" from the Bulawayan Group (~ 2.7 Gyr) of Zimbabwe (Fig. 6) that figure among the earliest descriptions of Precambrian stromatolites (MacGregor, 1940). In the wake of the establishment of extensive marine shelves during the Proterozoic, stromatolite development reached a climax (Hofmann, 1973; Monty, 1984), followed by the onset of a conspicuous decline at the dawn of the Phanerozoic due to the takeover of the oldest metazoan faunas (cf. Fig.1). This takeover marked the end of the dominion over the Earth's biosphere of the early microbial world that had lasted for 3 billion years of recorded geological history.

Summing up, the discovery of fossil laminated microbial mat communities in sediments approaching an age of 3.5 Gyr constitutes *prima facie* proof that benthic prokaryotes were widespread already in suitable aquatic habitats of the Archaean Earth. The morphological inventory of these oldest stromatolites, and the observed microfossil content of the ambient host rock or coeval sequences, allow a fairly detailed reconstruction of the Earth's earliest microbial ecosystems, indicating specifically that Archaean stromatolite builders were not markedly different from their counterparts in the younger record (inclusive of contemporary analogues). It appears well established that the principal microbial mat builders in the Archaean were filamentous and unicellular prokaryotes capable of both phototropic responses and photoautotrophic carbon fixation (Walter, 1983). The virtually unbroken stromatolite record from Archaean to present times attests, accordingly, to an astounding degree of uniformity in the physiological performance and social organisation of prokaryotic microbenthos over 3.5 Gyr of geological history.

2.2. THE GEOCHEMICAL RECORD OF LIFE

With carbon constituting the key element of life, the initiation of life processes on the juvenile Earth was bound to have profound geochemical consequences. Before the emergence of biology, the carbon dioxide degassed from the Earth's mantle could leave the atmosphere only in the form of carbonate as a result of the operation of the equilibrium system

$$CO_{2(gas)} \rightleftharpoons CO_{2(aq)} \rightleftharpoons HCO^{-}_{3(aq)} \rightleftharpoons CO^{2-}_{3(solid)} \qquad (1).$$

Here, the hydration of dissolved CO_2 gives bicarbonate (HCO_3^-) and subsequently carbonate (CO_3^{2-}) ions, the latter being susceptible to precipitation as solid carbonate assigned to the rock section of the carbon cycle.

Drawing heavily on the carbon content of the environment, the advent of life established a second effective sink for atmospheric carbon dioxide apart from the "carbonate pump" of Eq. (1). As living systems have an intrinsic drive to proliferate until the encounter of externally imposed limits, we may reasonably expect that the surface of the Earth was covered, before long, by a quasi-continuous veneer of life or "biosphere" that formed a geochemically reactive film at the interface between the solid crust and atmosphere/hydrosphere. Specifically, the first large-scale emplacement of CO_2-fixing ("autotrophic") microbial ecosystems must have exercised a significant impact on the carbon cycle by diverting a sizeable proportion of the surficial carbon flux from the carbonate sink to the organic carbon sink. As both carbonate and biologically generated (organic) carbon have, through the ages, been continuously relegated to the crust as part of newly-formed sediments, relevant information should have been encoded in the geologic record as from its very onset. The following text will briefly review the principal categories of biogeochemical and isotopic evidence that bear on the emergence of autotrophic CO_2-fixation, and specifically photosynthesis as the quantitatively most important process of biological carbon fixation.

2.2.1. *Sedimentary Organic Carbon as a Recorder of Ancient Life Processes.* Carbon occurs in the Earth's sedimentary shell in two forms, namely (i) as *oxidized* or carbonate carbon (C_{carb}), such as limestone [$CaCO_3$] or dolomite [$CaMg(CO_3)_2$], and (ii) as *reduced* or organic carbon (C_{org}) that constitutes the fossil residue of biogenic substances. The far bulk of the organic moiety is made up of so-called "kerogen", the acid-insoluble end-product of the degradation of primary organic matter transformed into a polycondensed aggregate of aliphatic and aromatic hydrocarbons (cf. Durand, 1980). In the average sedimentary rock, C_{org} is present in amounts between 0.5 and 0.6 % (Ronov, 1980); with a total mass of the sedimentary shell on the order of 2.4 x 10^{24}g (Garrels and Mackenzie, 1971), this percentage would translate into a C_{org}-reservoir between 1.2 and 1.4 x 10^{22}g.

While carbonate formation is driven by the equilibria of the carbon dioxide – bicarbonate – carbonate system, the far bulk of organic matter is produced by a limited number of photoautotrophic pathways entailing the light-powered reduction of CO_2 to the carbohydrate level by suitable reductants such as hydrogen sulfide or water, i.e.,

$$0.5\,H_2S + H_2O + CO_2 \xrightarrow{h\nu} CH_2O + H^+ + 0.5\,SO_4^{2-} \quad \Delta G_0' = +117.2\,\text{kJ} \quad (2)$$

$$2\,H_2O + CO_2 \xrightarrow{h\nu} CH_2O + H_2O + O_2 \quad \Delta G_0' = +470.7\,\text{kJ} \quad (3).$$

Because of its markedly lower energy demand, H_2S-based (bacterial) photosynthesis

(Eq. 2) has almost certainly preceded the H_2O-splitting process (utilized by cyanobacteria, eukaryotic algae and all higher plants; cf. Eq. 3) in the evolution of photoautotrophy. As most autotrophs have subsequently come to embark on the energetically more expensive water-splitting variant, there must have been a very strong evolutionary pressure to overcome the existing energy threshold by the integration into the photosynthetic pathway of a second light reaction ("photosystem II"). It is reasonable conjecture that any major proliferation of biomass on Earth was contingent on the utilization of water for the reduction of CO_2 as it permitted the exploitation of a virtually inexhaustible pool of readily accessible reducing power

The fossil residues of the organic matter thus generated can be traced back in sedimentary rocks to the very beginning of the record 3.8 Gyr ago (Allaart, 1976; Schidlowski et al., 1979). Systematic assays for C_{org} carried out on Phanerozoic rocks (Trask and Patnode, 1942; Ronov, 1958) as well as observed oscillations of the carbon isotope age functions (see Section 2.2.2) indicate that C_{org} deposition rates have moderately oscillated around a mean of perhaps 0.5% in the average sediment during this time span. Although both the preserved rock record and the available data base are progressively attenuated as we proceed into the older geological past, it seems fairly evident that the scatter of C_{org} in Precambrian sediments is virtually the same as in Phanerozoic rocks (Fig. 7). Hence, the average content of organic carbon in the oldest sediments does not appear to be basically different from that of geologically younger formations. It is worth noting that the (largely graphitized) organic carbon burden of the 3.8 Gyr-old Isua metasedimentary suite of West Greenland at the beginning of the record may amount to >0.6% (cf. Fig. 7) in the case of carbon-rich members of the suite.

Though at first astounding, the surprisingly modern C_{org}-content of Archaean sediments may not be unusual at all. It is by now firmly established that microbial communities figure among the biosphere's most productive ecosystems, with specifically benthic prokaryotes capable of sustaining prodigious rates of primary productivity between 8-12 g C_{org}/m^2 day (Krumbein and Cohen, 1977; Cohen et al., 1980). It is, moreover, generally accepted that the Precambrian was the "Golden Age" of prokaryotic ecosystems (Cloud, 1976; Schopf, 1983). If such high rates of primary production can be sustained by microbial photoautotrophs that operate at the prokaryotic level, a good case can be made that subsequent evolutionary improvements in the efficiency of the photosynthetic process cannot have been very impressive. As any production of biomass is *ultimately* nutrient-limited, it seems reasonable to envisage for the Earth's earliest biosphere a scenario characterized by a state of global plenitude or "biotic saturation" (Stanley, 1981) in which microbial ecosystems had proliferated in existing aquatic habitats to ultimate limits set by the availability of environmental resources, notably phosphorus and nitrogen as critical nutrients.

Summing up the evidence, we may state that organic carbon in the form of kerogen

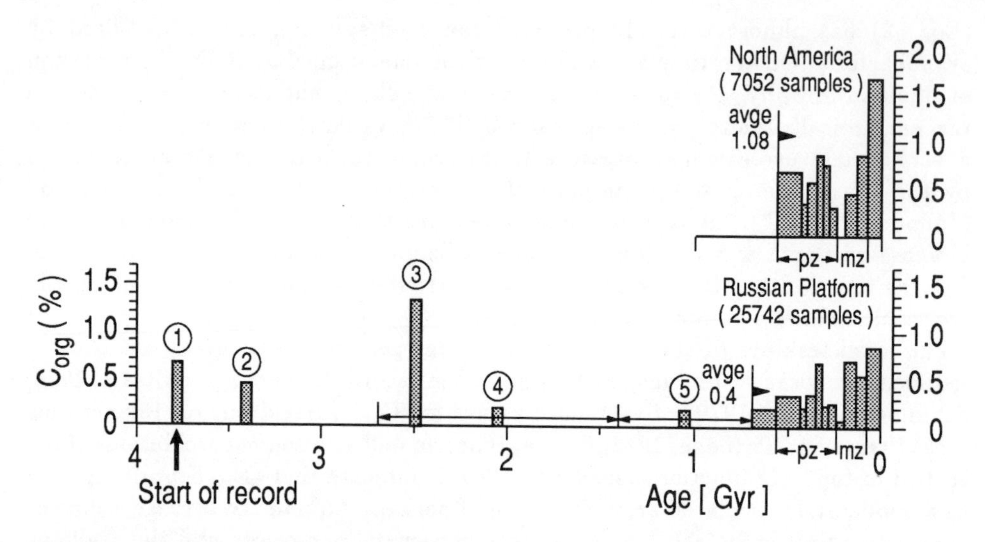

Figure 7. Organic carbon (C_{org}) content of average sedimentary rocks through geologic time. Note excellent data coverage for the last 0.6 Gyr (pz=Paleozoic, mz = Mesozoic) based on comprehensive samplings of both the North American and Russian Platforms (Trask and Patnode, 1942; Ronov, 1958, 1980). In contrast, the older record is poorly documented, but the general scatter of C_{org} in Precambrian sediments appears to be the same as in geologically younger rocks. (1) Early Archaean metasediments of the Isua suite, West Greenland; (2) Swaziland System, South Africa; (3) Hamersley Group, Australia; (4) Proterozoic 1-2 of Russian Platform; (5) Proterozoic 3 of Russian Platform. Adapted from Schidlowski, 1982 (see also detailed references therein).

and its graphitic derivatives is a common constituent of sedimentary rocks over the whole of the presently known record. Based on hitherto available data, a good case can be made that the C_{org}-content of the average sediment has stayed fairly uniform from Archaean to Recent, ranging mostly between 0.4 and 0.6%. The abundance of highly graphitized kerogenous materials in the 3.8 Gyr-old Isua suite is likely to attest to the operation of life processes already during the time of formation of the Earth's oldest sediments.

It should be noted that, in close association with kerogenous materials, there may occur quasi-pristine organic molecules (mostly pigments or single discrete hydrocarbon chains) that have preserved their identity over the whole pathway of kerogen evolution from dead organic matter to the polycondensed aggregates of solid hydrocarbons that figure as end-products of the diagenetic reconstitution of sedimentary organic matter. The record of such "biomarker" molecules or "chemofossils", respectively, goes far back into the Precambrian, gaining particular significance in connection with the geologically oldest (Proterozoic) petroleum occurrences (cf. McKirdy and Imbus, 1992; Summons and Powell, 1992). In general, the topic of Precambrian biomarkers represents a highly specialised and rapidly evolving field that holds considerable potential for a detailed approach to the history of ancient life processes (Hoering and Navale, 1987).

2.2.2. *Sedimentary Carbon Isotope Record as Index of Autotrophic Carbon Fixation.* Additional proof for the biogenicity of the organic carbon (kerogen) constituents of sedimentary rocks comes from observed $^{13}C/^{12}C$ fractionations between carbonates and kerogen constituents as the two principal sedimentary carbon species. As is known since the pioneering studies by Nier and Gulbransen (1938), Murphey and Nier (1941) and Rankama (1948), conversion of inorganic carbon into biological substances entails a discrete bias in favour of the light isotope (^{12}C), with the heavy species (^{13}C) preferentially held back in the inorganic feeder pool and consequently precipitated as carbonate. Subsequent work (cf. Craig, 1953; Park and Epstein, 1960; O'Leary, 1981) has confirmed that all common pathways of autotrophic carbon fixation discriminate against ^{13}C, principally as a result of two isotope-selecting steps inserted in the assimilatory sequence (Fig. 8), the quantitatively most important one being a kinetic isotope effect besetting the first irreversible enzymatic CO_2-fixing carboxylation reaction. This latter reaction is responsible for the incorporation of CO_2 into the carboxyl (COOH) group of an organic acid which, in turn, lends itself to further processing in ensuing metabolic pathways. In concert, these isotope effects have led to a conspicuous enrichment of ^{12}C in all forms of biogenic (reduced) carbon as compared to the inorganic (oxidized) carbon reservoir of the surficial environment consisting mainly of dis-

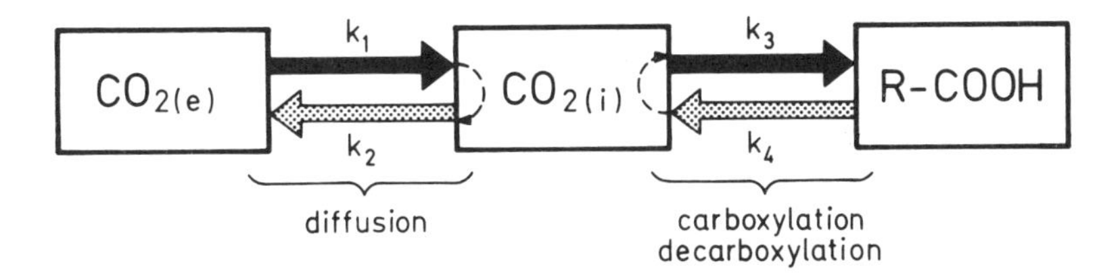

Figure 8. Principal isotope-discriminating steps in biological CO_2-fixation (black: assimilatory reactions; stippled: dissimilatory and other reverse processes; $k_1 - k_4$: corresponding rate constants). Carbon dioxide from the external (atmospheric) feeder pool [$CO_{2(e)}$] enters the living tissue to become "internal" $CO_{2(i)}$ on its way to the photosynthetically active sites, and is subsequently transformed to R-COOH that stands for the product of the first CO_2-fixing enzymatic carboxylation reaction. In sum, these transformations bring about a preferential accumulation of the light carbon isotope (^{12}C) in the synthesized biomass in the right-hand box.

solved bicarbonate ion (HCO_3^-) in seawater and atmospheric carbon dioxide. In terms of the conventional δ-notation, $\delta^{13}C$ values of average biomass usually turn out to be between 20 and 30°/oo more negative than those of marine bicarbonate, the most abundant inorganic carbon species in the environment (cf. Fig. 9, right-hand box).

The isotopic difference thus established between biogenic (organic) carbon and the surficial bicarbonate-carbonate pool is largely retained when organic and carbonate carbon enter newly formed sediments. Although diagenetic alteration in the sediment has been shown to induce secondary shifts in the original isotopic compositions of both C_{org} and C_{carb}, these shifts are usually small, hardly exceeding 2–3 °/oo in the case of diagenetically mature kerogens (for details see Hayes et al., 1983; Schidlowski, 1987; Schidlowski et al., 1983; and others). Also in the case of carbonates, secondary changes in the carbon isotope composition over the diagenetic pathway from parent mud to solid rock tend to stay well below 2 °/oo. Consequently, the original isotopic difference between organic and inorganic carbon as established in the surficial environment remains basically fixed after entry into the sedimentary cycle of both carbon species.

Students of the sedimentary carbon record had been always duly impressed by the near-constancy on the Gyr-scale of the isotope age functions of both C_{org} and C_{carb} (Fig. 9). As is obvious from these graphs, the carbon isotope spreads of both extant primary producers and of recent marine carbonate and bicarbonate have been basically transcribed into the sedimentary record back to 3.5, if not 3.8 Gyr ago. This would confirm that C_{org} and C_{carb} had always been transferred from the surficial environment to sedimentary rocks with little postdepositional change in their isotopic compositions. The narrow spread of the C_{carb} function suggests that the $\delta^{13}C$ values of the marine bicarbonate precursor of ancient carbonates had been always closely tethered to the zero permil line over 3.8 Gyr of Earth history, allowing little departure in either direction [for possible exceptions see Schidlowski et al. (1976) and Baker and Fallick (1989)]. In the case of the corresponding C_{org} function, it is obvious that the $\delta^{13}C_{org}$ range found in recent marine sediments (Fig. 9), integrates faithfully over the respective isotope spreads of the principal contributors to the contemporary biomass with just the extremes eliminated. This also confirms that the effect of a later diagenetic overprint on the primary isotope values is rather limited and, for the most part, gets lost within the broad scatter of primary values. Consequently, the kinetic isotope effect inherent in photosynthetic carbon fixation is propagated from the biosphere into the rock section of the carbon cycle almost unaltered, which opens up the possibility of tracing the isotopic signature of this process back into the geologic past.

With these relationships established, decoding of the vast body of isotopic information hitherto retrieved from the sedimentary record (cf. Schidlowski et al., 1975, 1979, 1983; Eichmann and Schidlowski, 1975; Veizer and Hoefs, 1976; Hayes et al., 1983) is fairly straightforward. There is little doubt that the conspicuous

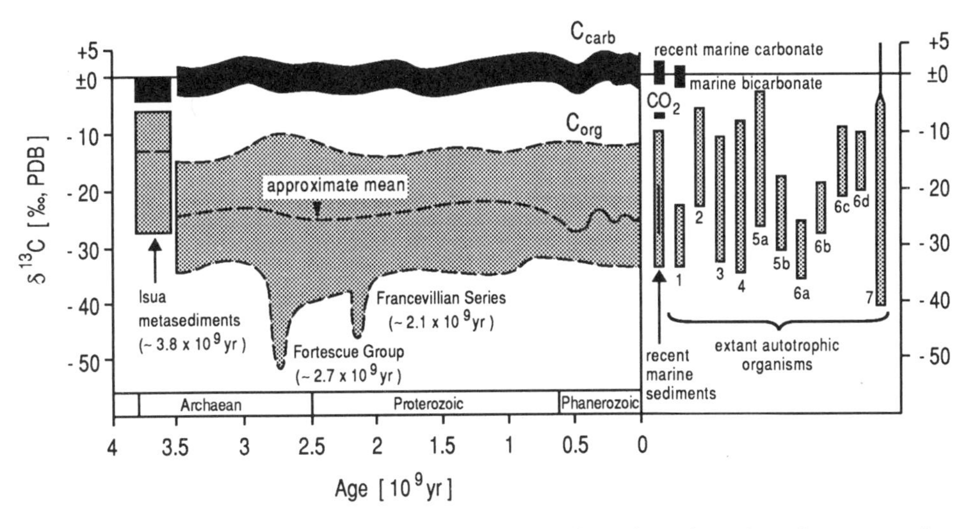

Figure 9. Isotope age functions of organic carbon (C_{org}) and carbonate carbon (C_{carb}) over 3.8 Gyr of recorded Earth history (left-hand box) as compared with the isotopic compositions of their progenitor substances in the present environment (marine bicarbonate and biogenic matter of various parentage, right-hand box). Isotopic compositions are given as $\delta^{13}C$ values indicating either an increase (+) or decrease (−) in the $^{13}C/^{12}C$ ratio of the respective substance (in permil difference) as compared to that of the PDB (Peedee belemnite) standard which defines the zero permil line on the δ-scale. Note that the $\delta^{13}C_{org}$ spread of the extant biomass is basically transcribed into recent marine sediments and the subsequent record back to 3.8 Gyr ago, with the Isua values moderately reset by amphibolite-grade metamorphism (cf. Fig. 10). The negative spikes at 2.7 and 2.1 Gyr indicate a large-scale involvement of methane in the formation of the respective kerogen precursors. Extant autotrophs contributing to the contemporary biomass are (1) C3 plants, (2) C4 plants, (3) CAM plants, (4) eukaryotic algae, (5a+b) natural and cultured cyanobacteria, (6) groups of photosynthetic bacteria other than cyanobacteria, (7) methanogens. $\delta^{13}C_{org}$ range in recent marine sediments according to Deines (1980) based on some 1.600 data points (black insert covers > 90% of the data base). – From Schidlowski (1988).

^{12}C-enrichment displayed by the data envelope for fossil organic carbon (Fig. 9) constitutes a coherent signal of autotrophic carbon fixation over almost 4 Gyr of recorded Earth history as it ultimately rests with the process that gave rise to the biological precursor materials. Moreover, the long-term uniformity of this signal attests to an extreme degree of conservatism of the basic biochemical mechanisms of carbon fixation. In fact, the mainstream of the envelope for $\delta^{13}C_{org}$ depicted

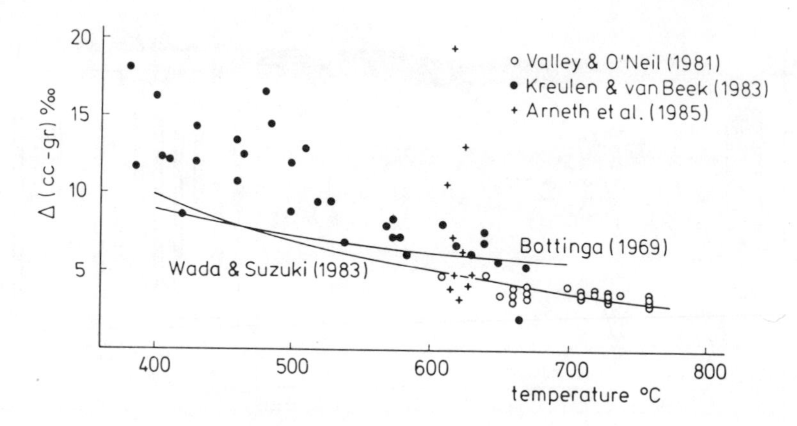

Figure 10. Decrease of isotopic fractionation Δ(cc-gr) $= \delta^{13}C_{cc} - \delta^{13}C_{gr}$ between coexisting calcite (cc) and graphite (gr) as a function of increasing temperature of metamorphism (sedimentary organic carbon preferentially occurs as graphite at higher metamorphic grades). Observed Δ(cc-gr) values reported by several authors from various metamorphic terranes are shown to scatter around both Bottinga's (1969) function of thermodynamically calculated isotope equilibria and an empirical fractionation curve by Wada and Suzuki (1983) calibrated by dolomite-calcite solvus temperatures. Note that the markedly decreased average fractionation of 10-11 °/oo between organic carbon and carbonate in the 3.8 Gyr-old Isua metasediments (cf. Fig. 9) broadly falls into the temperature range of the lower amphibolite facies (450-550°C).

in Fig. 9 can be most readily interpreted as the geochemical manifestation of the isotope-discriminating properties of one single enzyme, namely, ribulose-1,5-bisphosphate (RuBP) carboxylase, the key enzyme of the Calvin cycle.

It is well known today that most of the carbon transfer from the inorganic to the organic world proceeds via the RuBP carboxylase reaction that feeds CO_2 directly into the Calvin cycle as a 3-carbon compound (phosphoglycerate). Most autotrophic microorganisms and all green plants operate along this pathway of carbon assimilation; higher plants relying on it entirely are termed C3 plants. As a result, the bulk of the Earth's biomass (both extant and fossil) bears the isotopic signature of C3 (or Calvin cycle) photosynthesis characterized by the sizeable fractionations of the RuBP carboxylase reaction that assigns a mean $\delta^{13}C_{org}$ range of -26 ± 7°/oo to the bulk of terrestrial biogenic matter. Hence, the $\delta^{13}C_{org}$ age function shown in Fig. 9 may be taken as an index line of autotrophic carbon fixation over 3.8 Gyr of recorded Earth history. It should be noted that first powerful conjectures about the antiquity of photosynthesis had been based on simple precursor curves of the present carbon isotope age function (Junge et al., 1975).

Occasional negative offshoots from the long-term average of this function (Fig. 9) are commonly restricted to the Precambrian (Schoell and Wellmer, 1981; Hayes et al., 1983; Schidlowski et al., 1983; Weber et al., 1983) and suggest the involvement of methanotrophic pathways in the formation of the respective kerogen precursors. Basically, these excursions appear to be oddities confined to side stages of the carbon cycle that have never affected the economy of the global cycle as a whole.

The only demonstrable discontinuity in the $\delta^{13}C_{org}$ record depicted in Fig. 9 is the break between the ~3.8 Gyr-old Isua metasediments from West Greenland and the whole of the post-Isua record. The observed isotope shift is, however, fully consistent with the predictable effects of an isotopic re-equilibration between co-existing organic carbon and carbonate in response to the amphibolite-grade metamorphism experienced by the Isua suite. Both currently available thermodynamic data on $^{13}C/^{12}C$ exchange between C_{org} and C_{carb} as a function of increasing metamorphic temperatures and observational evidence from a host of geologically younger metamorphic terranes (Fig. 10) make it virtually certain that the "normal" sedimentary $\delta^{13}C_{org}$ and $\delta^{13}C_{carb}$ records had originally extended back to 3.8 Gyr ago, and that the Isua anomaly is evidently due to a metamorphic overprint (Schidlowski et al., 1979, 1983). This would imply that the isotopic signature of autotrophic carbon fixation as preserved in ancient kerogens and kerogen-derived graphitic materials persists through almost 4 Gyr of geological history.

3. Summary and Conclusions

Based on the currently available evidence and moderate extrapolations thereof, it is safe to say that the surface of the Archaean Earth had hosted a veneer of microbial (prokaryotic) life as from the onset of the sedimentary record 3.8 Gyr ago (and probably before). This conclusion rests on the following arguments:

- There exists a confirmed record of laminated biosedimentary structures ("stromatolites") back to 3.5 Gyr. With such structures having preserved the matting behaviour of microbial (specifically cyanobacterial) ecosystems thriving at the sediment-water interface, we may decree that communities of benthic prokaryotes were extant at the time of formation of the oldest stromatolite-bearing sediments.

- Cellularly preserved filamentous and coccoidal microfossils of prokaryotic (cyanobacterial and flexibacterial) affinity occur in sedimentary chert facies approaching an age of 3.5 Gyr. While metamorphic alteration of the older record has generally transformed these and related morphotypes into an undifferentiated "sticks and balls" assemblage qualifying at best for dubiofossil status we cannot, however, exclude at this stage that individual cell-like morphologies reported from the 3.8 Gyr-old Isua rocks may represent structurally degenerated members of an original *bona-fide* Archaean microflora.

- The record of sedimentary organic carbon in the form of kerogen (the fossil residue of primary biogenic substances) and its graphitic derivatives can be traced back to 3.8 Gyr ago, with the average C_{org}-content of Precambrian rocks staying well in the range of geologically younger sediments (0.5–0.6%). It seems worth noting that the C_{org}-content of the oldest terrestrial sediments from Isua, West Greenland, may even surpass the 0.6 percent mark in selected carbon-rich lithologies.

- Currently documented isotope age functions for sedimentary carbonate and organic carbon show observed $^{13}C/^{12}C$ fractionations between both carbon species to be diagnostic of the isotope-discriminating properties of ribulose-1,5-bisphosphate (RuBP) carboxylase, the key enzyme of the Calvin cycle that channels most of the carbon transfer from the nonliving to the living world. Accordingly, the isotope age curve of sedimentary organic carbon can be best interpreted as an index line of photosynthetic activity over the geological past. Allowing for a metamorphic overprint of the oldest record, biologically mediated carbon isotope fractionations have persisted rather uniformly as from 3.8 Gyr ago, attesting to an astounding degree of evolutionary conservatism in the biochemistry of autotrophic carbon fixation.

In concert, these lines of evidence document beyond reasonable doubt that life had emerged very early on the juvenile Earth, acting since then as a powerful geochemical agent with the concomitant establishment of a biogeochemical carbon cycle.

References

Allaart, J.H. (1976) 'The pre-3760 Myr old supracrustal rocks of the Isua area, central West Greenland, and the associated occurrence of quartz-banded ironstone', in B.F. Windley (ed.), The Early History of the Earth, Wiley, London, pp. 177-189.

Arneth, J.D., Schidlowski, M., Sarbas, B., Goerg, U. and Amstutz, G.C. (1985) 'Graphite content and isotopic fractionation between calcite-graphite pairs in metasediments from the Mgama Hills, Southern Kenya', Geochim. Cosmochim. Acta 49, 1553-1560.

Awramik, S.M. (1982) 'The pre-Phanerozoic fossil record', in H.D. Holland

and M. Schidlowski (eds.), Mineral Deposits and the Evolution of the Biosphere, Springer, Berlin, pp. 67-81.

Awramik, S.M., Schopf, J.W. and Walter, M.R. (1983) 'Filamentous fossil bacteria from the Archaean of Western Australia', in B. Nagy, R. Weber, J.C. Guerrero and M. Schidlowski (eds.), Developments and Interactions of the Precambrian Atmosphere, Lithosphere and Biosphere (Developments in Precambrian Geology 7), Elsevier, Amsterdam, pp. 249-266.

Baker, A.J. and Fallick, A.E. (1989) 'Evidence from Lewisian limestones for isotopically heavy carbon in two-thousand-million-year-old sea water', Nature 337, 352-354.

Barghoorn, E.S. and Tyler, S.A. (1965) 'Microorganisms from the Guntflint chert', Science 147, 563-577.

Bottinga, Y. (1969) 'Calculated fractionation factors for carbon and hydrogen isotope exchange in the system calcite - carbon dioxide - graphite - methane - hydrogen - water vapor', Geochim. Cosmochim. Acta 33, 49-64.

Bridgwater, D., Allaart, J.H., Schopf, J.W., Klein, C., Walter, M.R., Barghoorn, E.S., Strother, P., Knoll, A.H. and Gorman, B.E. (1981) 'Microfossil-like objects from the Archaean of Greenland: A cautionary note', Nature 289, 51-53.

Buick, R. (1984) 'Carbonaceous filaments from North Pole, Western Australia: Are they fossil bacteria in Archaean stromatolites?', Precambrian Res. 24, 157-172.

Burne, R.V. and Moore, L.S. (1987) 'Microbialites: Organosedimentary deposits of benthic microbial communities', Palaios 2, 241-254.

Byerly, G.R., Lowe, D.R. and Walsh, M.M. (1986) 'Stromatolites from the 3.300-3.500-Myr Swaziland Supergroup, Barberton Mountain Land, South Africa', Nature 319, 489-491.

Cloud, P.E. (1965) 'Significance of the Guntflint (Precambrian) microflora', Science 148, 27-45.

Cloud, P.E. (1976) 'Beginnings of biospheric evolution and their biogeochemical consequences', Paleobiology 2, 351-387.

Cohen, Y., Aizenshtat, Z., Stoler, A. and Jorgensen, B.B.(1980) 'The microbial geochemistry of Solar Lake, Sinai', in J.B. Ralph, P.A. Trudinger and M.R. Walter (eds.), Biogeochemistry of Ancient and Modern Environments, Springer, Berlin, pp. 167-172.

Corliss, J.B., Baross, J.A. and Hoffman, S.E. (1981) ' A hypothesis concerning the relationship between submarine hot springs and the origin of life on Earth', Oceanol. Acta 4 (Suppl.), 59-69.

Craig, H. (1953) 'The geochemistry of stable carbon isotopes', Geochim. Cosmochim. Acta 3, 53-92.

Deines, P. (1980) 'The isotopic composition of reduced organic carbon', in P. Fritz and J.C. Fontes (eds.), Handbook of Environmental Isotope

Geochemistry 1, Elsevier, Amsterdam, pp. 329-406.

Dunlop, J.S.R., Muir, M.D., Milne, V.A. and Groves, D.I. (1978) 'A new microfossil assemblage from the Archaean of Western Australia', Nature 274, 676-678.

Durand, B. (ed.) (1980) Kerogen–Insoluble Organic Matter from Sedimentary Rocks, Editions Technip, Paris, 519 pp.

Eichmann, R. and Schidlowski, M. (1975) 'Isotopic fractionation between coexisting organic carbon-carbonate pairs in Precambrian sediments', Geochim. Cosmochim. Acta 39, 585-595.

Garrels, R.M. and Mackenzie, F.T. (1971) Evolution of Sedimentary Rocks, Norton, New York, 397 pp.

Glaessner, M.F. (1983) 'The emergence of Metazoa in the early history of life', in B. Nagy, R. Weber, J.C. Guerrero and M. Schidlowski (eds.), Developments and Interactions of the Precambrian Atmosphere, Lithosphere and Biosphere (Developments in Precambrian Geology 7), Elsevier, Amsterdam, pp. 319-333.

Glaessner, M. (1984) The Dawn of Animal Life, Cambridge University Press, Cambridge, XI + 244 pp.

Greenberg, J.M. (1984) 'The structure and evolution of interstellar grains', Scient. Am. 250, 124-135.

Greenberg, J.M., Zhao, N. and Hage, J. (1989) 'Chemical evolution of interstellar dust, comets and the origins of life', Ann. Phys. Fr. 14, 103-133.

Hayes, J.M., Kaplan, I.R. and Wedeking, K.W. (1983) 'Precambrian organic geochemistry: Preservation of the record', in J.W. Schopf (ed.), Earth's Earliest Biosphere: Its Origin and Evolution, Princeton University Press, Princeton, N.J., pp. 93-134.

Hoering, T.C. and Navale, V. (1987) 'A search for molecular fossils in the kerogen of Precambrian sedimentary rocks', Precambrian Res. 34, 247-267.

Hofmann, H.J. (1973) 'Stromatolites: Characteristics and utility', Earth Science Rev. 9, 339-373.

Irvine, W.M. and Knacke, R.F. (1989) 'The chemistry of interstellar gas and grains', in S.K. Atreya, J.B. Pollack and M.S. Matthews (eds.), Origin and Evolution of Planetary and Satellite Atmospheres, University of Arizona Press, Tucson, pp. 3-34.

Junge, C.E., Schidlowski, M., Eichmann, R. and Pietrek, H. (1975) 'Model calculations for the terrestrial carbon cycle: Carbon isotope geochemistry and evolution of photosynthetic oxygen', J. Geophys. Res. 80, 4542-4552.

Kazmierczak, J. (1979) 'The eukaryotic nature of Eosphaera-like ferriferous structures from the Precambrian Gunflint Iron Formation, Canada: A comparative study', Precambrian Res. 9, 1-22.

Kissel, J. and Krueger, F.R. (1987) 'The organic component in dust from

comet Halley as measured by the PUMA mass spectrometer on board Vega 1', Nature 326, 755-760.
Knoll, A.H. and Barghoorn, E.S. (1977) 'Archean microfossils showing cell division from the Swaziland System of South Africa', Science 198, 396-398.
Kreulen, R. and Van Beek, P.C.J.M. (1983) 'The calcite-graphite isotope thermometer; data on graphite bearing marbles from Naxos, Greece', Geochim. Cosmochim. Acta 47, 1527-1530.
Krumbein, W.E. and Cohen, Y. (1977) 'Primary production, mat formation and lithification changes of oxygenic and facultative anoxygenic cyanophytes (cyanobacteria)', in E. Flügel (ed.), Fossil Algae, Springer, Berlin, pp. 37-56.
Lowe, D.R. (1980) 'Stromatolites 3.400-Myr old from the Archean of Western Australia', Nature 284, 441-443.
MacGregor, A.M. (1940) 'A Precambrian algal limestone in Southern Rhodesia', Trans. Geol. Soc. S. Afr. 43, 9-15.
Mason, T.R. and von Brunn, V. (1977) '3-Gyr-old stromatolites from South Africa', Nature 266, 47-49.
McKirdy, D.M. and Imbus, S.W. (1992) 'Precambrian petroleum: A decade of changing perceptions, in M. Schidlowski, S. Golubic, M.M. Kimberley, D.M. McKirdy and P.A. Trudinger (eds.), Early Organic Evolution: Implications for Mineral and Energy Resources, Springer, Berlin, pp. 176-192.
Miller, S.L. (1955) 'Production of some organic compounds under possible primitive Earth conditions', J. Am. Chem. Soc. 77, 2351-2361.
Miller, S.L. Urey, H.C. and Oró, J. (1976) 'Origin of organic compounds on the primitive Earth and in meteorites', J. Mol. Evol. 9, 59-72.
Monty, C. (1984) 'Stromatolites in Earth history', Terra cognita 4, 423-430.
Moorbath, S., O'Nions, R.K. and Pankhurst, R.J. (1973) 'Early Archaean age for the Isua iron-formation, West Greenland', Nature 245, 138-139
Muir, M.D. and Grant, P.R. (1976) 'Micropaleontological evidence from the Onverwacht Group, South Africa', in B.F. Windley (ed.), The Early History of the Earth, Wiley, London, pp. 595-604.
Murphey, B.F. and Nier, A.O. (1941) 'Variations in the relative abundance of the carbon isotopes', Phys. Rev. 59, 771-772.
Nier, A.O. and Gulbransen, E.A. (1939) 'Variations in the relative abundance of the carbon isotopes', J. Am. Chem. Soc. 61, 697-698.
O'Leary, M.H. (1981) 'Carbon isotope fractionation in plants', Phytochemistry 20, 553-567.
Oró, J., Miller, S.L. and Lazcano, A. (1990) 'The origin and early evolution of life on Earth', Ann. Rev. Earth Planet. Sci. 18, 317-356.
Orpen, J.L. and Wilson, J.F. (1981) 'Stromatolites at $\sim$3.500 Myr and a greenstone-granite unconformity in the Zimbabwean Archaean', Nature 291, 218-220.
Park, R. and Epstein, S. (1960) 'Carbon isotope fractionation during

photosynthesis', Geochim. Cosmochim. Acta 21, 110-126.

Pflug, H.D. (1978) 'Yeast-like microfossils detected in the oldest sediments of the Earth', Naturwissenschaften 65, 611-615.

Pflug, H.D. (1987) 'Chemical fossils in early minerals', Topics in Current Chemistry 139, 1-55.

Pflug, H.D. and Jaeschke-Boyer, H. (1979) 'Combined structural and chemical analysis of 3.800-Myr-old microfossils', Nature 280, 483-486.

Pflug, H.D. and Reitz, E. (1985) 'Earliest phytoplankton of eukaryotic affinity, Naturwissenschaften 72, 656-657.

Pflug, H.D. and Reitz, E. (1992) 'Palynostratigraphy in Phanerozoic and Precambrian metamorphic rocks, in M. Schidlowski, S. Golubic, M.M. Kimberley, D.M. McKirdy and P.A. Trudinger (eds.), Early Organic Evolution: Implications for Mineral and Energy Resources, Springer, Berlin, pp. 508-518.

Rankama, K. (1948) 'New evidence of the origin of Pre-Cambrian carbon', Geol. Soc. Amer. Bull. 59, 389-416.

Robbins, E.I. (1987) '*Appelella ferrifera*, a possible new iron-coated microfossil in the Isua iron-formation, southwestern Greenland', in P.W.U. Appel and G.L.LaBerge (eds.), Precambrian Iron-Formations, Theophrastus Publications, Athens, pp. 141-154.

Roedder, E. (1981) 'Are the 3.800-Myr-old Isua objects microfossils, limonite-stained fluid inclusions, or neither?', Nature 293, 459-462.

Ronov, A.B. (1958) 'Organic carbon in sedimentary rocks (in relation to the presence of petroleum)', Geochemistry 1958, 510-536.

Ronov, A.B. (1980) 'Osadochnaya Obolochka Zemli (Earth's Sedimentary Shell). 20th Vernadsky Lecture', Izdatel'stvo Nauka, Moscow, 97 pp. (in Russian).

Schidlowski, M. (1978) 'Evolution of the Earth's atmosphere: Current state and exploratory concepts', in H. Noda (ed.), Origin of Life, Center Acad. Publ. Japan, Tokyo, pp. 3-20.

Schidlowski, M. (1982) 'Content and isotopic composition of reduced carbon in sediments', in H.D. Holland and M. Schidlowski (eds.), Mineral Deposits and the Evolution of the Biosphere, Springer, Berlin, pp. 103-122.

Schidlowski, M. (1987) 'Application of stable carbon isotopes to early biochemical evolution on Earth', Ann. Rev. Earth Planet. Sci. 15, 47-72.

Schidlowski, M. (1988) 'A 3.800-million-year isotopic record of life from carbon in sedimentary rocks', Nature 333, 313-318.

Schidlowski, M. (1990) 'Life on the early Earth: Bridgehead from Cosmos or autochthonous phenomenon?', in K. Goplalan, V.K. Gaur, B.L.K. Somayajulu and J.D. MacDougall (eds.), From Mantle to Meteorites (Festschrift for Devendra Lal), Indian Academy of Sciences, Bangalore, pp. 189-199.

Schidlowski, M., Eichmann, R. and Junge C.E. (1975) 'Precambrian sedimentary carbonates: Carbon and oxygen isotope geochemistry and implications for the terrestrial oxygen budget', Precambrian Res. 2, 1-69.

Schidlowski, M., Eichmann, R. and Junge, C.E. (1976) 'Carbon isotope geochemistry of the Precambrian Lomagundi carbonate province, Rhodesia', Geochim. Cosmochim. Acta 40, 449-455.

Schidlowski, M., Appel, P.W.U., Eichmann, R. and Junge, C.E. (1979) 'Carbon isotope geochemistry of the 3.7 x 10^9 yr old Isua sediments, West Greenland: Implications for the Archaean carbon and oxygen cycles', Geochim. Cosmochim. Acta 43, 189-199.

Schidlowski, M., Hayes, J.M. and Kaplan, I.R. (1983) 'Isotopic inferences of ancient biochemistries: Carbon, sulfur, hydrogen and nitrogen', in J.W. Schopf (ed.), Earth's Earliest Biosphere: Its Origin and Evolution, Princeton University Press, Princeton, N.J., pp. 149-186.

Schoell, M. and Wellmer, F.W. (1981) 'Anomalous ^{13}C depletion in Early Precambrian graphites from Superior Province, Canada', Nature 290, 696-699.

Schopf, J.W. (1968) 'Microflora of the Bitter Springs Formation, Late Precambrian, Central Australia', J. Paleont. 42, 651-688.

Schopf, J.W. (ed.) (1983) Earth's Earliest Biosphere: Its Origin and Evolution, Princeton University Press, Princeton, N.J., XXV + 543 pp.

Schopf, J.W. and Oehler, D.Z. (1976) 'How old are the eukaryotes?', Science 193, 47-49.

Schopf, J.W. and Walter, M.R. (1983) 'Archean microfossils: New evidence of ancient microbes', in J.W. Schopf (ed.), Earth's Earliest Biosphere: Its Origin and Evolution, Princeton University Press, Princeton, N.J., pp. 214-239.

Schopf, J.W. and Packer, B.M. (1987) 'Early Archaean (3.3-billion to 3.5-billion-year-old) microfossils from Warrawoona Group, Australia', Science 237, 70-73.

Stanley, S.M. (1981) The New Evolutionary Timetable, Basic Books Inc., New York, 222 pp.

Summons, R.E. and Powell, T.G. (1992) 'Hydrocarbon composition of the Late Proterozoic oils of the Siberian Platform: Implications for the depositional environment of source rocks', in M. Schidlowski, S. Golubic, M.M. Kimberley, D.M. McKirdy and P.A. Trudinger (eds.), Early Organic Evolution: Implications for Mineral and Energy Resources, Springer, Berlin, pp. 296-307.

Trask, P.D. and Patnode, H.W. (1942) Source Beds of Petroleum, American Association of Petroleum Geologists, Tulsa, 566 pp.

Valley, J.W. and O'Neil, J.R. (1981) '^{13}C/^{12}C exchange between calcite and graphite: A possible thermometer in Grenville marbles', Geochim. Cosmochim. Acta 45, 411-419.

Veizer, J. and Hoefs, J. (1976) 'The nature of $^{18}O/^{16}O$ and $^{13}C/^{12}C$ secular trends in sedimentary carbonate rocks', Geochim. Cosmochim. Acta 40, 1387-1395.

Wada, H. and Suzuki, K. (1983) 'Carbon isotopic thermometry calibrated by dolomite-calcite solvus temperatures', Geochim. Cosmochim. Acta 47, 697-706.

Walker, J.C.G., Klein, C., Schidlowski, M., Schopf, J.W., Stevenson, D.J. and Walter, M.R. (1983) ' Environmental evolution of the Archean-Early Proterozoic Earth', in J.W. Schopf (ed.), Earth's Earliest Biosphere: Its Origin and Evolution, Princeton University Press, Princeton, N.J., pp. 260-290.

Walsh, M.M. (1992) 'Microfossils and possible microfossils from the Early Archean Onverwacht Group, Barberton Mountain Land, South Africa', Precambrian Res. 54, 271-293.

Walter, M.R. (1977) 'Interpreting stromatolites', Am. Sci. 65, 563-571.

Walter, M.R. (1983) 'Archean stromatolites: Evidence of the Earth's earliest benthos', in J.W. Schopf (ed.), Earth's Earliest Biosphere: Its Origin and Evolution, Princeton University Press, Princeton, N.J., pp. 187-213.

Walter, M.R., Buick, R. and Dunlop, J.S.R. (1980) 'Stromatolites 3.400 - 3.500 Myr old from the North Pole area, Western Australia', Nature 284, 443-445.

Weber, F., Schidlowski, M. Arneth, J.D. and Gauthier-Lafaye, F. (1983) 'Carbon isotope geochemistry of the Lower Proterozoic Francevillian Series of Gabon (Africa)', Terra cognita 3, 220.

INDEX

U

V

W

Y

Z

INDEX OF CHEMICAL SPECIES